T0406450

Genome Mapping and Genomics in Animals
Volume 3

Series Editor: Chittaranjan Kole

Noelle E. Cockett, Chittaranjan Kole
(Editors)

Genome Mapping and Genomics in Domestic Animals

With 49 Illustrations, 12 in Color

NOELLE E. COCKETT
Department of Animal, Dairy & Veterinary
   Sciences
Utah State University
Logan, UT 84322-4700
USA

e-mail: Noelle.Cockett@usu.edu

CHITTARANJAN KOLE
Department of Genetics & Biochemistry
Clemson University
Clemson, SC 29634
USA

e-mail: ckole@clemson.edu

ISBN 978-3-540-73834-3   e-ISBN 978-3-540-73835-0
DOI 10.1007/978-3-540-73835-0

Library of Congress Control Number: 2008929578

© Springer-Verlag Berlin Heidelberg 2009

This work is subject to copyright. All rights are reserved, whether the whole or part of the material is concerned, specifically the rights of translation, reprinting, reuse of illustrations, recitation, broadcasting, reproduction on microfilm or in any other way, and storage in data banks. Duplication of this publication or parts thereof is permitted only under the provisions of the German Copyright Law of September 9, 1965, in its current version, and permissions for use must always be obtained from Springer. Violations are liable for prosecution under the German Copyright Law.
The use of general descriptive names, registered names, trademarks, etc. in this publication does not imply, even in the absence of a specific statement, that such names are exempt from the relevant protective laws and regulations and therefore free for general use.

Cover design: WMXDesign GmbH, Heidelberg, Germany

Printed on acid-free paper

9 8 7 6 5 4 3 2 1

springer.com

Preface to the Series

The deciphering of the sequence of a gene for the first time, the gene for bacteriophage MS2 coat protein to be specific, by Walter Fiers and his coworkers in 1972 marked the beginning of a new era in genetics, popularly known as the genomics era. This was followed by the complete nucleotide sequence of the same bacteriophage in 1976 by the same group; DNA sequencing of another bacteriophage (Φ-X174) in 1977 by Fred Sanger, Walter Gilbert, and Allan Maxam, working independently; and first use of any DNA marker in gene mapping in 1980 for the human system by David Botstein. These landmark discoveries were immediately embraced by the life science community and were followed by an array of elegant experiments leading to the development of several novel concepts, tools, and strategies for elucidation of genes and genomes of living organisms of academic and economic interests to mankind.

The last two decades of the twentieth century witnessed the invention of the polymerase chain reaction; several types of molecular markers; techniques of cloning large DNA segments in artificial chromosomes; approaches to isolate and characterize genes; and tools for high-throughput sequencing, to name just a few. Another noteworthy development had been the formulation of different computer software to analyze the huge amount of data generated by genome mapping experiments, and above all deployment of information technology to store, search, and utilize enormous amounts of data particularly of cloned genes, transcripts, ESTs, proteins, and metabolites. This sweet and swift marriage of biology and information technology gave birth to bioinformatics and the new "omics" disciplines such as genomics, transcriptomics, proteomics, and metabolomics.

The tide of genome mapping and genomics flooded all phyla of the animal kingdom and all taxa of the plant kingdom and most obviously the prokaryotes. In the animal systems, we already had the gene sequence for the CFTR protein in humans in 1989; genome sequence of the model organism *Caenorhabditis elegans* in 1998; genetic maps of many higher animals with map positions of genes and gene-clusters during the nineties. We also happily witnessed the beginning of genome sequencing projects of three domestic animals (cow, dog, and horse) and poultry in 1993. All these achievements and endeavors culminated in the whole-genome sequence of the fruit-fly *Drosophila*, the garden pea of the animal system, in 2000 declaring a successful and pleasant ending of the genome science efforts of the twentieth century. The new millennium in 2001 started with the publication of the draft sequence of the human genome on February 15th by The International Human Genome Mapping Consortium and on February 16th by The Celera Genomics Sequencing Team.

A flurry of new concepts and tools in the first few years of the first decade of the twenty-first century has enriched the subject of genomics and the field has broadened to include the young and fast-growing disciplines of structural genomics, functional genomics, comparative genomics, evolutionary genomics, and neutraceutical genomics, to name just a few. We now have more, faster, cheaper, and cleverer mapping and sequencing strategies, association mapping and the 454 for example; several tools, such as microarrays and cDNA-AFLP to isolate hundreds of known and unknown genes within a short period, elegantly assisted by transcript-profiling and metabolic-profiling; identifying new genes from the knowledge-base of homologous genomes; and precise depiction of the road map of evolution of human and other members of the animal kingdom and their phylogenetic relationships with members of other species or genera. Within less than a decade of the deciphering of the first complete genome sequence for an animal species in 1998, we have complete sequences of some seventeen species of the animal kingdom including nematodes (2), arthropods (4), domestic animals and poultry (2), marsupial (1), wild animals (2), aquatic animals (4), human (1), and non-humanprimate (1). Many more genome mapping projects are progressing rapidly and their results are expected to be published soon.

The list of achievements in the fields of genome mapping and genomics in human and other members of the animal kingdom is enormous. It is also true that in today's world, in the global village of the new millennium, we have access to almost all information regarding the initiation, progress, and completion of all endeavors of animal genome sciences and can enrich our knowledge of the concepts, strategies, tools, and outcomes of the efforts being made in animal genome mapping and genomics. However, all this information is dispersed over the pages of periodicals, reviews on particular types of animals or their specific groups in hard copy versions, and also in electronic sources at innumerable links of web pages for research articles, reports, and databases. But we believe that there should be a single compilation, in both hard copy and electronic versions, embodying the information on the work already done and to be done in the fields of genome mapping and genomics of all members of the animal kingdom that are of diverse interests to mankind: academic, health, company, or environment.

We, therefore, planned for this series on Genome Mapping and Genomics in Animals with five book volumes dedicated to Arthropods; Fishes and Aquatic Animals; Domestic Animals; Laboratory Animals; and Human and Non-Human Primates. We have included chapters on the species for which substantial results have been obtained so far. Genomes of many of these species have been sequenced or are awaiting completion of sequencing soon. Overview on the contents of these volumes will be presented in the prefaces of the individual volumes.

It is an amazingly interesting and perplexing truth that only four nucleotides producing only twenty amino acids in their triplet combination could create anywhere between five to thirty million species of living organisms on the earth. An estimated number of about a half million vertebrate animal species have been described so far! Genomes of the few animal species from this enormous list that we know today are also too diverse to elucidate. It is therefore daring to edit a series on depiction of the diverse genomes we are presenting in over sixty chapters in the five volumes. Seven globally celebrated scientists with knowledge and expertise on different groups of animal systems, and human and non-human primates provided me with the inspiration and encouragement to undertake the job of the series editor. Wayne (Wayne B. Hunter), Tom (Thomas D. Kocher), Noelle (Noelle E. Cockett), Paul (Paul Denny), Ravi (Ravindranath Duggirala), Tony (Anthony G. Comuzzie), and Sarah (Sarah Williams-Blangero) were always available for consultations and clarifications on any aspect while editing the manuscripts of this series. During working on this series, I have been a student first, a scientist second, and an editor third and last, with the mission to present a comprehensive compilation of animal genome mapping and genomics to the students, scientists, and industries currently involved and to be involved in the study and practice of animal genome sciences.

I express my thanks and gratitude as a humble science worker to these seven volume editors for giving me an opportunity to have an enriching and pleasant view of the wide canvas of animal genome mapping and genomics. I also extend my thanks and gratitude to all the scientists who have generously collaborated with their elegant and lucid reviews on the rationale, concepts, methodologies, achievements, and future prospects of the particular systems they are working on, and for the subtle touches of their own experiences and philosophies.

As expected, the editing jobs of this series comprised communication with the volume editors, authors, and publishers; maintenance of the files in hard and soft copies; regular internet searches for verification of facts and databases; and above all maintenance of an environment to practice and enjoy science. My wife Phullara, our son Sourav, and our daughter Devleena were always with me on my travels as a small science worker on a long road of "miles to go before I sleep," not only for the successful completion of this series but also in all my efforts for teaching, research, and extension wherever I worked and stayed in my life.

We have already completed a seven-volume series on Genome Mapping and Molecular Breeding in Plants with Springer that has been very popular among students, scientists, and industries. We are also working on a series on Genome Mapping and Genomics in Microbes with Springer. It was, is, and will be enriching and entertaining to work with the experienced and wonderful people involved in the production of this series, including Sabine (Dr. S. Schwarz) and Jutta (Dr. J. Lindenborn) among many from the Springer family. I record my thanks and gratitude to them, here (and also submit in the databanks for future retrieval) for all their timely co-operation and advice when publishing these volumes.

I trust and believe that we must have missed deliberations on many significant animal species and left many mistakes on the pages of these volumes. All these lapses are surely mine, and all the credits must go to the volume editors, the authors, and the publisher. In the future these errors will be rectified on receipt of suggestions from the readers, and also there will be further improvement of the contents and general set-up of the volumes of this series.

Clemson
January 10, 2008

Chittaranjan Kole

Preface to the Volume

Over the past century, humans have used an expanding knowledge of genetics to improve the functionality and wellbeing of animals. The field of quantitative genetics has led to the selection and breeding of domesticated animals possessing superior genes for desirable traits. Enhanced animal selection, as well as a clearer understanding of the genes and genetic regulation underlying traits, is now possible through the study of genomics.

The nine chapters in this volume focus on genome mapping and genomics research that has been conducted in domesticated and farm species. Topics include the development of genome maps, descriptions of available genomic resources, phylogenetic analyses, domestication patterns, and genetic control of traits. While each chapter serves as a stand-alone description of genomics for that particular species, when read as a whole, the breadth of the research in domesticated and farm species is remarkable, particularly in the light of the limited funding, resources, and personnel as compared to the investment on humans and laboratory species. These limitations have resulted in the development of collaborations and consortiums that cross the globe. Clearly, the pooling of funding and expertise has expanded genomic resources for these species, and allowed prioritization of needs through a collective and iterative process.

While not all domesticated and farm species are included in the volume, the ones that are described here allow a comparison of the outcomes and approaches that were used across the various species. To encapsulate, the amount of genomics research that has occurred to date differs dramatically across the species. For example, there are only a limited number of molecular markers and a rudimentary genome map in cervids and water buffalo while full genome sequences are publicly available for cattle, chickens and dogs. And while there are limited outcomes from the research in some species, the impact of genomics research in livestock and domestic species has resulted in critical information. Direct outcomes have been the identification of genetic regions and in some cases, the causative mutation, that control a spectrum of traits including fertility, reproduction, growth rate and efficiency, milk production, carcass quality and composition, fitness, immune function, and disease traits. This progress is remarkable given that it has only been since the early 1990s, when genome linkage maps containing molecular markers were developed, that genome-wide studies for economic trait loci became feasible.

Comparative genomics is a critical component in the advancement of genomics for livestock and domesticated species. Anchoring the genome of one species to another has allowed an exchange of information and resources, particularly important when one of the species has limitations in funding and researchers, or is at an earlier stage in the discovery process. In addition to leveraging resources across species, an equally important outcome of the comparative genomic efforts is the comparison of locus order in mammalian species, providing additional information for understanding chromosomal evolution.

As mentioned, whole genome sequences for several species are now available or soon to be available, including cattle, dogs, chickens, horses and swine. Certainly a fully sequenced genome will advance the progress of research in every species but the utilization and impact of genomics resources is tempered by the number of active scientists working on that particular species. The economic impact of a species is also a major determinant of the emphasis placed on genomics resources. Several farm species, such as sheep and rabbits, are now used as models for biomedical studies, which has increased attention on securing genomics information.

It is important to note that the genomics research being conducted on these species is not static. With each passing week, more information is added. Therefore, the chapters included in this volume serve as a "snapshot" of the existing information available at the time that the chapter was written. Each author has included a section that highlights areas of future work and needs.

It has been a pleasure to work with the 23 authors who have contributed the chapters in this volume. These authors were invited to participate because they are experts within their field of study. As expected, they have added their own style and interpretation to the work conducted in their assigned species. The authors are affiliated with institutions from around the world, highlighting the global impact of the work being conducted. We appreciate the hard work and perseverance of the authors in the preparation of their chapters and contribution to the volume.

Logan, UT, USA
Clemson, SC, USA
April 15, 2008

Noelle E. Cockett
Chittaranjan Kole

Contents

Contributors XV

1 Cattle
M. D. MacNeil, J. M. Reecy, D. J. Garrick 1
1.1 Introduction 1
    1.1.1 History 1
    1.1.2 Economic Importance 1
1.2 Molecular Genetics 3
    1.2.1 Genetic-Mapping Resources 3
    1.2.2 Quantitative Trait Loci 5
    1.2.3 Using Genotypes in Breeding Cattle 8
1.3 Future Scope of Work 9
References 10

2 Water Buffalo
L. Iannuzzi and G. P. Di Meo 19
2.1 Introduction 19
2.1.1 Taxonomic Description 19
    2.1.2 Economic Importance 21
    2.1.3 Breeding Objectives 22
2.2 Molecular Genetics 23
    2.2.1 Classical Mapping Efforts 23
    2.2.2 Construction of Genetic Maps and Comparative Mapping 24
2.3 Future Scope of Work 28
References 28

3 Sheep
C. A. Bidwell, N. E. Cockett, J. F. Maddox and J. E. Beever. 33
3.1 Introduction 33
    3.1.1 Economic and Biomedical Importance 33
    3.1.2 History 34
    3.1.3 The Sheep Karyotype 34
3.2 Molecular Genetics 35
    3.2.1 Linkage Maps 36
    3.2.2 Mapped Traits in Sheep 36
    3.2.3 Resources for Mapping and Sequencing the Ovine Genome 39
3.3 Future Scope of Work 41
References 41

4 Deer
Richard J. Hall 47
4.1 Introduction 47
    4.1.1 Taxonomic Description 47
    4.1.2 Economic Importance 48
    4.1.3 Karotype of Cervids 48

XII Contents

4.2 Molecular Genetics 48
    4.2.1 Genetic Analysis Using Molecular Markers 48
    4.2.2 The Deer Genetic Map 60
    4.2.3 Genes 61
    4.2.4 Quantitative Trait Loci in Deer 62
    4.2.5 Chronic Wasting Disease of North American Cervids 65
4.3 Future Scope of Work 66
References 67

5 Poultry

Michael N. Romanov Alexei A. Sazanov, Irina Moiseyeva,
and Aleksandr F. Smirnov 75
5.1 Introduction 75
    5.1.1 Brief History and Zoological Description 75
    5.1.2 Chickens 77
    5.1.3 Economic Importance and Nutritional Value 82
    5.1.4 Breeding Objectives 85
5.2 Classic Genetics 86
    5.2.1 Brief History of Poultry Genetics 86
    5.2.2 Early Classical Mapping Efforts 86
    5.2.3 First Chicken Map 99
    5.2.4 Subsequent Classical Mapping 100
5.3 Molecular Genetics and Whole-Genome Sequence 101
    5.3.1 First-generation Molecular Maps 101
    5.3.2 Physical Maps 103
    5.3.3 Whole-Genome Sequence 105
    5.3.4 Chicken Genome and Sequence Features 107
    5.3.5 Genetics and Molecular Mapping in Other Birds 110
5.4 QTL and Functional Genomics 120
    5.4.1 QTL Analysis 120
    5.4.2 QTL: Growth, Meat Quality, and Productivity 120
    5.4.3 QTL: Egg Quality and Productivity 122
    5.4.4 QTL: Disease Resistance 122
    5.4.5 QTL: Behavior 123
    5.4.6 Toward Functional Genomics of Poultry 123
    5.5 Other Molecular Applications 124
    5.5.1 Biodiversity Studies 124
    5.5.2 Molecular Sexing 125
5.6 Conclusions 126
References 126

6 Turkey

Kent M. Reed 143
6.1 Introduction 143
    6.1.1 Origin and Domestication 143
    6.1.2 Taxonomy and Zoological Description 144
    6.1.3 Modern Breeding Objectives 144
    6.1.4 Economic Importance 145
    6.1.5 Karyotype and Genome 147
    6.1.6 Classical Mapping Efforts and Limitations 147

6.2	Construction of Genetic Maps	148
6.2.1	Genetic Markers	148
6.2.2	Primary Genetic Linkage Maps	151
6.2.3	Second-generation Linkage Map	151
6.2.4	Integrative Mapping	153
6.2.5	Comparative Mapping	154
6.2.6	Other Comparative Studies	155
6.2.7	Physical Mapping	157
6.2.8	The Next-Generation Physical Maps	157
6.3	Advanced Works, Functional Genomics	158
6.3.1	ESTs, Microarrays, and SAGE	158
6.3.2	Candidate Gene Mapping	159
6.4	Conclusion	160
References		160

7 Rabbit

Claire Rogel-Gaillard, Nuno Ferrand, and Helene Hayes 165

7.1	Introduction	165
7.1.1	History	165
7.1.2	Taxonomic Position	165
7.1.3	Physical Characteristics	166
7.1.4	Breeds	166
7.1.5	Domestication, Phylogeny, and Genetic Diversity	167
7.1.6	Economic Importance	171
7.2	Molecular Genetics	175
7.2.1	Cytogenetics	175
7.2.2	Genetic Molecular Markers	176
7.2.3	Physical Mapping Tools	204
7.2.4	Sequencing Data	204
7.2.5	Bioinformatics Tools	205
7.2.6	Expected Tools and Development	206
7.2.7	Genome Mapping	206
7.3	Future Scope of Work	220
7.3.1	Using Rabbits to Study the Domestication Process	220
7.3.2	Using Rabbits to Study Color Patterns	220
7.3.3	Using Rabbits to Study Early Embryonic Development	220
7.3.4	Using Rabbits to Produce Embryonic Stem Cells and Validate Candidate Genes	221
7.3.5	Perspectives of the Rabbit as a Farm Animal	221
References		223

8 Dog

D.S. Mosher, T.C. Spady, and E.A. Ostrander 231

8.1	Introduction	231
8.1.1	Dog Breeds	231
8.1.2	Genetic Diversity and Dog Breeds	233
8.1.3	The Superfamily Canoidae	234
8.1.4	Mitochondrial DNA (MtDNA) Analysis of Canids	234
8.1.5	The Domestic Dog Population	236

XIV   **Contents**

8.2	Molecular Genetics of Dogs	237
8.2.1	Canine Linkage and Radiation Hybrid Maps	237
8.2.2	Comparative Maps	238
8.2.3	Survey Sequence and a Canine Gene Map	239
8.2.4	The 7.5x Canine Genome Sequence	240
8.2.5	Single Gene Traits	241
8.2.6	Complex Traits	244
8.2.7	Resources	249
8.3	Future Scope of Work	250
References		251

9 Pig

C.W. Ernst and A.M. Ramos 257

9.1	Introduction	257
9.1.1	History and Description	257
9.1.2	Economic Importance	258
9.1.3	Breeding Objectives	258
9.2	Molecular Genetics of Pigs	259
9.2.1	Genetic Maps	259
9.2.2	Physical and RH Maps	260
9.2.3	Pig–Human Comparative Maps	260
9.2.4	Quantitative Trait Loci	261
9.2.5	Candidate Gene Discovery	262
9.2.6	Marker-Assisted Breeding	264
9.3	Functional Genomics	265
9.3.1	Expressed Sequence Tags	265
9.3.2	DNA Microarrays	265
9.4	Future Scope of Work	266
9.4.1	Swine Genome Sequencing	266
9.4.2	Expression QTL	266
9.4.3	Conclusion	268
References		268

Contributors

J.E. Beever
Department of Animal Sciences
University of Illinois
220 Edward R. Madigan Laboratory
1201 West Gregory Drive
Urbana, IL 61801, USA

C.A. Bidwell
Purdue University
Department of Animal Sciences
125 South Russell Street
West Lafayette
IN 47907–2042, USA
cbidwell@purdue.edu

N.E. Cockett
Department of Animal
Dairy & Veterinary Sciences
Utah State University
Logan, UT 84322–4700, USA

C.W. Ernst
Department of Animal Science
3385 Anthony Hall
Michigan State University
East Lansing, MI 48824, USA
ernstc@msu.edu

N. Ferrand
CIBIO, Centro de Investigação em
Biodiversidade e Recursos Genético
Campus Agrário de Vairão
4485-661 Vairão, Portugal
and
Departamento de Zoologia
e Antropologia
Faculdade de Ciências, Universidade
do Porto
Praça Gomes Teixeira, 4099-002 Porto
Portugal

D.J. Garrick
Department of Animal Science
Iowa State University
Ames, IA 50011, USA

R.J. Hall
AgResearch Ltd.
Invermay
Agricultural Centre
Private Bag 50034, Mosgiel 9053
New Zealand
Richard.Hall@esr.cri.nz

H. Hayes
INRA, UR339 Unité de Génétique
biochimique et Cytogénétique
F-78350 Jouy-en-Josas, France

L. Iannuzzi
National Research Council (CNR)
Institute of Animal Production Systems
in the Mediterranean Environments
(ISPAAM)
Laboratory of Animal Cytogenetics
and Gene Mapping
Naples, Italy
leopoldo.iannuzzi@ispaam.cnr.it

M.D. MacNeil
USDA
Agricultural Research Service
Miles City, MT 59301, USA
mike@larrl.ars.usda.gov

J.F. Maddox
Department of Veterinary Science
University of Melbourne, Victoria 3010
Australia

I. Moiseyeva
N.I. Vavilov Institute of General
Genetics
Russian Academy of Sciences
Gubkin Street 3
Moscow, GSP-1, 119991, Russia

D.S. Mosher
National Human Genome Research
Institute,
National Institutes of Health
50 South Drive
MSC 8000, Building 50, Room 5334
Bethesda, MD 20892-8000, USA

E.A. Ostrander
National Human Genome Research
Institute,
National Institutes of Health
50 South Drive MSC 8000,
Building 50, Room 5334
Bethesda, MD 20892-8000, USA
eostrand@mail.nih.gov

G. Pia Di Meo
National Research Council (CNR)
Institute of Animal Production
Systems in the Mediterranean
Environments (ISPAAM)
Laboratory of Animal Cytogenetics
and Gene Mapping
Naples, Italy

A.M. Ramos
Department of Animal Science
3385 Anthony Hall
Michigan State University
East Lansing, MI 48824, USA

J.M. Reecy
Department of Animal Science
Iowa State University
Ames, IA 50011, USA

K.M. Reed
Department of Veterinary and
Biomedical Sciences

College of Veterinary Medicine
University of Minnesota
St. Paul, MN 55108, USA
reedx054@tc.umn.edu

C. Rogel-Gaillard
INRA CEA, UMR314 Laboratoire
de Radiobiologie et Etude du Génome
F-78350 Jouy-en-Josas, France
claire.rogel-gaillard@jouy.inra.fr

M.N. Romanov
CRES – Conservation and Research
for Endangered Species
Zoological Society of San Diego
Arnold and
Mabel Beckman Center for
Conservation Research
15600 San Pasqual Valley Road
Escondido, CA 92027-7000, USA
mromanov@sandiegozoo.org

A.A. Sazanov
All-Russian Institute of Animal
Genetics and Breeding
Russian Academy of Agricultural
Science
Moskovskoye shosse
55A, St Petersburg, Pushkin 189620
Russia

A.F. Smirnov
Biological Research Institute
St Petersburg State University
Oranienbaumskoye shosse 2
St Petersburg, Stary Petergof 198504
Russia

T.C. Spady
National Human Genome Research
Institute,
National Institutes of Health
50 South Drive
MSC 8000, Building 50, Room 5334
Bethesda, MD 20892-8000, USA

CHAPTER 1

Cattle

Michael D. MacNeil[1] (✉), James M. Reecy, and Dorian J. Garrick[2]

[1] USDA, Agricultural Research Service, Miles City, MT 59301, USA, mike.macneil@ars.usda.gov
[2] Department of Animal Science, Iowa State University, Ames, IA 50011, USA

1.1
Introduction

1.1.1
History

All contemporary cattle are thought to have been domesticated from the now extinct aurochs, *Bos primigenius*. Evidence from mitochondrial DNA indicates divergence of two cattle taxa, *Bos indicus and Bos taurus*, more than 100,000 years ago (Loftus et al. 1994; Bradley et al. 1996). These two taxa were likely domesticated independently (Grigson 1980; Loftus et al. 1994) some 10,000 years before present. Mehrgrarh, in modern-day Pakistan, is a strong candidate for the site of domestication of *Bos indicus* cattle. Catal Hüyük, Anatolia, in modern-day Turkey, is a likely site of the Near-Eastern domestication of *Bos taurus* cattle. There is further but less secure evidence for additional domestications of aurochs in the Nile Valley (Bradley et al. 1996; Troy et al. 2001) and in North East Asia (Mannen et al. 2004). These domestications are postulated to give rise to the *Bos taurus* cattle of Africa and contribute to the gene pool of cattle in Mongolia, North China, Korea, and Japan, respectively.

Immigration of *Bos indicus* cattle into Africa appears to radiate from the Horn and East Coast (Hanotte et al. 2002). This immigration may result from local Arabian contacts. Alternatively, it may have been a consequence of long-distance trade on the Indian Ocean that was also responsible for introducing other domestic species into Africa (Clutton-Brock 1993). *Bos primigenius* was distributed throughout Europe at the end of the last Ice Age, and it remains unclear whether or not all European cattle are derived from the stock domesticated in western Asia (Bailey et al. 1996). However, the dominant mitochondrial DNA haplo-

type in European cattle is consistent with their being of Anatolian origin (Troy et al. 2001; Kühn et al. 2005). Genetic relationships among European *Bos taurus*, African *Bos taurus*, African *Bos indicus*, and Indian *Bos indicus* are illustrated in Fig. 1 (Bradley et al. 1998).

Cattle are not indigenous to Australia, North America, and South America. Importations of cattle to these countries followed the respective patterns of human exploration and colonization. Early importations to Australia and North America were predominantly British breeds, whereas early importations to South America were mainly of Spanish origin (Rouse 1970). Subsequently, *Bos indicus* cattle were brought to the more tropical areas of these continents (Rouse 1970) and still later there were substantial importations of *Bos taurus* cattle from continental Europe (Willham et al. 1993).

1.1.2
Economic Importance

Cattle have had a central role in the evolution of human cultures with significant numbers of cattle produced in every continent except Antarctica. From an economic perspective cattle are the most important domestic animal species (Cunningham 1992). In various parts of the world, cattle provide traction, milk, and meat. Worldwide production in 2004 of beef and veal, nonfat dry milk, butter, and cheese were predicted to be 51,191; 3,486; 6,676; and 13,373 thousand metric tons, respectively (USDA 2005).

Following Lush (1945), modern paradigms for genetic improvement of livestock are commonly traced to Robert Bakewell (1725–1795) whose admonitions included: "Like produces like or the likeness of some ancestor; inbreeding produces prepotency and refinement; breed the best to the best." Historical

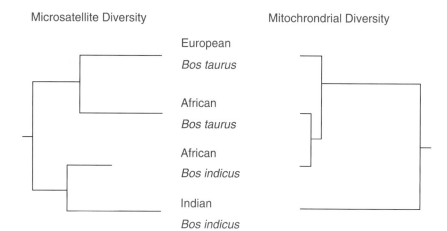

Fig. 1 Phylogenic tree derived from microsatellite allelic diversity and mitochondrial sequence diversity (Bradley et al. 1998). Note that African *Bos indicus* seemingly derives a majority of their genome from introgression of Indian *Bos indicus* alleles into the African *Bos taurus* maternal background

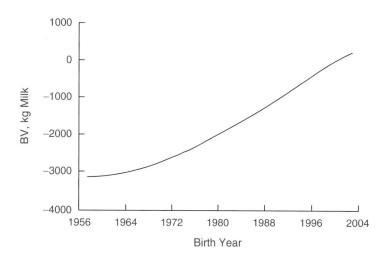

Fig. 2 Genetic trend for fluid milk production of US Holstein cows as of May 2005. From: http://aipl.arsusda.gov/eval/summary/trend.cfm

breeding objectives for cattle have focused primarily on increasing yield of milk or milk components in dairy breeds and growth rate, carcass weight, and composition in beef breeds. Success of these efforts in bringing about genetic change can be illustrated by genetic trends in fluid milk production by Holstein cattle in the US (Fig. 2) and growth to weaning by Hereford cattle in Canada (Fig. 3).

Genetic selection for increased production usually, although not always, leads to increased consumption of feed and gross efficiency, while some aspect of fertility is usually impaired (Roberts 1979; MacNeil et al. 1984; Wall et al. 2003). Technology for efficient multiple-trait selection was developed by Hazel (1934) more than 70 years ago and classical applications of multiple-trait selection for dairy and beef production were put forth in the early 1970s (Norman and Dickinson 1971; Cunningham and McClintock 1974; Dickerson et al. 1974). In regions, where feed costs and land prices are high, breeding objectives have been developed and implemented for dual-purpose cattle that are raised for both milk and meat production (e.g., Niebel 1986; Bekman and van Arendonk 1993). VanRaden (2004) reviewed the changing application of multiple-trait breeding objectives to US dairy production from the early 1970s to the present. Application of multiple-trait breeding objectives and selection indexes to beef production has been relatively more recent (e.g., Ponzoni and Newman 1989; Graser et al. 1994; MacNeil et al. 1994).

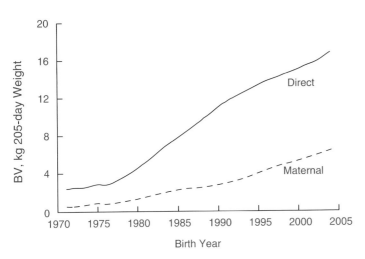

Fig. 3 Genetic trends for direct (solid line) and maternal (dashed line) 205-day weight of Canadian Hereford cattle as of the Spring 2005 Sire Summary. From: http://www.hereford.ca/pdf_files_2008/EPD%20Averages%20Tools%20-and%20Trends%202008.pdf

Genetic improvement of cattle is complicated by relatively late attainment of puberty and low reproductive rate, long generation interval, genetic antagonisms, manifestation of economically important phenotypes late in life, and valuable phenotypes that can only be observed post-harvest. Genomic science and its application to selection have the potential to partially overcome several of these obstacles by facilitating selection decisions being made earlier in life without the usual erosion of accuracy. Objectives of this chapter are to review: (1) the current state of genetic mapping in cattle, (2) assignment of economically important phenotypes to those maps, and (3) ongoing development of breeding programs that have been implemented using molecular genetic technologies.

1.2 Molecular Genetics

1.2.1 Genetic-Mapping Resources

The first bovine genetic maps were generated using somatic cell hybrid (SCH) panels and fluorescent in situ hybridization (FISH). Heuertz and Hors-Cayla (1981) and Womack and Moll (1985, 1986) were the first reports on the use of somatic cell hybrids to develop maps of conserved synteny between bovine genes and previously mapped human homologs of these genes. A bovine/hamster somatic cell radiation hybrid panel was recently developed and used to assign 1,303 previously unassigned expressed sequence tags (ESTs) to chromosome segments (Itoh et al. 2003). Itoh et al. (2005) also used a bovine radiation hybrid panel to map 3,216 microsatellites and 2,377 ESTs. ZOO-FISH mapping has been completed, in which human chromosome-specific painting probes were hybridized to cattle chromosomes (Hayes 1995; Solinas-Toldo et al. 1995; Chowdhary et al. 1996) to produce comparative maps. However, as with somatic cell hybrid panels, gene order could not be addressed with these chromosome pairs.

There are numerous reports on the development of chromosome-specific linkage maps (e.g., Barendse et al. 1993). However, Bishop et al. (1994) were the first to report a genetic linkage map for a majority of the bovine genomes. That map was shortly followed by second-generation medium density maps (Barendse et al. 1997; Kappes et al. 1997). At present, over 3,800 simple tandem repeat polymorphisms have been mapped in cattle by linkage analysis (Ihara et al. 2004). This most recent map comprised 29 sex-averaged autosomal linkage groups and a sex-specific X-chromosome linkage group covering 3,160 centiMorgans (cM). The average interval between markers was 1.4 cM.

Radiation hybrid (RH) mapping (Goss and Harris 1975) has recently been rediscovered as an effective approach to building ordered maps of sequence-tagged sites, regardless of allelic variation. Womack et al. (1997) reported on the generation of a 5,000-rad bovine whole-genome RH panel. This RH panel has served as a resource for mapping the expanding pool of bovine EST, for integrating the existing bovine linkage maps, and for the construction of ordered comparative maps relating the cattle genome to those of the human and mouse (Yang and Womack 1998). The main advantage of RH mapping over other methods is the straightforward positive/negative polymerase chain reaction (PCR) genotyping of RH panel clones. In addition, one can map genes without the need for polymorphisms, which is important when mapping sequences with a low rate of polymorphism (O'Brien et al. 1993). Numerous cattle chromosome-specific RH maps have been published (e.g., Womack et al. 1997; Yang et al. 1998; Rexroad and Womack 1999; Amarante et al. 2000). Subsequently, there have been several efforts to develop whole-genome radiation hybrid maps for cattle. To date, more than 4,000 microsatellites and 2,400 genes have been placed on bovine radiation hybrid maps (Band et al. 2000; Williams et al. 2002; Everts-van der Wind et al. 2004; Itoh et al. 2005). These resources will prove invaluable in the assembly of the bovine genome, which is ongoing in 2008. In addition, efforts are underway to integrate RH and linkage maps (Snelling et al. 2004).

In order to facilitate bovine gene sequencing, both bacterial artificial chromosome (BAC) and yeast artificial chromosome (YAC) libraries have been generated. Cai et al. (1995) reported construction of the first bovine BAC library. Subsequently, several other libraries have been generated (Zhu et al. 1999; Buitkamp et al. 2001; Eggen et al. 2001; http://bacpac.chori.org). Similarly, the first bovine YAC library was produced by Libert et al. (1993). Since that time, at least three other YAC libraries have been generated (Smith et al. 1996; Takeda et al. 1998; Hills et al. 1999). Over the course of the last couple of years the International Bovine BAC Map Consortium has developed a BAC fingerprint map that contains 257,912 clones in 655 contigs, and 32,885 singleton clones (http://www.bcgsc.ca/project/bovine/seqfiles). In addition, the National Institute of Agronomic Research (INRA) BAC map (Schibler et al. 2004; http://dga.jouy.inra.fr/cgi-bin/lgbc/main.pl?BASE=cattle) contains 96,261 clones in 5,096 contigs with 8,017 singletons. These maps share 18,890 CHORI-240 bovine BAC clones (W. Snelling, personal communication).

In March 2003, it was announced that the bovine genome would be sequenced at Baylor College of Medicine and a cow, L1 Dominette 01449 (http://animalgene.umn.edu/pedigraph/sequenced_cow.pdf), from the inbred Line 1 Hereford population maintained by USDA-ARS at Miles City, Montana (MacNeil et al. 1992), was selected for sequencing. Sequencing was initiated in December 2003 (http://www.nih.gov/news/pr/mar2003/nhgri-04.htm.) with assembly and annotation of the completed 7x bovine genome sequence expected by the end of 2006 (Steven Kappes, personal communication). Sequencing was facilitated by availability of a BAC library (CHORI 240; http://bacpac.chori.org) that had been constructed using DNA from her sire, L1 Domino 99375, with whom she had a pedigree-based additive relationship of >92%. On October 6, 2004, less than 11 months after the start of the project, the first draft of the bovine genome was released to the public (http://www.genome.gov/12512874). The current (release 2) assembly provides 6.2x coverage of the bovine genome. Sequence data are publicly available at the following databases: GenBank (www.ncbi.nih.gov/Genbank), EMBL Bank (www.ebi.ac.uk/embl/index.html), and DNA Data Bank of Japan (www.ddbj.nig.ac.jp). The data can be viewed via NCBI's Map Viewer (www.ncbi.nlm.nih.gov), UCSC Genome Browser (www.genome.gucsc.edu), and the Ensembl Genome Browser (www.ensembl.org). Complementary single nucleotide polymorphism (SNP) discovery is ongoing by sequencing DNA from one cow from each of the Holstein, Jersey, Norwegian Red, Angus, Limousin, and Brahman breeds. As of 2008, chips to assay more than 50,000 SNP are commercially available to estimate genetic variation within and between populations of cattle. These resources will allow bovine researchers to leverage infrastructure and expertise of the broader genome research community and provide greater opportunities for biomedical research (S. Kappes, personal communication).

To aid in the assembly of the bovine genome, a composite RH/linkage map is being constructed using information from the Shirakawa-Meat Animal Research Center (MARC) linkage map (Ihara et al. 2004), Shirakawa RH map (Itoh et al. 2005), ComRad RH map (Williams et al. 2002), and the ILTX-2004 RH map (Everts-van der Wind et al. 2004). The RH/linkage composite map is constructed using CarthaGene (de Givry et al. 2005), following procedures described by Snelling et al. (2004). The current composite map represents

9,112 markers, of which 3,919 markers are in at least two data sets, and 297 markers are common to all four data sets. The composite RH/Linkage map will be integrated with the BAC maps and assembled genomic sequence (W. Snelling, personal communication).

1.2.2
Quantitative Trait Loci

A primary objective of quantitative trait loci (QTL) detection studies in cattle is to identify markers that can be used in breeding programs through marker-assisted selection. Most QTL identification efforts have focused on milk production in dairy cattle and on carcass traits in beef cattle. There have been numerous whole-genome scans that have been implemented in both dairy and beef cattle. Investigation of candidate genes based on physiological pathways that control phenotypic expression has provided an alternative to whole-genome scans to detect genes corresponding to QTL (Rothschild and Soller 1997). Results for QTL-mapping studies may also identify critical biochemical pathways affecting various phenotypes for further investigation and manipulation.

Dairy Cattle

Geldermann (1975) and Weller et al. (1990) proposed daughter and granddaughter designs for QTL identification. These designs have been widely employed in dairy cattle, beginning with Georges et al. (1995). Populations studied originate from Finland (Elo et al. 1999; Viitala et al. 2003; Schulman et al. 2004), France (Biochard et al. 2003), Germany (Freyer et al. 2003a; Kuhn et al. 2003), the Netherlands (Spelman et al. 1996), Israel (Weller et al. 2003; Ron et al. 2004), New Zealand (Spelman et al. 1999), Norway (Klungland et al. 2001; Olsen et al. 2002), Sweden (Holmberg and Andersson-Eklund 2004), and North America (e.g., Georges et al. 1995; Zhang et al. 1998; Ashwell and Van Tassell 1999; Heyen et al. 1999; Ashwell et al. 2001, 2004). Recently, methods have been used that relax the pedigree constraints of the daughter and granddaughter designs (Chamberlain et al. 2002).

In a comprehensive review, Khatkar et al. (2004) found QTL that affected milk production identified on 20 of the 29 bovine chromosomes. Significant numbers of the detected QTL had pleiotropic effects, as may be expected given well-established genetic correlations among the phenotypes (Freyer et al. 2002, 2003b; Schrooten et al. 2004). Whether the pleiotropic QTL arise from pleiotropy at the gene level or from tight linkage is the subject of recent investigation (Freyer et al. 2004). In addition, any genetic change in milk yield without concomitant changes in protein and fat yields will result in changes in the percentages of these components. Thus, QTL identified for milk yield may have concomitant effects on composition.

Georges et al. (1995) observed that bovine chromosome 6 (BTA6) contained a QTL that influenced milk production. The presence of QTL on BTA6 associated with milk production and composition was subsequently confirmed by Kuhn et al. (1996), Spelman et al. (1996) (Fig. 4), Velmala et al. (1999), and

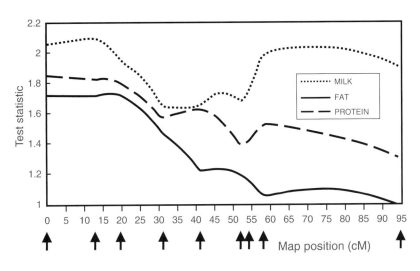

Fig. 4 Interval mapping of BTA6 marker effect on milk, fat and protein yields for six families of Dutch Holstein-Friesian cattle (Spelman et al. 1996). Arrows on the X-axis indicate positions of markers

others. As a consequence, BTA6 has been the subject of intense investigation to define the genetic basis for the QTL. These investigations have established the presence of multiple QTL regions indicating potential for several functional genes located on BTA6 to influence milk production (Ron et al. 2001; Freyer et al. 2002). Fine-mapping studies have further localized QTL on BTA6 affecting milk production (Olsen et al. 2004, 2005), and positional and functional candidate genes have been postulated (*osteopontin*, Schnabel et al. 2005; *peroxysome proliferator-activated receptor-gamma coactivator-1 alpha*, Weikard et al. 2005; *ABCG2*, Cohen-Zinder et al. 2005).

Coppieters et al. (1998) identified a QTL near the centromere on BTA14 with major effects on fat and protein percentages as well as milk yield (Fig. 5). The location of this QTL was subsequently resolved to a chromosome segment of approximately 5 cM (Riquet et al. 1999) and confirmed in independent populations (Heyen et al. 1999; Ashwell et al. 2001). Looft et al. (2001) found strong linkage between the QTL on BTA14 and an expressed sequence tag (EST) derived from mammary gland. Grisart et al. (2002) constructed a BAC contig corresponding to the BTA14 QTL marker interval and identified a nonconservative missense mutation in the positional candidate gene *AcylCoA:diacylglycerol acyltransferase (DGAT1)*. Winter et al. (2001) described the association of a lysine/alanine polymorphism (*K232A*) in *DGAT1* with milk fat content and postulated that this mutation was directly responsible for variation in milk fat content at the QTL. The effect of this polymorphism on milk production and/or composition was subsequently validated in New Zealand Jersey, Holstein-Friesian, and Ayrshire by Spelman et al. (2002), in Israeli Holstein by Weller et al. (2003), and in German Fleckvieh and Holstein by Thaller et al. (2003a). Recently, Grisart et al. (2004) presented genetic and functional data that confirmed the causality of the *DGAT1 K232A* mutation.

Georges et al. (1995) identified a centrally located QTL affecting milk yield on BTA20. This QTL was subsequently confirmed, the interval that contained the QTL was refined, and genes coding for the receptors of growth hormone and prolactin were proposed as positional candidate genes (Arranz et al. 1998). Additional conformation of a QTL on BTA20 that affects milk yield was provided by Mäki-Tanila et al. (1998) and Olsen et al. (2002). Subsequent dissection of the QTL revealed a substitution of tyrosine for phenylalanine in the transmembrane domain of the bovine growth hormone receptor protein that is associated with a strong effect on milk yield (Blott et al. 2003). There is also evidence to support additional QTLs for milk yield and composition that are linked to the *growth hormone receptor (GHR)* gene (Arranz et al. 1998; Blott et al. 2003).

Recent work has suggested additional sources of genetic variation in fat and protein yields and percentages, independent of the lysine/alanine polymorphism in *DGAT1*, but that are closely linked to it (Bennewitz et al. 2004). Kuhn et al. (2004) showed that alleles from the *DGAT1* promoter derived from a variable number tandem repeat (VNTR) polymorphism were associated

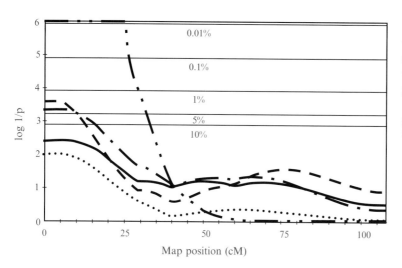

Fig. 5 Nonparametric multipoint multimarker regression mapping of BTA14 marker effect on ●: milk yield; ■: protein yield; □: protein percentage; ▲: fat yield; △: fat percentage (Coppieters et al. 1998). Horizontal reference lines indicate genome-wide probability of observed QTL effect

with milk fat content in animals that were homozygous for the 232A allele and suggested that variation in the number of tandem repeats might result in variability in the transcription level of *DGAT1*.

Mastitis is an inflammatory disease of the mammary gland that has considerable economic impact on dairying through treatment cost, lost production, decreased longevity, disposal of milk due to treatment with antibiotics, and additional labor. Numerous studies have mapped and confirmed QTL for somatic cell score and(or) mastitis resistance on BTA11 (Holmberg and Andersson-Eklund 2004; Schulman et al. 2004), BTA14 (Klungland et al. 2001; Schulman et al. 2002), BTA18 (Ashwell et al. 1997; Schrooten et al. 2000; Rodriguez-Zas et al. 2002; Schulman et al. 2002; Bennewitz et al. 2003; Kuhn et al. 2003), BTA21 (Rodriguez-Zas et al. 2002; Khatib et al. 2005), BTA23 (Ashwell et al. 1997; Heyen et al. 1999), and BTA27 (Klungland et al. 2001; Schulman et al. 2004). Additional QTL affecting either somatic cell score or mastitis resistance have been identified on BTA1, 3, 4, 5, 6, 7, 10, 15, 20, and 22. Schwerin et al. (2003) used mRNA differential display to identify DNA sequences that were differentially expressed in noninfected and infected quarters of a cow. The results of mRNA differential display combined with RH mapping led to the identification of four candidate genes affecting resistance to mastitis: *OSTF1* on BTA8, *AHCY* on BTA13, *PRKDC* on BTA14, and *HNRPU* on BTA16 (Schwerin et al. 2003).

In addition, other studies have evaluated potential for QTL to influence phenotypes related to reproductive rate (Kirpatrick et al. 2000; Lien et al. 2000; Ashwell et al. 2004; Cruickshank et al. 2004; Gonda et al. 2004; Ron et al. 2004), dystocia (Grupe et al. 1998), longevity (Ron et al. 2004), dairy type (Spelman et al. 1999; Connor et al. 2004; Van Tassell et al. 2004), behavior (Hiendleder et al. 2003), udder conformation (Hiendleder et al. 2003), and other diseases (Georges et al. 1993; Hanotte et al. 2003; Zhang et al. 2004). Ashwell and Van Tassell (1999) and Schrooten et al. (2000) evaluated 30 and 27 phenotypes descriptive of conformation and function, respectively. In a majority of these cases, there has yet to be substantial independent confirmation of the reported QTL. Exceptions include a putative QTL on BTA5 affecting ovulation rate or twinning which has been fine mapped by Meuwissen et al. (2002).

Beef Cattle

Backcross and F_2 designs using diverse breeds have predominated QTL identification studies in beef cattle, and several studies using the genome scan approach have successfully identified QTL that affect body composition traits, carcass yield and quality, and growth (Keele et al. 1999; Stone et al. 1999; Schmutz et al. 2000; Casas et al. 2000, 2003, 2004; MacNeil and Grosz 2002; Kim et al. 2003a, b; Li et al. 2004). These designs were developed for use with inbred lines, each homozygous for different alleles at marker loci and assumed QTL (Soller et al. 1976; Zhuchenko et al. 1979). Until recently, it was unknown to what extent these experiments were relevant within breeds. Thallman et al. (2003) reported that QTL influenced meat quality and carcass composition traits, which were identified in a two-breed backcross also segregated within breeds. Further, the resolution of these QTL was only on the order of several tens of centiMorgans. Unfortunately, to date, only one causal mutation for QTL in beef cattle (discussed later) has been identified. While comparative maps between cattle and humans can be used to identify candidate genes for testing, current maps obtained from linkage and radiation hybrid panel analyses contain numerous gaps that limit their utility. Determining the bovine genome sequence (http://www.usda.gov/news/releases/2003/12/0420.htm) will alleviate this problem.

The phenotype commonly referred to as double muscling, or more correctly as muscular hypertrophy, was first documented by Culley (1807) and has been the subject of considerable investigation in cattle (see reviews by Hanset 1981; Arthur 1995; Bellinge et al. 2005). This phenotype was mapped by Charlier et al. (1995) to within 2 cM of a marker locus on BTA2 using microsatellites genotyped across a backcross family. Smith et al. (1997) mapped the myostatin gene to a locus in the interval that was previously shown to contain the muscular hypertrophy locus and that was cytogenetically indistinguishable from it, suggesting that myostatin may be the gene causing muscular hypertrophy in cattle. Mutations in myostatin alleles common to the Belgian Blue and Piedmontese breeds were shown to confer the characteristic increase in muscle mass (Grobet et al. 1997; Kambadur et al. 1997; McPherron and Lee 1997). Pleiotropic effects on other agronomic phenotypes have also been reported (Casas et al. 1998; Short et al. 2002).

Building on previous results related to QTL for fat content of milk, a role in triglyceride synthesis (Cases et al. 1998), expression of sequence tags in bovine adipose tissue (Fries and Winter 2002), and radiation hybrid mapping to BTA14 (Womack et al. 1997), *DGAT1* was suggested as a positional and functional candidate gene for intramuscular fat deposition in cattle (Thaller et al. 2003b). The chromosome regions flanking bovine *DGAT1* are gene rich (Winter et al. 2004), and results from fine-mapping analyses are suggestive of yet to be identified polymorphisms affecting fat deposition (Moore et al. 2003). Also linked with *DGAT1* is the gene encoding thyroglobulin (*TG*), whose product is a precursor of hormones affecting lipid metabolism (Barendse 1999). In the beef cattle industry, commercial tests have been designed to evaluate polymorphisms in TG and used to aid genetic improvement of marbling.

Inconsistent tenderness of beef is frequently identified by US consumers as a primary reason for dissatisfaction with beef (e.g., Huffman et al. 1996). Casas et al. (2000) and Morris et al. (2001) independently identified a QTL for shear force, an objective measure that is highly correlated with tenderness as perceived by consumers, to BTA29. The gene encoding micromolar calcium-activated neutral protease (*CAPN1*), which degrades myofibrillar proteins under postmortem conditions and appears to be the primary enzyme in the postmortem tenderization process (Koohmaraie 1992, 1994, 1996), was colocated with the QTL (Smith et al. 2000). Page et al. (2002) presented evidence that SNPs predicted to result in amino acid changes in the *CAPN1* gene were associated with variation in meat tenderness. The utility of these markers was subsequently validated and has been developed into a commercial test to improve tenderness of beef (Page et al. 2004).

1.2.3
Using Genotypes in Breeding Cattle

Numerous simulation studies have documented theoretical potential for use of genotypic information for marker-assisted selection (MAS) to increase rate of response to selection (e.g., Soller 1978; Soller and Beckmann 1983; Lande and Tompson 1990). Results of MAS are expected to be especially favorable in multiple ovulation and embryo transfer schemes

where selection can be practiced within full-sib families (Kashi et al. 1990; Ruane and Colleau, 1996; Gomez-Raya and Klemetsdal 1999). However, negative covariance between QTL and polygenes results from selection (Bulmer 1971) and may compromise response resulting from MAS in comparison with traditional selection (Gomez-Raya and Gibson 1993; Gibson 1994; Spelman and Garrick 1997). Schulman and Dentine (2005) suggested within-family two-stage selection using MAS will result in more rapid genetic gain than traditional selection and that this advantage will persist over several generations. To date, demonstrated gain in rate of genetic improvement from use of DNA-based tests for QTL is limited, because selection without markers is already quite accurate, very few QTL have been identified as of yet, and genotyping is relatively expensive (Goddard 2003).

The B-variant of the milk protein β-lactoglobulin is associated with a DNA polymorphism resulting in decreased synthesis of this protein variant in milk, leading to a decrease in whey protein concentration and an increase in casein concentration, which results in an increase in cheese yield. Tests have long been available for milk protein variants, originally based on testing for the presence of each variant in milk protein and more recently from direct DNA tests. From 1995 onward, 28 suppliers (representing 8,500 cows) belonging to a small cheese manufacturing cooperative preferentially bred females to bulls that were homozygous for the B allele at the β-lactoglobulin locus (Boland et al. 2000).

Some studies have found association between consumption of the A1 variant of β-casein and onset of insulin-dependent diabetes. In addition, there is epidemiological evidence to suggest a relationship between consumption of milk (and therefore the A1 variant of β-casein) and coronary heart disease. Subsequent studies have been unable to consistently confirm these associations (Hill et al. 2002). The studies showing a benefit of milk containing the A2 variant of β-casein have led to the formation of a New Zealand company to market milk obtained from cows that are homozygous for the A2 allele, presumably with claims as to the health benefits with regard to diabetes and heart disease. A recent agreement was reached between the New Zealand Company and a US corporation to test some 100,000 cows in the US with the aim of marketing A2 milk in 5,000 US health food stores.

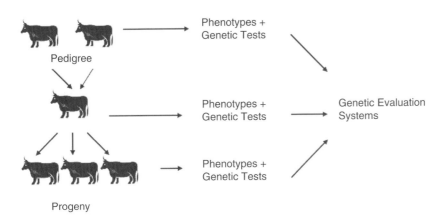

Fig. 6 Schematic representation of the genetic evaluation system including both phenotypic and genotypic data for prediction of breeding values (after Thallman 2004)

Causal mutations have been identified for more than 28 genetic disorders that affect cattle (Nicholas 2003). Molecular genetic tests have been developed for several of these disorders including: bovine leukocyte adhesion deficiency (Schuster et al. 1992), uridine monophosphate synthase deficiency (Schwenger et al. 1993), and complex vertebral malformation (Grzybowski 2003). Given that every animal likely carries a few deleterious recessive alleles, it is unlikely that culling carrier bulls will be an effective solution for eliminating all genetic defects. The general philosophy has shifted from trying to eliminate carrier animals as rapidly as possible to one of a controlled reduction in allele frequency at the population level and avoidance of matings between carriers. Unpublished simulations show net industry benefit from continued use of proven carrier bulls if their genetic merit is sufficiently high, with future carrier males being discarded before progeny testing in favor of noncarrier sibs.

It has been envisioned that phenotypic and genotypic data will be combined in future genetic evaluation systems to produce marker-assisted estimates of genetic merit (Fig. 6; Thallman 2004). Cornell University currently produces a marker-assisted genetic evaluation for shear force, a quantitative measure of meat tenderness, for the American Simmental Association using phenotypes from progeny testing and genotypes at *SNP316* in the *CAPN1* gene (E. J. Pollak and R. L. Quaas, personal communication). In an evaluation of 120 progeny tested sires, average accuracy of the resulting estimated predicted difference (EPD) was increased from 0.26 to 0.28 by including the genotypic information. However, accuracy of a genetic evaluation computed using only the genotypic information from this single marker would be on the order of 0.05. Thus, collection of phenotype data is still a critical requirement for accurate evaluation of genetic potential.

1.3 Future Scope of Work

Over the past several years, tremendous progress has been made with respect to cloning/obtainment of putative embryonic stem cells and targeted genetic modification/homologous recombination in somatic cells (e.g., see reviews by Tian et al. 2003; Wang and Zhou 2003; Saito et al. 2004). When coupled with the sequencing of the bovine genome, it will be possible to manipulate the bovine genome in ways only previously possible in laboratory species. It is expected that these advancements will dramatically facilitate our understanding of the biology of beef and dairy cattle.

Hand in hand with these advancements are numerous bovine-specific functional genomic resources that are either available or being developed (e.g., full-length complementary DNA or cDNA clones; cDNA and long-oligo microarrays). Recent development of these resources has dramatically increased the need for bioinformatic resources, and several efforts are underway (e.g., http://www.ncbi.nlm.nih.gov/genome/guide/cow/index.html; http://bovinegenome.org/; http://www.livestockgenomics.csiro.au/; http://locus.jouy.inra.fr/cgi-bin/bovmap/intro.pl; http://www.tigr.org/microarray/; http://www.animalgenome.org/

cattle/). Together these genomic, functional genomic, and bioinformatic resources will facilitate defining molecular mechanisms underlying phenotypes of interest.

Ultrahigh-throughput SNP genotyping with modest cost and mapping of haplotypes may allow association-based approaches to be applied to candidate genes, QTL regions, or serve as a basis for genome-wide scanning of cattle in a manner analogous to that envisioned by The International HapMap Consortium (2003). However, there is only marginal improvement in precision of traditional QTL mapping by linkage analysis through the use of a dense marker map (Darvasi et al. 1993). Thus, a shift from the present paradigm of searching for causative mutations underlying QTL to a breeding strategy based on molecular biology may be anticipated. Meuwissen et al. (2001) demonstrated that accurate estimates of breeding value could be obtained for animals that have no phenotypic record of their own and no progeny using a dense and complete marker map and phenotypic information from the parental and grandparental generations. This approach also facilitates shortening the generation interval and thus accelerating genetic improvement.

Dekkers and Hospital (2002) concluded that genotype-environment interaction, pleiotropy, and epistasis would make selection for complex traits solely on the basis of molecular information risky unless consequent response was confirmed by phenotypic evaluation. Mapping traits to chromosomal locations is a concomitant requirement for accurately and precisely assessing phenotypic variation. Daughter and granddaughter designs use the law of averages to reduce environmental variance in mapping QTL in dairy cattle. QTL-mapping strategies for beef cattle have relied on crosses of phenotypically diverse breeds to increase statistical power by magnifying the smallest true difference to be detected. Functional approaches are also inherently dependent on accurate and precise phenotypes.

References

Amarante MRV, Yang YP, Kata SR, Lopes CR, Womack JE (2000) RH maps of bovine chromosomes 15 and 29: conservation of human chromosomes 11 and 5. Mamm Genome 11:364–368

Arthur PF (1995) Double muscling in cattle: a review. Aust J Agri Res 46:1493–1515

Arranz JJ, Coppieters W, Berzi P, Cambisano N, Grisart B, Karim L, Marcq F, Moreau L, Mezerm C, Riquet J, Simon P, Vanmanshoven P, Wagenaar D, Georges M (1998) A QTL affecting milk yield and composition traits maps to bovine chromosome 20: a confirmation. Anim Genet 29:107–115

Ashwell MS, Rexroad CE, Miller RH, VanRaden PM, Da Y (1997) Detection of loci affecting milk production and health traits in an elite U.S. Holstein population using microsatellite markers. Anim Genet 28:216–222

Ashwell MS, Van Tassell CP (1999) Detection of putative loci affecting milk, health, and type traits in a US Holstein population using 70 microsatellite markers in a genome scan. J Dairy Sci 82:2497–2502

Ashwell MS, Van Tassell CP, Sonstegard TS (2001) A genome scan to identify quantitative trait loci affecting economically important traits in a US Holstein population. J Dairy Sci 84:2535–2542

Ashwell MS, Heyen DW, Sonstegard TS, Van Tassell CP, Da Y, VanRaden PM, Ron M, Weller JI, Lewin HA (2004) Detection of quantitative trait loci affecting milk production, health, and reproductive traits in Holstein cattle. J Dairy Sci 87:468–475

Bailey JF, Richards MB, Macauley VA, Colson IB, James IT, Bradley DG, Hedges REM, Sykes BC (1996) Ancient DNA suggests a recent expansion of European cattle from a diverse wild progenitor species. Proc R Soc Lond B 263:1467–1473

Band MR, Larson JH, Rebeiz M, Green CA, Heyen DW, Donovan J, Windish R, Steining C, Mahyuddin P, Womack JE, Lewin HA (2000) An ordered comparative map of the cattle and human genomes. Genome Res 10:1359–1368

Barendse WJ (1999) Assessing lipid metabolism. Patent. International Publication Number: WO 99/23248. World International Property Organization

Barendse W, Armitage SM, Ryan AM, Moore SS, Clayton D, Georges M, Womack JE, Hetzel J (1993) A genetic-map of DNA loci on bovine chromosome-1. Genomics 18:602–608

Barendse W, Vaiman D, Kemp SJ, Sugimoto Y, Armitage SM, Williams JL, Sun HS, Eggen A, Agaba M, Aleyasin SA, Band M, Bishop MD, Buitkamp J, Byrne K, Collins F, Cooper L, Coppetiers W, Denys B, Drinkwater RD, Easterday K, Elduque C, Ennis S, Erhardt G, Ferretti L, Flavin N, Gao Q, Georges M, Gurung R, Harlizius B, Hawkins G, Hetzel J, Hirano T, Hulme D, Jorgensen C, Kessler M, Kirkpatrick BW, Konfortov B, Kostia S, Kuhn C, Lenstra JA, Leveziel H, Lewin HA, Leyhe B, Lil L, Burriel IM, McGraw RA, Miller JR, Moody DE, Moore SS, Nakane S, Nijman IJ, Olsaker I, Pomp D, Rando A, Ron M, Shalom A, Teale AJ, Thieven U, Urquhart BGD, Vage DI, VandeWeghe A, Varvio S, Velmala R, Vikki J, Weikard R, Woodside C, Womack JE, Zanotti M, Zaragoza P (1997) A medium-density genetic linkage map of the bovine genome. Mamm Genom 8:21–28

Bekman H, van Arendonk JAM (1993) Derivation of economic values for veal, beef and milk production traits using profit equations. Livest Prod Sci 34:35–56

Bellinge RHS, Liberles DA, Iaschi SPA, O'Brien PA, Tay GK (2005) Myostatin and its implications on animal breeding: a review. Anim Genet 36:1–6

Bennewitz J, Reinsch N, Grohs C, Leveziel H, Malafosse A, Thomsen H, Xu N, Looft C, Kuhn C, Brockmann GA, Schwerin M, Weimann C, Hiendleder S, Erhardt G, Medjugorac I, Russ I, Forster M, Brenig B, Reinhardt F, Reents R, Averdunk G, Blumel J, Boichard D, Kalm E (2003) Combined analysis of data from two granddaughter designs: a simple strategy for QTL confirmation and increasing experimental power in dairy cattle. Genet Sel Evol 35:319–338

Bennewitz J, Reinsch N, Paul S, Looft C, Kaupe B, Weimann C, Erhardt G, Thaller G, Kuhn C, Schwerin M, Thomsen H, Reinhardt F, Reents R, Kalm E (2004) The *DGAT1 K232A* mutation is not solely responsible for the milk production quantitative trait locus on the bovine chromosome 14. J Dairy Sci 87:431–442

Bishop MD, Kappes SM, Keele JW, Stone RT, Sunden SLF, Hawkins GA, Toldo SS, Fries R, Grosz MD, Yoo JY, Beattie CW (1994) A genetic linkage map for cattle. Genetics 136:619–639

Blott S, Kim J, Moisio S, Schmidt-Kuntzel A, Cornet A, Berzi P, Cambisano N, Ford C, Grisart B, Johnson D, Karim L, Simon P, Snell R, Spelman R, Wong J, Vilkki J, Goerges M, Farnir F, Coppieters W (2003) Molecular dissection of a quantitative trait locus: a phenylalanine-to-tyrosine substitution in the transmembrane domain of the bovine growth hormone receptor is associated with a major effect on milk yield and composition. Genetics 163:253–266

Boichard D, Grohs C, Bourgeois F, Cerqueira F, Faugeras R, Neau A, Rupp R, Amigues Y, Boscher MY, Leveziel H (2003) Detection of genes influencing economic traits in three French dairy cattle breeds. Genet Sel Evol 35:77–101

Boland M, Hill J, O'Connor P (2000) Changing the milk supply to increase cheese yield: the Kaikoura experience. In: Milk Composition. British Soc Anim Sci Occasional Publ 25:313–316

Bradley DG, MacHugh DE, Cunningham P, Loftus RT (1996) Mitochondrial diversity and origins of African and European cattle. Proc Natl Acad Sci USA 93:5131–5135

Bradley DG, Loftus RT, Cunningham P, MacHugh DE (1998) Genetics and domestic cattle origins. Evol Anthro 6:79–86

Buitkamp J, Kollers S, Durstewitz G, Fries R, Welzel K, Schafer K, Kellermann A, Lehrach H (2001) Construction and characterization of a gridded cattle BAC library. Anim Genet 31:347–351

Bulmer MG (1971) The effect of selection on genetic variability. Am Nat 105:201–211

Cai L, Taylor JF, Wing RA, Gallagher DS, Woo SS, Davis SK (1995) Construction and characterization of a bovine bacterial artificial chromosome library. Genomics 29:413–425

Casas E, Keele JW, Shackelford SD, Koohmaraie M, Sonstegard TS, Smith TPL, Kappes SM, Stone RT (1998) Association of the muscle hypertrophy locus with carcass traits in beef cattle. J Anim Sci 76:468–473

Casas E, Shackelford SD, Keele JW, Stone RT, Kappes SM, Koohmaraie M (2000) Quantitative trait loci affecting growth and carcass composition of cattle segregating alternate forms of myostatin. J Anim Sci 78:560–569

Casas E, Shackelford SD, Keele JW, Koohmaraie M, Smith TPL, Stone RT (2003) Detection of quantitative trait loci for growth and carcass composition in cattle. J Anim Sci 81:2976–2983

Casas E, Keele JW, Shackelford SD, Koohmaraie M, Stone RT (2004) Identification of quantitative trait loci for growth and carcass composition in cattle. Anim Genet 35:2–6

Cases S, Smith SJ, Zhang YW, Myers HM, Lear SR, Sande E, Novak S, Collins C, Welch CB, Lusis AJ, Erickson SK, Farese RV (1998) Identification of a gene encoding an acyl CoA: diacylglycerol acyltransferase, a key enzyme in triacylglycerol synthesis. Proc Natl Acad Sci USA 95:13018–13023

Chamberlain A, McParlan H, Balasingham T, Carrick M, Bowman P, Robinson N, Goddard M (2002) Mapping QTL affecting milk composition traits in dairy cattle using a complex pedigree. Proc 7th World Congr on Genet Appl Livest Prod, Montpellier, France, August 19–23, paper 09–08

Charlier C, Coppieters W, Farnir F, Grobet L, Leroy PL, Michaux C, Mni M, Schwers A, VanManshoven P, Hanset R, Georges M (1995) The MH gene causing double-muscling in cattle maps to bovine chromosome 2. Mamm Genom 6:788–792

Chowdhary BP, Fronicke L, Gustavsson I, Scherthan H (1996) Comparative analysis of the cattle and human genomes: detection of ZOO-FISH and gene mapping-based chromosomal homologies. Mamm Genom 7:297–302

Clutton-Brock J (1993) The spread of domestic animals in Africa. In: Shaw T, Sinclair PJJ, Andah B, Okpoko A (eds) The Archaeology of Africa: Food, Metals, and Towns. Routledge, London, UK, pp 61–70

Cohen-Zinder M, Seroussi E, Larkin DM, Loor JJ, Everts-van der Wind A, Lee JH, Drackley JK, Band MR, Hernande AG, Shani M, Lewin HA, Weller JI, Ron M (2005) Identification of a missense mutation in the bovine ABCG2 gene with a major effect on the QTL on chromosome 6 affecting milk yield and composition in Holstein cattle. Genome Res 15:936–944

Connor EE, Sonstegard TS, Ashwell MS, Bennett GL, Williams JL (2004) An expanded comparative map of bovine chromosome 27 targeting dairy form QTL regions. Anim Genet 35:265–269

Coppieters W, Riquet J, Arranz JJ, Berzi P, Cambisano N, Grisart B, Karim L, Marcq F, Moreau L, Nezer C, Simon P, Vanmanshoven P, Wagenaar D, Georges M (1998) A QTL with major effect on milk yield and composition maps to bovine Chromosome 14. Mamm Genom 9:540–544

Cruickshank J, Dentine MR, Berger PJ, Kirkpatrick BW (2004) Evidence for quantitative trait loci affecting twinning rate in North American Holstein cattle. Anim Genet 35:206–212

Culley G (1807) Observations on Livestock, 4th edn. G. Woodfall, London, UK

Cunningham EP (1992) Selected Issues in Livestock Industry Development. The World Bank, Washington, DC, USA, pp 1–12

Cunningham EP, McClintock AE (1974) Selection in dual-purpose populations: effect of beef crosses and cow replacement rates. Ann Genet Sel Anim 6:227–239

Darvasi A, Weinreb A, Minke V, Weller JI, Soller M (1993) Detecting marker-QTL linkage and estimating QTL gene effect and map location using a saturated genetic map. Genetics 134:943–951

de Givry S, Bouchez M, Chabrier P, Milan D, Schiex T (2005) CarthaGene: multipopulation integrated genetic and radiation hybrid mapping. Bioinformatics 21:1703–1704

Dekkers JCM, Hospital F (2002) The use of molecular genetics in the improvement of agricultural populations. Nat Rev Genet 3:22–32

Dickerson GE, Künzi N, Cundiff LV, Koch RM, Arthaud VH, Gregory KE (1974) Selection criteria for efficient beef production. J Anim Sci 39:659–673

Eggen A, Gautier M, Billaut A, Petit E, Hayes H, Laurent P, Urban C, Pfister-Genskow M, Eilertsen K, Bishop MD (2001) Construction and characterization of a bovine BAC library with four genome-equivalent coverage. Genet Sel Evol 33:543–548

Elo KT, Vikki J, deKoning DJ, Velmala RJ, Maki-Tanila AV (1999) A quantitative trait locus for live weight maps to bovine chromosome 23. Mamm Genom 10:831–835

Everts-van der Wind A, Kata SR, Band MR, Rebeiz M, Larkin DM, Everts RE, Green CA, Liu L, Natarajan S, Goldammer T, Lee JH, McKay S, Womack JE, Lewin HA (2004) A 1463 gene cattle-human comparative map with anchor points defined by human genome sequence coordinates. Genome Res 14:1424–1437

Freyer G, Kuhn C, Weikard R, Zhang Q, Mayer M, Hoeschele I (2002) Multiple QTL on chromosome six in dairy cattle affecting yield and content traits. J Anim Breed Genet 119:69–82

Freyer G, Kuhn C, Weikard R (2003a) Comparisons of different statistical approaches of QTL detection by evaluating results from a real dairy cattle data set. Archiv fur Tierzucht 46:413–423

Freyer G, Sorensen P, Kuhn C, Weikard R, Hoeschele I (2003b) Search for pleiotropic QTL on chromosome BTA6 affecting yield traits of milk production. J Dairy Sci 86:999–1008

Freyer G, Sorensen P, Kuhn C, Weikard R (2004) Investigations in the character of QTL affecting negatively correlated milk traits. J Anim Breed Genet 121:40–51

Fries R, Winter A (2002) Method of testing a mammal for its predisposition for fat content of milk and/or its predisposition for meat marbling. Intl Patent Appl No. PCT/EP 02/07520. World Intl Property Org

Geldermann H (1975) Investigations on inheritance of quantitative characters in animals by gene markers. I. Methods. Theor Appl Genet 46:319–330

Georges M, Dietz AB, Mishra A, Nielsen D, Sargeant LS, Sorensen A, Steele MR, Zhao XY, Leipold H, Womack JE, Lathrop M (1993) Microsatellite mapping of the gene causing Weaver Disease in cattle will allow the study of and associated quantitative trait locus. Proc Natl Acad Sci USA 90: 1058–1062

Georges M, Nielsen D, MacKinnon M, Mishra A, Okimoto R, Pasquino AT, Sargeant LS Sorensen A, Steele MR, Zhao X, Womack JE, Hoeschele I (1995) Mapping quantitative trait loci controlling milk production in dairy cattle by exploiting progeny testing. Genetics 139:907–920

Gibson JP (1994) Short-term gain at the expense of long-term response with selection of identified loci. Proc 5th World Congr Genet Appl Livest Prod 21:201–204

Goddard ME (2003) Animal breeding in the (post-) genomic era. Anim Sci 76:353–365

Gomez-Raya L, Gibson JP (1993) Within family selection at an otherwise unselected locus in dairy cattle. Genome 36:433–439

Gomez-Raya L, Klemetsdal G (1999) Two-stage selection strategies utilizing marker-quantitative trait locus information and individual performance. J Anim Sci 77:2008–2018

Gonda MG, Arias JA, Shook GE, Kirkpatrick BW (2004) Identification of an ovulation rate QTL in cattle on BTA14 using selective DNA pooling and interval mapping. Anim Genet 35:298–304

Goss SJ, Harris H (1975) New method for mapping genes in human chromosomes. Nature 255:680–684

Graser H-U, Nitter G, Barwick SA (1994) Evaluation of advanced industry breeding schemes for Australian beef cattle. II. Selection on combinations of growth, reproduction and carcase criteria. Aust J Agri Res 45:1657–1669

Grigson C (1980) The craniology and relationships of four species of Bos. 5. *Bos Indicus* L. J Archeological Sci 7:3–32

Grisart B, Coppieters W, Farnir F, Karim L, Ford C, Berzi P, Cambisano N, Mni M, Reid S, Simon P, Spelman R, Georges M, Snell R (2002) Positional candidate cloning of a QTL in dairy cattle: identification of a missense mutation in the bovine DGAT1 gene with major effect on milk yield and composition. Genome Res 12:222–231

Grisart B, Farnir F, Karim L, Cambisano N, Kim JJ, Kvasz A, Mni M, Simon P, Frere JM, Coppieters W, Georges M (2004) Genetic and functional confirmation of the causality of the *DGAT1 K232A* quantitative trait nucleotide in affecting milk yield and composition. Proc Natl Acad Sci USA 101:2398–2403

Grobet L, Martin LJR, Poncelet D, Pirottin D, Brouwers B, Riquet J, Schoeberlein A, Dunner S, Menissier F, Massabanda J, Fries R, Hanset R, Georges M (1997) A deletion in the bovine myostatin gene causes the double-muscled phenotype in cattle. Nat Genet 17:71–74

Grupe S, Panicke L, Dietl G, Kuhn C, Gulard V, Schwerin M (1998) Identification of loci with significant effects on stillbirth and calving difficulties in Holstein cattle. Archiv. Anim Breed 41:151–158

Grzybowski G (2003) Complex vertebral malformation and its implication for cattle breeding. Medycyna Weterynaryjna 59:107–111

Hanotte O, Bradley DG, Ochleng JW, Verjee Y, Hill EW, Rege JE (2002) African Pastoralism: genetic imprints of origins and migrations. Science 296:336–339

Hanotte O, Ronin Y, Agaba M, Nilsson P, Gelhaus A, Horstmann R, Sugimoto Y, Kemp S, Gibson J, Korol A, Soller M, Teale A (2003) Mapping of quantitative trait loci controlling trypanotolerance in a cross of tolerant West African N'Dama and susceptible East African Boran cattle. Proc Natl Acad Sci USA 100:7443–7448

Hanset R (1981) Double muscling in cattle – its inheritance – its meaning for beef production. Ann Med Vet 125:85–95

Hayes H (1995) Chromosome painting with human chromosome-specific DNA libraries reveals the extent and distribution of conserved segments in bovine chromosomes. Cytogenet Cell Genet 71:168–174

Hazel LN (1934) The genetic basis for constructing selection indexes. Genetics 28:476–490

Heuertz S, Hors-Cayla MC (1981) Cattle gene-mapping by somatic-cell hybridization study of 17 enzyme markers. Cytogenet Cell Genet 30:137–145

Heyen DW, Weller JI, Ron M, Band M, Beever JE, Feldmesser E, Da Y, Wiggans GR, VanRaden PM, Lewin HA (1999) A genome scan for QTL influencing milk production and health traits in dairy cattle. Physiol Genom 1:165–175

Hiendleder S, Thomsen H, Reinsch N, Bennewitz J, Leyhe-Horn B, Looft C, Xu N, Medjugorac I, Russ I, Kuhn C, Brockmann GA, Blumel J, Brenig B, Reinhardt F, Reents R, Averdunk G, Schwerin M, Forster M, Kalm E, Erhardt G (2003) Mapping of QTL for body conformation and behavior in cattle. J Hered 94:496–506

Hill JP, Crawford RA, Boland MJ (2002) Milk and consumer health: a review of the evidence for a relationship between the consumption of beta-casein A1 with heart disease and insulin-dependent diabetes mellitus. Proc NZ Soc Anim Prod 62:111–114

Hills D, Tracey S, Masabanda J, Fries R, Schalkwyk LC, Lehrach H, Miller JR, Williams LJ (1999) A bovine YAC library containing four- to five-fold genome equivalents. Mamm Genom 10:837–838

Holmberg M, Andersson-Eklund L (2004) Quantitative trait loci affecting health traits in Swedish dairy cattle. J Dairy Sci 87:2653–2659

Huffman KL, Miller MF, Hoover LC, Wu CK, Brittin HC, Ramsey CB (1996) Effect of beef tenderness on consumer satisfaction with steaks consumed in the home and restaurant. J Anim Sci 74:91–97

Ihara N, Takasuga A, Mizoshita K, Takeda H, Sugimoto M, Mizoguchi Y, Hirano T, Itoh T, Watanabe T, Reed KM, Snelling WM, Kappes SM, Beattie CW, Bennett GL, Sugimoto Y (2004) A comprehensive genetic map of the cattle genome based on 3802 microsatellites. Genome Res 14:1987–1998

Itoh T, Takasuga A, Watanabe T, Sugimoto YT (2003) Mapping of 1400 expressed sequence tags in the bovine genome using a somatic cell hybrid panel. Anim Genet 34:362–370

Itoh T, Watanabe T, Ihara N, Mariani P, Beattie CW, Sugimoto Y, Takasuga A (2005) A comprehensive radiation hybrid map of the bovine genome comprising 5593 loci. Genomics 85:413–424

Kambadur R, Sharma M, Smith TPL, Bass JJ (1997) Mutations in myostatin (GDF8) in double muscled Belgian Blue and Piedmontese cattle. Genome Res 7:910–915

Kappes SM, Keele JW, Stone RT, Sonstegard TS, Smith TPL, McGraw RA, LopezCorrales NL, Beattie CW (1997) A second-generation linkage map of the bovine genome. Genome Res 7:235–249

Kashi Y, Hallerman E, Soller M (1990) Marker assisted selection of candidate bulls for progeny testing programs. Anim Prod 51:63–74

Keele JW, Shackelford SD, Kappes SM, Koohmaraie M, Stone RT (1999) A region on bovine chromosome 15 influences beef longissimus tenderness in steers. J Anim Sci 77:1364–1371

Khatib H, Heifetz E, Dekkers JCM (2005) Association of the protease inhibitor gene with production traits in Holstein dairy cattle. J Dairy Sci. 88:1208–1213

Khatkar MS, Thomson PC, Tammen I, Raadsma HW (2004) Quantitative trait loci mapping in dairy cattle: review and meta-analysis. Genet Sel Evol 36:163–190

Kim J-J, Farnir F, Savell J, Taylor JF (2003a) Detection of quantitative trait loci for growth and beef carcass fatness traits in a cross between *Bos taurus* (Angus) and *Bos indicus* (Brahman) cattle. J Anim Sci 81:1933–1942

Kim JW, Park SI, Yeo JS (2003b) Linkage mapping and QTL on chromosome 6 in Hanwoo (Korean cattle). Asian-Aust J Anim Sci 16:1402–1405

Kirkpatrick BW, Byla BM, Gregory KE (2000) Mapping quantitative trait loci for bovine ovulation rate. Mamm Genom 11:136–139

Klungland H, Sabry A, Heringstad B, Olsen H, Gomez-Raya GL, Vage DI, Olsaker I, Odegard J, Klemetsdal G, Schulman N, Vikki J, Ruane J, Aasland M, Ronningen K, Lien S (2001)

Quantitative trait loci affecting clinical mastitis and somatic cell count in dairy cattle. Mamm Genom 12:837–842

Koohmaraie M (1992) The role of Ca²⁺-dependent proteases (calpains) in postmortem proteolysis and meat tenderness. Biochemie 74:239–245

Koohmaraie M (1994) Muscle proteinases and meat aging. Meat Sci 36:93–104

Koohmaraie M (1996) Biochemical factors regulating the toughening and tenderization process of meat. Meat Sci 43:S193–S201

Kuhn C, Weikard R, Goldammer T, Grupe S, Olsaker I, Schwerin M (1996) Isolation and application of chromosome 6 specific microsatellite markers for detection of QTL for milk-production traits in cattle. J Anim Breed Genet 113:355–362

Kuhn C, Bennewitz J, Reinsch N, Xu N, Thomsen H, Looft C, Brockman GA, Schwerin M, Weimann C, Hiendieder S, Erhardt G, Medjugorac I, Forster M, Brenig B, Reinhardt F, Reents R, Russ I, Averdunk G, Blumel J, Kalm E (2003) Quantitative trait loci mapping of functional traits in the German Holstein cattle population. J Dairy Sci 86:360–368

Kuhn C, Thaller G, Winter A, Bininda-Emonds ORP, Kaupe B, Erhardt G, Bennewitz J, Schwerin M, Fries R (2004) Evidence for multiple alleles at the *DGAT1* locus better explains a quantitative tip trait locus with major effect on milk fat content in cattle. Genetics 167:1873–1881

Kühn R, Ludt C, Manhart H, Peters J, Neumair E, Rottmann O (2005) Close genetic relationship of early Neolithic cattle from Ziegelberg (Freising, Germany) with modern breeds. J Anim Breed Genet 122 (Suppl 1):36–44

Lande R, Thompson R (1990) Efficiency of marker assisted selection in the improvement of quantitative traits. Genetics 124:743–756

Li C, Basarab J, Snelling WM, Benkel B, Kneeland J, Murdoch B, Hansen C, Moore SS (2004) Identification and fine mapping of quantitative trait loci for backfat on bovine chromosomes 2, 5, 6, 19, 21, and 23 in a commercial line of *Bos taurus*. J Anim Sci 82:967–972

Lien S, Karlsen A, Klemetsdal G, Vage DI, Olsaker I, Klungland H, Aasland M, Heringstad B, Ruane J, Gomez-Raya L (2000) A primary scan of the bovine genome for quantitative trait loci affecting twinning rate. Mamm Genom 10:877–882

Libert F, Lefort A, Okimoto R, Womack J, Georges M (1993) Construction of a bovine genomic library of large yeast artificial chromosome clones. Genomics 18:270–276

Loftus RT, MacHugh DE, Bradley DG, Sharp PM, Cunningham P (1994) Evidence for two independent domestications of cattle. Proc Natl Acad Sci USA 91:2757–2761

Looft C, Reinsch N, Karall-Albrecht C, Paul S, Brink M, Thomsen H, Brockmann G, Kuhn C, Schwerin M, Kalm E (2001) A mammary gland EST showing linkage disequilibrium to a milk production QTL on bovine chromosome 14. Mamm Genom 12:646–650

Lush JL (1945) Animal Breeding Plans. Iowa State College Press, Ames, USA

MacNeil MD, Cundiff LV, Dinkel CA, Koch RM (1984) Genetic correlations among sex-limited traits in beef cattle. J Anim Sci 58:1171–1180

MacNeil MD, Urick JJ, Newman S, Knapp BW (1992) Selection for postweaning growth in inbred Hereford cattle: The Fort Keogh, Montana Line 1 example. J Anim Sci 70:723–733

MacNeil MD, Newman S, Enns RM, Stewart-Smith J (1994) Relative economic values for Canadian beef production using specialized sire and dam lines. Can J Anim Sci 74:411–417

MacNeil MD, Grosz MD (2002) Genome-wide scans for QTL affecting carcass traits in Hereford x composite double backcross populations. J Anim Sci 80:2316–2324

Mäki-Tanila A, de Koning D-J, Elo K, Moisio S, Velmala R, Vikki J (1998) Mapping of multiple quantitative trait loci by regression in half sib designs. Proc 6th World Congr Genet Appl Livest Prod, Armidale, Australia, January 11–16, 26:269–272

Mannen H, Kohno M, Nagata Y, Tsuji S, Bradley DG, Yeo JS, Nyamsamba D, Zagdsuren Y, Yokohama M, Nomura K, Amono T (2004) Independent mitochondrial origin and historical genetic differentiation in North Eastern Asian cattle. Mol Phytogenet Evol 32:539–544

McPherron AC, Lee SJ (1997) Doubling muscling in cattle due to mutations in the myostatin gene. Proc Natl Acad Sci USA 94:12457–12461

Meuwissen THE, Hayes BJ, Goddard ME (2001) Prediction of total genetic value using genome-wide dense marker maps. Genetics 157:1819–1829

Meuwissen THE, Karlsen A, Lien S, Olsaker I, Goddard ME (2002) Fine mapping of a quantitative trait locus for twinning rate using combined linkage and linkage disequilibrium mapping. Genetics 161:373–379

Moore SS, Li C, Basarab J, Snelling WM, Kneeland J, Murdoch B, Hansen C, Benkel B (2003) Fine mapping of quantitative trait loci and assessment of positional candidate genes for backfat on bovine chromosome 14 in a commercial line of Bos taurus. J Anim Sci 81:1919–1925

Morris CA, Cullen NG, Hickey SM, Crawford AM, Hyndman DL, Bottema CDK, Pitchford WS (2001) Progress in DNA marker studies of beef carcass composition and meat quality in New Zealand and Australia. Proc Assoc Adv Anim Breed Genet, Queenstown, NZ 14:17–22

Nicholas FW (2003) Online Mendelian Inheritance in Animals (OMIA): a comparative knowledgebase of genetic disorders and other familial traits in non-laboratory animals. Nucl Acids Res 31:275–277

Niebel E (1986) Economic evaluation of breeding objectives for milk and beef production in temperate zones. Proc 3rd World Congr Genet Appl Livest Prod, Lincoln, NE 9:18–32

Norman HD, Dickinson FN (1971) An economic index for determining the relative value of milk and fat in predicted differences of bulls and cow index values of cows. ARS-44–223. Dairy Herd Improv Lett 47(1):1–34

O'Brien SJ, Womack JE, Lyons LA, Moore KJ, Jenkins NA, Copeland NG (1993) Anchored reference loci for comparative genome mapping in mammals. Nat Genet 3:103–12

Olsen HG, Gomez-Raya L, Vage DI, Olsaker I, Klungland H, Svendsen M, Adnoy T, Sabry A, Klemetsdal G, Schulman N, Kramer W, Thaller G, Ronningen K, Lien S (2002) A genome scan for quantitative trait loci affecting milk production in Norwegian dairy cattle. J Dairy Sci 85:3124–3130

Olsen HG, Lien S, Svendsen M, Nilsen H, Roseth A, Opsal MA, Meuwissen THE (2004) Fine mapping of milk production QTL on BTA6 by combined linkage and linkage disequilibrium analysis. J Dairy Sci 87:690–698

Olsen HG, Lien S, Gautier M, Nilsen H, Roseth A, Berg PR, Sundsaasen KK, Svendsen M, Meuwissen THE (2005) Mapping of a milk production quantitative trait locus to a 420-kb region on bovine chromosome 6. Genetics 169:275–283

Page BT, Casas E, Heaton MP, Cullen NG, Hyndman DL, Morris CA, Crawford AM, Wheeler TL, Koohmaraie M, Keele JW, Smith TPL (2002) Evaluation of single-nucleotide polymorphisms in CAPN1 for association with meat tenderness in cattle. J Anim Sci 80:3077–3085

Page BT, Casas E, Quaas RL, Thallman RM, Wheeler TL, Shackelford SD, Koohmaraie M, White SN, Bennett GL, Keele JW, Dikeman ME, Smith TPL (2004) Association of markers in the bovine *CAPN1* gene with meat tenderness in large crossbred populations that sample influential industry sires. J Anim Sci 82:3474–3481

Ponzoni RW, Newman S (1989) Developing breeding objectives for Australian beef cattle production. Anim Prod 49:35–47

Rexroad CE, Womack JE (1999) Parallel RH mapping of BTA1 with HSA3 and HSA21. Mamm Genom 10:1095–1097

Riquet J, Coppieters W, Cambisano N, Arranz JJ, Berzi P, Davis SK, Grisart B, Farnir F, Karim L, Mni M, Simon P, Taylor JF, Vanmanshoven P, Wagenaar D, Womack JE, Georges M (1999) Fine-mapping of quantitative trait loci by identity by descent in outbred populations: application to milk production in dairy cattle. Proc Natl Acad Sci USA 96:9252–9257

Roberts RC (1979) Side effects of selection for growth in laboratory animals. Livest Prod Sci 6:93–104

Rodriguez-Zas SL, Southey BR, Heyen DW, Lewin HA (2002) Interval and composite interval mapping of somatic cell score, yield, and components of milk in dairy cattle. J Dairy Sci 85:3081–3091

Ron M, Kliger D, Feldmesser E, Seroussi E, Ezra E, Weller JI (2001) Multiple quantitative trait locus analysis of bovine chromosome 6 in the Israeli Holstein population by a daughter design. Genetics 159:727–735

Ron M, Feldmesser E, Golik M, Tager-Cohen I, Kliger D, Reiss V, Domochovsky R, Alus O, Seroussi E, Erza E, Weller JI (2004) A complete genome scan of the Israeli Holstein population for quantitative trait loci by a daughter design. J Dairy Sci 87:476–490

Rothschild MF, Soller M (1997) Candidate gene analysis to detect genes controlling traits of economic importance in domestic livestock. Probe 8:13–22

Rouse JE (1970) World Cattle. University of Oklahoma Press, Norman, USA

Ruane J, Colleau JJ (1996) Marker assisted selection for a sex-limited character in a nucleus breeding population. J Dairy Sci 79:1666–1678

Saito S, Lui B, Yokoyama K (2004) Animal embryonic stem (ES) cells: self-renewal, pluropotency, trangenesis and nuclear transfer. Hum Cell 17:107–115

Schmutz SM, Buchanan FC, Plante Y, Winkelman-Sim DC, Aalhus J, Boles JA, Moker JS (2000) Mapping collagenase and a QTL to beef tenderness to cattle chromosome 29. Plant and Animal Genome VIII Conf, San Diego, CA, USA, p 143

Schibler L, Roig A, Mahe MF, Save JC, Gautier M, Taourit S, Boichard D, Eggen A, Cribiu EP (2004) A first generation bovine BAC-based physical map. Genet Sel Evol 36:105–122

Schnabel RD, Kim JJ, Ashwell MS, Sonstegard TS, VanTassell CP, Connor EE, Taylor JF (2005) Fine mapping milk production quantitative trait loci on BTA6: analysis of the bovine osteopontin gene. Proc Natl Acad Sci USA 102:6896–6901

Schrooten C, Bovenhuis H, Coppieters W, Van Arendonk JAM (2000) Whole genome scan to detect quantitative trait loci for conformation and functional traits in dairy cattle. J Dairy Sci 83:795–806

Schrooten C, Bink MCAM, Bovenhuis H (2004) Whole genome scan to detect chromosomal regions affecting multiple traits in dairy cattle. J Dairy Sci 87:3550–3560

Schulman NF, Dentine MR (2005) Linkage disequilibrium and selection response in two-stage marker assisted selection of dairy cattle over several generations. J Anim Breed Genet 122:110–116

Schulman NF, Moisio SM, de Koning DJ, Elo K, Maki-Tanila JA, Vilkki J (2002) QTL for health traits in Finnish Ayshire cattle. Proc 7th World Congr Genet Appl Livest Prod, Montpellier, France, August 19–23, paper 09–19

Schulman NF, Viitala SM, de Koning DJ, Virta J, Maki-Tanila JA, Vilkki JH (2004) Quantitative trait loci for health traits in Finnish Ayrshire cattle. J Dairy Sci 87:443–449

Schuster DE, Kehrli ME, Ackermann MR, Gilbert RO (1992) Identification and prevalence of a genetic defect that causes leukocyte adhesion deficiency in Holstein cattle. Proc Natl Acad Sci USA 89:9225–9229

Schwenger B, Schober S, Simon D (1993) DUMPS cattle carry a point mutation in the uridine monophosphate synthase gene. Genomics 16:241–244

Schwerin M, Czernek-Schäfer D, Goldhammer T, Kata SR, Womack JE, Pareek R, Pareek C, Walawski K, Brunner RM (2003) Application of disease-associated differentially expressed genes – Mining for functional candidate genes for mastitis resistance in cattle. Genet Sel Evol 35 (Suppl 1): S19–S34

Short RE, MacNeil MD, Grosz MD, Gerrard DE, Grings EE (2002) Pleiotrophic effects in Hereford, Limousin, and Piedmontese F_2 crossbred calves of genes controlling muscularity including the Piedmontese myostatin allele. J Anim Sci 80:1–11

Smith TPL, Alexander LJ, Sonstegard TS, Yoo J, Beattie CW, Broom MF (1996) Construction and characterization of a large insert bovine YAC library with five-fold genomic coverage. Mamm Genom 7:155–156

Smith TPL, Lopez-Corrales NL, Kappes SM, Sonstegard TS (1997) Myostatin maps to the interval containing the bovine mh locus. Mamm Genom 8:742–744

Smith TPL, Casas E, Rexroad CE III, Kappes SM, Keele JW (2000) Bovine CAPN1 maps to a region of BTA29 containing a quantitative trait locus for meat tenderness. J Anim Sci 78:2589–2594

Snelling WM, Gautier M, Keele JW, Smith TPL, Stone RT, Harhay GP, Bennett GL, Ihara N, Takasuga A, Takeda H, Sugimoto Y, Eggen A (2004) Integrating linkage and radiation hybrid mapping data for bovine chromosome 15. BMC Genomics 5: Art No 77

Solinas-Toldo S, Lengauer C, Fries R (1995) Comparative genome map of human and cattle. Genomics 27:489–496

Soller M (1978) The use of loci associated with quantitative effects in dairy cattle improvement. Anim Prod 27:133–139

Soller M, Beckmann JS (1983) Genetic polymorphisms in varietal identification and genetic improvement. Theor Appl Genet 67:25–33

Soller M, Genizi A, Brody T (1976) On the power of experimental designs for detection of linkage between marker loci and quantitative trait loci in crosses between inbred lines. Theor Appl Genet 47:35–39

Spelman RJ, Coppieters W, Karim L, vanArendonk JAM, Bovenhuis H (1996) Quantitative trait loci analysis for five milk production traits on chromosome six in the Dutch Holstein-Friesian population. Genetics 144:1799–1807

Spelman RJ, Garrick D (1997) Utilization of marker assisted selection in a commercial dairy cow population. Livest Prod Sci 47:139–147

Spelman RJ, Huisman AE, Singireddy SR, Coppieters W, Arranz J, Georges M, Garrick DJ (1999) Short communication: quantitative trait loci analysis on 17 nonproduction traits in the New Zealand dairy population. J Dairy Sci 82:2514–2516

Spelman RJ, Ford CA, McElhinney P, Gregory GC, Snell RG (2002) Characterization of the *DGAT1* gene in the New Zealand dairy population. J Dairy Sci 85:3514–3517

Stone RT, Keele JW, Shackelford SD, Kappes SM, Koohmaraie M (1999) A primary screen of the bovine genome for quantitative trait loci affecting carcass and growth traits. J Anim Sci 77:1379–1384

Takeda H, Yamakuchi H, Ihara N, Hara K, Watanabe T, Sugimoto Y, Oshiro T, Kishine H, Kano Y, Kohno K (1998) Construction of a bovine yeast artificial chromosome (YAC) library. Anim Genet 29:216–219

Thaller G, Kramer W, Winter A, Kaupe B, Erhardt G, Fries R (2003a) Effects of *DGAT1* variants on milk production traits in German cattle breeds. J Anim Sci 81:1911–1918

Thaller G, Kuhn C, Winter A, Ewald G, Bellmann O, Wegner J, Zuhlke H, Fries R (2003b) *DGAT1*, a new positional and functional candidate gene for intramuscular fat deposition in cattle. Anim Genet 34:354–357

Thallman RM (2004) DNA testing and marker assisted selection. Proc Beef Improv Fed, Sioux Fall, SD, May 25–28, pp 20–25

Thallman RM, Moser DW, Dressler EW, Totir LR, Fernando RL, Kachman SD, Rumph JM, Dikeman ME, Pollak EJ (2003) Carcass Merit Project: DNA marker validation: http://www.beefimprovement.org/GPW-CarcassMeritProject-Final.pdf

The International HapMap Consortium (2003) The International HapMap Project. Nature 426:789–796

Tian XC, Kubota C, Enright B, Yang X (2003) Cloning animals by somatic cell nuclear transfer – biological factors. Reprod Biol Endocrinol 1:98

Troy CS, MacHugh DE, Bailey JF, Mcgee DA, Loftus RT, Cunningham P, Chamberlain AT, Sykes BC, Bradley DG (2001) Genetic evidence for Near-Eastern origins of European cattle. Nature 401:1088–1091

USDA (2005) http://www.fas.usda.gov/psd/complete_tables/ downloaded 4 April (2005)

VanRaden PM (2004) Invited review: selection on net merit to improve lifetime profit. J Dairy Sci 87:3125–3131

Van Tassell CP, Sonstegard TS, Ashwell MS (2004) Mapping quantitative trait loci affecting dairy conformation to chromosome 27 in two Holstein grandsire families. J Dairy Sci 87:450–457

Velmala RJ, Vilkki HJ, Elo KT, de Koning DJ, Maki-Tanila AV (1999) A search for quantitative trait loci for milk production traits on chromosome 6 in Finnish Ayrshire cattle. Anim Genet 30:136–143

Viitala SM, Schulman NF, de Koning DJ, Elo K, Kinos R, Virta A, Virta J, Maki-Tanila A, Vikki JH (2003) Quantitative trait loci affecting milk production traits in Finnish Ayrshire dairy cattle. J Dairy Sci 86:1828–1836

Wall E, Brotherstone S, Woolliams JA, Banos G, Coffey MP (2003) Genetic evaluation of fertility using direct and correlated traits. J Dairy Sci 86:4093–4102

Wang B, Zhou J (2003) Specific genetic modifications of domestic animals by gene targeting and animal cloning. Reprod Biol Endocrinol 1:103

Weikard R, Kuhn C, Goldammer T, Freyer G, Schwerin M (2005) The bovine PPARGC1A gene: molecular characterization and association of an SNP with variation of milk fat synthesis. Physiol Genomics 21:1–13

Weller JI, Kashi Y, Soller M (1990) Power of daughter and granddaughter designs for determining linkage between marker loci and quantitative trait loci in dairy cattle. J Dairy Sci 73:2525–2537

Weller JI, Golik M, Seroussi E, Ezra E, Ron M (2003) Population-wide analysis of a QTL affecting milk-fat production in the Israeli Holstein population. J Dairy Sci 86:2219–2227

Willham RL, Baker F, Wallace R (1993) Ideas into Action. University Printing Services, Oklahoma State Univ, Stillwater, USA

Williams JL, Eggen A, Ferretti L, Farr CJ, Gautier M, Amati G, Ball G, Caramorr T, Critcher R, Costa S, Hextall P, Hills D, Jeulin A, Kiguwa SL, Ross O, Smith AL, Saunier K, Urquhart B, Waddington D (2002) A bovine whole-genome radiation hybrid panel and outline map. Mamm Genom 13:469–474

Winter A, Alzinger A, Fries R (2004) Assessment of the gene content of the chromosomal regions flanking bovine *DGAT1*. Genomics 83:172–180

Winter A, Kramer W, Werner FAO, Kollers S, Kata S, Durstewitz G, Buitkamp J, Womack JE, Thaller G, Fries R (2001) Association of a lysine-232/alanine polymorphism in a bovine gene encoding acyl-CoA: diacylglycerol acyltransferase (*DGAT1*) with variation at a quantitative trait locus for milk fat content. Proc Natl Acad Sci USA 99:9300–9305

Womack JE, Moll YD (1985) Gene-mapping in cattle – Extensive homology with the human map. Cytogenet Cell Genet 40:781–781

Womack JE, Moll YD (1986) Gene map of the cow – Conservation of linkage with mouse and man. J Hered 77:2–7

Womack JE, Johnson JS, Owens EK, Rexroad CE III, Schläpfer J, Yang Y-P (1997) A whole genome radiation hybrid panel for bovine genome mapping. Mamm Genom 8:854–856

Yang YP, Womack JE (1998) Parallel radiation hybrid mapping: a powerful tool for high-resolution genomic comparison. Genome Res 8:731–736

Yang YP, Rexroad CE, Schlapfer J, Womack JE (1998) An integrated radiation hybrid map of bovine chromosome 19 and ordered comparative mapping with human chromosome 17. Genomics 48:93–99

Zhang C, de Koning DJ, Hernandez-Sanchez J, Haley CS, Williams JL, Wiener P (2004) Mapping of multiple quantitative trait loci affecting bovine spongiform encephalopathy. Genetics 167:1863–1872

Zhang Q, Boichard D, Hoeschele I, Ernst C, Eggen A, Murkve B, Pfister-Genskow M, Witte LA, Grignola FE, Uimari P, Thaller G, Bishop MD (1998) Mapping quantitative trait loci for milk production and health of dairy cattle in a large outbred pedigree. Genetics 149:1959–1973

Zhu BL, Smith JA, Tracey SM, Konfortov BA, Welzel K, Schalkwyk LC, Lehrach H, Kollers S, Masabanda J, Buitkamp J, Fries R, Williams JL, Miller JR (1999) A 5x genome coverage bovine BAC library: production, characterization, and distribution. Mamm Genom 10:706–709

Zhuchenko AA, Korol AB, Andryushchenko VK (1979) Linkage between loci of quantitative characters and marker loci 1. Model Genetika 14:771–779

CHAPTER 2

Water Buffalo

Leopoldo Iannuzzi(✉) and Guilia Pia Di Meo

National Research Council (CNR), Institute of Animal Production Systems in the Mediterranean Environments (ISPAAM), Laboratory of Animal Cytogenetics and Gene Mapping, Naples, Italy, leopoldo.iannuzzi@ispaam.cnr.it

2.1
Introduction

2.1.1
Taxonomic Description

Buffalo belong to the Bovini tribe of the Bovidae family, which in turn belongs to the Ruminantia suborder of the Cetartiodactyla order. Two main species of buffalo are found in the world: the Asiatic buffalo (*Bubalus bubalis*) and the African buffalo (*Syncerus caffer*). The Asiatic buffalo originated in India where domestication took place in the third millennium BC (Cockrill 1981) and/or from China where domestication occurred in the fifth millennium BC (Chen and Li 1989). The African buffalo, originating from central Africa, is a wild species living mainly in Ethiopia, Somalia, Zambia, Zimbabwe, Namibia, Botswana, Mozambique, South Africa, Kenya, and Tanzania. While there are around 900,000 African buffalo currently (East 1999), declines have occurred in recent years. Two extreme forms or types of buffalo are known: the "savannah buffalo" (the most numerous one) and the "forest buffalo" (only 60,000 head), which are commonly split into two or three subspecies: the large black savannah or Cape buffalo, *Syncerus caffer caffer*; the small reddish forest buffalo, *S. c. nanus*; and the intermediate Sudan buffalo from West Africa, *S. c. brachyceros* (Buchholtz 1990; Kingdon 1997). A fourth subspecies, *S. c. mathewsi*, known as the relict "mountain buffalo" is recognized by some authorities (Kingdon 1997), while East (1999) splits the savannah buffalo into three subspecies: the west African *S. c. brachyceros*, the central African *S. c. aequinoctialis*, and the southern *S. c. caffer*, in addition to the *S. c. nanus* as a separate group.

Cytogenetic studies have revealed that *S. c. caffer* has $2n = 52$ and a fundamental number (FN) equals

to 60, while *S. c. nanus* has $2n = 54$ and FN = 60. These two subspecies interbreed, although the F_1 has $2n = 53$ and may have reduced fertility due to the presence of unbalanced gametes, giving rise to unbalanced zygotes (and embryos). These zygotes die in early embryonic life, as occurs in cattle heterozygous carriers of rob (1;29) (Gustavsson 1980). The main karyotypic difference between these two species is the presence of four (*S. c. caffer*) and three (*S. c. nanus*) bi-armed autosome pairs, while the other chromosomes are acrocentric, including X (the largest acrocentric) and Y (a small acrocentric) chromosomes. The bi-armed pairs in *S. c. caffer* correspond to centric fusion translocations of cattle (ancestral bovid) chromosomes (1;13, 2;3, 5;20, 11;29, respectively) (Gallagher and Womack 1992).

The Asiatic (water) buffalo (*B. bubalis*) has two main subspecies: the river buffalo with a karyotype $2n = 50$ and FN = 60 (Fig. 1) and the swamp buffalo ($2n = 48$, FN = 58). Apparently these two subspecies differ by one chromosome pair (and the resulting FN), where a tandem fusion translocation between river buffalo (BBU) chromosomes 4 and 9 (telomere of BBU4p and centromere of BBU9) is comparable to swamp buffalo chromosome 1 (Di Berardino and Iannuzzi 1981) (Fig. 2). Thus, all chromosome arms and bi-armed pairs are conserved between the two species. Crosses between the two species are fertile, although the hybrid has 49 chromosomes and may have a lower reproductive value.

The river buffalo has five bi-armed autosomes (Fig. 1) and all remaining chromosomes are acrocentric, including both the X (the largest acrocentric) and Y (small acrocentric) chromosomes. The five bi-armed river buffalo chromosomes (BBU1 to BBU5) correspond to five centric fusion translocations of cattle chromosomes (and bovine syntenic groups) according to CSKBB (1994) and ISCNDB

Genome Mapping and Genomics in Animals, Volume 3
Genome Mapping and Genomics in Domestic Animals
N.E. Cockett, C. Kole (Eds.)
© Springer-Verlag Berlin Heidelberg 2009

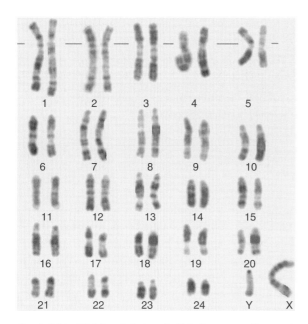

Fig. 1 Male RBG-banded river buffalo karyotype (2n = 50, XY), produced from a single cell and arranged according to the standard karyotype. Lines indicate centromeric positions in the five biarmed chromosomes

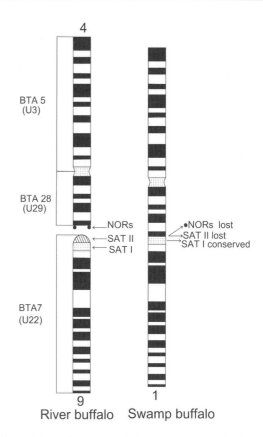

Fig. 2 Origin of swamp buffalo chromosome 1 by tandem fusion translocation of river buffalo chromosomes 9 (acrocentric) and 4 (submetacentric). The fusion occurred between the centromere and telomere of river buffalo chromosomes BBU9 and BBU4p, respectively, and resulted in the loss of both NORs and satellite II (SAT II) DNA. Only SAT I DNA was conserved. R-banded ideograms were drawn from standard river buffalo karyotype (CSKBB, 1994) and the reconstruction of the possible origin of swamp buffalo chromosome 1 was performed on the basis of published data (Di Berardino and Iannuzzi 1981; Tanaka et al. 1999). The homologous cattle (BTA) chromosomes and relative syntenic groups are also indicated

(2001): BBU1 (1;27-U10/U25), BBU2 (2;23-U17/U20), BBU3 (8;19-U18/U21), BBU4 (5;28-U3/U29), and BBU5 (16;29-U1/U7). The fusion of these bi-armed pairs is accompanied by a substantial loss of constitutive heterochromatin (HC), with very small C-bands found in the centromeres of the bi-armed pairs, compared to centromeres of all acrocentric chromosomes, including X, which shows the largest heterochromatin block (Fig. 3). The Y-chromosome shows variable C-banding patterns depending on the degree of chromosome denaturation. Indeed, the Y-chromosome appears completely heterochromatic or with a strong C-band that is distally located (Fig. 3). Thus, the C-banding technique (especially CBA-banding) distinguishes river buffalo sex chromosomes (especially the Y-chromosome) from the autosomes and is used for studying river buffalo sex chromosome abnormalities (Iannuzzi et al. 2000a, 2001a, 2004, 2005).

Because river buffalo BBU4 originated from centric fusion of cattle chromosomes 5 and 28 (CSKBB 1994), and river buffalo BBU9 is homologous to cattle chromosome 7 (CSKBB 1994), the following cattle homologous chromosomes (and bovine syntenic groups) are present in swamp buffalo chromosome 1: BTA5 (U3), BTA28 (U29), and BTA7 (U22) (Fig. 2).

During the tandem fusion event, the centromere of BBU9 was apparently deleted or inactivated while the nucleolus organizer regions (NORs) present at the telomeres of BBU4p (BTA28) (Iannuzzi et al. 1996) were lost (Di Berardino and Iannuzzi 1981). Indeed, there are six nucleolus organizer (NO) chromosomes in river buffalo, located at the telomeres of 3p, 4p, 6, 21, 23, and 24 (Iannuzzi et al. 1996), while the swamp buffalo has five NO-chromosomes (Di Berardino and Iannuzzi 1981). However, fluorescent in situ hybridization (FISH) mapping with two river buffalo probes containing satellite DNA similar to bovine SAT I and

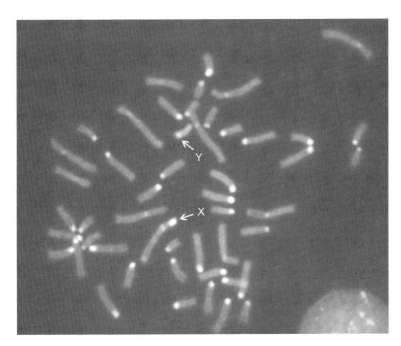

Fig. 3 CBA-banded male river buffalo metaphases. Note the strong C-bands in all acrocentric chromosomes (including the X chromosome, which shows the largest centromeric HC-block with a clear, additional proximal positive C-band) and the very small C-bands in biarmed chromosomes. The Y chromosome appears to be C-band positive only in the distal (almost telomeric) region, whereas the centromeric C-band is negative

SAT II (79% and 81% similarity, respectively) revealed signals in the proximal region of swamp buffalo chromosome 1q (at the presumed region of tandem fusion) with only the SAT I probe, although the signal intensity was stronger in the acrocentric chromosomes than in the bi-armed pairs (Tanaka et al. 1999). This result was also confirmed with C-banding (Di Meo et al. 1995) (Fig. 3). The SAT II probe produced a signal on all centromeric chromosome regions but not on the proximal end of swamp buffalo chromosome 1q (Tanaka et al. 1999). Thus, large portions of HC and satellite DNA present in river buffalo BBU9 and the NORs present at the BBU4p telomere were lost during the tandem fusion translocation that formed swamp buffalo chromosome 1 (Fig. 2).

Because no bi-armed chromosome pairs are shared between genuses *Syncerus* and *Bubalus* (Gallagher and Womack 1992; CSKBB 1994), crosses between them are not possible because the hybrid would have an unbalanced chromosome set. Thus, the designation of two separate genera is also supported by chromosomal morphology.

Another Asiatic buffalo species that is in danger of extinction is *Bubalus depressicornis* (2n = 48), which is found on the Sulawesi (Indonesia) island as two subspecies: *Bubalus depressicornis depressicornis* (Lowland Anoa) and *Bubalus depressicornis quarlesi* (Mountain Anoa). This species is the smallest buffalo in the world, with a mature size between 70 (Mountain Anoa) and 100 cm (Lowland Anoa) tall with a shape and size similar to that of a goat. This species is very similar to the river *B. bubalis*. Indeed, four of the six bi-armed chromosome pairs in *Anoa* are centric fusion translocations of cattle homologs similar to what occurred in *Bubalus* (1;27, 2;23, 8;19, 5;28), whereas the other *Anoa* bi-armed chromosomes are different combinations of cattle chromosomes (11;20 and 17;15) (Gallagher et al. 1999).

A standard karyotype using six banding techniques and G- and R-banded ideograms (CSKBB 1994) has been established for river buffalo. A comparative genetic comparison of river buffalo and cattle, goat and sheep standard karyotypes (ISC-NDB 2001) has been established by Iannuzzi et al. (2001d) using the same markers assigned to G/Q and R-banded cattle chromosomes (Hayes et al. 2000). These results were used to construct cattle G- and R-banded ideograms, as well as definitive chromosome homologies between river buffalo and related bovids (ISCNDB 2001).

2.1.2
Economic Importance

There are about 168 million water buffalo in the world, with 161 million in Asia (of which 95 million are in India), 3.7 million in Africa (essentially all are in Egypt), 3.3 million in South America, 500,000 in Europe, and 40,000 in Australia (Dhanda 2004; Borghese and Mazzi 2005). Although worldwide numbers of buffalo are about 1/9 the numbers of cattle, significantly more humans depend on buffalo, especially in South-East Asia, than any other domestic species (FAO and UNEP 2000). Therefore, buffalo have great economic importance in the world. Moreover, unlike other domesticated bovids, the number of river buffalo has grown in the last 20 years at an average rate of +2% per year worldwide, with the highest increases being in Europe (+4.5% per year), India (+5.3%), and Pakistan (4.8%) (Dhanda 2004). In Italy, a +1,600% increase in buffalo numbers occurred from 1957 to 2001 (Borghese 2004).

The most numerous river buffalo breeds are found in India–Pakistan and include Murrah, Nili-Ravi, Kundi, Surti, Mehsana, Jaffarabadi, Bradawari, Tarai, Nagpuri, Pandharpuri. Manda, Jerangi, Kalahandi, Sambalpur, Toda, and South Kanara breeds. The Mediterranean breed is found in Italy, Egypt, Iran, Turkey, Romania, Iraq, Bulgaria, Syria, and Greece. There is an offshoot Italian Mediterranean breed, with around 300,000 animals, found primarily in Campania. This breed has not been crossed with other breeds since its introduction from North Africa (Egypt) or central Europe during the fifth to seventh century AD (Kierstein et al. 2004), unlike other neighbouring Mediterranean countries where crosses are performed, primarily with the Indian Murrah breed.

There are 3.345 million river buffalo in South America, with 3 million found in Brazil alone (Loures 2001). Because buffalo were imported into South America from India and Mediterranean countries, as well as Australia and the Philippines where swamp buffalo are found, hybrids between river and swamp buffalo (the Carrabao buffalo) exist in many South American countries.

The swamp buffalo is mainly raised in the Philippines, Malaysia, Thailand, Australia, and China, the latter with about 23 million head (Liang et al. 2004). Crosses with river buffalo from India (Murrah breed) and Pakistan (Nili-Ravi) were performed in China during 1950–1970 (Liang et al. 2004). Although the swamp buffalo is considered to have only one breed, Chunxi and Zhongquan (2001) refer to 18 breeds of swamp buffaloes which evolved separately in different areas of China. However, no genetic analyses that support this suggestion have been performed.

Borghese and Mazzi (2005) report the following breeds in China: Binhu, Enshi, Xinyang, Fuan, Yanjin, Xinglong, Wenzhou, Haizi, Shanghai, Guizhou, Fuling, Dehong, Diandong, Dechang, Xilin, Fuzhong, and Dongliu.

River buffalo are raised for both milk and meat production. About 50% of the milk is used for drinking in many countries (India, Pakistan, Egypt, and Nepal, in particular), while in Italy buffalo milk is used to produce cheese, mainly "mozzarella" (Zicarelli 2004). Milk composition varies among breeds, with protein content varying between 4.2 and 4.6% and fat between 6.0 and 8.5% (Moioli and Borghese 2005). The highest milk production has been recorded in the Murrah, Jafarabadi, Nili-Ravi, and Mediterranean breeds (1,500–2,800 kg per lactation) (Moioli and Borghese 2005). However, milk records are primarily maintained in Italy where most female buffalo are registered in the Italian Mediterranean Buffalo Breed (Razza Bufala Mediterranea Italiana). River buffalo milk is richer in both proteins (4.65%) and fat (8.30%) than cattle milk (3.5% and 3.4%, respectively) (Zicarelli 2004), which may explain its use in cheese making (primarily mozzarella) in several countries, especially in Italy. Percentage of casein fractions in the milk vary between river buffalo and cattle. Specifically, alpha-s1 is 30.2% in buffalo and 38% in cow, alpha s2 casein is 17.6% in buffalo and 10.5% in cow, beta-casein is 33.9% in buffalo and 36.5% in cow, and K-casein is 15.4% in buffalo and 12.5% in cow (Zicarelli 2004).

The production of buffalo for meat is common in many countries and has recently increased in Italy. Buffalo meat is about 1/3 lower in cholesterol than beef, and therefore, recommended in low cholesterol diets.

2.1.3
Breeding Objectives

Because milk production is much lower in swamp buffalo than in river buffalo, a breeding program for crossing swamp buffalo with river buffalo (Murrah, Nili-Ravi, and Jafarabady, in particular) has been established in the Philippines, China, and Australia. However, the success of these breeding programs has been compromised because (i) the hybrids have $2n = 49$ chromosomes, leading to reproductive problems due to unbalanced gametes as reported above, and (ii) higher milk production requires larger quantities of food, which are very often not available in these countries, especially on the predominant small farms.

Within specific breeds, breeding programs are focused on genetic improvement of milk production using progeny tests and recorded milk traits. The highest proportion of buffalo milk recording is performed in Italy with 28.6% of cows, followed by Bulgaria (8.5%), and Iran (4.5%) (Moioli 2005).

A breeding program currently underway in Italy, in collaboration with buffalo breeder associations, includes the cytogenetic characterization of bulls referred to reproductive centres and females with reproductive problems. About 20% of females investigated for reproductive problems were found to carry sex chromosome aberrations such as X-monosomy, X-trisomy, sex reversal syndrome, and freemartin, and all female carriers were sterile because of damaged internal reproductive organs (Iannuzzi et al. 2000a, 2001a, 2004, 2005). Freemartinism was the most common syndrome (Iannuzzi et al. 2005) and almost all cases were from single births, suggesting that male co-twins died during early embryonic life. Producers often kept the Freemartin females for up to four years before having clinical and cytogenetic analyses completed. Similar to the Freemartin condition in cattle, internal sex organs showed variable damage (from complete lack of internal sex adducts to the atrophy of Muller ducts), which was not correlated with the percentage of male cells (Iannuzzi et al. 2005). It is likely that the variable Freemartin effect is due to the time of placental anastomosis, which occurs at 2–3 weeks in cattle and 3–4 weeks in buffalo, as well as to sex differentiation, which occurs at 40–50 days in cattle for male co-twins and about one week later for females (Ruvinsky and Spicer, 1999).

2.2
Molecular Genetics

2.2.1
Classical Mapping Efforts

Few studies of genetic variability or phylogenetic analyses have been performed in water buffalo. A set of 21 microsatellite loci in 11 populations of Asian water buffaloes (8 swamp and 3 river buffalo, respectively) was studied by Backer et al. (1997), who described significant differences in allelic variability between swamp and river buffalo types and within each buffalo type.

Genetic variation among eight Indian breeds was performed using 27 microsatellite loci (Kumar et al. 2006). This study identified 166 alleles in the Toda breed and up to 194 alleles in the Mehsana and Murrah breeds. The Toda, Jaffarabadi, and Pandharpuri breeds had one lineage each, while Bhadawari, Nagpuri, Surati, Mehsana, and Murrah breeds were identified as admixture. The results from this study can be used for developing rational breeding and conservation strategies for Indian buffalo (Kumar et al. 2006).

An analysis of 20 bovine microsatellites was used to compare bovid species, including the river buffalo (Ritz et al. 2000). Results from that study indicated that *Bos taurus* and *Bos indicus* grouped first, followed by *Bos frontalis* and *Bos grunniens*. The *Bison bison* branched off next and *B. bubalis* and *S. caffer* emerged as the two most divergent species from the *Bos* clade (Ritz et al. 2000). Phylogenetic analysis based on amplified fragment length polymorphism (AFLP) fingerprinting of bovid species, including African and water buffalo, revealed three tree reconstructions: African buffalo with water buffalo, ox with zebu, and bison with wisent (Buntjer et al. 2002).

Analysis of the mitochondrial D-loop DNA sequence determined on 80 water buffaloes, of which 19 were swamp buffaloes and 61 river buffaloes (Murrah, Jafarabadi, and Mediterranean breeds), revealed only one domestication event of water buffalo, which occurred on the Indian subcontinent 5,000 years ago (Kierstein et al. 2004). On the South-East Asian mainland, these populations interbred with wild buffaloes and/or domestic animals from China (Kierstein et al. 2004).

Polymorphisms in the major histocompatibility complex (MHC) *DRB (Bubu-DRB)* and *DRA (Bubu-DRA)* loci have been studied in one swamp buffalo and three river buffalo breeds. Eight alleles for *Bubu-DRB* were found in all river buffalo and swamp breeds (Sena et al. 2003). The presence of two alleles of *DRA* in water buffalo (and *Anoa depressicornis*) was considered important by Sena et al. (2003), given that *DRA* is a highly conserved polypeptide in mammals, especially in ruminants where no other substitutions in non-peptide-binding sites (PBS) have been found (Sena et al. 2003).

The complete coding region sequence of the river buffalo *SRY* gene has been determined (Parma et al. 2004). This sequence has been used to detect mutations in sex reversal syndrome (Iannuzzi et al. 2001a, 2004).

Analyses of sequences and expression profiles of buffalo interleukin-12 (*IL12*) revealed significant sequence identity with bovine *IL12* and functional cross-reactivity with bovine immune cells (Premraj et al. 2006). A full-length cDNA of interleukin-18 (*IL18*) of the Indian water buffalo was determined, revealing a very similar amino acid sequence (99% and 95% identity) to cattle and sheep, respectively (Chaudhury and Bera 2005).

2.2.2
Construction of Genetic Maps and Comparative Mapping

Assignments of both type I (known genes) and type II loci (generally microsatellies) have been performed in river buffalo using a somatic cell hybrid panel (El Nahas et al. 1996) and FISH-mapping techniques (Iannuzzi 1998; Iannuzzi et al. 2003). The first genetic map for river buffalo with only 54 loci, mostly assigned by FISH, was reported by Iannuzzi (1998). In this first genetic map, at least one bovine molecular marker (and associated syntenic group) was assigned to each river buffalo chromosome or chromosome arm. Improved genetic maps with 99 (El Nahas et al. 2001) and 293 (Iannuzzi et al. 2003) loci were later established. The later map included 171 type I loci and 122 type II loci, which were mostly microsatellites. Of the 293 assigned loci, 247 were assigned by FISH (Iannuzzi et al. 2003a) (Fig. 4).

An advanced river buffalo cytogenetic map is shown in Fig. 5. It includes loci previously mapped by Iannuzzi et al. (2003a) and FISH-mapped loci: *SLC26A2* to BBU9q26 (Kierstein et al. 2003); *SMN* to BBU19q13 (Iannuzzi et al. 2003b); *LEP* to BBU8q32 (Vallinoto et al. 2004); *FHIT* to BBU21q24 (Di Meo et al. 2005a); *GDF8, TTN, GCG, NEB, CXCR4, MYL1, ACADL, IGFBP2*, and *FN1* to BBU2q (Di Meo et al. 2006); *MUC1* to BBU6q13 (Perucatti et al. 2006); *TRG2* to BBU8q17 (Antonacci et al. 2007); *UMN0301, UMN0504* to BBUY (Di Meo et al. 2005b). The total number of loci assigned to river buffalo is now 309. Of these loci, 186 are type I and 124 type II. Although some chromosome bands are still without markers, specifically along chromosomes 1q, 3q, 7, 9, 12, 21, and 24, this cytogenetic map covers all chromosomes and chromosome regions, improving our knowledge on

the river buffalo genome, especially considering that a linkage map is still lacking in this species and preliminary radiation hybrid (RH) maps have only recently been performed for some river buffalo chromosomes (Amaral et al. 2007; Strafuzza et al. 2007). Particularly, dense maps are available for BBU1p, BBU2q, BBU3p, BBU6prox, BBU8dist, BBU14, BBU15prox, BBU16, BBU18, BBU20, BBU-X, and BBU-Ydist.

To detect conserved chromosome segments and syntenies, comparative mapping studies have been performed between river buffalo and other related species (cattle, sheep, and goat), as well as between river buffalo and humans. These studies revealed high levels of homology among autosomal chromosomes of bovids. Indeed, the same chromosome banding patterns and gene order among all autosomes (or chromosome arms) of cattle, river buffalo, sheep, and goat have so far been found (Iannuzzi et al. 1999, 2000b,c, 2001b–d; Di Meo et al. 2000, 2002, 2005a, 2006). The only exception is between "bovine" (cattle and river buffalo) and "caprine" (sheep and goat) chromosomes 9 and 14. Indeed, a simple translocation/inversion of a pericentromeric chromosome segment from bovine chromosome 9 to caprine chromosome 14 differentiates these two autosomes, supported by both linkage (de Gortari et al. 1998) and FISH-mapping (Iannuzzi et al. 2001c) analyses. Comparison of bovine chromosome 9 with human chromosomes revealed a more conserved synteny between bovine chromosome 9 and HSA9 than that between caprine chromosome 9 and HSA6q (Iannuzzi et al. 2001c). Thus, the caprine karyotype is a derivative of the bovine one (ancestral bovid).

In contrast to the strong similarity among bovid autosomal chromosomes, sex chromosomes diverged through more complex chromosome rearrangements. X-chromosome of bovids can mainly be found as one of three types: the submetacentric cattle type, the acrocentric eland (or river buffalo) type, and the acrocentric sheep (or goat) type with small and visible p-arms. Chromosome banding comparisons demonstrated that large portions of these chromosomes are conserved (Iannuzzi and Di Meo 1995), with the presence of large blocks of constitutive heterochromatin (HC) in BBU-X (Di Meo et al. 1995) and their absence in both BTA-X and OAR/CHI-X (Iannuzzi and Di Meo 1995). Detailed comparative cytogenetic maps representing the order of loci along sex chromosomes of cattle, river buffalo, and sheep/goat X-chromosomes revealed complex chromosome

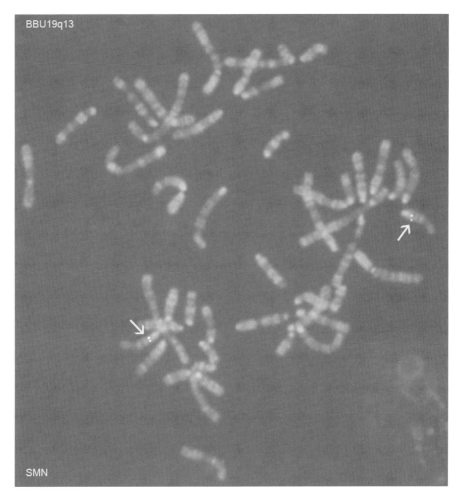

Fig. 4 Map assignment of the survival motor neuron (*SMR*) gene on river buffalo chromosome 19q13 using FISH. Bovine bacterial artificial chromosome (BAC) clones were biotinylated and precipitated in the presence of bovine COT-1 DNA, ethanol precipitated, and dissolved in hybridization solution. Detection of fluorescein FITC-signals staining with propidium iodide and mounting with antifade was then completed. Green FITC-signals were superimposed on red RBPI-(R-banding using late incorporation of BrdU and propidium iodide staining) banded chromosomes. The image is presented in black and white

rearrangements that occurred during evolution of the karyotypes of these species (Piumi et al. 1998; Robinson et al. 1998; Iannuzzi et al. 2000c). In particular, BTA-X and BBU-X share the same gene order but a different centromere position. Hence, a centromere transposition (or centromere repositioning) with loss of constitutive heterochromatin (HC) differentiates BTA-X from BBU-X (Iannuzzi et al. 2000c). When comparing bovine X (BTA-X and BBU-X) with caprine X (OAR-X and CHI-X), at least four chromosome transpositions including a centromere repositioning were found (Iannuzzi et al. 2000c).

A similar situation exists for the Y-chromosomes of bovids. Indeed, comparative FISH-mapping analyses performed among the Y-chromosomes of cattle (*Bos taurus*), zebu (*Bos indicus*, BIN), river buffalo, and sheep/goat revealed complex chromosome rearrangements. In particular, BTA-Y (submetacentric) and BIN-Y (acrocentric with small and visible p-arms) differ in a centromere transposition (or repositioning) or pericentric inversion, yet they retain the same gene order along the distal regions (Di Meo et al. 2005b). BTA-Y and BBU-Y differ in a pericentric inversion with loss (from BBU-Y to BTA-Y) or gain (from BTA-Y to BBU-Y) of HC, BBU-Y being larger than BTA-Y. OAR-Y/CHI-Y (very small metacentrics) differs from BBU-Y in a pericentric inversion and greater loss of HC and from BTA-Y and BIN-Y in a

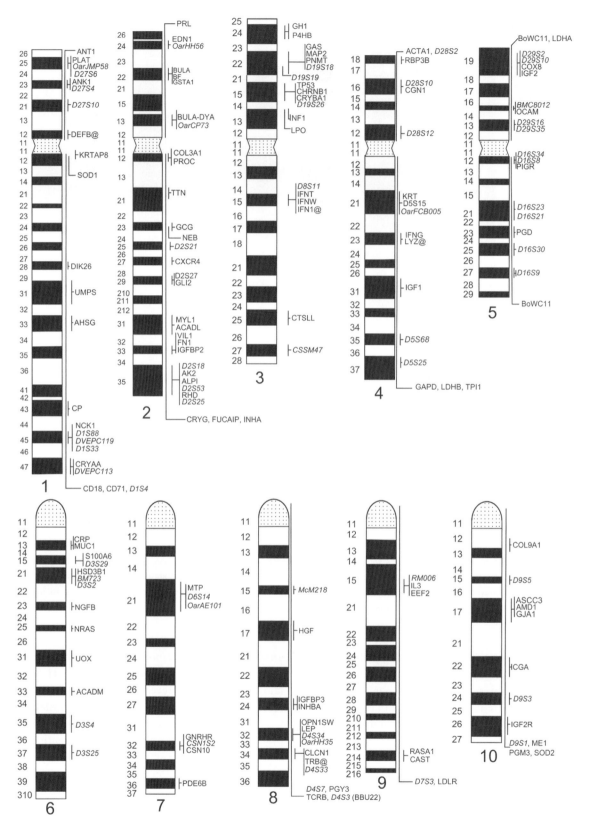

Fig. 5 An updated river buffalo cytogenetic map on an R-banded river buffalo standard karyotype. Type I (known genes) and type II (generally microsatellites) loci are normal characters and italicized characters, respectively

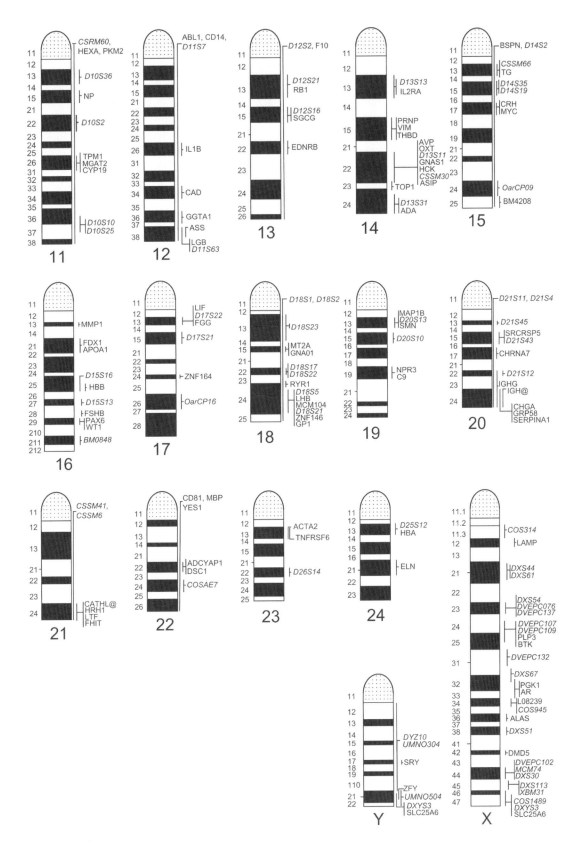

Fig. 5 (Continued)

centromere transposition with loss of HC (Di Meo et al. 2005b).

The use of human chromosome painting probes identified at least 41 human chromosome conserved segments between humans and river buffalo (Iannuzzi et al. 1998). Detailed cytogenetic maps revealed more human segments along river buffalo chromosomes than was resolved with Zoo–FISH mapping, indicating complex chromosome rearrangements within large conserved chromosome regions of river buffalo and human chromosomes (Iannuzzi et al. 1999, 2000b,c, 2001b–d; Di Meo et al. 2000, 2002, 2005a, 2006). For example, BBU2q was almost entirely contained by HSA2q and HSA1p probes (centromeric and telomeric regions) (Iannuzzi et al. 1998), resulting in only three human segments. However, high-resolution comparative FISH-mapping of type I loci on bovine (including BBU2q) chromosomes, and HSA2q and HSA1p, revealed at least ten human segments, nine along HSA2q and one along HSA1p. These complex chromosome rearrangements that differentiate bovids and primates occurred about 10 million years ago (Di Meo et al. 2006).

Another important parameter from an evolutionary point of view is represented by the NORs, which are all located at the telomeres of five (cattle, sheep, goat, swamp buffalo) and six (river buffalo) autosomes of domestic bovids. Considering the high degree of autosome conservation among bovids (with only centromeric regions affected by chromosome rearrangements), we expected to find the same nucleolus organizer chromosomes (NOCs) in bovids, but instead, due to simple NOR-translocations, only some NORs were conserved to homologous chromosomes or chromosome arms (Table 1) (Iannuzzi et al. 1996; Gallagher et al. 1999). In particular, only two homologous NOCs were common to river buffalo and cattle (BBU6 and BBU24, homologous to BTA3 and BTA25, respectively), as well as to river buffalo and goat/sheep (BBU4p and BBU6, homologous to CHI3 and CHI28, respectively, and to OAR1p and OAR25, respectively). These four bovids share only one NOC (BBU6, BTA3, CHI3, and OAR1p), while goat and sheep share the same NOC, confirming their evolutionary proximity.

2.3
Future Scope of Work

Linkage and RH maps are critical for the advancement of genomics in water buffalo. Particular emphasis should be placed on regions where quantitative trait loci (QTL) for production traits have been found in other species. Collaborative studies among groups involved in linkage-, RH-, and FISH-mapping analyses should be performed to enhance our understanding of the water buffalo genome. The cytogenetic map reported herein is an appropriate point of reference for these studies. The strong chromosome homology found between river buffalo and cattle and the imminent sequencing of the cattle genome will rapidly add to our knowledge of this economically important species.

References

Amaral MEJ, Owens KE, Elliot JS, Fichey C, Shaffer AA, Agarwale R, Womack JE (2007) Construction of a river buffalo (*Bubalus bubalis*) whole-genome radiation hybrid panel and preliminary RH mapping of chromosomes 3 and 10. Anim Genet 38:311–314

Antonacci R, Maccarelli G, Di Meo GP, Piccinni B, Miccoli MC, Cribiu EP, Perucatti A, Iannuzzi L, Ciccarese S (2007) Molecular *in situ* hybridization analysis of sheep and goat BAC clones identifies the transcriptional orientation of T cell receptor gamma genes on chromosome 4 in bovids. Vet Res Comm 31:977–983

Backer JSR, Moore SS, Hetzel DJS, Evans D, Tan SG, Byrne K (1997) Genetic diversity of Asian water buffalo (*Bubalus bubalis*): microsatellite variation and comparison with protein-coding loci. Anim Genet 28:103–115

Borghese A (2004) Recent development of buffaloes in Europe and near east. Proc 7th World Buffalo Congress, Manila, The Philippines, October 20–23, 1:10–16

Borghese A, Mazzi M (2005) Buffalo population and strategies in the world. In: Borghese A (ed) Buffalo Production and

Table 1 Nucleolus organizer chromosomes (NOCs) in river buffalo, cattle, sheep, and goat according to the latest standard chromosome nomenclature (ISCNDA 2001)[a]

Species	Nucleolus organizer chromosomes[b]							
River buffalo	3p	**4p**	**6**	21	23	**24**		
Cattle			**3**			25	2	4	11
Sheep		25	**1p**				2q	4	3q
Goat		28	**3**				2	4	5

[a] Data elaborated from Iannuzzi et al. (1996) and Gallagher et al. (1999)

[b] Homologous NOCs (in bold) are aligned among the four species

Research. REU Technical series 67. FAO Regional Office for Europe, pp 1–39

Buchholtz C (1990) Cattle. In: Parker SP (ed) Grzimek's Encyclopedia of Mammals. Vol 5. McGraw-Hill, New York, USA, pp 360–417

Buntjer JB, Otsen M, Nijman IJ, Kuiper MT, Lenstra JA (2002) Phylogeny of bovine species based on AFLP fingerprinting. Heredity 88:46–51

Chaudhury P, Bera BC (2005) Cloning and sequencing of Indian water buffalo interleukin-18 cDNA. Intl Immunogenet 32:75–78

Chen YS, Li XH (1989) New evidence of the origin and domestication of the Chinese swamp buffalo (*Bubalus bubalis*). Buff J 1:51–55

Chunxi Z, Zhongquan L (2001) Buffalo genetic resources in China. Buff News l 16:1–7

Cockrill WR (1981) The water buffalo: a review. Brit Vet J 137:8–16

CSKBB (1994) Standard karyotype of the river buffalo (*Bubalus bubalis* L., 2n = 50). Report of the Committee for the Standardization of Banded Karyotypes of the River Buffalo (Iannuzzi L, coordinator). Cytogenet Cell Genet 67:02–113

Dhanda OP (2004) Developments in water buffalo in Asia and Oceania. Proc 7th World Buffalo Congress, Manila, The Philippines, October 20–23. 1:17–27

de Gortari MG, Freking BA, Cuthbertson RP, Kappes SM, Keele JW, Stone RT, Leymaster KA, Dodds KG, Crawford AM, Beattie CW (1998) A second-generation linkage map of the sheep genome. Mamm Genom 9:204–209

Di Berardino D, Iannuzzi L (1981) Chromosome banding homologies in swamp and murrah buffalo. J Hered 72:183–188

Di Meo GP, Perucatti A, Iannuzzi L, Rangel-Figueiredo MT, Ferrara L (1995) Constitutive heterochromatin polymorphism in river buffalo chromosomes. Caryologia 48:137–145

Di Meo GP, Perucatti A, Schibler L, Incarnato D, Ferrara L, Cribiu EP, Iannuzzi L (2000) Thirteen type I loci from HSA4q, HSA6p, HSA7q and HSA12q were comparatively FISH-mapped in four river buffalo and sheep chromosomes. Cytogenet Cell Genet 90:102–105

Di Meo GP, Perucatti A, Incarnato D, Ferretti L, Di Berardino D, Caputi Jambrenghi A, Vonghia G, Iannuzzi L (2002) Comparative mapping of twenty-eight bovine loci in sheep (*Ovis aries*, 2n = 54) and river buffalo (*Bubalus bubalis*, 2n = 50) by FISH. Cytogenet Genom Res 98:262–264

Di Meo GP, Perucatti A, Uboldi C, Roperto S, Incarnato D, Roperto F, Williams J, Eggen A, Ferretti L, Iannuzzi L (2005a) Comparative mapping of the fragile histidine triad (FHIT) gene in cattle, river buffalo, sheep and goat by FISH and assignment to BTA22 by RH-mapping: a comparison with HSA3. Anim Genet 36:363–364

Di Meo GP, Perucatti A, Floriot S, Incarnato D, Rullo R, Caputi Jambrenghi A, Ferretti L, Vonghia G, Cribiu E, Eggen A, Iannuzzi L (2005b) Chromosome evolution and improved

cytogenetic maps of the Y chromosome in cattle, zebu, river buffalo, sheep and goat. Chrom Res 13:349–355

Di Meo GP, Gallagher DS, Perucatti A, Wu X, Incarnato D, Mohammadi G, Taylor JF, Iannuzzi L (2006) Mapping of 11 genes by FISH to BTA2, BBU2q, OAR2q and CHI2, and comparison with HSA2q. Anim Genet 37:299–300

East R (1999) African Antelope Database 1998. IUCN/SSC Antelope Specialist Group. Gland, Switzerland and Cambridge, UK: IUCN

El Nahas SM, Oraby HA, de Hondt HA, Medhat AM, Zahran MM, Mahfouz ER, Kari AM (1996) Synteny mapping in river buffalo. Mamm Genome 7:831–834

El Nahas SM, de Hondt HA, Womack JE (2001) Current status of the river buffalo (*Bubalus bubalis*) gene map. J Hered 92:221–225

FAO, UNEP (2000) World Watch List for Domestic Animal Diversity, 3rd ed. Information Division, FAO, Rome, Italy

Gallagher DS, Womack JE (1992) Chromosome conservation in the Bovidae: J Hered 83:287–298

Gallagher DS, Davis SK, De Donato M, Burzlaff JD, Womack JE, Taylor JF, Kumamoto AT (1999) A molecular cytogenetic analysis of the tribe Bovini (Artiodactyla: Bovidae: Bovinae) with emphasis on the sex chromosome morphology and NOR distribution. Chrom Res 7:481–492

Gustavsson I (1980) Chromosome aberrations and their influence on the reproductive performance of domestic animals – a review. Z Tierzuchtg Zuchtgsbiol 97:176–195

Hayes H, Di Meo GP, Gautier M, Laurent P, Eggen A, Iannuzzi L (2000) Localization by FISH of the 31 Texas nomenclature type I markers to both Q- and R-banded bovine chromosomes. Cytogenet Cell Genet 90:315–320

Iannuzzi L (1998) A genetic physical map in river buffalo (*Bubalus bubalis*, 2n = 50). Cayologia 51:311–318

Iannuzzi L, Di Meo GP (1995) Chromosomal evolution in bovids: a comparison of cattle, sheep and goat G- and R-banded chromosomes and cytogenetic divergences among cattle, goat and river buffalo sex chromosomes. Chrom Res 3:291–299

Iannuzzi L, Di Meo GP, Perucatti A (1996) Identification of nucleolus organizer chromosomes and frequency of active NORs in river buffalo (*Bubalus bubalis* L.). Caryologia 49:27–34

Iannuzzi L, Di Meo GP, Perucatti A, Bardaro T (1998) ZOO-FISH and R-banding reveal extensive conservation of human chromosome regions in euchromatic regions of river buffalo chromosomes. Cytogenet Cell Genet 82:210–214

Iannuzzi L, Gallagher DS, Di Meo GP, Yang Y, Womack JE, Davis SK, Taylor JF (1999) Comparative FISH-mapping of six expressed gene loci to river buffalo and sheep. Cytogenet Cell Genet 84:161–163

Iannuzzi L, Di Meo GP, Perucatti A, Zicarelli L (2000a) A case of sex chromosome monosomy (2n = 49, X0) in the river buffalo (*Bubalus bubalis*). Vet Record 147:690–691

Iannuzzi L, Di Meo GP, Perucatti A, Schibler L, Incarnato D, Ferrara L, Bardaro T, Cribiu EP (2000b) Sixteen type I loci from six chromosomes were comparatively fluorescent in-situ mapped to river buffalo (*Bubalus bubalis*) and sheep (*Ovis aries*) chromosomes. Chrom Res 8:45–47

Iannuzzi L, Di Meo GP, Perucatti A, Incarnato D, Schibler L, Cribiu EP (2000c) Comparative FISH-mapping of bovid X chromosomes reveals homologies and divergences between the subfamilies *Bovinae* and *Caprinae*. Cytogenet Cell Genet 89:171–176

Iannuzzi L, Di Meo GP, Perucatti A, Di Palo R, Zicarelli L (2001a) 50, XY gonadal dysgenesis (Swier's syndrome) in a female river buffalo (*Bubalus bubalis*). Vet Record 148:634–635

Iannuzzi L, Gallagher DS, Di Meo GP, Schlapfer J, Perucatti A, Amarante MRV, Incarnato D, Davis SK, Taylor JF, Womack JE (2001b) Twelve loci from HSA10, HSA11 and HSA20 were comparatively FISH-mapped on river buffalo and sheep chromosomes. Cytogenet Cell Genet 93:124–126

Iannuzzi L, Di Meo GP, Perucatti A, Schibler L, Incarnato D, Cribiu EP (2001c) Comparative FISH-mapping in river buffalo and sheep chromosomes: assignment of forty autosomal type I loci from sixteen human chromosomes. Cytogenet Cell Genet 94:43–48

Iannuzzi L, Di Meo GP, Hayes H, Perucatti A, Incarnato D, Gautier M, Eggen A (2001d) FISH-mapping of the 31 bovine Texas markers to river buffalo chromosomes. Chrom Res 9:339–342

Iannuzzi L, Di Meo GP, Perucatti A, Schibler L, Incarnato D, Gallagher D, Eggen A, Ferretti L, Cribiu EP, Womack J (2003) The river buffalo (*Bubalus bubalis*, 2n = 50) cytogenetic map: assignment of 64 loci by fluorescence *in situ* hybridisation and R-banding. Cytogenet Genom Res 102:65–75

Iannuzzi L, Di Meo GP, Perucatti A, Incarnato D, Di Palo R, Zicarelli L (2004) Reproductive disturbances and sex chromosome abnormalities in river buffaloes. Vet Record 154:823–824

Iannuzzi L, Di Meo GP, Perucatti A, Ciotola F, Incarnato D, Di Palo R, Peretti V, Campanile G, Zicarelli L (2005) Free-martinism in river buffalo: clinical and cytogenetic observations. Cytogenet Genom Res 108:335–358

Iannuzzi L, Di Meo GP, Perucatti A, Rullo R, Incarnato D, Longevi M, Bongioni A, Molteni L, Galli A, Canotti M, Eggen A (2003b) Comparative FISH-mapping of the survival of motor neuron gene (SMN) in domestic bovids. Cytogenet Genome Res 102:39–41

ISCNDB (2001) International system for chromosome nomenclature of domestic bovids. Di Berardino D, Di Meo GP, Gallagher DS, Hayes H, Iannuzzi L (eds) Cytogenet Cell Genet 92:283–299

Kierstein G, Vallinoto M, Silva A, Schneider MP, Iannuzzi L, Brenig B (2004) Analysis of mitochondrial D-loop region casts new light on domestic water buffalo (*Bubalus bubalus*) phylogeny. Mol Phylogenet Evol 30:308–324

Kingdon J (1997) The Kingdon Field Guide to African Mammals. Academic Press, London and New York

Kierstein G, Iannuzzi L, Silva A, Schneider MP, Baumgartner BG, Brenig B (2003) Assignment of solute carrier family 26 (sulfate transporter), member 2 (SLC26a2) to river buffalo (*Bubalus bubalis*, 2n = 50) chromosome 9q26 (BBU9q26) by fluorescence in situ hybridisation and R-banding. Cytogenet Genome Res 103:202A

Kumar S, Gupta J, Kumar N, Dikshit K, Navani N, Jain P, Nagarajan M (2006) Genetic variation and relationships among eight Indian riverine buffalo breeds. Mol Ecol 15:593–600

Liang XW, Yang BZ, Zhang XF, Zou CX, Huang YJ (2004) Progress of scientific research on buffaloes in China. Proc 7th World Buffalo Congress, Manila, The Philippines, October 20–23. 1:29–35

Loures R (2001) Buffalo production systems in Americas. Proc 6th World Buffalo Congress, Maracaibo, Zulia, Venezuela, May 20–23. pp 74–86

Moioli B (2005) Breeding and selection of dairy buffaloes. In: Borghese A (ed) Buffalo Production and Research. REU Tech Sr. FAO Regional Office for Europe, pp 41–50

Moioli B, Borghese A (2005) Buffalo breeds and management system. In: Borghese A (ed) Buffalo Production and Research. REU Tech Sr 67. FAO Regional Office for Europe, pp 51–76

Parma P, Feligini M, Greppi G, Enne G (2004) The complete coding region sequence of river buffalo (*Bubalus bubalis*) SRY gene. DNA Seq 15:77–80

Perucatti A, Di Meo GP, Vallinoto M, Kierstein G, Schneider MPC, Incarnato D, Caputi Jambrenghi A, Mohammadi G, Vonghia G, Silva A, Brenig B, Iannuzzi L (2006) FISH-mapping of *LEP* and *SLC26A2* genes in sheep, goat and cattle R-banded chromosomes: comparison between bovine, ovine and caprine chromosome 4 (BTA4/OAR4/CHI4) and human chromosome 7 (HSA7). Cytogenet Genom Res 115:7–9

Piumi F, Schibler L, Vaiman D, Oustry A, Cribiu EP (1998) Comparative cytogenetic mapping reveals chromosome rearrangements between the X chromosomes of two closely related mammalian species (cattle and goats). Cytogenet Cell Genet 81:36–41

Premraj A, Sreekumar E, Jain M, Rasool TJ (2006) Buffalo (*Bubalus bubalis*) interleukin-12: analysis of expression profiles and functional cross-reactivity with bovine system. Mol Immunol 43:822–829

Ritz LR, Glowatzki-Mullis ML, MacHugh DE, Gailard C (2000) Phylogenetic analysis of the tribe Bovini using microsatellites. Anim Genet 31:178–185

Robinson TJ, Harrison WR, Ponce de Leon FA, Davis SK, Elder FFB (1998) A molecular cytogenetic analysis of the X chromosome repatterning in the Bovidae: transpositions, inversions, and phylogenetic inference. Cytogenet Cell Genet 80:179–184

Ruvinsky A, Spicer LJ (1999) Developmental genetics. In: Fries R, Ruvinsky A (eds) The Genetics of Cattle. CABI Publ, Wallingford, Oxon, UK

Sena L, Schneider MPC, Brenig B, Honeycutt RL, Womack JE (2003) Polymorphism in MHC-DRA and –DRB alleles of water buffalo (*Bubalus bubalis*) reveal different features from cattle DR alleles. Anim Genet 34:1–10

Strafuzza NB, Iarnella P, Miziara MN, Agarwala R, Schaffer AA, Riggs PK, Womack JE, Amaral MEJ (2007) Preliminary RH map for river buffalo chromosome 6 and comparison to bovine chromosome 3. Anim Genet 38:406–409

Tanaka K, Matsuda Y, Masangkay JS, Solis CD, Anunciado RV, Namikawa T (1999) Characterization and chromosomal distribution of satellite DNA sequences of the water buffalo (*Bubalus bubalis*). J Hered 90:418–22

Vallinoto M, Scheider MP, Silva A, Iannuzzi L, Brenig B (2004) Molecular cloning and analysis of the bubaline leptin gene. Anim Genet 35:362–363

Zicarelli L (2004) Buffalo milk: its proprieties, dairy yield and mozzarella production. Vet Res Comm 28:127–135

CHAPTER 3

Sheep

Christopher A. Bidwell[1] (✉), Noelle E. Cockett[2], Jill F. Maddox[3], and Jon E. Beever[4]

[1] Purdue University, Department of Animal Sciences, 125 South Russell Street, West Lafayette, IN 47907–2042, USA, cbidwell@purdue.edu

[2] Department of Animal, Dairy & Veterinary Sciences, Utah State University, Logan, UT 84322–4700, USA

[3] Department of Veterinary Science, University of Melbourne, Victoria 3010, Australia

[4] Department of Animal Sciences, University of Illinois at Urbana-Champaign, 220 Edward R. Madigan Laboratory, 1201 West Gregory Drive, Urbana, IL 61801, USA

3.1
Introduction

Animal geneticists have been searching for the molecular basis of production traits in livestock species, including sheep, for over 40 years. Phenotypes of interest in sheep include fertility, reproduction, growth rate and efficiency, milk production, carcass quality and composition, wool characteristics, and disease resistance. The development of an ovine genome map containing molecular markers and genes has greatly advanced the identification of genetic regions influencing and controlling traits in sheep. Other genomic resources available for researchers include an ovine radiation hybrid panel, large-insert genomic libraries, large-scale sequencing projects, and most recently, the virtual sheep genome, a whole-genome physical map orientated against the human, dog, and cow genome assemblies. In order to continue the identification of genes controlling important phenotypes in sheep, development of the comparative maps should continue. Complete sequencing and annotation of the ovine genome will further facilitate studies in sheep, other livestock species, and humans.

3.1.1
Economic and Biomedical Importance

Sheep are critical to the worldwide production of food and fiber, as well as serving an expanding role in biomedical research. Total meat production from lambs and sheep rose to a record of 13 million metric tons in 2005. Sheep production contributes US$50 billion (farm gate prices) annually, and they are particularly important to the economies of less developed nations. Worldwide, there are more than 1 billion sheep belonging to over 1,300 breeds (Scherf 2000), demonstrating the vast genetic variability that exists in sheep and the impact of this species in agricultural settings.

Animal geneticists have successfully elucidated the molecular basis for many sheep production traits over the past decade using current genetic maps and genomic tools, as described in Sect. 3.2.2. The identification of genetic regions controlling economically important traits in sheep will lead to the selection of animals with enhanced production.

The analysis of traits unique to sheep has significantly extended our knowledge of mammalian gene function and regulation beyond what is known is humans and mice (Cockett et al. 2001; Maddox and Cockett 2007). The moderate size, low maintenance requirements, and the docile temperament of sheep make them a practical large-animal model. There are over 100,000 PubMed references for sheep, making it one of the most quoted medical species. The Online Mendelian Inheritance in Animals database (http://omia.angis.org.au/) lists 67 potential sheep models for human disorders, which is nearly the same number (70) as for the pig. Sheep have been used to study inherited diseases such as rickets (Thompson et al. 2007), cystic fibrosis (Harris 1997), muscular dystrophy (Johnsen et al. 1997), osteoarthritis (Scott 2001), and lysosomal storage diseases (Ellinwood et al. 2004). The feasibility of in utero gene therapy (Porada et al. 2004), the safety of retroviral-based transmission systems for gene

Genome Mapping and Genomics in Animals, Volume 3
Genome Mapping and Genomics in Domestic Animals
N.E. Cockett, C. Kole (Eds.)
© Springer-Verlag Berlin Heidelberg 2009

therapy (Van den Broeke and Burny 2003), the possibility of stem cell transplantation (Mackenzie and Flake 2001), and therapies for osteoporosis (Turner 2002) have all been tested using sheep as the large-animal model for humans. Currently there are 244 projects in the National Institutes of Health (NIH) CRISP database (http://crisp.cit.nih.gov/) using sheep as a model, including 77 in the area of fetal development, 34 for surgery modeling, 27 in metabolism, 11 immunity projects, and 10 studies on prion diseases.

The first animal produced by nuclear transfer was a sheep named Dolly (Campbell et al. 1996), and ovine embryonic stem cell lines have been produced (Zhu et al. 2007), establishing sheep as a potentially valuable model for enhancing reproductive and therapeutic technologies. Sheep models have been developed to examine human medical conditions such as brain injury (Finnie et al. 2005), the effects of fetal alcohol exposure (Parnell et al. 2007), intervertebral disc degeneration in the spinal column (Zhou et al. 2007), and the physiological effects of smoke inhalation (Sakurai et al. 2007). In addition, therapies for osteoporosis (Turner 2002), the feasibility of in utero gene therapy (Porada et al. 2004), the safety of retroviral-based transmission systems for gene therapy (Van den Broeke and Burny 2003), and the possibility of stem cell transplantation (Mackenzie and Flake 2001) have all been tested using sheep as the large-animal model for humans.

3.1.2
History

Sheep first appeared in the fossil records about 2.5 million years ago (Maijala 1997). These Ice Age specimens were larger in stature than sheep of today. It is likely that modern domestic sheep and the European mouflon were derived from a common ancestor (Hiendleder et al. 1990). Domestication of sheep began about 9000 BC in southwestern Asia and was completed around 4000 BC with the westward migration across Asia and Europe to the Atlantic Ocean (Maijala 1997). Originally, sheep were sought after for their meat, hide, and bones, but with domestication the value of secondary products such as wool and milk was enhanced.

Analysis of mitochondrial DNA from animals in modern sheep breeds identified separate ancestral maternal origins for domestic sheep in Europe and Asia (Hiendleder et al. 1990). Sequence divergence of 2.7–4.4% has been estimated between the two lineages (Hiendleder et al 1998; Meadows et al. 2005), suggesting significant gene flow across breeds of European and Asian origins. A third maternal lineage was recently identified in Chinese breeds (Guo et al. 2005), suggesting a novel origin for sheep in China.

3.1.3
The Sheep Karyotype

Although numerous sheep karyotypes have been published, a standard G-band karyotype for sheep was agreed upon at the ninth North American Colloquium on Domestic Animal Cytogenetics and Gene Mapping in 1995 (Ansari et al. 1999). The chromosomal organizations of cattle and sheep show a high degree of conservation that has been characterized through cytogenetic analysis (Hayes et al. 1991; Hediger et al. 1991; Ansari et al. 1993, 1999), linkage mapping (Montgomery et al. 1995; Lord et al. 1996; Maddox et al. 2001), gene order (Crawford et al. 1995; de Gortari et al. 1998), and recently, large-scale DNA sequence comparisons (Dalrymple et al. 2007).

Sheep ($2n = 54$) have 26 pairs of autosomes, including three pairs of submetacentric chromosomes and 23 pairs of acrocentric chromosomes, in contrast to cattle and goats that both have 29 pairs of acrocentric chromosomes. Sheep and goats have acrocentric sex chromosomes, whereas cattle have metacentric sex chromosomes. The three submetracentric autosomes in sheep (ovine chromosomes OAR1, OAR2, and OAR3) are likely the result of Robertsonian translocations that occurred during the divergence of caprine and ovine species. The highly conserved nature of the sheep and cattle chromosome organization can be seen in Fig. 1, which is based on 1,937 loci that are in common between the sheep and cattle genetic maps. The centric fusions of two ancestral chromosomes to form the sheep metacentric chromosomes are seen in Fig. 1 as diagonals with opposite slopes for OAR1 (bovine chromosomes BTA1/3), OAR2 (BTA2/8), and OAR3 (BTA5/11). Chromosomal banding patterns of goat and sheep chromosomes show that OAR1, OAR2, and OAR3 correspond to centric fusions of caprine chromosomes (CHI) 1/3, CHI2/8, and CHI4/9, respectively (Hayes et al. 1991).

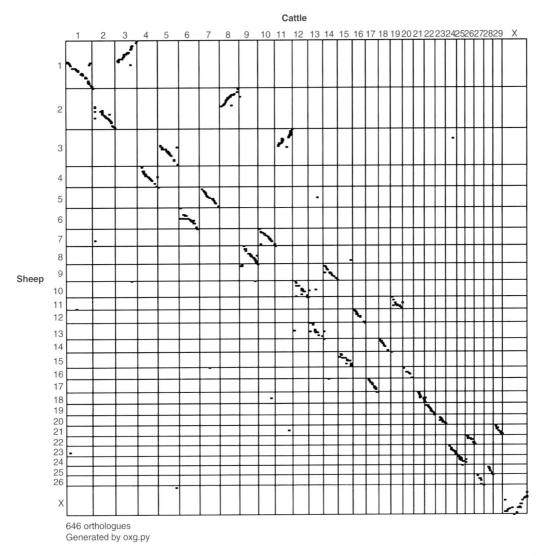

Fig. 1 An Oxford grid of the genetic maps of sheep and cattle. The genetic map of sheep for each chromosome (*vertical axis*) was aligned with the corresponding cattle chromosome (*horizontal axis*) using common loci. The correspondence of the three sheep metacentric chromosomes (OAR 1, OAR 2, and OAR 3) to six cattle acrocentric chromosomes is evident. The other sheep acrocentric chromosomes show a high degree of conservation to bovine chromosomes in chromosomal structure and gene order. The Oxford grid was generated by OXGRID – The Oxford Grid Project (Table 1) using a total of 1,937 sheep loci, which included 551 protein coding sequences

Conserved chromosomal segments between sheep and humans have been determined using human chromosome painting probes on R-banded sheep chromosomes (Iannuzzi et al. 1999) and by probing Indian muntjac deer chromosomal preparations with human and sheep painting probes (Burkin et al. 1997). A total of 48 human segments were found across sheep chromosomes at the resolution achievable with the Burkin et al. (1997) approach.

3.2
Molecular Genetics

Because of significant limits in funding, development of the sheep genome map has relied heavily on resources from other species, particularly cattle. Gene-coding sequences are highly conserved between sheep and cattle, sharing approximately 96% identity

(Maddox et al. 2006). However, these sequences comprise only 1% of the genome. Identity levels of about 90% are expected for noncoding sequences. For nonrepetitive regions, experimental evidence suggests that only 58% of bovine genomic primers will amplify a product from ovine genomic DNA, no doubt because of divergence in the primer sequences (de Gortari et al. 1998). Of those bovine markers that amplify ovine DNA, about 60% are polymorphic in sheep (Crawford et al. 1995; de Gortari et al. 1998).

3.2.1
Linkage Maps

Development of the ovine linkage map has been critical for genomic research in sheep. The first published linkage map was the outcome of a genome scan that was initiated to map the Booroola fecundity gene (Montgomery et al. 1993) using 12 pedigrees segregating for the Booroola gene. This map contained 52 markers, including microsatellites and candidate gene restriction fragment length polymorphisms (RFLPs), assigned to 19 linkage groups (Crawford et al. 1994). A more extensive ovine linkage map was later published by this research group (Crawford et al. 1995) and contained 246 markers (86 ovine microsatellites, 126 bovine microsatellites, 1 deer microsatellite, and 33 known genes) across all 26 sheep autosomes. Marker spacing was between 10 and 30 cM and total coverage of the map was 2,070 cM (about 75% of the genome). This map was constructed using the AgResearch International Mapping Flock (IMF), which included 127 sheep from nine three-generation full-sibling families derived from Texel, Coopworth, Perendale, Romney, and Merino founders. The 98 F_2 progeny all shared a common paternal grandsire, for a total of 222 informative meioses. The IMF has been an important resource for continued development of the ovine linkage map (Maddox et al. 2001, 2006).

Another ovine genetic map was published in 1998 by the USDA-ARS group in Nebraska (de Gortari et al. 1998). The map contained 519 markers (402 bovine microsatellites, 101 ovine microsatellites, and 16 known genes) and spanned 3,063 cM across the autosomes, with an average marker spacing of 6.5 cM. The USDA pedigree included 247 backcross progeny produced by mating four F_1 rams to 44 Romanov ewes. This map and information on the markers can be accessed at the USDA-ARS sheep genome site (http://www.marc.usda.gov/genome/sheep/sheep.html).

A third-generation map containing 1,062 loci (941 anonymous loci and 121 genes) mapped across the IMF population was generated using data from 15 laboratories (Maddox et al. 2001). The map spanned 3,400 cM (sex-averaged) for the autosomes and 132 cM (female) on the X-chromosome.

The sheep linkage map is periodically updated. The most current map (version 4.6) can be viewed on the Australian Gene Mapping Web Site (http://rubens.its.unimelb.edu.au/jillm/jill.htm), which is maintained by Jill Maddox, University of Melbourne, Australia. The current map has nearly 1,400 markers spanning over 3,500 cM (Table 1). Submetacentric chromosomes OAR1, OAR2, and OAR3 have between 136 and 117 markers spanning over 300 cM for each chromosome. The remaining acrocentric chromosomes have between 65 and 22 markers with a range of 159 cM (OAR5) to 7 cM (OAR26) coverage.

Ovine linkage maps contain relatively few expressed genes because of the difficulty in identifying allelic variation needed for linkage analyses. Assignments to the ovine physical map have been made using somatic cell hybrids (Saidi-Mehtar et al. 1979; Saidi-Mehtar et al. 1981; Burkin et al. 1998) and in situ hybridization. The physical map was reviewed by Broad et al. (1997), but an updated physical map, including map assignments of 90 additional loci using FISH and R-banding, is now available (Di Meo et al. 2007).

3.2.2
Mapped Traits in Sheep

Sheep possess many heritable traits that are of economic importance in modern animal agriculture. The contribution of sheep to food and fiber markets through meat, milk, and wool production is of global significance. To date, quantitative trait loci (QTL) for multiple traits including milk production (Diez-Tascon et al. 2000; Barillet et al. 2005; Gutiérrez-Gil et al. 2007), parasite resistance (Beh et al. 2002; Crawford et al. 2006; Davis et al. 2006; Beraldi et al. 2007), wool traits (McLaren et al. 1997; Purvis and Franklin 2005), fecal soiling of the wool (MacDonald et al. 1998), growth and carcass traits (Walling et al. 2004; Karamichou et al. 2006), and bone density (Campbell et al. 2003) have been identified. SheepQTLdb

Table 1 Summary data for the sheep linkage map (Version 4.6)[1]

Chromosome (OAR)	Markers	Sex ave (cM)	Female (cM)	Male (cM)
1	136	324.4	290.3	359.3
2	117	301.3	280.4	323.4
3	122	303.2	289.9	313.2
4	47	147.8	131.0	155.5
5	50	158.9	158.9	169.2
6	64	155.7	132.2	176.9
7	65	148.9	139.6	160.9
8	45	126.9	117.7	135.5
9	45	126.9	117.7	135.5
10	36	100.2	96.5	104.2
11	45	109.6	103.2	118.3
12	36	106.4	88.4	116.6
13	36	128.3	128.0	128.7
14	42	120.0	97.9	142.1
15	56	123.8	101.5	142.8
16	38	84.7	77.1	91.4
17	50	130.0	113.3	146.6
18	48	127.7	109.3	150.1
19	30	72.1	65.4	85.0
20	49	103.3	89.0	118.8
21	27	75.5	65.6	84.9
22	22	82.9	60.6	98.7
23	40	85.0	73.0	96.8
24	36	102.1	99.3	105.3
25	26	68.3	56.2	76.6
26	22	71.1	57.3	85.9
X	61	90.6	131.3	58.6
Total	1391	3575.6	3270.6	3880.8

[1]Data compiled from Australian Gene Mapping Web Site (http://rubens.its.unimelb.edu.au/~jillm/jill.htm)

(http://sphinx.vet.unimelb.edu.au/QTLdb/) was recently launched (Hu et al. 2007) and will be a valuable tool for cataloging and comparing QTL studies in sheep.

In addition, the chromosomal assignments of many single-gene traits in sheep are known, including those for wool color (Sponenberg 1997), coat color in Soay (Gratten et al. 2007), inherited ovine arthrogryposis or IOA (Murphy et al. 2007a), hairlessness (Finocchiaro et al. 2003), meat traits such as *callipyge* (Cockett et al. 1994, 1996; Freking et al. 1998; Freking et al. 2002; Smit et al. 2003) and *rib-eye muscling* (Nicholl et al. 1998), and horns (Montgomery et al. 1996). Causative mutations for heritable defects such as spider lamb syndrome (Cockett et al. 1999; Beever et al. 2006), glycogen storage disease (Tan et al. 1997), and scrapie susceptibility (reviewed by Smit et al. 2002) have been identified. Interestingly, sheep carry the most identified genetic mutations associated with reproduction of any livestock species, and many of these have been mapped to specific chromosomal regions (Montgomery et al. 1994; Cockett et al. 2001; Rohrer 2004; McNatty et al. 2005). While several causative mutations for fecundity traits such as Booroola (Mulsant et al. 2001; Souza et al. 2001; Wilson et al. 2001) and Inverdale (Galloway et al. 2000) have been identified, causative mutations for many other traits are not yet known.

The callipyge trait and the QTL for muscle hypertrophy near the *myostatin* locus in Texel sheep are examples of novel mutations that are contributing to our understanding of transcriptional and post-transcriptional gene regulation. The *callipyge* locus has been mapped to the distal end of ovine chromosome 18 within an imprinted cluster of genes (Cockett et al. 1994; Charlier et al. 2001a). The callipyge phenotype is inherited in a unique manner termed polar overdominance (Cockett et al. 1996; Freking et al. 1998). Animals inheriting a wild-type allele from the dam and a mutant *CLPG* allele from the sire ($+^{Mat}/CLPG^{Pat}$) have the callipyge phenotype, whereas maternal heterozygous ($CLPG^{Mat}/+^{Pat}$) and homozygous ($CLPG^{Mat}/CLPG^{Pat}$) lambs have muscling and carcass compositions similar to wild-type (+/+) animals (Cockett et al. 1996). Six genes have been identified in the *callipyge* region (Fig. 2). *DLK1, DAT*, and *PEG11* have paternal allele-specific expression, whereas *GTL2, PEG11AS*, and *MEG8* are expressed from the maternal allele (Charlier et al. 2001a). The causative mutation has been identified as a single base pair transition from an A (wild type) to a G (*CLPG*) that occurs in the intergenic region between *DLK1* and *GTL2* (Freking et al. 2002; Smit et al. 2003). It has been demonstrated that the *callipyge* mutation alters the expression of some genes in proximity to the *callipyge* locus (Bidwell et al. 2001, 2004; Charlier et al. 2001b; Murphy et al. 2005; Takeda et al. 2006; White et al. 2007), but not all (Smit et al. 2005). The increased expression occurs when the mutation is inherited in *cis* and maintains parent-of-origin imprinting (Charlier et al. 2001b). Other down-stream changes in expression resulting from the *callipyge* mutation have been investigated using microarrays (Vuocolo et al. 2006).

Both *DLK1* and *PEG11* genes exhibit polar overdominant gene expression patterns in callipyge sheep based on transcript abundance (Fig. 3). The expression

Fig. 2 The callipyge region of ovine chromosome 18. A diagram of the callipyge region based on GenBank accession no. AF345168 is shown. Six known transcripts are indicated along with the direction of transcription (*arrows*). Transcripts expressed from the paternal allele are shown as gray arrows and those expressed from the maternal allele are shown as black arrows. The thin vertical line indicates the position of the causative mutation. The wild-type allele (+) has an A and the mutant callipyge allele (*CLPG*) has a G at this position. Reprinted from Bidwell et al. (2004)

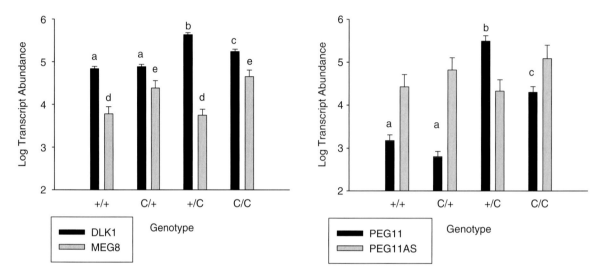

Fig. 3 Transcript abundance of four genes in the *callipyge* region, measured in the gluteus medius at 8 weeks of age. Least square means and standard errors for transcript abundance by genotype are shown for the gluteus medius that undergoes hypertrophy in callipyge lambs. The genotypes are given with the maternal allele first followed by the paternal allele and the *CLPG* allele has been abbreviated to C. Quantification of *DLK1* and *MEG8* transcripts is shown on the left, and *PEG11* and *PEG11AS* transcripts are shown on the right. Different superscripts indicate significant differences ($P < 0.05$) in genotypic means for a given transcript. Reprinted from Bidwell et al. (2004)

of both genes is the greatest in callipyge animals (+^{Mat}/*CLPG*^{Pat}), but they have intermediate levels of expression in *CLPG*/*CLPG* animals. Therefore, both genes are candidates for muscle hypertrophy associated with the callipyge trait. The *PEG11AS* transcript produces microRNA from intron sequences that have been shown to interact with the *PEG11* transcript (Seitz et al. 2003; Davis et al. 2005). This trans interaction between transcripts from the paternal and the maternal chromosomes is a potential mechanism for the polar overdominance inheritance of the callipyge phenotype (Georges et al. 2003). This result leads to the suggestion that an RNA interference mechanism may normally keep *PEG11* gene expression suppressed in postnatal skeletal muscle. The paternal heterozygous animals have a unique combination of *PEG11* and *PEG11AS* expression where *PEG11* is substantially more abundant than *PEG11AS* (Fig. 3). Therefore, the *cis* effect of the mutation on *PEG11AS* in *CLPG*/*CLPG* animals leads to a reduced level of *PEG11* expression (Bidwell et al. 2004; Perkins et al. 2006; Fleming-Waddell et al. 2007).

Genome scans for quantitative trait loci associated with muscle hypertrophy in Texel sheep have consistently shown a QTL near the myostatin locus on OAR2 (Walling et al. 2004; Johnson et al. 2005). Recently, a single nucleotide polymorphism has been identified in the 3′ untranslated region of the *myostatin* mRNA that makes the mRNA a target of post-transcriptional regulation (Clop et al. 2006). This *myostatin* mutation in Texels creates an 8 bp motif that is known to be a target sequence for *miR*-1 and *miR*-206 microRNA, which are expressed in skeletal muscle. The mutation causes a reduced translational efficiency of *myostatin* RNA leading to reduced levels of circulating myostatin, which results in muscle hypertrophy in animals with the Texel allele (Clop et al. 2006). This muscle hypertrophy trait in Texel sheep has led to the identification of new mechanisms for generating phenotypic variation (Clop et al. 2006).

3.2.3
Resources for Mapping and Sequencing the Ovine Genome

While relatively few resources have been devoted to the sheep genome as compared to humans and rodents, tools necessary for genomic research in sheep are being accumulated. In 2002, the International Sheep Genomics Consortium (ISGC) was formed, with emphasis on the development of public genome resources that contribute to the sheep genome map and ultimately lead to a completely sequenced ovine genome (Table 2). The Consortium is composed of scientists, commodity organizations, and funding agencies from Australia, France, Kenya, New Zealand, United Kingdom, and United States. Several successful projects of the ISGC are described later.

Ovine Radiation Hybrid Panels

A 5,000 rad ovine radiation hybrid (RH) panel that was developed through a collaborative project between Utah State University and Texas A&M University (Eng et al. 2004) is now available. Ninety clones with retention frequencies between 15 and 40% have been selected for inclusion in the USUoRH5000 panel and DNA from the panel has been distributed to five international laboratories. In a preliminary experiment, 257 markers covering all autosomes were screened across the panel; 131 (51%) of these markers produce resolvable patterns (Wu et al. 2006). Recent studies using the USUoRH5000 include preliminary RH maps for OAR2, 6, 23, and 26, for a total of 580 loci (Goldammer et al. 2006; Nomura et al. 2006; Tetens et al. 2006; Wu et al. 2007), as well as regions on OAR8, 9, and 20 that are homologous to HSA6 (Wu et al. 2008). The panel is now being typed with 500 microsatellite markers also mapped on the ovine and bovine linkage maps, 500 expressed sequence tags (ESTs) developed from ovine and bovine complementary DNA (cDNA) sequences with known locations on the human genome map, and 500 primers developed from the ovine bacterial artificial chromosome (BAC) end sequencing project

Table 2 Online resources for sheep gene mapping and genomics

Source	Site	Internet address
Genetic Maps		
University of Melbourne	Australian Gene Mapping Web Site	http://rubens.its.unimelb.edu.au/~jillm/jill.htm
USDA/ARS MARC	Sheep genetic map	http://www.marc.usda.gov/genome/sheep/sheep.html
Roslin institute	ArkDB	www.thearkdb.org
The Oxford grid project	OXGRID	http://oxgrid.angis.org.au/sheep/
Genomics Resources		
International Sheep Genomics Consortium	ISGC projects	http://www.sheephapmap.org
National Center for Biotechnology Information	Sheep genome resources	http://www.ncbi.nlm.nih.gov/genome/guide/sheep/index.html
Children's Hospital Oakland Research Institute	BACPAC Resource Center	http://bacpac.chori.org/home.htm
BACPAC Resources	CHORI 243 ovine BAC library	http://bacpac.chori.org/library.php?id=162
CSIRO Livestock Genomics	Sheep BAC end alignment	http://www.livestockgenomics.csiro.au/perl/gbrowse.cgi/sheepbacend/

(see later). Data from these loci will be used to develop a framework/comprehensive RH map for sheep. An online, real-time comparative database will be developed for web-based transmission of mapping data on the distributed ovine RH panels (N Cockett, personal communication).

In addition, an ovine 12,000 rad RH panel containing 90 clones with an average retention of 31.8% and a resolution of 15 kb/cR has been constructed at INRA (Laurent et al. 2007). Sixty-seven markers have been mapped to a 23 Mb region on OAR18 using the INRA panel. Because a high radiation dose creates relatively small DNA fragments resulting in enhanced mapping resolution, this 12,000-rad panel will be useful for physical contig development and fine mapping of genetic regions. However, development of a whole-genome map using this panel will be difficult.

Sheep Bacterial Artificial Chromosome Libraries

Two ovine BAC libraries were constructed in 1999 (Gill et al. 1999; Vaiman et al. 1999). These two libraries are remarkably similar, with 90,000 and 60,000 clones, respectively; an average clone size of 123 kb and 103 kb, respectively; and covering three and two genome equivalents, respectively. DNAs from the libraries have been pooled, allowing polymerase chain reaction (PCR)-based screening for sequences of interest.

In 2003, a memorandum of understanding among AgResearch (New Zealand), Meat and Livestock Australia, the USDA-ARS Meat Animal Research Center (Nebraska, USA), and Utah State University (USA) resulted in the construction of another ovine BAC library. The CHORI-243 library was made by Pieter de Jong's group at BACPAC Resources Center (http://bacpac.chori.org/) using DNA from a Texel ram contributed by the USDA-ARS group. This library consists of 202,752 clones with an average insert size of 184 kb, and has 10-fold genome coverage (Nefedov et al. 2003). It is organized in two segments; bacterial colonies for each segment have been arrayed on filters which can be used for screening the library. The library, filter arrays, and individual BAC clones are available through the BACPAC Resources Center (Table 1).

BAC End Sequences and the Virtual Sheep Genome

A joint collaboration was established between Utah State University (USA), The Institute for Genomic Research (TIGR, USA), Australian Wool Innovation, Meat and Livestock Australia, AgResearch (New Zealand), and Genesis-Faraday (UK), with support of the Alliance for Animal Genome Research, to obtain large-scale BAC end-sequences from the CHORI-243 BAC library. The sequencing portion of the BES project is now complete, with 376,493 sequences from 193,073 BAC clones. The average BAC end-sequence length was 687 bp, with a range of 64–1044 bp. Over 99% of the clones provided at least 100 bp sequence. A total of 258,650,691 bp sequence (approximately 6% of the genome) was produced from this project and can be accessed at the National Center for Biotechnology Information (NCBI) Sheep Genome Resource site (http://www.ncbi.nlm.nih.gov/genome/guide/sheep/index.html).

The BAC-end sequences have provided an important resource for generating new marker sequences to enhance the ovine linkage and radiation hybrid maps. Furthermore, the paired end sequences from the BAC have been incorporated into a "virtual sheep genome" using comparative analysis with sequenced genomes of cattle, dogs, and humans (Dalrymple et al. 2007). Alignment of 84,624 ovine BACs with paired end sequences in the correct orientation and spacing to the bovine and dog genome sequences resulted in 1,172 BAC mega-contigs, which covered 91.2% of the human genome. Further extension of the virtual genome using interchromosomal breaks and the sheep marker map resulted in 70% coverage for the sheep genome. Matches to the bovine, dog, and human sequences; orientation of the contigs; and identity of the BAC clones in the virtual sheep genome can be obtained through a database established by CSIRO (http://www.livestockgenomics.csiro.au/sheep.shtml).

The International Sheep Genome Consortium has further leveraged the ovine BAC end sequence information through a 2005 grant from the Australian International Sciences Linkage Program directed toward single nucleotide polymorphism (SNP) discovery using a resequencing approach and a first-generation HapMap (http://www.sheephapmap.org/). A pilot study was started in mid-2006 using Sanger sequencing of regions of the genome corresponding to selected BAC ends positioned on the virtual sheep genome and some ESTs across a panel of diverse sheep and in pools of DNA from these animals. Samples from nine breeds (Polled

Dorset, Merino, Awassi, Lacaune, Red Masai, Texel, Romney, Katahdin, and Gulf Coast Native) were contributed by researchers from across the world (J Kijas, personal communication). Frequency of SNPs in the resequencing project was about one in every 200 bases (188 bases/SNP for the BAC end sequences and 428 bases/SNP for the EST sequence), in contrast to an expected SNP frequency of one in every 500 bases for BAC sequencing (assuming two haplotypes within the BAC library).

Using SNPs from the pilot project, a 1.5K SNP GoldenGate chip has been constructed by the John Hopkins SNP Center (http://snpcenter.grcf.jhmi.edu/), which can be assayed using Illumina's BeadArray technology (http://www.illumina.com/pages.ilmn?ID=5). Selection of the SNPs was based on the following criteria: minor allele frequency (MAF) ≥ 0.2; Illumina conversion score ≥ 0.6; at least 500 bp of contiguous unmasked sequence; representation of each BAC-CGC in the virtual sheep genome; and targets from ovine ESTs with selected intron and 3'UTR attributes. Fifteen animals each from Tibetan, Scottish Blackface, Suffolk, Charolais, Soay, Texel, Romney, German Brown, Red Masai, African Dorper, Italian Sarda, Poll Dorset, Australian Merino, Javanese, Sumatran Thin Tail, US Rambouillet, US Suffolks, bighorn sheep, and Dall's sheep will be genotyped with the chip, as well as the USUo5000RH panel and the IMF pedigree. Data collected on these samples will be used to develop a first-generation ovine HapMap (J Kijas, personal communication).

3.3
Future Scope of Work

Utilization of genomic resources now available in sheep will lead to more efficient studies on specific chromosomal regions of interest, thereby extending functional studies in sheep. These resources will greatly facilitate current positional cloning efforts to identify causal mutations that underlie economic trait loci. Scientists involved in sheep molecular genetics will be better able to exploit comparative information from the fully sequenced, information-rich genomes such as human, mouse, and cattle. Genomic resources in sheep will also provide the scaffold for sequencing

the ovine genome, a project that the sheep community is working toward for the future.

References

Ansari HA, Pearce PD, Maher DW (1993) Resolving ambiguities in the karyotype of domestic sheep (*Ovis aries*). Chromosoma 102:97–106

Ansari HA, Bosma AA, Broad TE, Bunch TD, Long SE, Maher DW, Pearce PD, Popescu CP (1999) Standard G-, Q-, and R-banded ideograms of the domestic sheep (*Ovis aries*) and homology with cattle (*Bos taurus*): report of the committee for the standardization of the sheep karyotype. Cytogenet Cell Genet 87:134–42

Barillet F, Arranz JJ, Carta A (2005) Mapping quantitative trait loci for milk production and genetic polymorphisms of milk proteins in dairy sheep. Genet Sel Evol 37(Suppl 1): S109–S123

Beever JE, Smit MA, Meyers SN, Hadfield TS, Bottema C, Albretsen J, Cockett NE (2006) A single-base change in the tyrosine kinase II domain of ovine FGFR3I causes hereditary chondrodysplasia in sheep. Anim Genet 37:66–71

Beh KJ, Hulme DJ, Callaghan MJ, Leish Z, Lenane I, Windon RG, Maddox JF (2002) A genome scan for quantitative loci affecting resistance to *Trichostrongylus columbriformis* in sheep. Anim Genet 33:97–106

Beraldi D, McRae AF, Gratten J, Pilkington JG, Slate J, Visscher PM, Pemberton JM (2007) Quantitative trait loci (QTL) mapping of resistance to strongyles and coccidian in the free-living Soay sheep (*Ovis aries*). Intl J Parasitol 37:121–129

Bidwell CA, Shay TL, Georges M, Beever JE, Berghmans S, Cockett NE (2001) Differential expression of the GTL2 gene within the callipyge region of ovine chromosome 18. Anim Genet 32:248–256

Bidwell CA, Kramer LN, Perkins AC, Hadfield TS, Moody DE, Cockett NE (2004) Expression of *PEG11* and *PEG-11AS* transcripts in normal and callipyge sheep. BMC Biol 2:17

Broad TE, Hayes H, Long SE (1997) Cytogenetics: physical chromosome maps. In: Piper L, Ruvinsky A (eds) The Genetics of Sheep. CABI, Oxon, UK, pp 241–295

Burkin DJ, Yang F, Broad T, Wienberg J, Hill DF, Ferguson-Smith MA (1997) Use of the Indian muntjac idiogram to align conserved chromosomal segments in sheep and human genomes by chromosomal painting. Genomics 46:143–147

Burkin DJ, Broad TE, Lambeth MR, Burkin HR, Jones C (1998) New gene assignments using a complete, characterized sheep-hamster somatic cell hybrid panel. Anim Genet 29:48–54

Campbell AW, Bain WE, McRae AF, Broad TE, Johnstone PD, Dodds KG, Veenvliet BA, Greer GJ, Glass BC, Beattie AE, Jopson NB, McEwan JC (2003) Bone density in sheep: genetic variation and quantitative trait localisation. Bone 33:540–548

Campbell KH, McWhir J, Ritchie WA, Wilmut I (1996) Sheep cloned by nuclear transfer from a cultured cell line. Nature 380:64–66

Charlier C, Segers K, Karim L, Shay T, Gyapay G, Cockett N, Georges M (2001a) The callipyge mutation enhances the expression of coregulated imprinted genes in cis without affecting their imprinting status. Nat Genet 27:367–369

Charlier C, Segers K, Wagenaar D, Karim L, Berghmans S, Jaillon O, Shay T, Weissenbach J, Cockett N, Gyapay G, Georges M (2001b) Human-ovine comparative sequencing of a 250-kb imprinted domain encompassing the callipyge (clpg) locus and identification of six imprinted transcripts: DLK1, DAT, GTL2, PEG11, antiPEG11, and MEG8. Genom Res 11:850–862

Clop A, Marcq F, Takeda H, Pirottin D, Tordoir X, Bibe B, Bouix J, Caiment F, Elsen JM, Eychenne F, Larzul C, Laville E, Meish F, Milenkovic D, Tobin J, Charlier C, Georges M (2006) A mutation creating a potential illegitimate microRNA target site in the myostatin gene affects muscularity in sheep. Nat Genet 38:813–818

Cockett NE, Jackson SP, Shay TL, Farnir F, Berghmans S, Snowder GD, Nielsen DM, Georges M (1996) Polar overdominance at the ovine callipyge locus. Science 273:236–238

Cockett NE, Jackson SP, Shay TL, Nielsen D, Steele MR, Green RD, Georges M (1994) Chromosomal localization of the callipyge gene in sheep (Ovis aries) using bovine DNA markers. Proc Natl Acad Sci USA 91:3019–3023

Cockett NE, Shay TL, Beever JE, Nielsen D, Albertsen J, Georges M, Peterson K, Stephens A, Vernon W, Timofeevskaia O, South S, Mork J, Maciulis A, Bunch TD (1999) Localization of the locus causing spider lamb syndrome to the distal end of ovine chromosome 6. Mamm Genom 10:35–38

Cockett NE, Shay TL, Smit M (2001) Analysis of the sheep genome. Physiol Genom 7:69–78

Crawford AM, Montgomery GW, Pierson CA, Brown T, Dodds KG, Sunden SL, Henry HM, Ede AJ, Swarbrick PA, Berryman T, Penty JM, Hill DF (1994) Sheep linkage mapping: nineteen linkage groups derived from the analysis of paternal half-sib families. Genetics 87:271–277

Crawford AM, Dodds KG, Ede AE, Pierson CA, Montgomery GW, Garmonsway G, Beattie AE, Davies K, Maddox JF, Kappes SM, Stone RT, Nguyen TC, Penty JM, Lord EA, Broom JE, Buitkamp J, Schwaiger W, Epplen JT, Matthew P, Matthews ME, Hulme DJ, Beh KJ, McGraw RA, Beattie CW (1995) An autosomal genetic linkage map of the sheep genome. Genetics 140:703–724

Crawford AM, Paterson KA, Dodds KG, Diez Tascon C, Williamson PA, Thomson MR, Bisset SA, Beattie AE, Greer GJ, Green RS, Wheeler R, Shaw RJ, Knowler K, McEwan JC (2006) Discovery of quantitative trait loci for resistance to parasitic nematode infection in sheep: I. Analysis of outcross pedigrees. BMC Genomics 7:178–187

Dalrymple BP, Kirkness EF, Nefodov M, McWilliam S, Ratnakumar A, Barris W, Zhao S, Shetty J, Maddox JF, O'Grady M, Nicholas F, Crawford AM, Smith T, de Jong P, McEwan JC, Oddy VH, Cockett NE (2007) Constructing the virtual sheep genome. Genome Res 8:R152 (doi:10.1186/gb-2007-8-7-r152)

Davis E, Caiment F, Tordoir X, Cavaille J, Ferguson-Smith A, Cockett C, Georges M, Charlier C (2005) RNAi-mediated allelic trans-interaction at the imprinted Rtl1/Peg11 locus. Curr Biol 15:743–749

Davis G, Stear MJ, Benothman M, Abuagob O, Kerr A, Mitchell S, Bishop SC (2006) Quantitative trait loci associated with parasitic infection in Scottish blackface sheep. Heredity 96:252–258

de Gortari MJ, Freking BA, Cuthbertson RP, Kappes SM, Keele JW, Stone RT, Leymaster KA, Dodds KG, Crawford AM, Beattie CW (1998) A second-generation linkage map of the sheep genome. Mamm Genom 9:204–209

Diez-Tascon C, Bayon Y, Arranz JJ, De La Fuente F, San Primitivo F (2000) Mapping quantitative trait loci for milk production traits on ovine chromosome 6. J Dairy Res 68:389–397

Di Meo GP, Perucatti A, Floriot S, Hayes H, Schibler L, Rullo R, Incarnato D, Ferretti L, Cockett N, Cribiu E, Williams JL, Eggen A, Iannuzzi L (2007). An advanced sheep (Ovis aries, 2n = 54) cytogenetic map and assignment of 88 new autosomal loci by fluorescence in situ hybridization and R-banding. Anim Genet 38:233–240

Ellinwood NM, Vite CH, Haskins ME (2004) Gene therapy for lysosomal storage diseases: the lessons and promise of animal models. J Gene Med 6:481–506

Eng SL, Owens E, Womack JE, Cockett NE (2004) Development of an ovine whole-genome radiation hybrid panel. Plant & Animal Genome XII Conf, San Diego, CA, USA, P 650

Finocchiaro R, Portolano B, Damiani G, Caroli A, Budelli E, Bolla P, Pagnacco G (2003) The hairless (hr) gene is involved in the congenital hypotrichosis of Valle del Belice sheep. Genet Sel Evol 35(Suppl 1):S147–S156

Finnie J, Manavis J, Jones N, Blumbergs P (2005) Topography of c-fos immunoreactivity in an ovine head impact model. J Clin Neurosci 12:799-803

Fleming-Waddell JN, Wilson LM, Olbricht GR, Vuocolo T, Byrne K, Craig BA, Tellam RL, Cockett NE, Bidwell CA (2007) Analysis of gene expression during the onset of muscle hypertrophy in callipyge lambs. Anim Genet 38:28–36

Freking BA, Keele JW, Beattie CW, Kappes SM, Smith TP, Sonstegard TS, Nielsen MK, Leymaster KA (1998) Evaluation of the ovine callipyge locus: I. Relative chromosomal position and gene action. J Anim Sci 76:2062–2071

Freking BA, Murphy SK, Wylie AA, Rhodes SJ, Keele JW, Leymaster KA, Jirtle RL, Smith TP (2002) Identification of the single base change causing the callipyge muscle hypertrophy phenotype, the only known example of polar overdominance in mammals. Genome Res 12:1496–1506

Galloway SM, McNatty KP, Cambridge LM, Laitinen MP, Juengel JL, Jokiranta TS, McLaren RJ, Luiro K, Dodds KG, Montgomery GW, Beattie AE, Davis GH, Ritvos O (2000) Mutations in an oocyte-derived growth factor gene (BMP15) cause increased ovulation rate and infertility in a dosage-sensitive manner. Nat Genet 25:279–283

Georges M, Charlier C, Cockett N (2003) The callipyge locus: evidence for the trans interaction of reciprocally imprinted genes. Trends Genet 19:248–252

Gill CA, Davis SK, Taylor JF, Cockett NE, Bottema CDK (1999) Construction and characterization of an ovine bacterial artificial chromosome library. Mamm Genom 10: 1108–1111

Goldammer T, Hadfield TS, Kühn C, Gill CA, Weikard R, Brunner RM, Wimmers K, Womack JE, Cockett NE (2006) A comparative radiation hybrid map of sheep chromosome OAR26. Proc 30th Intl Conf Anim Genet, Porto Seguro, Brazil, August 20–25, B212

Guo J, Du LX, Ma YH, Guan WJ, Li HB, Zhao QJ, Lis X, Rao SQ (2005) A novel maternal lineage revealed in sheep (*Ovis aries*). Anim Genet 36:331–336

Gratten J, Beraldi D, Lowder BV, McRae AF, Visscher PM, Pemberton JM, Slate J (2007) Compelling evidence that a single nucleotide substitution in *TYRP1* is responsible for coat-colour polymorphism in a free-living population of Soay sheep. Proc Biol Sci 274:619–626

Gutiérrez-Gil B, El-Zarei MF, Bayón Y, Alvarez L, de la Fuente LF, San Primitivo F, Arranz JJ (2007) Detection of quantitative trait loci influencing somatic cell score in Spanish Churra sheep. J Dairy Sci 90:422–426

Harris A (1997) Towards an ovine model of cystic fibrosis. Hum Mol Genet 13:2191–2193

Hayes H, Petit E, Dutrillaux (1991) Comparison of RBG-banded karyotypes of cattle sheep and goats. Cytogenet Cell Genet 57:51–55

Hediger R, Ansari HA, Stranszinger G (1991) Chromosome banding and gene localizations support extensive conservation of chromosome structure between sheep and cattle. Cytogenet Cell Genet 57:127–134

Hiendleder S, Mainz K, Plante Y, Lewalski H (1998) Analysis of mitochondrial DNA indicates that domestic sheep are derived from two different ancestral maternal sources: no evidence for contributions from urial and argali sheep. J Hered 89:113–120

Hu Z, Fritz ER, Reecy JM (2007) AnimalQTLdb: a livestock QTL database tool set for positional QTL information mining and beyond. Nucl Acids Res 35:D604–D609

Iannuzzi L, Di Meo GP, Perucatti A, Incarnato D (1999) Comparison of the human with the sheep genomes by use of human chromosome-specific painting probes. Mamm Genom 10:719–723

Johnsen RD, Laing NG, Huxtable CR, Kakulas BA (1997) Normal expression of adhalin and merosin in ovine congenital progressive muscular dystrophy. Aust Vet J 75:215–216

Johnson PL, McEwan JC, Dodds KG, Purchas RW, Blair HT (2005) A directed search in the region of GDF8 for quantitative trait loci affecting carcass traits in Texel sheep. J Anim Sci 83:1988–2000

Karamichou E, Richardson RI, Nute GR, Gibson KP, Bishop SC (2006) Genetic analyses and quantitative trait loci detection, using a partial genome scan for intramuscular fatty acid composition in Scottish Blackface sheep. J Anim Sci 84:3228–3238

Laurent P, Schibler L, Vaiman A, Laubier J, Delcros C, Cosseddu G, Vaiman D, Cribiu EP, Yerle M (2007) A 12,000-rad whole-genome radiation hybrid panel in sheep: application to the study of the ovine chromosome 18 region containing a QTL for scrapie susceptibility. Anim Genet 38:358–363

Lord EA, Lumsden JM, Dodds KG, Henry HM, Crawford AM, Ansari HA, Pearce PD, Maher DW, Stone RT, Kappes SM, Beattie CW, Montgomery GW (1996) The linkage map of sheep Chromosome 6 compared with orthologous regions in other species. Mamm Genom 7:373–376

MacDonald PA, McEwan JC, Goosen GJ, Dodds KG, Green RS, Wheeler RW, Knowler KJ, Greer GJ, Crawford AM (1998) A dagginess QTL in sheep. Proc 26th Intl Conf Anim Genet, Auckland, New Zealand, August 9–14. pp 102

Mackenzie TC, Flake AW (2001) Human mesenchymal stem cells persist, demonstrate site-specific multipotential differentiation, and are present in sites of wound healing and tissue regeneration after transplantation into fetal sheep. Blood Cells Mol Dis 27:601–604

Maddox JF, Cockett NE (2007) An update on sheep and goat linkage maps and other genomics resources. Small Rumin Res 70:4–20

Maddox JF, Dalrymple B, McWilliam S, McColloch A, Kirkness E, Townley D, Barris W, Kijas J, Ratnakumar A, McEwen J, Nicholas F, International Sheep Genome Consortium (2006) Enhanced mapping tools for the sheep genome – BACs, microsatellites, SNP's and the virtual map. Proc 30th Intl Conf Anim Genet, Porto Seguro, Brazil, August 20–25, E507

Maddox JF, Davies KP, Crawford AM, Hulme DJ, Vaiman D, Cribiu EP, Freking BA, Beh KJ, Cockett NE, Kang N, Riffkin CD, Drinkwater R, Moore SS, Dodds K, Lumsden JK, Adelson D, Birkin H, Broom JE, Buitkamp J, Cambridge E, Cushwa WT, Gerard G, Galloway S, Harrison B, Hawken RJ, Hiendleder S, Henry H, Medrano J, Paterson K, Phua SH, Schibler L, Stone RT, van Hest B (2001) An enhanced linkage map of the sheep genome comprising more than 1000 loci. Genom Res 11:1275–1289

Maijala K (1997) Genetic aspects of domestication, common breeds and their origins. In: Piper I, Ruvinsky A (eds) The Genetics of Sheep. CABI, Oxon, UK, pp 13–49

McLaren RJ, Rogers GR, Davies KP, Maddox JF, Montgomery GW (1997) Linkage mapping of wool keratin and keratin-associated protein genes in sheep. Mamm Genom 8: 938–940

McNatty KP, Galloway SM, Wilson T, Smith P, Hudson NL, O'Connell A, Bibby AH, Heath DA, Davis GH, Hanrahan JP, Juengel JL (2005) Physiological effects of major genes affecting ovulation rate in sheep. Genet Sel Evol 37 (Suppl 1):S25–S38

Meadows JR, Li K, Kantanen J, Tapio M, Sipos W, Pardeshi V, Gupta V, Calvo JH, Whan V, Norris B, Kijas JW (2005) Mitochondrial sequence reveals high levels of gene flow between breeds of domestic sheep from Asia and Europe. J Hered 96:494–501

Montgomery GW, Crawford AM, Penty JM, Dodds KG, Ede AJ (1993) The ovine Booroola fecundity gene (FecB) is linked to markers from a region of human chromosome 4q. Nat Genet 4:410–414

Montgomery GW, Henry HM, Dodds KG, Beattie AE, Wuliji T, Crawford AM (1996) Mapping the Horns (Ho) locus in sheep: a further locus controlling horn development in domestic animals. J Hered 87:358–363

Montgomery GW, Lord EA, Penty JM, Dodds KG, Broad T, Cambridge L, Sunden S, Stone R, Crawford A (1994) The Booroola fecundity (FecB) gene maps to sheep chromosome 6. Genomics 22:148–53

Montgomery GW, Penty JM, Henry HM, Sise JA, Lord EA, Dodds KG, Hill DF (1995) Sheep linkage mapping: RFLP markers for comparative mapping studies. Anim Genet 26:249–259

Mulsant P, Lecerf F, Fabre S, Schibler L, Monget P, Lanneluc I, Pisselet C, Riquet J, Monniaux D, Callebaut I, Cribiu E, Thimonier J, Teyssier J, Bodin L, Cognie Y, Elsen JM (2001) Mutation in bone morphogenetic protein receptor-1B is associated with increased ovulation rate in Booroola Merino ewes. Proc Natl Acad Sci USA 98:5104–5109

Murphy AM, MacHugh DE, Park SD, Scraggs E, Haley CS, Lynn DJ, Boland MP, Doherty ML (2007) Linkage mapping of the locus for inherited ovine arthrogryposis (IOA) to sheep chromosome 5. Mamm Genome 18:43–52

Murphy SK, Freking BA, Smith TP, Leymaster K, Nolan CM, Wylie AA, Evans HK Jirtle RL (2005) Abnormal postnatal maintenance of elevated DLK1 transcript levels in callipyge sheep. Mamm Genom 16:171–183

Nefedov M, Zhu B, Thorsen J, Shi CL, Cao Q, Osoegawa K, de Jong P (2003) New chicken, turkey, salmon, bovine, porcine, and sheep genomic BAC libraries to complement world wide effort to map farm animals genomes. In: Plant and Animal Genome XI Conf, San Diego, CA, USA, p 87

Nicholl GB, Burkin HR, Broad TE, Jopson NB, Greer GJ, Bain WE, Wright CS, Dodds KG, Fennessy PF, McEwan JC (1998) Genetic linkage of microsatellite markers to the Carwell locus for rib-eye muscling in sheep. Proc 6th World Congr on Genet Appl to Livestock Prod, Armidale, Australia, January 11–16, 26:529–532

Nomura K, Wu CH, Hadfield TS, Womack JE, Cockett NE (2006) Radiation hybrid maps of sheep chromosome 2 and 6. Proc 30th Intl Conf on Anim Genet, Porto Seguro, Brazil, August 20–25, B220

Parnell SE, Ramadoss J, Delp MD, Ramsey MW, Chen WJ, West JR, Cudd TA (2007) Chronic ethanol increases fetal cerebral blood flow specific to the ethanol sensitive cerebellum under normoxaemic, hypercapnic and acidaemic conditions: ovine model. Exp Physiol, 92:933–943

Perkins AC, Kramer LN, Moody DE, Hadfield TS, Cockett NE, Bidwell CA (2006) Postnatal changes in gene expression from the callipyge region in sheep skeletal muscle. Anim Genet 37:535–542

Porada CD, Park P, Almeida-Porada G, Zanjani ED (2004) The sheep model of in utero gene therapy. Fetal Diagn Ther 19:23–30

Purvis IW, Franklin IR (2005) Major genes and QTL influencing wool production and quality: a review. Genet Sel Evol 37 (Suppl 1):S97–S107

Rohrer GA (2004) An overview of genomics research and its impact on livestock production. Reprod Fertility Dev 16:47–54

Saidi-Mehtar N, Hors-Cayla MC, Van Cong M, Benne F (1979) Sheep gene mapping by somatic cell hybridization. Cytogen Cell Genet 25:200–210

Saidi-Mehtar N, Hors-Cayla MC, Van Cong N (1981) Sheep gene mapping by somatic cell hybridization: four syntenic groups ENO1-PGP, ME1-PGM3, LDHB-PEPB-TP1, and G6PD-GALA. Cytogenet Cell Genet 30:193–204

Sakurai H, Soejima K, Nozaki M, Traber LD, Traber DL (2007) Effect of ablated airway blood flow on systemic and pulmonary microvascular permeability after smoke inhalation in sheep. Burns 3(7):885–891 epub 08 May 2007 (PMID 17493760)

Scherf BD (2000) World watch list for domestic animal diversity. FAO/UNEP, Domestic Animal Diversity Information System (http://http://www.fao.org/dad-is/index.asp)

Scott PR (2001) Osteoarthritis of the elbow joint in adult sheep. Vet Rec 149:652–654

Seitz H, Youngson N, Lin SP, Dalbert S, Paulsen M, Bachellerie JP, Ferguson-Smith AC, Cavaillé J (2003) Imprinted microRNA genes transcribed antisense to a reciprocally imprinted retrotransposon-like gene. Nat Genet 34:261–262

Smit MA, Cockett NE, Beever JE, Shay TL, Eng SL (2002) Scrapie in sheep: a transmissible spongiform encephalopathy. Sheep Goat Res J 17:21–32

Smit M, Segers K, Shay T, Baraldi F, Gyapay G, Snowder G, Georges M, Cockett N, Charlier C (2003) Mosaicism of

Solid Gold supports the causality of the SNP in determinism of the callipyge phenotype. Genetics 163:453–456

Smit MA, Tordoir X, Baraldi F, Davis E, Gyapay G, Cockett N, Georges M, Charlier C (2005) BEGAIN: A novel imprinted gene in the *DLK1-GTL2* domain unaffected by the callipyge mutation. Mamm Genom 16:801–814

Souza CJ, MacDougall C, Campbell BK, McNeilly AS, Baird DT (2001) The Booroola (FecB) phenotype is associated with a mutation in the bone morphogenetic receptor type 1 B (BMPR1B) gene. J Endocrinol 169:R1–R6

Sponenberg DP (1997) Genetics of colour and hair texture. In: Piper L, Ruvinsky A (eds) The Genetics of Sheep. CABI, Oxon, UK, pp 51–86

Takeda H, Caiment T, Smit M, Hiand S, Tordoir X, Cockett N, Georges M, Charlier C (2006) The callipyge mutation enhances bidirectional long-range *DLK1-GTL2* intragenic transcription in *cis*. Proc Natl Acad Sci USA 103: 8119–8124

Tan P, Allen JG, Wilton SD, Akkari PA, Huxtable CR, Laing NG (1997) A splice-site mutation causing ovine McArdle's disease. Neuromuscular Disorders 7:336–342

Tetens J, Goldammer T, Maddox JF, Cockett N, Drögemüller C (2006) A high-resolution, integrated, comprehensive, and comparative radiation hybrid map of sheep chromosome. Proc 30th Intl Conf Anim Genet, Porto Seguro, Brazil, August 20–25, B180

Thompson KG, Dittmer KE, Blair HT, Fairley RA, Sim DF (2007) An outbreak of rickets in Corriedale sheep: evidence for a genetic aetiology. NZ Vet J 55:137–142

Turner AS (2002) The sheep as a model for osteoporosis in humans. Vet J 163:232–239

Vaiman D, Billault A, Tabet-Aoul K, Schibler L, Vilette D, Oustry-Vaiman A, Soravito C, Cribiu EP (1999) Construction and characterization of a sheep BAC library of three genome equivalents. Mamm Genom 10:585–587

Van den Broeke A, Burny A (2003) Retroviral vector biosafety: lessons from sheep. J Biomed Biotechnol 1:9–12

Vuocolo T, Byrne K, White J, McWilliam S, Reverter T, Cockett NE, Tellam RL (2006) Identification of a gene network contributing to muscle hypertrophy in callipyge skeletal muscle. Physiol Genom 28:253–272

Walling GA, Visscher PM, Wilson AD, McTeir BL, Simm G, Bishop SC (2004) Mapping of quantitative trait loci for growth and carcass traits in commercial sheep populations. J Anim Sci 82:2234–2245

Wilson T, Wu X-Y, Juengel JL, Ross IK, LUmsden JM, Lord EA, Dodds KG, Walling GA, McEwan JC, O'Connell AR, McNatty KP, Montgomery GW (2001) Highly prolific Booroola sheep have a mutation in the intracellular kinase domain of bone morphogenetic protein IB receptor (ALK-6) that is expressed in both oocytes and granulose cell. Biol Reprod 64:1225–1235

White JD, Vuocolog T, McDonagh M, Grounds MD, Harper G, Cockett NE, Tellam R (2007) Cellular analysis of the *callipyge* phenotype through skeletal muscle development: association of DLK1 with developing satellite cells. Differentiation 76:283–298

Wu CH, Nomura K, Goldammer T, Hadfield T, Womack JE, Cockett NE (2007) An ovine whole-genome radiation hybrid panel used to construct an RH map of ovine chromosome 9. Anim Genet 38:534–536

Wu CH, Nomura K, Goldammer T, Hadfield T, Dalrymple BP, McWilliam S, Maddox JF, Womack JE, Cockett NE (2008) A high-resolution comparative radiation hybrid map of ovine chromosomal regions that are homologous to human chromosome 6 (HSA6). Anim Genet In Press

Zhou H, Hou S, Shang W, Wu W, Cheng Y, Mei F, Peng B (2007) A new in vivo animal model to create intervertebral disc degeneration characterized by MRI, radiography, CT/discogram, biochemistry, and histology. Spine 32:864–872

Zhu SX, Sun Z, Zhang JP (2007) Ovine (*Ovis aries*) blastula from an in vitro production system and isolation of primary embryonic stem cells. Zygote 15:35–41

CHAPTER 4

Deer

Richard J. Hall

AgResearch Ltd., Invermay Agricultural Centre, Private Bag 50034, Mosgiel 9053, New Zealand, Current address: Richard.Hall@esr.cri.nz

4.1
Introduction

Perhaps no other animal upon this planet has attracted so much attention and mysticism as surrounds the deer that inhabit our forests, plains and mountains. Deer have achieved an archetypal status in the collective mind of both modern and ancient human cultures, embodying many of our views of the natural world in their innocence and wildness. Despite the intimacy of the complicated relationship between humans and deer, and despite our attempts to understand these animals, there will always be a degree of mystery surrounding these magical creatures. The intensity of human interest in deer is evident from the extent of hunting, domestication and conservation of their kind. Indeed, perhaps one of the most well-recognized artistic depictions of an animal in our history is the 'Monarch of the Glen' by Edwin Landseer (Fig. 1). With advances in scientific understanding, it has now become possible to delve deeper into the biology of these creatures, and in particular into their genetics and genomic biology, which is the focus of this chapter.

4.1.1
Taxonomic Description

Deer are ruminants that exist in a wide variety of habitats across the globe and are considerably diverse in their morphology. Decisions upon the taxonomic classification of deer species have been the subject of intense discussion in more recent times, often due to developments and findings from molecular genetic data. Most living deer species have been assigned to the Cervidae family, encapsulating at least 40 species of deer, and an additional five species have been assigned to the Moschidae family (Grubb and Groves 2003). The subfamily Cervinae are found in Eurasia, with the exception of the North American wapiti (*Cervus elaphus*), and are referred to as the Old World deer species. The Odocoileinae, or New World deer species, can be found in both North America, such as the white-tailed deer (*Odocoileus viginianus*), in South America such as the brocket deer (*Mazama*), and in Eurasia, such as the roe deer (*Capreolus capreolus*). Tribal status within the Cervidae has been assigned to the holoarctic deer species of reindeer (caribou; *Rangifer* sp.) and moose (*Alces alces*). More difficult to place have been the muntjac deer, which now have subfamily status within the Cervidae (Muntiacinae), and even more so the Chinese water deer (*Hydropotes inermis*) which is the only extant species in the subfamily of Hydropotinae. Complicating the taxonomic designations are the various hybridizations that occur between deer species, such as the well-described hybridizations between members of the genera *Cervus*, particularly for sika (*C. nippon*), red deer and North American wapiti (*C. elaphus*). Further complications have arisen due to the various morphotypes of *C. elaphus*, where the subspecies status has been much debated (Ludt et al. 2004).

Anthropogenic influence on deer has been profound and includes the management of deer populations as a game species (particularly the roe deer and red deer of Europe), habitat destruction or modification, and the gross over-exploitation of deer that has often adversely affected the genetic diversity of endemic deer species. The introduction of foreign deer species has often been shown to have negative effects on endemic deer populations either through increased competition or through hybridization. Indeed, introductions of foreign deer species can be highly destructive in a 'naive' ecosystem. Even in an

Genome Mapping and Genomics in Animals, Volume 3
Genome Mapping and Genomics in Domestic Animals
N.E. Cockett, C. Kole (Eds.)
© Springer-Verlag Berlin Heidelberg 2009

Fig. 1 Landseer, Edwin H., Monarch of the Glen (oil-on-canvas), 1851. Diageo Archives, Scotland. (Reproduced with kind permission from Diageo Archives)

existing ecological niche of an endemic deer species, the extinction its natural predators (and thus population control) can be highly disruptive to the ecological balance (Côte et al. 2004).

4.1.2
Economic Importance

The semi-domestication of the reindeer by humans (*Rangifer tarandus* spp.) has occurred over many thousands of years, over a greater range than just the holoarctic region that it inhabits today (Gordon 2003). Various ancient human cultures have tended herds of reindeer for the supply of meat, milk, fur, spiritual influences and transportation (Gordon 2003), a relationship that still continues today for the indigenous peoples of the arctic regions. Domestication of the other deer species has not necessarily been achieved, although deer are kept in captive populations for a number of purposes, and the fallow deer has flourished in a close semi-domestic relationship with humans (Hemmer 1990).

During the last 100 years, there has been increased activity in the farming of species other than the traditional livestock. Various attributes of the products produced by deer such as meat (venison), pelt and antler have attracted a great deal of interest in certain market places. Farming of deer species mitigates the environmental impact of taking deer from wild populations and has also allowed for genetic selection to be applied to traits of interest (such as antler development). Deer are now farmed worldwide, and the red deer and North American wapiti (*Cervus elaphus* sp.) are the prominent species. Deer farming has been particularly successful in New Zealand where approximately 800,000 deer are processed annually for export of venison and velvet antler products (O'Conner 2006). The development of genetic knowledge in deer will assist selection strategies, although the application of this technology is still very much in its infancy.

4.1.3
Karotype of Cervids

The karyotype of the cervid genome has been well described for a number of species (Huang et al. 2006) where the ancestral cervid karyotype has been proposed to be $2n = 70$ (Neitzl 1987), with the diversity of chromosomal rearrangements observed between the species being attributable to Robertsonian translocations in the Cervinae and Odocoileinae subfamilies (Fontana and Rubini 1990), and tandem chromosome fusions in the Muntiacinae (Fontana and Rubini 1990).

In addition to cytogenetic studies, there has been extensive examination of cervid genetic variation at the molecular level, including the analysis of specific genes and molecular polymorphism in the mitochondrial and nuclear genomes. The genetic analysis of deer and current body of knowledge about the cervid genome are reviewed here.

4.2
Molecular Genetics

4.2.1
Genetic Analysis Using Molecular Markers

The development of genetic markers in deer has followed the development of biochemistry and molecular biology, from the initial studies that utilized allozymes to studies that incorporate molecular markers derived from both mitochondrial DNA and the nuclear genome.

The majority of studies that have utilized molecular markers in the family Cervidae have addressed the conservation and taxonomic issues that are peculiar to each species. Within Cervidae there has been much debate over the assignment of deer at each taxonomic level. This is especially true for the *Cervus elaphus* species (Polziehn and Strobeck 1998; Ludt et al. 2004).

Population genetic studies in deer have included molecular data to determine effective population sizes and other diversity statistics, with important consequences for the management of both threatened indigenous populations and in the control and management of introduced deer populations.

Allozymes

Before the advent of genetic analysis, the taxonomic classification of deer species was based upon morphological characteristics, such as stature, antler characteristics, pelage and voice (reviewed in Grubb and Groves 2003). Allozymes and blood serum proteins provided the first means of detecting genetic variation in deer. Such markers are ideally suited to use in a species that has not been the subject of molecular genetic analysis in the past, as these methods require no a priori information on protein or gene sequence (Hartl and Clark 1997). However, there are several disadvantages in the use of protein variants as markers. Most obvious is the limited number of markers available for assay and the limited number of alleles that are observed for a given allozyme (Pemberton and Slate 1998). The analysis of allozymes precludes high-throughput technologies, and even modest sample sizes will pose a significant amount of practical work in the laboratory. It has also been shown in *Cervus elaphus* that these markers may indeed be under selection, and therefore estimates of genetic diversity may be biased (Hartl et al. 1991). Collection of samples for allozyme assays may also be much more difficult than the collection of DNA for molecular studies, due to their often secretive and elusive nature of deer, and difficulties in navigating the terrain they inhabit.

Blood group and serum polymorphisms have proven useful in the distinction of species and subspecies (Herzog 1990). However, such markers suffer from similar difficulties as allozymes, which include the limited number of available markers and alleles, and difficulties in processing large numbers of animals.

Mitochondrial DNA Variation

Genetic variation in the cervid mitochondrial genome has been utilized largely in population genetic analysis or phylogenetic studies. Mitochondrial DNA (mtDNA) sequence variation is highly appropriate for phylogenetic analysis amongst closely related species, as compared to nuclear DNA markers. This is because mtDNA shows more rapid evolution (especially the hypervariable D-loop region), maternal inheritance and the absence of recombination (Brown et al. 1979), although there is some evidence that recombination may occur in the mtDNA of ruminants (Slate and Phua 2003).

In recent years, there has been a rapid expansion in the sequencing of mitochondrial genomes (Feijao et al. 2006). Complete mitochondrial genome sequences are now available for red deer (*Cervus elaphus*: GenBank Accession: NC_007704), reindeer (*Rangifer tarandus*: GenBank Accession: NC_007703) and sika deer (*Cervus nippon taiouanus*: GenBank Accession: NC_008462). As yet there has been no formal analysis or publication of these genomic sequences, although many studies have already utilized mtDNA sequence in the past.

Extensive sequencing of specific mtDNA regions has occurred to characterize genetic variation for population genetic studies or to determine phylogenies for deer. Genes that have been sequenced include the ribosomal RNA (rRNA) and transfer RNA (tRNA) (Miyamoto et al. 1990; Kraus and Miyamoto 1991), the hypervariable D-loop region (Douzery and Randi 1997; Polziehn and Strobeck 1998) and the cytochrome-b gene (Randi et al. 1998a), although the list of authors who have utilized mtDNA sequence is far greater than those shown here. The specific use of mtDNA sequence variation will be referred to in the following sections for the many applications of this molecular genetic data in cervids.

Nuclear DNA Variation in Cervids

Much of the work in the development of nuclear DNA markers has relied upon comparative work already completed for other species, in particular other ruminants. It is, therefore, appropriate to make mention of the distinction between the type I and type II markers. Type I markers refer to those markers that are conserved between species (or perhaps subspecies), but are usually observed to be invariant within each species and are often found in coding sequence. These markers show greatest utility in comparative analysis between species (O'Brien et al. 1993). Type II markers are usually only found to be variant within a species, or perhaps between only closely related species (O'Brien et al. 1993). These markers usually reside within non-coding sequence and most that are currently known in deer are microsatellites (Slate et al. 2002a), a variable dinucleotide or trinucleotide repeat motif flanked by unique sequence.

A number of nuclear genetic markers have been described in cervids, including minisatellite DNA (see for examples, Li et al. 2002; Lin and Li 2006), restriction fragment-length polymorphisms (RFLPs) and amplified fragment-length polymorphisms (AFLPs). The RFLPs in deer have usually relied upon the hybridization of a known probe to gene sequence where the digestion with restriction endonuclease reveals polymorphism between individuals (Tate et al. 1995). The RFLP variation is most likely to represent differences in coding sequence where a change of amino acid may occur, and thus RFLPs are often likely to have functional significance, which is particularly relevant for gene-mapping studies. The AFLPs utilize semi-redundant primer sets that allow for the amplification of random segments of DNA that appear variable in length. Their greatest utility lies in the creation of markers for a species in which few molecular markers have been discovered (Slate et al. 2002a). These AFLPs were used intensively in the creation of the deer genetic map (Slate et al. 2002a).

The first reported observation of microsatellites in *Cervinae* was made using 'DNA fingerprinting' in the roe deer (*Capreolus capreolus* L.; Mörsch and Leibenguth 1993). In more recent years, the isolation and characterization of microsatellite loci in deer has flourished, although not to the extent as for species that have a complete genome sequence. The discovery of microsatellites has been facilitated by the transfer of microsatellite assays from other species, especially within the Artiodactyla, where the sequence surrounding the repeat motif is conserved to a great enough extent to allow successful amplification by PCR (Slate et al. 1998). In particular, the livestock species of cattle, goat and sheep have provided many transferable microsatellite loci that have been used in a variety of cervid species, which include moose, red deer, reindeer and roe deer (Engel et al. 1996; Kuhn et al. 1996; Roed 1998; Slate et al. 1998). The transferral of microsatellites from caprine and ovine lineages is usually successful (50–90% successful), where less success is reported for Bovidae (16–50% successful), with approximately 50% of the successful markers being polymorphic (Engel et al. 1996; Kuhn et al. 1996; Roed 1998; Slate et al. 1998). Variation in success rates between studies depends upon the species selected and PCR optimization protocols. The current availability of microsatellite markers is described for each species in the section on population genetics (see later).

At present, there has been no large-scale identification of single nucleotide polymorphism (SNP) in cervids beyond the sequencing of individual genes such as the major histocompatibility complex, the κ-casein gene, reproductive hormones and the prion-protein gene. A review of this type of nuclear DNA variation is described later herein. It should also be noted that most RFLPs identified in the deer genetic map are likely to be SNPs and thus represent the largest collection of SNP described for deer (Slate et al. 2002a).

Phylogenetic Analysis

Cladistic analyses have helped to clarify the complicated phylogenetic relationships between members of Cervidae; however, such analyses may be confounded due to parallelism and convergent evolution in the morphological characters that are used (Janis and Scott 1987; Grubb and Groves 2003). Mitochondrial DNA has been used to resolve various discrepancies in the relationships of Cervidae with other members of pecora (Gatesy et al. 1992) and has also proven useful for resolving the phylogeny within the Cervidae. For example, sequencing of the mtDNA cytochrome-b gene in 11 cervids found that the Cervidae are monophyletic within pecoran Artiodactyla (Randi et al. 1998a).

Specific attributes of the phylogeny of Cervidae have been hotly debated, but a consensus viewpoint is beginning to emerge that incorporates findings from a variety of molecular genetic sources that include mtDNA, microsatellites, RFLPs and nuclear gene sequence. An example of a recent phylogenetic analysis that utilized mtDNA of Cervidae is shown in Fig. 2 (Pitra et al. 2004). A close association between members of the Cervinae (Old World deer) and

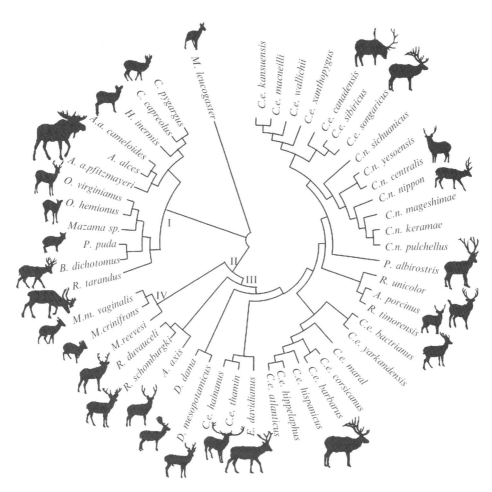

Fig. 2 Phylogenetic tree showing the relationships amongst members of Cervidae, based upon cytochrome b gene sequence data from mitochondrial DNA. Telemetacarpalian deer (I) were identified as distinct from plesiometacarpalian deer (II). Notice also that the 'Old World' deer are a sister group of the muntjac (IV). (Reproduced with kind permission from Pitra et al. 2004 and Brent Haufmann at www.ultimateungulate.com)

Muntiacinae has been identified by molecular studies (Miyamoto et al. 1990; Kraus and Miyamoto 1991; Cronin et al. 1996; Pitra et al. 2004). Molecular studies have confirmed the split between the Cervinae and the Odocoileinae (New World deer) which are likely to have shared a common ancestor in Europe about 10 million years ago (Mya) (Miyamoto et al. 1990; Pitra et al. 2004). The Odocoileinae have been shown to be monophyletic (Cronin et al. 1996; Douzery and Randi 1997) with sister groups of *Alces* (moose), *Rangifer* (reindeer) and *Capreolus* (roe deer; Cronin 1996). Previous evidence for the split of Hydropotinae from other subfamilies in Cervidae is due to the absence of antlers, but this may have been correlated with a decrease in body size (Gould 1974) and has been brought into question by molecular studies. Molecular studies suggested that *Hydropotes inermis* resides within Odocoileinae and is most closely related to *Capreolus* (Kraus and Miyamoto 1991; Douzery and Randi 1997; Lee et al. 1997; Pitra et al. 2004).

Paraphyly of *Cervus elaphus* has been observed in the molecular data (Polziehn and Strobeck 1998; Ludt et al. 2004; Pitra et al. 2004), where many of the current subspecies designations were not supported. Indeed, it has been suggested that the species *C. elaphus* be divided into two species, consisting of a western group (mostly wapiti) and eastern group (mostly red deer; Ludt et al. 2004). North American wapiti had previously been regarded as a separate species, *Cervus canadensis*, but this status was later reviewed and wapiti were included as a subspecies in *Cervus elaphus* due to their ability to interbreed and their proximity to the polar region (Grubb and Groves 2003). A similar amount of sequence divergence was observed between wapiti and red deer, as compared to that observed between sika and red deer (Polziehn and Strobeck 1998). The authors suggest that a review of the subspecies status for North American wapiti needs to be undertaken (Polziehn and Strobeck 1998). This contention is further supported by an analysis that found that European red deer, wapiti and sika were all monophyletic with wapiti being closer to sika than red deer (Kuwayama and Ozawa 2000). Sika have been shown to be monophyletic in relation to red deer, from mtDNA sequence data of both the control region and the cytochrome-b gene (Cook et al. 1999).

The recovery of ancient DNA from fossil remains of *Megaloceros giganteus* (the so-called Irish Elk) has assisted in the resolution of the phylogenetic position of this now extinct deer species within the Cervidae (Kuehn et al. 2005). Cytochrome-b sequences amplified from samples of DNA recovered from antler and bone from *Megaloceros giganteus* were compared to 11 extant deer taxa. This extinct deer species was shown to be most closely related to *Cervus elaphus* (Kuehn et al. 2005), but it had previously been suggested that the Megalocerini had shared a common ancestor with *Dama dama* as the closest extant relative due to the palmation of antlers that is common to both species (Gould 1974). However, more recent studies that incorporate molecular and morphological data place *Megaloceros* in the same clade as fallow deer, as distinct from members of the genus *Cervus*, thus highlighting the importance of using a number of characters for phylogenetic analyses including the use of morphological data (Lister et al. 2005; Hughes et al. 2006).

Population Genetics, Conservation and Management in Cervidae

A massive amount of work has been expended in the analysis of population genetics in Cervidae; much of this activity used molecular markers, especially mitochondrial DNA variation and microsatellites. Many reviewers have already succinctly covered the large array of literature that is pertinent to each species of deer, and readers will be referred to these works where appropriate. In the following section, it is not the intention to completely describe in detail all facets of the population genetics pertaining to the various species of deer. Rather, this section aims to portray the utility of molecular markers in population genetic studies in deer, and also in comparison to those studies that have already been achieved using classical genetic markers such as allozymes and serum protein polymorphism. The vast majority of population genetic studies have focussed on three species, the white-tailed deer, the red deer and its wapiti subspecies, and the roe deer. A number of factors have led to the intensive study of these three species over other deer. Each of these species is present within developed nations, making them more amenable to research. Indeed, perhaps the greatest driver of research into these species has been the enormous amount of pressure they have been placed under as a consequence of human development. Many of the advances in molecular and population genetics that have been made in these three species have had benefits for other lesser studied members of the Cervidae, although this is by no means always the case. Equally determined efforts have been made in the reindeer, moose, sika deer, amongst others that have certainly contributed to the advancement of knowledge of deer populations and genetic diversity.

In this section, each species is considered. The reader's attention is drawn to the unique studies within each species that have gone beyond the description of population genetic parameters for the species and populations studied.

White-tailed deer (*Odocoileus virginianus*). The white-tailed deer was the first species of deer in which individual microsatellites were characterized by sequencing (DeWoody et al. 1995). More recently, a set of microsatellites that are suitable for high-throughput genotyping in white-tailed deer have been reported (Anderson et al. 2002), suitable for studies of parentage and population genetics in this species.

The application of molecular markers in white-tailed deer has largely been in population genetic studies, focussing on the conservation of this indigenous North American deer species. The exploitation of indigenous animals and habitat degradation that occurred during the colonization of North America by Europeans pressed many regional populations of species such as the white-tailed deer and wapiti to critically low levels. Various efforts have been made, many of which have been very successful, to reintroduce and restore populations of white-tailed deer in threatened areas. Restoration plans often involve the reintroduction of animals or the remediation of habitat, but the potential effects on genetic diversity were often not considered during early attempts. Both founder effects and a concomitant loss of genetic diversity have the potential to arise as a consequence of restoration programmes, both of which may be deleterious to the conservation of the species. In addition, protracted periods of small population size may also reduce genetic diversity through inbreeding. Many regional studies on genetic diversity in white-tailed deer have been carried out at both micro- and macro-geographical scales. A variety of genetic markers have been utilized, with much of the early work completed using allozyme assays, which were later combined with mitochondrial DNA data. Allozymes have also proven useful in more recent analyses as a means of determining temporal trends in genetic diversity. This is possible over a greater time frame than molecular markers, due to the earlier discovery of allozymes (Kollars et al. 2004).

There is considerable variation in estimates of genetic diversity in white-tailed deer using allozyme data, where both high and low levels of genetic heterogeneity have been observed within and between populations (Manlove et al. 1975; Ramsey et al. 1979; Chesser et al. 1982; Sheffield et al. 1985; Breshears et al. 1988; Purdue et al. 2000). Particular emphasis has been placed upon areas such as the south-eastern USA, where white-tailed deer were nearly completely extirpated during the latter part of the 19th century. The effect of restocking on genetic diversity appears to be highly dependent upon the populations under study, where studies that utilize mtDNA and/or allozyme data infer either a reduction in genetic diversity (Kennedy et al. 1987; Karlin et al. 1989; Ellsworth et al. 1994a; Leberg et al. 1994; Kollars et al. 2004) or no significant effect (Ellsworth et al. 1994b; Leberg and Ellsworth 1999). The use of allozymes as markers of genetic diversity is complicated, as such genes are often under selective pressure, a factor which has been shown to contribute to diversity estimates in restored populations of white-tailed deer (Kennedy et al. 1987). A study of genetic variation at the major histocompatibility complex (MHC) class II locus revealed high levels of diversity despite the potential population bottlenecks that may have occurred

during European colonization (Van Den Bussche et al. 2002). This is also interesting given that the MHC locus is known to be under selective pressure (Ditchkoff et al. 2001). More recently, the use of microsatellite loci to examine the effects of restoration on white-tailed deer populations in the south-eastern USA has been reported (DeYoung et al. 2003). Seventeen microsatellites were genotyped in >500 deer from 16 populations from Mississippi, a population where severe depletion of white-tailed deer populations was known to have occurred. Despite evidence for genetic bottlenecks and population sub-structure, there were relatively high levels of genetic variation in the populations studied (DeYoung et al. 2003).

Within the genus *Odocoileus*, genetic comparisons of white-tailed deer with closely related subspecies black-tailed deer (*Odocoileus hemionus sitkensis; columbianus*) and mule deer (*Odocoileus hemionus hemionus*) have been made. Only limited transfer of genetic material between populations from these three taxa was shown to occur, with the most significant hybridization occurring between subspecies of *O. hemionus* (Scribner et al. 1991). Various spatial and demographic effects on genetic variation within mule deer populations have also been explored (Scribner et al. 1991).

Roe deer (*Capreolus capreolus* L.). Phylogeographic studies in roe deer had been conducted, initially using DNA fingerprinting analysis (Volmer et al. 1995) and then by some extensive work in mtDNA (Jaeger et al. 1992; Randi et al. 1998b; Wiehler and Tiedemann 1998). A wealth of genetic studies on roe deer have recently been published, after an initial paucity of research in the early 1990s as compared to the other species such as white-tailed deer and red deer. These studies have used allozymes, and both mtDNA and microsatellites in the exploration of the phylogeography of roe deer in Europe (Coulon et al. 2004, 2006; Randi et al. 2004; Nies et al. 2005; Lorenzini and Lovari 2006; Zachos et al. 2006).

The roe deer was the first species of deer in which microsatellites were observed by 'DNA fingerprinting' techniques (Mörsch and Leibenguth 1993). But it was the use of conserved microsatellites derived from Bovidae that provided the first means of testing microsatellite variation in roe deer (10 microsatellites; Roed 1998). A very small-scale study of the population genetics of roe deer in Scandinavia was achieved using four microsatellites derived from other species

(Pemberton and Slate 1998). The actual isolation and characterization of microsatellites derived in roe deer did not occur until 2000 when seven microsatellites were isolated from the roe deer genome (Fickel and Reinsch 2000). The set of microsatellites available to researchers working in roe deer has expanded more recently, with the development of more than 50 polymorphic microsatellites based on cross-amplification from other cervids and Bovidae (Galan et al. 2003; Vial et al. 2003). Similar to the transference of microsatellites between bovids and other cervids, just over 50% of markers that were investigated in these studies amplified genomic DNA from roe deer. In both studies, ~30% of these microsatellites showed polymorphism, and a multiplex system was developed for a subset of these in each study (11-plex, Vial et al. 2003; 12-plex, Galan et al. 2003).

The first large-scale study of roe deer populations in Northern Europe was achieved using 57 microsatellites derived from the Bovidae and Caprinae (Wang and Schreiber 2001). Roe deer showed a high level of polymorphism when compared to other cervids, and no evidence for spatial heterogeneity was observed with either microsatellite or allozyme loci. Typically, the microsatellites showed much higher diversity estimates than did the allozyme data.

Similar to studies of temporal genetic variability that have been conducted in white-tailed deer, a large-scale study spanning 34 years from was conducted on a roe deer population in the Vosges du Nord Mountains in France (Wang et al. 2002). This was possible due to the collection of antlers as hunting trophies, where a specially adapted protocol for the isolation of DNA was used. Many of the sampling strategies utilized in population genetic analyses in deer, and indeed other species, incorporate samples that have been collected over different years. It is, however, possible for the genetic parameters in a population to change over a relatively short frame which would affect the results reported in population genetic studies. Indeed, temporal instability in all microsatellite loci examined in the roe deer population was observed over the 34-year time period (Wang et al. 2002). Reduced allelic diversity was correlated with a decrease in population density that occurred during the study period. Whilst it was not possible to estimate the temporal trends in matrilineal lines (due to the absence of antlers in female roe deer), it was hypothesized that the philopatry observed in females would have a stabilizing effect on allelic diversity in a local population, as has been observed in both red deer (Schreiber et al. 1994) and white-tailed deer (Purdue et al. 2000). These data place importance upon the temporal sampling strategy used to infer genetic diversity, either in roe deer, or other deer species.

European red deer and North American wapiti (*Cervus elaphus* ssp.). The diverse and widely distributed deer *Cervus elaphus* has been intensively studied using molecular markers. The diversity of populations and subspecies, and the hybridizations between these, have allowed for a wide range of genetic analyses to be undertaken using the full suite of molecular markers. *Cervus elaphus* is distributed across Europe and is known as the red deer, and in Asia and North America as wapiti (Grubb and Groves 2003).

As for most other deer species, allozymes have been used extensively as markers of genetic variation in *Cervus elaphus*. Much of the work completed using allozymes has sought to distinguish the relationships within and amongst the various subspecies of *Cervus elaphus*. Using 35 enzyme loci, the successful distinction of four common subspecies of *Cervus elaphus* present in Europe was possible (Gyllensten et al. 1983), *C. e. atlanticus*, *C. e. scoticus*, *C. e. elaphus* and *C. e. hippelaphus* (previously *C. e. germanicus*). Allozymes have also been able to resolve genetic structure in populations of *Cervus elaphus* spp. in Europe (Hartl et al. 1990; Stratil et al. 1990; Strandgaard and Simonsen 1993; Stroehlein et al. 1993; Lorenzini et al. 1998; Martinez et al. 2002).

Allozyme studies have largely been limited to red deer, but the distinction of wapiti from red deer (Tate and McEwan 1992; He et al. 1997) and genetic analysis within wapiti populations (Kucera 1991) have been done. Individual allozymes have been used to study the hybridization between subspecies of *Cervus elaphus* that include wapiti, red deer and the various hybridizations that may occur with *Cervus nippon* (Sika; Herzog 1988; Herzog 1990; Linnell and Cross 1991).

An expansion in both the scale and breadth of genetic research in *Cervus elaphus* occurred with the advent of molecular markers in mitochondrial DNA (mtDNA) due to the ease of marker development, genotyping and the informativeness of mtDNA polymorphisms. The first characterization of mtDNA variation in Cervidae was undertaken to describe the phylogeny of members of the Cervidae family, with emphasis on New World deer species (Cronin 1991), with a similar study of the Old World species of deer (that encompasses the entire Cervinae family) having since been completed (Ludt et al. 2004). Similarly, more detailed comparisons that have examined the relationship between red deer and wapiti have been made using mtDNA (Polziehn and Strobeck 2002). These studies have challenged the classification of North American wapiti as a subspecies (*C. e. canadensis*), suggesting perhaps a return to a species classification (*C. canadensis*).

Phylogeographic analysis within the North American wapiti using RFLP assays of mtDNA, revealed no evidence for genetic differentiation in subspecies or populations (Cronin 1992). As mtDNA is maternally inherited, it was suggested that the absence of variation in North American wapiti may be due to the presence of founder effects when wapiti colonized North America during the late Pleistocene era (Cronin 1992). A similar study was later conducted for European red deer, from a number of different populations in western, central and south-eastern Europe. Evidence for population substructure in European red deer was revealed, where haplotypes were shown to be monomorphic within populations but variable between them (Hartl et al. 1995b).

Population genetic analyses in *Cervus elaphus* have also been achieved using nuclear genetic markers. The type of marker selected for population genetic studies can have a profound influence on the observed diversity and conclusions that can be drawn from the study. The use of allozyme variation, mtDNA and nuclear genetic variation has often been used simultaneously in genetic analyses for comparison (e.g., Nussey et al. 2006).

Detailed studies of local geographic populations, particularly in the European red deer, have been made such as the Corsican red deer (Zachos et al. 2003; Hmwe et al. 2006), Bavarian and Czech (Hartl et al. 2005), French (Hartl et al. 2005), the Isle of Rhum (Pemberton et al. 1996; Pemberton et al. 1998; Nussey et al. 2005), Carpathian (Feulner et al. 2004) and red deer in Switzerland (Kuehn et al. 2004) using combinations of the various genetic markers. Microsatellites have also been utilized in studies of population genetics in the wapiti (Polziehn et al. 2000; Williams et al. 2002) and for parentage analysis (Talbot et al. 1996).

Sika deer (*Cervus nippon*). Genetic markers have not been used extensively in sika deer, until recently.

Initially such studies were limited to introduced populations and utilized allozyme data, such as for the sika found in Ireland (Linnell and Cross 1991). Allozyme data provided a means to identify hybrid animals and provided insights into the relationship of sika to other members of *Cervus* (Herzog 1988; Linnell and Cross 1991). Subsequent studies of allozyme variation in natural populations distinguished between sika subspecies within Japan (Yokohama et al. 1995; Fukui et al. 1996). Despite an initial paucity of studies on natural populations of sika deer, many have since been published that have utilized mtDNA, and more recently, nuclear DNA genetic markers. This is often required to evaluate the genetic diversity of threatened species such as the Vietnamese sika deer (*C. n. pseuaxis*) for the purposes of conservation (Thevenon et al. 2003, 2004).

Mitochondrial DNA sequence variation has provided some clarity over issues regarding sika subspecies in Japan, where *C. n. yesoensis* and *C. n. yakoshimae* have been shown to be distinct subspecies (Tamate and Tsuchiya 1995; Yokohama et al. 1995), but the separation of *C. e. centralis* and the sika deer from the island of Hokkaido *C. e. yesoensis* has not been supported by mtDNA analysis (Tamate and Tsuchiya 1995).

Sika deer in Japan have endured large fluctuations in populations due to over-exploitation by hunting activities, and population explosion and collapse where numbers of deer have been higher or lower than can be supported by the available levels of vegetation in the forests. Analysis at the subspecific level in Japanese sika was made using sequence data from the mtDNA D-loop region for *C. e. centralis* from the Japanese mainland island of Hokkaido (Nagata et al. 1998). Population structure was shown to occur, with seven matrilineal haplotypes being identified in the mtDNA sequence data. This structure was attributed at least, in part, to the historical climatic and anthropogenic effects that are known to have occurred. The occurrence of population bottlenecks was supported by the data and could be correlated with the historical accounts of factors affecting sika deer populations, such as harsh winters (Nagata et al. 1998). The effect of population bottlenecks has been explored by the excavation of ancient mtDNA from sika deer remains on the Hokkaido Island of Japan (Nabata et al. 2004).

Sika deer subspecies from the greater Asia region have also been compared using mtDNA sequence from the D-loop region and the cytochrome-b gene (Cook et al. 1999). *Cervus nippon* was shown to be monophyletic in comparison to *Cervus elaphus*, in support of the designation of sika deer as a separate species. However, it was not possible to infer the dispersal of sika deer throughout Asia from these data, although it was suggested that a north eastern Asiatic origin for sika deer was unlikely (Cook et al. 1999). It has also been possible to estimate the divergence time of Japanese sika from the mainland Asian sika deer, which agrees with palaeontogical evidence at around 0.3 Mya (Nagata et al. 1999). In this same analysis, a large divergence between southern and northern Japanese sika was revealed, suggesting that a single colonization event (across land bridges) was not likely to have accounted for the arrival of sika deer in Japan (Nagata et al. 1999).

Many introduced populations of sika deer occur throughout the world (Whitehead 1993), and some of these populations may pose a threat to native deer species by hybridization (Goodman et al. 1999). Microsatellite data have been employed to assess the extent of hybridization in the Argyll region of Scotland where sika deer have been introduced and have hybridized with the indigenous red deer population. Evidence for the introgression of sika genetics into the red deer is presented, but no evidence for a selective advantage in either hybrid population could be identified, possibly due to the relatively short time frame since the introduction of sika deer into this region (~80 years; Goodman et al. 1999).

Reindeer and caribou (*Rangifer tarandus*). As for the other deer species, protein polymorphisms were initially used to study the population genetics and phylogeography of reindeer. A number of subspecies of reindeer have been named that also include caribou in North America. Polymorphisms in the electrophoretic mobility of serum transferrin, which is exceptionally polymorphic in this species, were used to investigate the effect of the introduction of domestic reindeer on North American populations of the native caribou (Roed and Whitten 1986) and to elucidate the phylogenetic relationships between subspecies of caribou in eastern Canada (Roed et al. 1991).

With the advent of molecular markers, a wealth of studies have been published that have considered various aspects of population genetics and phylogeography. The reindeer and the caribou of North America

provide a unique opportunity to examine the effects of admixture, large-scale population fluctuations and domestication, as all these phenomena are known to have occurred over a relatively large time scale, where the domestication of reindeer is thought to have begun in Eurasia more than 10,000 years ago (Gordon 2003). A variety of molecular markers have been employed in these analyses including nuclear DNA gene sequences (namely, the MHC and the κ-casein gene), mtDNA and microsatellites. Many of these genetic markers have been adapted from other species, in particular, the transference of microsatellites from other cervids or even other Artiodactyla. However, over 30 specific reindeer microsatellite markers have been developed (Wilson et al. 1997; Roed and Midthjell 1998).

Many studies have focussed upon the influence of the introduced domestic reindeer from Eurasia on the native caribou populations of North America. Allele frequencies in all genetic markers have been shown to differ between the introduced reindeer and native caribou (Cronin et al. 1995). Even though the introduction of reindeer into Alaska from Russia occurred over 100 years ago, there has been relatively little gene flow between subspecies (Roed and Whitten 1986; Cronin et al. 1995, 2003, 2006). A similar finding has also been observed in Greenland where introductions of Norwegian domestic reindeer occurred 50 years ago (Jepsen et al. 2002). The phylogenetic relationships between subspecies have been analysed using molecular data, which appear to be largely in agreement with the taxonomic classification of the *Rangifer tarandus* subspecies (Cronin et al. 2005).

Moose (*Alces alces*). The study of population differentiation in the moose (*Alces alces*) using allozymes proved difficult due to the limited variability observed at these loci (Baccus et al. 1983). However, if enough loci are included then evidence for genetic heterogeneity has been revealed (Ryman et al. 1980). Even within the usually highly polymorphic major histocompatibility complex genes, only low levels of polymorphism have been detected, in both the European and the North American moose populations (Mikko and Andersson 1995; Ellegren et al. 1996).

A detailed investigation of moose populations in Canada using microsatellites derived from other cervids and bovids revealed modest genetic heterogeneity (Broders et al. 1999). Population bottlenecks caused by founder effects were thought to have contributed to this loss of genetic variation (Broders et al. 1999). This was mostly attributed to the introduction, or re-introduction, of a small number of founder animals to geographic regions that were previously uninhabited or had suffered over-exploitation by hunting (Broders et al. 1999). Similar findings were made in a later study, but here the influence of co-habitation with other cervid species and the effect of parasites on the frequency of MHC genotypes and genetic heterogeneity were also investigated (Wilson et al. 2003). Discordant levels of genetic variability were found using microsatellites and MHC, and in the later, the parasite burden and the presence of other host animals were associated with non-balancing selection. One moose genotype was shown to predominate in areas that were inhabited by white-tailed deer (another possible host) and also in moose populations that were known to be infected with the parasite *Parelaphostrongylus tenuis* (Wilson et al. 2003). The implication that the MHC has been under functional selection at a point of time in the past has been made in studies of MHC alleles in moose populations from both Europe and North America (Mikko and Andersson 1995).

Fallow deer (*Dama dama* spp.). Despite the relative abundance of fallow deer populations in Europe and across the globe, low genetic variability is the hallmark of almost all extant fallow deer (*Dama dama*) populations, as revealed in the initial studies of biochemical variation using serum proteins and allozyme data where many of the proteins examined have been shown to be monomorphic (Pemberton and Smith 1985; Hartl et al. 1986; Randi and Apollonio 1988; Herzog 1989; Wehner et al. 1991; Schreiber and Fakler 1996). This is likely due to the origin of all extant populations from a Middle Eastern refuge, all other populations becoming extinct during the last Ice Age. Whilst studies of genetic variation in fallow deer using molecular markers have largely been neglected, Scandura and colleagues (1998) have used randomly amplified polymorphic DNA-polymerase chain reaction (RAPD-PCR) to identify nuclear DNA genetic variation in an Italian population of fallow deer. Microsatellite data have also been used to determine paternity and reproductive success in fallow deer (Say et al. 2003).

Other deer species. The development of molecular markers in deer has been applied recently to many of the less well-characterized deer species. Many of these species are endangered or threatened, and genetic analysis will likely prove instrumental in their management and rehabilitation. Most of these species have come from Asia and include the chital deer (*Cervus axis*), Indian swamp deer (*Cervus duvauceli branderi*), forest musk deer (*Moschus berezovskii*), tufted deer (*Elaphodus cephalophus*), Chinese water deer (*Hydropotes inermis*) and black muntjac (*Muntiacus crinifrons*).

In the forest musk deer, it has been necessary to isolate microsatellite sequence from the species due to non-transference of repeats from other cervids. Biotin capture of cloned DNA containing repeat sequence was used to increase the efficiency of selecting microsatellite positive clones, and a total of 14 polymorphic microsatellite loci have been isolated and characterized for use in this species (Zou et al. 2005; Xia et al. 2006). Initial studies of population genetic parameters in a single population from the Sichuan province in China have revealed that there is likely to be considerable genetic variability in this species, thus raising issues related to the conservation of diversity in this threatened species. The isolation and characterization of 10 microsatellite loci in the chital deer (*Cervus axis*) has also been achieved, and high levels of genetic variation have been found in these deer, even within a single zoological collection (Gaur et al. 2003).

Unlike allozyme assays, microsatellite loci can be adapted for high-throughput analysis that may often be required in genetic analysis. Various multiplex methods have been derived for genetic studies in deer populations. A successful multiplex assay is described for the transferral of 11 microsatellite loci from the bovine and ovine families to four cervid species, namely Eld's deer (*Cervus eldii* spp.), Indian Swamp deer, Rusa (*Cervus timorensis russa*) and Vietnamese sika (*Cervus nippon pseudaxis*) (Bonnet et al. 2002), and have also been shown to be applicable to other cervid species (Bonnet et al. 2002).

Ali and colleagues (1998) explored the genome of the central Indian barasinga (Indian swamp deer, *Cervus duvauceli branderi*) using probes for conserved repeat motifs. Only two samples were available for analysis due to the critically endangered status of this species. Several of the repeat motifs were shown to be absent from the genome of *C. d. branderi*, in addition to an observed lack of genetic diversity in this species, although this is perhaps not surprising given the small sample size.

Genetic diversity in both wild and captive populations of the endangered black muntjac deer (*Muntiacus crinifrons*) has been determined using the mtDNA control region (Wu and Fang 2005). From these data it was revealed that in the wild this species has had a small effective population size in the past, as revealed by the low nucleotide diversity found in the wild cohort. However, low haplotypic diversity in the captive population has important consequences for the management of this population, where the authors suggest the introduction of new animals from the wild may be necessary to increase haplotypic diversity as well as a readjusted breeding programme to maintain the diversity already present within the captive population.

Determination of Parentage

The assessment of parentage relationships in wild animal populations using molecular markers was pioneered in deer, specifically in the European red deer. Previously developed methods for resolving parentage in wild animal populations had utilized exclusion criteria in the determination of sire (Morin et al. 1994). In the exclusion approach, genotypes that are incompatible between sire and offspring are used to exclude candidate sires, with the intention of leaving one male non-excluded and therefore identified as the parent. However, even with the use of many genetic markers, it can be difficult to exclude an unrelated male. The exclusion approach becomes more difficult in situations where there are a large number of candidate sires, such as may be found in the wild, and erroneous exclusions can occur due to missing or incorrect genotypes. The computer program CERVUS was designed by Marshall and colleagues, for use in a wild red deer population on the island of Rum, Scotland (Marshall et al. 1998). This program utilizes a likelihood-based approach to identify the most-likely sire from a selection of candidates, based upon the inheritance of codominant marker genotypes. For greatest power, maternal genotypes are included, allowing fewer markers to be genotyped in order to resolve parentage.

Since the inception of CERVUS, it has been widely and successfully used to infer parentage both within the Cervidae and across the animal kingdom (Slate et al. 2000). It is robust to genotyping error and can be used for large-scale studies, such as those encountered in wild populations. Whilst intensive studies of

the inference of maternal parentage in farmed red deer have shown that it is possible to achieve good accuracy by visual observation (Vankova et al. 2001), this is highly labour-intensive. In a follow-up study on red deer from the Isle of Rum, 84 microsatellite loci were used to infer parentage using CERVUS in a retrospective assessment of the accuracy of this software (Slate et al. 2000). For comparison, behavioural observations were also used to infer parentage, and it was shown that CERVUS could accurately predict paternity (Slate et al. 2000).

The CERVUS program can be applied to domestic situations, so long as Hardy-Weinberg equilibrium is observed for at the loci that are used. In breeding operations for farmed deer, the assessment of parentage is critical to the assessment of sire and dam genetics, and thus for making genetic gain. Indeed, in New Zealand where red deer and wapiti are farmed in large numbers (O'Conner 2006), the use of DNA-based parentage testing in the stud breeding industry has been widely adopted (Crawford et al. 2006), resulting in a decrease in the labour units that are required to determine parentage.

A potential confounding effect in the analysis of parentage using microsatellite loci is the presence of null alleles (non-amplifying alleles). Instances of null alleles have certainly been documented in red deer, which may be further complicated by across-species use of microsatellites (Pemberton et al. 1995). Mismatches between PCR primers and target genomic sequence are likely to occur in the use of non-cervid primers on the cervid genome. Any additional mismatch errors that are due to cervid-specific polymorphism in the primer-binding region may present as 'null alleles' when in actual fact the microsatellite is still present and has just failed to amplify. The number of mismatches between primer and target sequence has certainly been shown to affect the successful amplification of PCR amplicons, where an ~7% decrease in success rate has been observed per mismatch in cross-species PCR (Housley et al. 2006).

Parentage analysis has also been undertaken in the white-tailed deer, where does often produce litters of two or more fawns. Multiple paternity in captive white-tailed deer was identified in approximately one-quarter of litters (litter size ≥ 2; DeYoung et al. 2002). The first evidence for the occurrence of multiple paternity in a large ungulate was provided using the program CERVUS incorporating 19 microsatellite loci. In a more recent study, the occurrence of multiple paternity has been shown to occur in wild and free-ranging populations of white-tailed deer at approximately the same rate as that observed in captive populations (22% in twins; Sorin 2004). These findings hold important consequences in the consideration of mating strategies in deer and other mammals where multiple paternity may occur.

Forensic Applications for Molecular Markers in Cervidae

The use of molecular markers has allowed for the identification of the species of origin when confronted with an unknown tissue sample from an animal of interest. A forensic application for such technologies has been explored in deer, in particular for the identification of 'poaching' or illegally procured deer, whether this is from a hunting estate or from herds that are protected for conservation purposes.

Mitochondrial DNA was first employed for the purposes of deer forensic analysis, due to its ability to distinguish between closely related species (Matsunaga et al. 1998). The distinction between meat originating from deer (venison), as compared to other livestock species such as sheep and goats, was initially achieved by amplification and restriction digest profiles of the cytochrome-b gene in mtDNA (Chikuni et al. 1994). A PCR-based assay on venison from red deer or sika was designed in the cytochrome-b gene, which is specific for the meat from deer, but not from other animals. Restriction fragment-length polymorphism (RFLP) within the amplified PCR product could then be used to distinguish between red deer and sika deer (Matsunaga et al. 1998). These findings allowed for the detection of gross hybridization or the identification of unknown meat samples. Indeed, an assurance of species-of-origin, most commonly for consumers who purchase game meat such as venison, has been demonstrated using PCR-RFLP of the mtDNA 12S rRNA gene, where it is possible to distinguish between meat from three cervid species (red deer, fallow deer and roe deer) from that of cattle, sheep or goat (Fajardo et al. 2006).

In China, only two endemic sika deer species are known to currently occur in wild populations, *C. n. sinchuanicus* and *C. n. kopschi*. However, other introduced populations of sika deer have been established and become common, such as *C. n. nippon* and *C. n. hortulorum*. Consumer desire for products from sika deer for use in traditional Chinese medicine may have negative effects upon threatened populations of the endemic Chinese species of sika deer. Sequencing of the mtDNA control region allows for the distinction of introduced and endemic sika deer in China, and forensic analysis of sika deer products can now be used to determine the subspecies of origin and thus provide a means to detect poaching of threatened Chinese sika (Wu et al. 2005).

It is not only the meat from deer that can be used as a sample for forensic or other genetic analyses. A method has been developed for the extraction and amplification of DNA from deer antlers (O'Connell and Denome 1999). Antlers are often important to hunters, as they are collected as trophies, and for deer it is possible to sample male animals that have had their antlers collected (Wang et al. 2002). Furthermore, methods have been developed within deer that allow for the extraction of DNA from scat samples collected in the field (Huber et al. 2003), and successful PCR from this DNA may allow for the identification of species, sex and individuals within wild populations (Huber et al. 2002). Although these techniques will perhaps find greater application in studies of population genetics, they may also be applicable to forensic analysis.

Similar to investigations of human criminal activity, the discovery of microsatellites has improved forensic analysis in deer. A number of studies have sought to identify microsatellites that are specifically suited to this application - those that show high degrees of polymorphism and provide reliable genotyping assays. Microsatellites have proven useful for the distinction between different deer species, especially where game management practices or conservation issues may differ between deer species that reside in the same geographic area. Six microsatellites derived in reindeer (Wilson et al. 1997; Roed and Midthjell 1998) were shown to amplify in red deer, roe deer and fallow deer, which are all species present in Germany (Poetsch et al. 2001). Similarly in North America, 11 microsatellites were specifically developed in wapiti (*C. e. canadensis*) for forensic application. The development of microsatellites within this subspecies avoids the potential artefacts that may occur in cross-species amplification, such as 'stutter bands' (Jones et al. 2002).

A novel extension of microsatellite use in forensic analysis that has already shown application in conservation genetics is the use of population assignment tests that have recently been developed (Excoffier and Heckel 2006). The uniqueness of endemic deer species, such as certain regional populations of red deer found in Europe, can be threatened by the introduction of deer from alien populations. Bayesian clustering methods have been applied to multilocus microsatellite data to determine the population of origin of four red deer that were suspected to have been illegally introduced into the red deer population of Luxembourg (Hmwe et al. 2006). It was concluded that the four candidate deer under investigation had been illegally introduced (Frantz et al. 2006).

4.2.2
The Deer Genetic Map

Until recently, most genomic markers in red deer and indeed all other extant deer species have been anonymous, and the only inferences about marker position were made by comparative analyses to other species. For organisms that have yet to have their genome sequenced, the construction of a genetic map still remains a critical juncture in the advancement of genomic knowledge. Both genetic and cytogenetic maps have been compiled for some taxa of Cervinae (Liming et al. 1980; Bonnet et al. 2001; Slate et al. 2002a, b; Huang et al. 2006). These will undoubtedly prove useful in the construction of genetic maps for other members of this family, and also for gene-mapping efforts, comparative genome studies and finally for the eventual sequencing of the deer genome.

The deer genetic map was derived from an interspecific hybrid between *Cervus elaphus* (red deer) and *Elaphurus davidianus* (Pere David's deer), Pere David's deer being one of the most distant relatives to *Cervus elaphus* in the Cervinae subfamily (Groves and Grubb 1987). It has still been possible to create fertile interspecific hybrids by natural mating and even more successfully using artificial breeding technologies, despite the large degree of morphological and genetic divergence between the two species (Asher et al. 1988).

Such an outcross between two divergent species means that species-specific variation will occur in virtually any DNA sequence. Because very few anonymous type II molecular markers were available, it was possible to create a genetic map using DNA probes from coding sequence that are usually invariant within a species (the so-called type I markers).

Pere David's deer is extinct in the wild, but has survived in captivity in zoological collections around the world. The species at present remains listed as 'critically endangered' on the IUCN red list (http://www.iucnredlist.org/search/details.php/7121/all), but reintroductions in the wild in China and the establishment of breeding programmes in captivity are likely to ensure the survival of this species. Viable hybrids derived from Pere David's deer and red deer have been known to occur since the late 19th century, although the success of natural mating has been limited (Beck and Wemmer 1983). A comparatively large-scale hybridization programme was established in New Zealand (AgResearch at the Invermay Agricultural Centre) after approximately 70 Pere David's deer were imported into New Zealand in 1984 and 1985, both for conservation purposes and for application to the farming of deer in New Zealand (Tate et al. 1998). Due to the seasonal nature of breeding, the breeding cycle of red deer (and wapiti) is not perfectly matched to pasture growth or market demand for venison (Nicol and Barry 2002). It was hoped that the introgression of genes from the short-day breeding Pere David's deer would assist in altering the reproductive season of the resultant hybrids to better match these facets of agricultural production. At the same time, a project to establish a genetic map was initiated. Several F_1 male hybrids were produced by artificial insemination and natural mating of red deer hinds (Asher et al. 1988). Three of the F_1 males were used as sires to produce 143 backcross calves of ¼ Pere David: ¾ red deer. A prototypic genetic linkage map was assembled for a small number of genetic markers that were protein coding type I loci showing variation between these species (Tate et al. 1995). A total of 17 genetic markers were mapped to four linkage groups, consisting of five proteins analysed for electrophoretic isoforms and 12 RFLP of 12 cDNA probes derived from human, mouse and cattle (Tate et al. 1995). The term RFLV, or restriction fragment-length variant, was used in this study to denote the characteristic of these markers as type I, variants between species, as opposed to polymorphic within the species. While this prototypic genetic map had limited utility for gene-mapping experiments, interesting comparisons of gene order were made with genetic maps from humans and cattle that revealed some chromosomal rearrangements in syntenic groups within ruminants as compared to humans.

Using Pere David's deer x red deer hybrid-mapping resource, Slate and colleagues (2002a) compiled a more detailed genetic map consisting of 621 markers that spanned 2,532.2 cM across 34 linkage groups. An expanded mapping herd was created over a six-year mating period from seven F_1 sires crossed to red deer dams, giving rise to 351 backcross calves of ¼ Pere David's: ¾ red deer (Tate et al. 1997). Approximately half of these markers were mapped in a subset of 89 backcross calves derived from two of the F_1 sires. Both type I and type II genetic markers were used in this study including 229 AFLPs, 153 microsatellites, 150 RFLVs, 73 ESTs and 16 proteins. Ninety-three additional markers were shown to be variant between the two species but were not assigned to any linkage group. An independent study of 93 loci from the initial deer map has since verified marker location and marker order, part of a quantitative trait loci (QTL) study of wild free-living red deer on the Isle of Rum (Slate et al. 2002b).

A cytogenetic map in the Vietnamese sika deer (*Cervus nippon pseudaxis*; Bonnet et al. 2001) allows for the assignment of linkage groups on the deer genetic map to their respective chromosomes (Slate et al. 2002a). This was achieved using fluorescence in situ hybridization (FISH) mapping of 59 molecular markers (both type I and microsatellites). Whilst the chromosome number in *C. n. pseudaxis* is greater than that of Pere David's deer or red deer, these results will still be applicable to other Cervinae, as alterations in chromosome number have been caused by Robertsonian translocations (Fontana and Rubini 1990). Comparison between banding patterns in cattle chromosomes versus those for sika was also determined, and complete concordance between linkage group assignment and marker order was observed when the cytogenetic map was compared to the deer genetic map (Bonnet et al. 2001).

Comparative genomic analysis, examining karyotypic evolution and chromosomal structure in Mammalia and Artiodactyla, has been possible since the creation of the deer genetic map. It was noted that the deer genetic map (>2,500 cM) was significantly shorter (Slate et al. 2002a) than those of cattle (>3,500 cM) and sheep (>3,000 cM), yet was of similar length to that observed in goat (~2,500 cM). A number of technical artefacts may have contributed

to this reduced length, such as the reduced recombination rate that may occur in males, given the map was derived from F_1 sires (Slate et al. 2002a). From comparison of the deer genetic map with other ruminants, it appears that repeated chromosomal fissions of the Pecoran ancestral karyotype ($2n = 6$) gave rise to the ancestral deer karyotype ($2n = 70$ in Cervinae (Bonnet et al. 2001; Slate et al. 2002a). These data support the assertion that chromosomal rearrangements of a similar type tend to reoccur during the evolution of a karyotype, such as chromosomal fission in deer (Bonnet et al. 2001; Slate et al. 2002a) and also that greater amounts of chromosomal rearrangements have occurred in rodents and in humans as compared to the ruminants (Slate et al. 2002a).

4.2.3
Genes

At present, there is only one study that reports the likely segregation of a single gene that affects a phenotypic character, which is the number of antler points in a white-tailed deer pedigree (Templeton et al. 1982). Templeton and colleagues report monogenic inheritance of antler point number in captive populations of white-tailed deer in half-sib analysis (Templeton et al. 1982). Mendelian inheritance of a dominant and a recessive allele with major effects on antler point number was the only model not rejected in their analysis, which allowed for 6–10 antler points or 2–5 antler points, respectively (in a goodness-of-fit test, similar results were reported using maximum likelihood). An additive model of genetic variance for this trait was rejected. In red deer, allozyme loci have been associated with antler traits and other morphological characters, in which selective hunting pressure for antlers as trophies may have resulted in changes to allele frequencies, but these correlative associations have not been verified by robust mapping techniques (Hartl et al. 1991).

Given the highly seasonal nature of breeding in a number of the Cervid species, there have been a number of functional genomic studies that have investigated the role of single genes on aspects of seasonal reproduction. These genes may be likely to represent candidates for gene-mapping studies in deer. For example, the red deer (*Cervus elaphus*) is a seasonal breeder, with a circannual breeding cycle that is triggered by a decrease in day-length. Key hormones have been identified as playing a role in this seasonal breeding pattern, which include prolactin, which is secreted from the pituitary under melatonin control in response to photoperiod. In the male (stag), plasma concentrations of the hormone prolactin increase in autumn in response to a decreasing day-length (Barrell et al. 1985). As a consequence, gonadotrophins are released from the pituitary, and receptors in the testis effect an increase in testosterone production, testis growth and spermatogenesis. Three RNA transcripts have been cloned and sequenced, and include a partial 1.7 kb transcript (non-membrane bound) that is relatively abundant in the testis (Clarke et al. 1995; Jabbour et al. 1996, 1998), and two full-length transcripts of 2.5 kb and 3.5 kb (both membrane bound) that are present in many tissues (Clarke et al. 1995). A premature stop codon in the 1.7 kb RNA transcript occurs as the result of a single base pair frame-shift mutation, and it has been suggested that this occurs to down-regulate the expression of the full-length receptor in testis in a red deer as a means to controlling seasonal breeding. A change in the structure of the expressed protein derived from the 1.7 kb transcript compared to the full-length transcripts may affect the function of this protein (Jabbour et al. 1998).

Several other deer reproductive hormones have been cloned and sequenced, and are found in both published literature and within public databases. The glycoprotein α-subunit and follicle-stimulating hormone (FSH) and Luteinizing hormone (LH) have been cloned and sequenced in both the North American wapiti (Clark et al. 2005) and sika (GenBank accession numbers: AY156688, and AY066018). Detailed studies of the relationship to gene sequences from other mammals were achieved, and the patterns of expression in the pituitary and muscle were examined (Clark et al. 2005). Pituitary growth hormone has been characterized in red deer (Wallis et al. 2006), and various growth factors have been cloned and sequenced in the roe deer (Wagener et al. 2000). Each of these studies specifically examines the expression of these hormones as they relate to the seasonal breeding.

The genes that affect fitness-related traits in deer have also been the subject of molecular investigation. Research in this area has been possible as most cervids are predominately wild animals and are therefore subject to natural selection. Genetic determinants of traits such as immune system competence are likely to be under selective pressure in the wild. The MHC genes have been studied in free-living populations of white-tailed deer in North America, not only for the determination of phylogenetic or population genetic parameters but also to determine whether these genes are involved in fitness in the wild (Ditchkoff et al. 2001). The MHC genes are candidates for the 'advertisement of good genes' hypothesis, as they are involved in the presentation of antigens and subsequent immune response. Individuals that were heterozygous for alleles at the *MHC-DRB* locus had an increased antler size and body mass, and showed greater resistance to abomasal parasites (Ditchkoff et al. 2001), therefore attesting to the supposition that antler size is an indicator of the genetic quality of the male animal, at least for white-tailed deer. Associations of allozyme loci with morphological traits in red deer have also been reported (Hartl et al. 1991, 1995a, 2005).

4.2.4
Quantitative Trait Loci in Deer

With the development of a genetic map in deer, it has been possible to undertake genetic mapping to identify quantitative trait loci (QTL), and this has been accomplished in both a domestic and a wild deer population. The initial genetic mapping resource in deer was produced from an outcross between Pere David's deer (*Elaphurus davidianus*) and red deer (*Cervus elaphus*), that included more than 300 backcross progeny (¾ red deer: ¼ Pere David's deer; Tate et al. 1997). For the purposes of gene/QTL mapping, this resource is especially interesting given the large divergence in a variety of developmental, morphological and reproductive traits. A number of studies reporting QTL have been published since the first description of genetic mapping in these interspecies hybrids (Tate et al. 1995). Many of the traits between these two species are highly divergent, including gestation length, body weight, antler casting date and various morphological traits, making the backcross progeny resource a valuable tool for identifying QTL that affect these traits (Tate et al. 1995, 1998). Photographs of Pere David's deer, red deer and the F_1 progeny from this backcross are shown in Fig. 3, where

Fig. 3 A female Père David's deer (**a**) as compared to a female red deer (**b**). Note the elongated facial features, longer legs and heavier musculature of the Père David's deer. Additional features such as splayed hooves and antler morphology also differ between the two species. An interspecific hybrid between the Père David's deer and red deer was used to create a genetic map in deer. This resource was also used for the identification of numerous QTL that affect morphological and growth traits between these two species. A female F1 hybrid is also pictured shown for here for comparison (**c**). (Photos courtesy of Geoff Asher, AgResearch, Invermay, New Zealand)

many of the morphological differences become immediately apparent.

Linkage mapping for growth rate and live weight are described by Goosen et al. (1999) using 250 genetic markers that included RFLV, protein variants and microsatellites at an average marker spacing of ~7 cM (total genome length 1,240 cM). Seasonal changes in growth rate and live weight are known to occur in red deer, but differ from those observed for Pere David's deer (Tate et al. 1997). The two mapping techniques employed were analysis of variance (ANOVA) and an interval linkage-mapping approach. Significant support (genome-wide level of significance) for QTL affecting both live weight and growth rate were identified. Five regions were shown to contain significant QTLs that affected live weight (LG 1, LG 12 and LG 20) and growth rate (LG 5, LG 12 and LG20). The allele substitution effect was not necessarily always intuitive, where both positive and negative effects on weight-related traits were reported for the Pere David's deer allele, given that Pere David's deer is larger than the red deer. It was suggested that the large genetic distance between these two species resulted in the unpredictable allele substitution effects, and that epistatic effects or recessive alleles may account for the observed conflict. Candidate genes were identified within the QTL intervals that included the insulin-like growth factor genes and their receptors (*IGF1, IGF2, IGF1R*), the growth hormone gene and growth hormone-releasing hormone (*GHRH*), although no fine-mapping efforts have been undertaken to resolve the genes that affect these weight-related traits.

Another QTL study on this same Pere David's deer x red deer-mapping resource searched for QTL affecting seasonal antler traits, which include the date and live weight at pedicle initiation, and the dates of antler cleaning and antler casting (Goosen et al. 2000). A significant QTL affecting pedicle initiation and live weight at this time was identified on linkage group 12 that explained approximately 13% of the phenotypic variance in this trait. The allele substitution effect accounted for a 21-day difference in pedicle initiation date, where Pere David's deer allele delayed pedicle initiation when substituted for a red deer allele. This is contrary to expectations, given that Pere David's deer shows markedly earlier seasonal breeding (and thus for other seasonal traits such as antler initiation). Similarly, other QTLs

have been identified in this resource population that have been shown to affect seasonal gestation length (LG 16) and birth weight (LG 4; Tate et al. 1998). Validation of all QTLs identified in Pere David's deer x red deer-mapping resource has yet to be achieved in an independent population, but these studies will undoubtedly prove useful when more genome information becomes available for fine mapping in the cervid genome.

No positional candidate genes have been cloned. This is perhaps in part due to the limited statistical power and resolution of the published QTL-mapping studies, and also the limited number of markers presently available on the deer genetic map. For any given QTL interval identified (many of which are tens of centiMorgans in length), there will be many genes at each locus, where each presents a potential candidate gene. The positional cloning of genes has already proven difficult in species that have a much greater level of genomic information available, such as the mapping of complex disease loci in humans.

Such QTLs are also of interest to evolutionary biologists studying natural populations, because the existence of QTLs for traits associated with fitness would suggest that quantitative genetic variation for fitness is not necessarily eroded in the way that Fisher had proposed (Fisher 1958). To this end, an attempt to map QTL underlying birth weight, a trait associated with lifetime breeding success (Kruuk et al. 1999), was undertaken in the red deer population on the Isle of Rum, Scotland. These deer provide a large intergenerational pedigree resource with detailed life-history records, and where parentage has been recorded for some time (Clutton-Brock et al. 1982; Kruuk et al. 2000), making them suitable for QTL-mapping studies. Mapping QTL in red deer from the Isle of Rum has suggested that genes of larger effect can account for fitness traits (Slate et al. 2002b). Both interval mapping and variance components analysis were used to search for QTL affecting birth weight, using a distribution of microsatellite markers (and three allozyme markers) covering ~ 60% of the deer genome (Slate et al. 2002b). Three linkage groups showed suggestive evidence for linkage, with one on LG 21 surpassing the genome-wide level of significance (VC analysis, $P = 0.006$). Each QTL had major effects on birth weight of between 0.75 and 1.0 phenotypic standard deviations (from interval-mapping data, similar results for VC analysis). However, the QTL effects may be over-estimated due to the 'Beavis' effect, where <500 progeny may lead to upwardly biased estimates of QTL effect size (Xu 2003), perhaps by even more than two-fold in this instance (Slate et al. 2002b). Nonetheless, a relatively large effect is still likely for this QTL given its detection in a limited number of progeny (Slate et al. 2002b), suggesting that major genes can account for a proportion of the additive genetic variation that is observed in fitness traits.

A great deal of research has been expended into the genetics of species differences (particularly for those that contribute to reproductive isolation) using QTL mapping in interspecific hybrids. These studies have been conducted in model organisms such as *Drosophila* sp. (Liu et al. 1996; Zeng et al. 2000), the mouse (Cheverud et al. 2004) and in plants such as monkeyflower (*Mimulus*; Bradshaw et al. 1998). Each of these studies adds to the body of evidence suggesting that genes of major effect can contribute significantly to the morphological differences observed between species, although evidence for Fisher's infinitesimal model of many genes with small effect is also apparent in these studies. Attempts to determine the genetic basis of speciation in deer have been made using Pere David's deer x red deer hybrid resource developed for QTL studies (Maqbool et al. 2007). A number of morphological traits were measured that include the dimensions of leg, cranial, and tail traits at two stages of development, both pre- and post-puberty. Three QTLs of major effect were revealed in the study by Maqbool et al. (2007) using a composite interval mapping with ~75% coverage of the deer genome. These QTLs were located on linkage groups 3, 5 and 12, and represent putative pleiotropic QTL that each affect more than one trait. As only 210 progeny were included in the analysis, it is likely that the effect size was over-estimated due to the 'Beavis' effect, but these QTLs were shown to account for between 29 and 58% of the phenotypic variance in nasal and foot morphology in the interspecies hybrids. Each of these QTLs affected more than one trait, and the authors suggest that pleiotropy might play a role in the determination of species-specific morphologies as observed between large mammals, such as deer. However, it is also possible that different QTLs are co-located and could not be resolved in this study.

4.2.5
Chronic Wasting Disease of North American Cervids

Chronic wasting disease (CWD) is a fatal neurodegenerative disease that affects North American cervids. It is endemic in wild and farmed populations of wapiti, white-tailed deer and mule deer (Williams and Young 1982; Spraker et al. 1997). Chronic wasting disease belongs to a group of disorders known as the transmissible spongiform encephalopathies (TSE) that include scrapie in sheep, Creutzfeldt-Jakob disease (CJD) in humans and bovine spongiform encephalopathy (BSE) in cattle (Aguzzi 2006). The TSEs are unique in that they may occur as infectious, genetic or sporadic disease, where a common link between them is the involvement of the prion protein (PrP; Aguzzi 2006). Conversion of the normal cellular prion-protein (PrP^c) to a protease-resistant form (PrP^{TSE}, PrP^{Sc} or other abbreviations) allows for the accumulation of this protein in the central nervous system, leading to spongiform encephalopathy and neurodegeneration (Cohen and Prusiner 1998; Horiuchi and Caughey 1999). The aberrant form of PrP can also be found in the peripheral nervous system, lymphoid and other tissues, with each of these sites of accumulation allowing for different modes of both horizontal and inter-species transmission to occur for each TSE (Brown et al. 2000). Variability in the *PRNP* gene sequence that occurs in a species may affect the genetic susceptibility of an individual (Supattapone et al. 2001). The aggregation of PrP^c and PrP^{TSE} is thought to allow for the conversion of α-helical protein to β-sheet conformation (Cohen and Prusiner 1998) and thus the formation of amyloid fibres which are an important part of the pathogenesis of TSEs (Come et al. 1993). It has been hypothesized that variant forms of *PRNP* may play a role in the aetiology of CWD in deer, and more recently a number of studies have sought to characterize this gene, and variation within it, in the North American cervid populations.

Owing to the highly conserved nature of the *PRNP* gene between species, it has been possible to clone and sequence most of the *PRNP* gene in the Rocky Mountain elk (wapiti; Cervenakova et al. 1997; Schatzl et al. 1997) and mule deer (Cervenakova et al. 1997) using degenerate primers based on *PRNP* sequences from other species. The wapiti sequence was found to be invariant, but an amino acid substitution was found at codon 138 (Ser138Asp). Additional variation has been observed in the *PRNP* gene in wapiti, mule deer and white-tailed deer.

A polymorphism in human *PRNP* gene at codon 129 has been shown to contribute to susceptibility to sporadic CJD, where individuals that are homozygous for methionine at this position are at an increased risk of developing the condition (Collinge et al. 1991; Palmer et al. 1991). At the corresponding position in the wapiti *PRNP* gene, codon 132, an M132L substitution (O'Rourke et al. 1998) has been identified and a synonymous substitution at codon 104 has been discovered (AAG to AAA; O'Rourke et al. 1999). A significant difference in the frequency of the M132L polymorphism has been observed in free-ranging Rocky Mountain elk (wapiti) when comparing CWD-affected versus unaffected populations, thus suggesting a role for this polymorphism in genetic susceptibility to CWD in wapiti (O'Rourke et al. 1999).

PRNP genetic variation in the genus *Odocoileus*, which includes both mule deer and white-tailed deer, appears to be greater than in wapiti, although this is likely due to the greater extent of sequencing undertaken in the *Odocoileus* sp. Variation in the *PRNP* coding sequence, promoter region and 3'UTR has been documented (Heaton et al. 2003; Johnson et al. 2003; O'Rourke et al. 2004; Jewell et al. 2005), that includes a number of non-synonymous amino acid substitutions (D20G, Q95H, G96S, A116G, S225F). Initial studies in the *Odocoileus* sp. were confounded by the presence of a *PRNP* pseudogene (*PRNPψ*), that was later identified by cloning and sequencing of *PRNP*-positive clones in a BAC library derived from mule-deer genomic sequence (Brayton et al. 2004). The pseudogene is known to occur in both subspecies of *O. hemionus* (mule deer and black-tailed deer) and white-tailed deer, but has not been shown to occur in Old World deer species such as wapiti. A number of non-synonymous substitutions in *PRNPψ*, some of which had previously been documented as occurring in the functional gene, were observed in mule deer (G65E, R151C) and variation in an octapeptide repeat (O'Rourke et al. 2004; Johnson et al. 2006) and a Q226K substitution (Johnson et al. 2006) has also been observed in *PRNPψ*

of white-tailed deer. Common to both species, the functional gene encodes a serine (S) at codon 138 with all pseudogene alleles encoding an asparagine (N) at this position (O'Rourke et al. 2004), this had previously been documented as a polymorphism of *PRNP*.

In white-tailed deer, an association between non-synonymous amino acid substitutions in *PRNP* and susceptibility to CWD has been reported. There appears to be no common resistant genotype that provides complete protection from CWD, although certain rare genotypes (in combination) appear to confer resistance (Johnson et al. 2003; O'Rourke et al. 2004) and the presence of at least one copy of the more common alleles at Q95H and G96S confers moderate resistance to CWD (Johnson et al. 2006). The pseudogene *PRNPψ* has shown no apparent involvement in CWD (O'Rourke et al. 2004).

In the mule deer (*O. hemionus*), a large-scale study of free-ranging populations revealed that the S225F polymorphism conferred a significant protective effect from disease where the odds of a CWD-infected deer having an 225SS genotype was 30 times greater than for a CWD-negative animal (Jewell et al. 2005). An increase in the S225SF genotype in wild populations may assist in reducing CWD prevalence, but an analysis of population structure will be necessary to distinguish genuine allelic association from the spurious associations that may arise due to population structure (Jewell et al. 2005).

Continued research into the genetic susceptibility of CWD in cervids is required if the effective management of this disease in both wild and captive populations of North American cervids is to be achieved. More recently, a transgenic mouse model for the study of CWD has been created by the insertion of the cervid *PRNP* gene into the mouse genome (Browning et al. 2004), and it may therefore soon be possible to empirically test the effects of various polymorphisms of the cervid PrP protein on the development of CWD.

4.3
Future Scope of Work

Many of the cervid species now have a well-described set of molecular markers that are available for various genetic analyses. The path leading to this has followed the development of molecular genetics in other mammals. Much of the genetic analysis in cervids has utilized molecular markers as descriptors of genetic variability for the inference of population genetic parameters. This has largely held applications for conservation and game management, and once a suite of markers has been developed, whether from mtDNA or nuclear DNA, the need for additional genomic information is not necessarily of paramount importance in such studies. However, a sequenced deer genome is a large and obvious next step in the natural progression of genome-based research, as has been observed for numerous other species. The need for a sequenced deer genome is likely to have greatest application for those researchers whose interests lie in farmed captive populations of deer, where the link between genotype and desirable phenotypes will be more easily resolved with access to a fully sequenced genome. As the cost of sequencing declines due to increases in efficiency, achieving a sequenced deer genome becomes more feasible. As has been seen in the other species with a sequenced genome, insights into genome organization and comparative analysis with other sequenced species will become possible.

For the time being, and until the deer genome sequence (of at least one cervid) becomes available, much progress is likely to be made through the use of genome sequence data from other ruminants, especially cattle and sheep. It has already been demonstrated that microsatellite loci are conserved between cattle, sheep, goat, deer and other ruminants (e.g., Pepin et al. 1995; Slate et al. 1998) and for researchers who seek to map genes and QTL in deer, the availability of large numbers of transferrable microsatellite loci from cattle and sheep genome projects offers a way forward. The identification and characterization of genes in deer has also become more easily achieved using a bioinformatic approach in the design of sequencing primers. To this effect, the absence of a complete (or draft) genomic sequence in deer need not be an immediate hindrance to genome mapping in these species. In the time it takes to achieve a complete deer genome sequence, it is likely that a number of successful gene-mapping studies will have been achieved by using a comparative genomics approach.

Acknowledgements. I am very grateful and wish to express my thanks to Josephine Pemberton and Noelle Cockett for their suggestions and review of the manuscript. I also thank Diageo Archives for their permission to reproduce 'The Monarch of the Glen' and also Joerns Fickel and Colin Groves for permission to reproduce the figure from their manuscript, and also to Brent Haufmann from www.ultimateungulate.com for the images contained therein. I also thank Geoff Asher for his photos of Pere David's deer, red deer and the F_1 hybrid.

References

Aguzzi, A (2006) Prion diseases of humans and farm animals: epidemiology, genetics, and pathogenesis. J Neurochem 97:1726–39

Ali, S, Ansari, S, Ehtesham, NZ, Azfer, MA, Homkar, U, Gopal, R, Hasnain, SE (1998) Analysis of the evolutionarily conserved repeat motifs in the genome of the highly endangered central Indian swamp deer *Cervus duvauceli branderi*. Gene 223:361–367

Anderson, JD, Honeycutt, RL, Gonzales, RA, Gee, KL, Skow, LC, Gallagher, RL, Honeycuti, DA, DeYoung, RW (2002) Development of microsatellite DNA markers for the automated genetic characterization of white-tailed deer populations. J Wildlife Manag 66:67–74

Asher, GW, Adam, JL, Otway, W, Bowmar, P, Van Reenan, G, Mackintosh, CG, Dratch, P (1988) Hybridization of Pere David's deer (*Elaphurus davidianus*) and red deer (*Cervus elaphus*) by artificial insemination. J Zool Lond 215:197–203

Baccus, R, Ryman, N, Smith, MH, Reuterwall, C, Cameron, D (1983) Genetic variability and differentiation of large grazing mammals. J Mamm 64:109–120

Barrell, GK, Muir, PD, Sykes, AR (1985) Seasonal profiles of plasma testosterone, prolactin and growth hormone in red deer stags. In: Fennessy, P. F., Drew, K. R. (eds) The Biology of Deer Production. The Royal Society of New Zealand, New Zealand, pp 185–190

Beck, BB, Wemmer, C (1983) The Biology and Management of an Extinct Species – Pere David's Deer. Noyes Publications, New Jersey

Bonnet, A, Thevenon, S, Claro, F, Gautier, M, Hayes, H (2001) Cytogenetic comparison between Vietnamese sika deer and cattle: R-banded karyotypes and FISH mapping. Chromosome Res 9:673–687

Bonnet, A, Thevenon, S, Maudet, F, Maillard, JC (2002) Efficiency of semi-automated fluorescent multiplex PCRs with 11 microsatellite markers for genetic studies of deer populations. Anim Genet 33:343–350

Bradshaw, HD, Jr., Otto, KG, Frewen, BE, McKay, JK, Schemske, DW (1998) Quantitative trait loci affecting differences in floral morphology between two species of monkeyflower (Mimulus). Genetics 149:367–382

Brayton, KA, O'Rourke, KI, Lyda, AK, Miller, MW, Knowles, DP (2004) A processed pseudogene contributes to apparent mule deer prion gene heterogeneity. Gene 326:167–173

Breshears, DD, Smith, MH, Cothran, EG, Johns, PE (1988) Genetic variability in white-tailed deer. Heredity 60:139–146

Broders, HG, Mahoney, SP, Montevecchi, WA, Davidson, WS (1999) Population genetic structure and the effect of founder events on the genetic variability of moose, Alces alces, in Canada. Mol Ecol 8:1309–1315

Brown, KL, Ritchie, DL, McBride, PA, Bruce, ME (2000) Detection of PrP in extraneural tissues. Microsc Res Tech 50:40–45

Brown, WM, George, M, Jr., Wilson, AC (1979) Rapid evolution of animal mitochondrial DNA. Proc Natl Acad Sci USA 76:1967–1971

Browning, SR, Mason, GL, Seward, T, Green, M, Eliason, GA, Mathiason, C, Miller, MW, Williams, ES, Hoover, E, Telling, GC (2004) Transmission of prions from mule deer and elk with chronic wasting disease to transgenic mice expressing cervid PrP. J Virol 78:13345–13350

Cervenakova, L, Rohwer, R, Williams, S, Brown, P, Gajdusek, DC (1997) High sequence homology of the PrP gene in mule deer and Rocky Mountain elk. Lancet 350:219–220

Chesser, RK, Smith, MH, Johns, PE, Manlove, MN, Straney, DO, Baccus, R (1982) Spatial temporal and age dependent heterozygosity of beta hemoglobin in white-tailed deer *Odocoileus-virginianus*. J Wildlife Management 46:983–990

Cheverud, JM, Ehrich, TH, Vaughn, TT, Koreishi, SF, Linsey, RB, Pletscher, LS (2004) Pleiotropic effects on mandibular morphology II: differential epistasis and genetic variation in morphological integration. J Exp Zoolog B Mol Dev Evol 302:424–435

Chikuni, K, Tabata, T, Saito, M, Monma, M (1994) Sequencing of mitochondrial cytochrome b genes for the identification of meat species. Anim Sci Technol 65:571–57

Clark, RJ, Furlan, MA, Chedrese, PJ (2005) Cloning of the elk common glycoprotein alpha-subunit and the FSH and LH beta-subunit cDNAs. J Reprod Dev 51:607–616

Clarke, LA, Edery, M, Loudon, AS, Randall, VA, Postel-Vinay, MC, Kelly, PA, Jabbour, HN (1995) Expression of the prolactin receptor gene during the breeding and non-breeding seasons in red deer (*Cervus elaphus*): evidence for the expression of two forms in the testis. J Endocrinol 146:313–321

Clutton-Brock, TH, Guinness, FE, Albon, SD (1982) Red deer behaviour and ecology of the two sexes. Edinburgh University Press, Edinburgh

Cohen, FE, Prusiner, SB (1998) Pathologic conformations of prion proteins. Annu Rev Biochem 67:793–819

Collinge, J, Palmer, MS, Dryden, AJ (1991) Genetic predisposition to iatrogenic Creutzfeldt-Jakob disease. Lancet 337:1441–2

Come, JH, Fraser, PE, Lansbury, PT, Jr. (1993) A kinetic model for amyloid formation in the prion diseases: importance of seeding. Proc Natl Acad Sci USA 90:5959–5963

Cook, CE, Wang, Y, Sensabaugh, G (1999) A mitochondrial control region and cytochrome b phylogeny of sika deer (*Cervus nippon*) and report of tandem repeats in the control region. Mol Phylogenet Evol 12:47–56

Côte, SD, Rooney, TP, Tremblay, J-P, Dussault, C, Waller, DM (2004) Ecological impacts of deer overabundance. Annu Rev Ecol Evol Syst 35:113–147

Coulon, A, Cosson, JF, Angibault, JM, Cargnelutti, B, Galan, M, Morellet, N, Petit, E, Aulagnier, S, Hewison, AJ (2004) Landscape connectivity influences gene flow in a roe deer population inhabiting a fragmented landscape: an individual-based approach. Mol Ecol 13:2841–2850

Coulon, A, Guillot, G, Cosson, JF, Angibault, JM, Aulagnier, S, Cargnelutti, B, Galan, M, Hewison, AJ (2006) Genetic structure is influenced by landscape features: empirical evidence from a roe deer population. Mol Ecol 15:1669–1679

Crawford, AM, Bixley, MJ, Anderson, RM, McEwan, JC (2006) Uptake of DNA testing by the livestock industries of New Zealand. Proc 8th World Congr on Genet Appl Livestock Prod, Belo Horizonte, Brazil, August, pp 13–18

Cronin, MA (1991) Mitochondrial-DNA phylogeny of deer (Cervidae). J Mamm 72:533–566

Cronin, MA (1992) Intraspecific variation in mitochondrial DNA of North American cervids. J Mamm 73:70–82

Cronin, MA, Renecker, L, Pierson, BJ, Patton, JC (1995) Genetic variation in domestic reindeer and wild caribou in Alaska. Anim Genet 26:427–34

Cronin, MA, Stuart, R, Pierson, BJ, Patton, JC (1996) K-casein gene phylogeny of higher ruminants (Pecora, Artiodactyla). Mol Phylogenet Evol 6:295–311

Cronin, MA, Patton, JC, Balmysheva, N, MacNeil, MD (2003) Genetic variation in caribou and reindeer (*Rangifer tarandus*). Anim Genet 34:33–41

Cronin, MA, MacNeil, MD, Patton, JC (2005) Variation in mitochondrial DNA and microsatellite DNA in caribou (*Rangifer tarandus*) in North America. J Mamm 86:495–505

Cronin, MA, Macneil, MD, Patton, JC (2006) Mitochondrial DNA and microsatellite DNA variation in domestic Reindeer (*Rangifer tarandus tarandus*) and relationships with wild Caribou (*Rangifer tarandus granti, Rangifer tarandus groenlandicus, and Rangifer tarandus caribou*). J Hered 97:525–530

DeWoody, JA, Honeycutt, RL, Skow, LC (1995) Microsatellite markers in white-tailed deer. J Hered 86:317–319

DeYoung, RW, Demarais, S, Gonzales, RA, Honeycutt, RL, Gee, KL (2002) Multiple paternity in white-tailed deer (*Odocoileus virginianus*) revealed by DNA microsatellites. J Mamm 83:884–892

DeYoung, RW, Demarais, S, Honeycutt, RL, Rooney, AP, Gonzales, RA, Gee, KL (2003) Genetic consequences of white-tailed deer (*Odocoileus virginianus*) restoration in Mississippi. Mol Ecol 12:3237–3252

Ditchkoff, SS, Lochmiller, RL, Masters, RE, Hoofer, SR, Van Den Bussche, RA (2001) Major-histocompatibility-complex-associated variation in secondary sexual traits of white-tailed deer (*Odocoileus virginianus*): evidence for good-genes advertisement. Evolution Int J Org Evolution 55:616–625

Douzery, E, Randi, E (1997) The mitochondrial control region of Cervidae: evolutionary patterns and phylogenetic content. Mol Biol Evol 14:1154–1166

Ellegren, H, Mikko, S, Wallin, K, Andersson, L (1996) Limited polymorphism at the major histocompatibility complex (MHC) loci in the Swedish moose. Mol Ecol 5:3–9

Ellsworth, DL, Honeycutt, RL, Silvy, NJ, Bichman, JW, Klimstra, WD (1994a) Historical biogeography and contemporary patterns of mitochondrial DNA variation in white-tailed deer from southeastern United States. Evolution 48:122–136

Ellsworth, DL, Honeycutt, RL, Silvy, NJ, Smith, MH, Bickham, JW, Klimstra, WD (1994b) White-tailed deer restoration to the southeastern United States: evaluating genetic variation. J Wildlife Manage 58:686–697

Engel, SR, Linn, RA, Taylor, JF, Davis, SK (1996) Conservation of microsatellite loci across species of artiodactyls: implications for population studies. J Mamm 77:504–518

Excoffier, L, Heckel, G (2006) Computer programs for population genetics data analysis: a survival guide. Nat Rev Genet 7:745–758

Fajardo, V, Gonzalez, I, Lopez-Calleja, I, Martin, I, Hernandez, PE, Garcia, T, Martin, R (2006) PCR-RFLP authentication of meats from red deer (*Cervus elaphus*), fallow deer (*Dama dama*), roe deer (*Capreolus capreolus*), cattle (*Bos taurus*), sheep (*Ovis aries*), and goat (*Capra hircus*). J Agric Food Chem 54:1144–1150

Feijao, PC, Neiva, LS, de Azeredo-Espin, AM, Lessinger, AC (2006) AMiGA: the arthropodan mitochondrial genomes accessible database. Bioinformatics 22:902–903

Feulner, PG, Bielfeldt, W, Zachos, FE, Bradvarovic, J, Eckert, I, Hartl, GB (2004) Mitochondrial DNA and microsatellite analyses of the genetic status of the presumed subspecies *Cervus elaphus montanus* (Carpathian red deer). Heredity 93:299–306

Fickel, J, Reinsch, A (2000) Microsatellite markers for the European Roe deer (*Capreolus capreolus*). Mol Ecol 9:994–995

Fisher, RA (1958) The genetical theory of natural selection. Dover, New York

Fontana, F, Rubini, M (1990) Chromosomal evolution in Cervidae. Biosystems 24:157–174

Frantz, AC, Pourtois, JT, Heuertz, M, Schley, L, Flamand, MC, Krier, A, Bertouille, S, Chaumont, F, Burke, T (2006) Genetic structure and assignment tests demonstrate illegal translocation of red deer (*Cervus elaphus*) into a continuous population. Mol Ecol 15:3191–3203

Fukui, E, Kojima, T, Yoshizawa, M, Muramatsu, S, Yoshikawa, T, Tanji, T (1996) Genetic polymorphism of electrophoretic type of serum proteins and red cell enzymes in Japanese sika deer. Anim Sci Technol 67:574–578

Galan, M, Cosson, JF, Aulagnier, S, Maillard, JC, Thevenon, S, Hewison, AJM (2003) Cross-amplification tests of ungulate primers in roe deer (*Capreolus capreolus*) to develop a multiplex panel of 12 microsatellite loci. Mol Ecol Notes 3:142–146

Gatesy, J, Yelon, D, DeSalle, R, Vrba, ES (1992) Phylogeny of the Bovidae (Artiodactyla, Mammalia), based on mitochondrial ribosomal DNA sequences. Mol Biol Evol 9:433–446

Gaur, A, Singh, A, Arunabala, V, Umapathy, G, Shailaja, K, Singh, L (2003) Development and characterization of 10 novel microsatellite markers from Chital deer (*Cervus axis*) and their cross-amplification in other related species. Mol Ecol Notes 3:607–609

Goodman, SJ, Barton, NH, Swanson, G, Abernethy, K, Pemberton, JM (1999) Introgression through rare hybridization: a genetic study of a hybrid zone between red and sika deer (genus Cervus) in Argyll, Scotland. Genetics 152:355–371

Goosen, GJ, Dodds, KG, Tate, ML, Fennessy, PF (1999) QTL for live weight traits in Pere David's x red deer interspecies hybrids. J Hered 90:643–647

Goosen, GJ, Dodds, KG, Tate, ML, Fennessy, PF (2000) QTL for pubertal and seasonality traits in male Pere David's x red deer interspecies hybrids. J Hered 91:397–400

Gordon, B (2003) Rangifer and man: an ancient relationship. Rangifer 14:15–28

Gould, SJ (1974) The origin and function of bizarre structures, antler size and skull size in the Irish elk, *Megaloceros gianteus*. Evolution 28:191–220

Groves, C, Grubb, P (1987) Relationships of living deer. In: Wemmer, C. (ed) Biology and Management of the Cervidae. Smithsonian Institution Press, Washington DC

Grubb, P, Groves, CP (2003) Relationships of living deer. In: Wemmer, C. M. (ed) Biology and Management of the Cervidae. Smithsonian Institution Press, Washington, DC pp. 21–59

Gyllensten, U, Ryman, N, Reuterwall, C, Dratch, P (1983) Genetic differentiation in four European subspecies of red deer (*Cervus elaphus* L.). Heredity 51:561–580

Hartl, DL, Clark, AG (1997) Principles of Population Genetics. Sinauer Associates, Sunderland

Hartl, GB, Schleger, A, Slowak, M (1986) Genetic variability in fallow deer, Dama dama L. Anim Genet 17:335–341

Hartl, GB, Willing, R, Lang, G, Klein, F, Koller, J (1990) Genetic variability and differentiation in red deer (*Cervus elaphus* L) of Central Europe. Genet Sel Evol 22:289–306

Hartl, GB, Lang, G, Klein, F, Willing, R (1991) Relationships between allozymes, heterozygosity and morphological characters in red deer (*Cervus elaphus*), and the influence of selective hunting on allele frequency distribution. Heredity 66 (Pt 3):343–350

Hartl, GB, Apollonio, M, Mattioli, L (1995a) Genetic determination of cervid antlers in relation to their significance in social interactions. Acta Theriol 40:199–205

Hartl, GB, Nadlinger, K, Apollonio, M, Markov, G, Klein, F, Lang, G, Findo, S, Markowski, J (1995b) Extensive mitochondrial-DNA differentiation among European Red deer (*Cervus elaphus*) populations: implications for conservation and management. Z Säugetierkunde 60:41–52

Hartl, GB, Zachos, FE, Nadlinger, K, Ratkiewicz, M, Klein, F, Lang, G (2005) Allozyme and mitochondrial DNA analysis of French red deer (*Cervus elaphus*) populations: genetic structure and its implications for management and conservation. Mamm Biol 70:24–34

He, K, Wilton, SD, Tate, ML, Murphy, MP (1997) Characterization of the erythrocyte superoxide dismutase allozymes in the deer *Cervus elaphus*. Anim Genet 28:299–301

Heaton, MP, Leymaster, KA, Freking, BA, Hawk, DA, Smith, TP, Keele, JW, Snelling, WM, Fox, JM, Chitko-McKown, CG, Laegreid, WW (2003) Prion gene sequence variation within diverse groups of U.S. sheep, beef cattle, and deer. Mamm Genome 14:765–777

Hemmer, H (1990) Domestication: The Decline of Environmental Appreciation. Cambridge University Press, Cambridge, UK

Herzog, S (1988) Polymorphism and genetic control of erythrocyte 6-phosphogluconate dehydrogenase in the genus Cervus. Anim Genet 19:291–294

Herzog, S (1989) Genetic polymorphism of transferrin in fallow deer, *Cervus dama L.* Anim Genet 20:421–426

Herzog, S (1990) Genetic analysis of erythrocyte superoxide dismutase polymorphism in the genus Cervus. Anim Genet 21:391–400

Hmwe, SS, Zachos, FE, Eckert, I, Lorenzini, R, Fico, R, Hartl, GB (2006) Conservation genetics of the endangered red deer from Sardinia and Mesola with further remarks on the phylogeography of *Cervus elaphus corsicanus*. Biol J Linn Soc 88:691–701

Horiuchi, M, Caughey, B (1999) Specific binding of normal prion protein to the scrapie form via a localized domain initiates its conversion to the protease-resistant state. Embo J 18:3193–203

Housley, DJ, Zalewski, ZA, Beckett, SE Venta, PJ (2006) Design factors that influence PCR amplification success of cross-species primers among 1147 mammalian primer pairs. BMC Genomics 7:253

Huang, L, Chi, J, Nie, W, Wang, J Yang, F (2006) Phylogenomics of several deer species revealed by comparative chromosome painting with Chinese muntjac paints. Genetica 127:25–33

Huber, S, Bruns, U Arnold, W (2002) Sex determination of red deer using polymerase chain reaction of DNA from feces. Wildlife Soc Bull 30:208–212

Huber, S, Bruns, U, Arnold, W (2003) Genotyping herbivore feces facilitating their further analyses. Wildlife Soc Bull 31:692–697

Hughes, S, Hayden, TJ, Douady, CJ, Tougard, C, Germonpre, M, Stuart, A, Lbova, L, Carden, RF, Hanni, C, Say, L (2006) Molecular phylogeny of the extinct giant deer, *Megaloceros giganteus*. Mol Phylogenet Evol 40:285–291

Jabbour, HN, Clarke, LA, Boddy, S, Pezet, A, Edery, M, Kelly, PA (1996) Cloning, sequencing and functional analysis of a truncated cDNA encoding red deer prolactin receptor: an alternative tyrosine residue mediates beta-casein promoter activation. Mol Cell Endocrinol 123:17–26

Jabbour, HN, Clarke, LA, Bramley, T, Postel-Vinay, MC, Kelly, PA, Edery, M (1998) Alternative splicing of the prolactin receptor gene generates a 1.7 kb RNA transcript that is linked to prolactin function in the red deer testis. J Mol Endocrinol 21:51–59

Jaeger, F, Hecht, W, Herzog, A (1992) Investigation of mitochondrial DNA from Hessian Roe Deer *Capreolus capreolus*. Z Jagdwiss 38:26–33

Janis, CM, Scott, KM (1987) The interrelationships of higher ruminant families, with special emphasis on the members of the Cervidae. Am Mus Novit 2893:1–85

Jepsen, BI, Siegismund, HR, Fredholm, M (2002) Population genetics of native caribou (*Rangifer tarandus groenlandicus*) and the semi-domestic reindeer (*Rangifer tarandus tarandus*) in Southwestern Greenland: evidence of introgression. Conserv Genet 3:401–409

Jewell, JE, Conner, MM, Wolfe, LL, Miller, MW, Williams, ES (2005) Low frequency of PrP genotype 225SF among free-ranging mule deer (*Odocoileus hemionus*) with chronic wasting disease. J Gen Virol 86:2127–34

Johnson, C, Johnson, J, Clayton, M, McKenzie, D, Aiken, J (2003) Prion protein gene heterogeneity in free-ranging white-tailed deer within the chronic wasting disease affected region of Wisconsin. J Wildl Dis 39:576–581

Johnson, C, Johnson, J, Vanderloo, JP, Keane, D, Aiken, JM, McKenzie, D (2006) Prion protein polymorphisms in white-tailed deer influence susceptibility to chronic wasting disease. J Gen Virol 87:2109–2114

Jones, KC, Levine, KF, Banks, JD (2002) Characterization of 11 polymorphic tetranucleotide microsatellites for forensic applications in California elk (*Cervus elaphus canadensis*). Mol Ecol Notes 2:425–427

Karlin, AA, Heidt, GA, Sugg, DW (1989) Genetic variation and heterozygosity in white-tailed deer in southern Arkansas USA. Am Midl Nat 121:273–284

Kennedy, PK, Kennedy, ML, Beck, ML (1987) Genetic variability in white-tailed deer (*Odocoileus virginianus*) and its relationship to environmental parameters and herd origin (Cervidae). Genetica 74:189–201

Kollars, PG, Beck, ML, Mech, SG, Kennedy, PK, Kennedy, ML (2004) Temporal and spatial genetic variability in white-tailed deer (*Odocoileus virginianus*). Genetica 121:269–276

Kraus, F, Miyamoto, MM (1991) Rapid cladogenesis among the pecoran ruminants, evidence from mitochondrial DNA sequences. Syst Zool 40:117–130

Kruuk, LE, Clutton-Brock, TH, Slate, J, Pemberton, JM, Brotherstone, S, Guinness, FE (2000) Heritability of fitness in a wild mammal population. Proc Natl Acad Sci USA 97:698–703

Kruuk, LEB, Clutton-Brock, TH, Rose, KE, Guinness, FE (1999) Early determinants of lifetime reproductive success differ between the sexes in red deer. Proc R Soc Lond B Biol Ser. B. 266:1655–1661

Kucera, TE (1991) Genetic variability in Tule Elk. California Fish Game 77:70–78

Kuehn, R, Haller, H, Schroeder, W, Rottmann, O (2004) Genetic roots of the red deer (*Cervus elaphus*) population in Eastern Switzerland. J Hered 95:136–143

Kuehn, R, Ludt, CJ, Schroeder, W, Rottmann, O (2005) Molecular phylogeny of *Megaloceros giganteus*—the giant deer or just a giant red deer? Zoolog Sci 22:1031–1044

Kuhn, R, Anastassiadis, C, Pirchner, F (1996) Transfer of bovine microsatellites to the cervine (*Cervus elaphus*). Anim Genet 27:199–201

Kuwayama, R, Ozawa, T (2000) Phylogenetic relationships among European red deer, wapiti, and sika deer inferred from mitochondrial DNA sequences. Mol Phylogenet Evol 15:115–123

Leberg, PL, Stangel, PW, Hillstead, HO, Marchinton, RL, Smith, MH (1994) Genetic structure of reintroduced wild turkey and white-tailed deer populations. J Wildlife Manage 58:698–711

Leberg, PL, Ellsworth, DL (1999) Further evaluation of the genetic consequences of translocations on southeastern white-tailed deer populations. J Wildlife Manage 63:327–334

Lee, C, Court, DR, Cho, C, Haslett, JL, Lin, CC (1997) Higherorder organization of subrepeats and the evolution of cervid satellite I DNA. J Mol Evol 44:327–335

Li, YC, Lee, C, Chang, WS, Li, SY, Lin, CC (2002) Isolation and identification of a novel satellite DNA family highly conserved in several Cervidae species. Chromosoma 111:176–183

Liming, S, Yingying, Y, Xingsheng, D (1980) Comparative cytogenetic studies on the red muntjac, Chinese muntjac, and their F1 hybrids. Cytogenet Cell Genet 26:22–27

Lin, CC, Li, YC (2006) Chromosomal distribution and organization of three cervid satellite DNAs in Chinese water deer (*Hydropotes inermis*). Cytogenet Genome Res 114:147–154

Linnell, JC, Cross, TF (1991) The biochemical systematics of red and sika deer (genus Cervus) in Ireland. Hereditas 115:267–273

Lister, AM, Edwards, CJ, Nock, DA, Bunce, M, van Pijlen, IA, Bradley, DG, Thomas, MG, Barnes, I (2005) The phylogenetic position of the 'giant deer' *Megaloceros giganteus*. Nature 438:850–853

Liu, J, Mercer, JM, Stam, LF, Gibson, GC, Zeng, ZB, Laurie, CC (1996) Genetic analysis of a morphological shape difference in the male genitalia of *Drosophila simulans* and *D. mauritiana*. Genetics 142:1129–1145

Lorenzini, R, Mattioli, S, Fico, R (1998) Allozyme variation in native red deer *Cervus elaphus* of Mesola Wood, northern Italy: Implications for conservation. Acta Theriol Suppl. 5:63–74

Lorenzini, R, Lovari, S (2006) Genetic diversity and phylogeography of the European roe deer: the refuge area theory revisited. Biol J Linn Soc 88:85–100

Ludt, CJ, Schroeder, W, Rottmann, O, Kuehn, R (2004) Mitochondrial DNA phylogeography of red deer (*Cervus elaphus*). Mol Phylogenet Evol 31:1064–1083

Maqbool, NJ, Tate, ML, Dodds, KG, Anderson, RM, McEwan, KM, Mathias, HC, McEwan, JC, Hall, RJ (2007) A QTL study of growth and body shape in the inter-species hybrid of Père David's deer (*Elaphurus davidianus*) and red deer (*Cervus elaphus*). Anim Genet 38:270–276

Manlove, MN, Avise, JC, Hillstead, HO, Ramsey, PR, Smith, MH, Straney, DO (1975) Starch gel electrophoresis for the study of population genetics in white-tailed deer. Proc, Ann Conf S E Assoc Game Fish Comm 29:392–403

Marshall, TC, Slate, J, Kruuk, LE, Pemberton, JM (1998) Statistical confidence for likelihood-based paternity inference in natural populations. Mol Ecol 7:639–655

Martinez, JG, Carranza, J, Fernandez-Garcia, JL, Sanchez-Prieto, CB (2002) Genetic variation of red deer populations under hunting exploitation in southwestern Spain. J Mamm 66:1273–1282

Matsunaga, T, Chikuni, K, Tanabe, R, Muroya, S, Nakai, H, Shibata, K, Yamada, J, Shinmura, Y (1998) Determination of mitochondrial cytochrome B gene sequence for red deer (*Cervus elaphus*) and the differentiation of closely related deer meats. Meat Sci 49:379–385

Mikko, S, Andersson, L (1995) Low major histocompatibility complex class II diversity in European and North American moose. Proc Natl Acad Sci USA 92:4259–4263

Miyamoto, MM, Kraus, F, Ryder, OA (1990) Phylogeny and evolution of antlered deer determined from mitochondrial DNA sequences. Proc Natl Acad Sci USA 87:6127–6131

Morin, PA, Wallis, J, Moore, JJ, Woodruff, DS (1994) Paternity exclusion in a community of wild chimpanzees using hypervariable simple sequence repeats. Mol Ecol 3:469–477

Mörsch, G, Leibenguth, F (1993) DNA fingerprinting of the roe deer, *Capreolus capreolus* L. Comp Biochem Physiol B 104:229–233

Nabata, D, Masuda, R, Takahashi, O, Nagata, J (2004) Bottleneck effects on the sika deer *Cervus nippon* population in Hokkaido, revealed by ancient DNA analysis. Zool Sci 21:473–481

Nagata, J, Masuda, R, Kaji, K, Kaneko, M, Yoshida, MC (1998) Genetic variation and population structure of the Japanese sika deer (*Cervus nippon*) in Hokkaido Island, based on mitochondrial D-loop sequences. Mol Ecol 7:871–877

Nagata, J, Masuda, R, Tamate, HB, Hamasaki, S, Ochiai, K, Asada, M, Tatsuzawa, S, Suda, K, Tado, H, Yoshida, MC (1999) Two genetically distinct lineages of the sika deer, *Cervus nippon*, in Japanese islands: comparison of mitochondrial D-loop region sequences. Mol Phylogenet Evol 13:511–519

Neitzl, H (1987) Chromosome evolution of Cervidae: karyotypic and molecular aspects. In: Obe, G., Basler, A. (eds) Basic and Applied Aspects in Cytogenetics. Springer-Verlag, Berlin, pp 90–112

Nicol, AM, Barry, TN (2002) Pastures and forages for deer growth. In: Casey, M. J. (ed) Proc NZ Grasslands Assoc

Nies, G, Zachos, FE, Hartl, GB (2005) The impact of female philopatry on population differentiation in the European roe deer (*Capreolus capreolus*) as revealed by mitochondrial DNA and allozymes. Mamm Biol 70:130–134

Nussey, DH, Coltman, DW, Coulson, T, Kruuk, LE, Donald, A, Morris, SJ, Clutton-Brock, TH, Pemberton, J (2005) Rapidly declining fine-scale spatial genetic structure in female red deer. Mol Ecol 14:3395–405

Nussey, DH, Pemberton, J, Donald, A, Kruuk, LE (2006) Genetic consequences of human management in an introduced island population of red deer (*Cervus elaphus*). Heredity 97:56–65

O'Brien, SJ, Womack, JE, Lyons, LA, Moore, KJ, Jenkins, NA, Copeland, NG (1993) Anchored reference loci for comparative genome mapping in mammals. Nat Genet 3:103–112

O'Connell, A, Denome, RM (1999) Microsatellite locus amplification using deer antler DNA. Biotechniques 26:1086–1088

O'Conner, MJ (2006) Current deer industry issues. Proc Deer Branch of the New Zealand Veterinary Association, pp 26–28

O'Rourke, KI, Baszler, TV, Miller, JM, Spraker, TR, Sadler-Riggleman, I, Knowles, DP (1998) Monoclonal antibody F89/160.1.5 defines a conserved epitope on the ruminant prion protein. J Clin Microbiol 36:1750–1755

O'Rourke, KI, Besser, TE, Miller, MW, Cline, TF, Spraker, TR, Jenny, AL, Wild, MA, Zebarth, GL, Williams, ES (1999) PrP genotypes of captive and free-ranging Rocky Mountain elk (*Cervus elaphus nelsoni*) with chronic wasting disease. J Gen Virol 80 (Pt 10):2765–2769

O'Rourke, KI, Spraker, TR, Hamburg, LK, Besser, TE, Brayton, KA, Knowles, DP (2004) Polymorphisms in the prion precursor functional gene but not the pseudogene are associated with susceptibility to chronic wasting disease in white-tailed deer. J Gen Virol 85:1339–1346

Palmer, MS, Dryden, AJ, Hughes, JT, Collinge, J (1991) Homozygous prion protein genotype predisposes to sporadic Creutzfeldt-Jakob disease. Nature 352:340–242

Pemberton, JM, Smith, RH (1985) Lack of biochemical polymorphism in British fallow deer. Heredity 55 (Pt 2): 199–207

Pemberton, JM, Slate, J, Bancroft, DR, Barrett, JA (1995) Non-amplifying alleles at microsatellite loci: a caution for parentage and population studies. Mol Ecol 4:249–52

Pemberton, JM, Smith, JA, Coulson, TN, Marshall, TC, Slate, J, Paterson, S, Albon, SD, Clutton-Brock, TH (1996) The maintenance of genetic polymorphism in small island populations: large mammals in the Hebrides. Philos Trans R Soc Lond B Biol Sci 351:745–752

Pemberton, JM, Coltman, DW, Coulson, TN, Paterson, S, Smith, JA, Slate, J, Marshall, TC (1998) Microsatellites as markers of fitness in free-living populations. Anim Genet 29 Suppl 1:1–7

Pemberton, JM, Slate, J (1998) Genetic studies of wild deer populations: a technical revolution. In: Milne, J. A. (ed) Proc 3rd Internat Congr Biol Deer. Macauley Land Use Research Institute and Moredun Research Institute

Pepin, L, Amigues, Y, Lepingle, A, Berthier, JL, Bensaid, A, Vaiman, D (1995) Sequence conservation of microsatellites between *Bos taurus* (cattle), *Capra hircus* (goat) and related species. Examples of use in parentage testing and phylogeny analysis. Heredity 74 (Pt 1):53–61

Pitra, C, Fickel, J, Meijaard, E, Groves, PC (2004) Evolution and phylogeny of old world deer. Mol Phylogenet Evol 33:880–95

Poetsch, M, Seefeldt, S, Maschke, M, Lignitz, E (2001) Analysis of microsatellite polymorphism in red deer, roe deer, and fallow deer — possible employment in forensic applications. Forensic Sci Int 116:1–8

Polziehn, RO, Strobeck, C (1998) Phylogeny of wapiti, red deer, sika deer, and other North American cervids as determined from mitochondrial DNA. Mol Phylogenet Evol 10:249–258

Polziehn, RO, Hamr, J, Mallory, FF, Strobeck, C (2000) Microsatellite analysis of North American wapiti (*Cervus elaphus*) populations. Mol Ecol 9:1561–1576

Polziehn, RO, Strobeck, C (2002) A phylogenetic comparison of red deer and wapiti using mitochondrial DNA. Mol Phylogenet Evol 22:342–356

Purdue, JR, Smith, MH, Patton, J (2000) Female philopatry and extreme spatial genetic heterogeneity in white-tailed deer. J Mamm 81:179–185

Ramsey, PR, Avise, JC, Smith, MH, Urbston, DF (1979) Biochemical variation and genetic heterogeneity in South Carolina USA deer populations. J Wildlife Manage 43:136–142

Randi, E, Apollonio, M (1988) Low biochemical variability in European fallow deer (*Dama dama L.*): natural bottlenecks and the effects of domestication. Heredity 61 (Pt 3):405–410

Randi, E, Mucci, N, Pierpaoli, M, Douzery, E (1998a) New phylogenetic perspectives on the Cervidae (Artiodactyla) are provided by the mitochondrial cytochrome b gene. Proc Biol Sci 265:793–801

Randi, E, Pierpaoli, M, Danilkin, A (1998b) Mitochondrial DNA polymorphism in populations of Siberian and European roe deer (*Capreolus pygargus and C. capreolus*). Heredity 80 (Pt 4):429–437

Randi, E, Alves, PC, Carranza, J, Milosevic-Zlatanovic, S, Sfougaris, A, Mucci, N (2004) Phylogeography of roe deer (*Capreolus capreolus*) populations: the effects of historical genetic subdivisions and recent nonequilibrium dynamics. Mol Ecol 13:3071–3083

Roed, KH, Whitten, KR (1986) Transferrin variation and evolution of Alaskan USA reindeer and caribou *Rangifer tarandus* L. Rangifer S1:247–252

Roed, KH, Ferguson, MAD, Crete, M, Bergerud, TA (1991) Genetic variation in transferrin as a predictor for differentiation and evolution of caribou from eastern Canada. Rangifer 11:65–74

Roed, KH (1998) Microsatellite variation in Scandinavian Cervidae using primers derived from Bovidae. Hereditas 129:19–25

Roed, KH, Midthjell, L (1998) Microsatellites in reindeer, *Rangifer tarandus*, and their use in other cervids. Mol Ecol 7:1773–1776

Ryman, N, Reuterwall, C, Nygren, K, Nygren, T (1980) Genetic variation and differentiation in Scandinavian moose (*Alces alces*): are large mammals monomorphic? Evolution 34:1037–1049

Say, L, Naulty, F, Hayden, TJ (2003) Genetic and behavioural estimates of reproductive skew in male fallow deer. Mol Ecol 12:2793–800

Scandura, M, Tiedemann, R, Apollonio, M, Hartl, GB (1998) Genetic variation in an isolated Italian population of fallow deer *Dama dama* as revealed by RAPD-PCR. Acta Theriol Suppl 5:163–169

Schatzl, HM, Wopfner, F, Gilch, S, von Brunn, A, Jager, G (1997) Is codon 129 of prion protein polymorphic in human beings but not in animals? Lancet 349:1603–1604

Schreiber, A, Klein, F, Lang, G (1994) Transferrin polymorphism of red deer in France: evidence for spatial genetic microstructure of an autochthonous herd. Genet Sel Evol 26:187–203

Schreiber, A, Fakler, P (1996) NADH diaphorase polymorphism in European fallow deer. Biochem Genet 34:61–65

Scribner, KT, Smith, MH, Garrott, RA, Carpenter, LH (1991) Temporal spatial and age-specific changes in genotypic composition of mule deer. J Mamm 72:126–137

Sheffield, SR, Morgan, RP, Feldhamer, GA, Harman, DM (1985) Genetic variation in white-tailed deer *Odocoileus-virginianus* populations in western Maryland USA. J Mamm 66:243–255

Slate, J, Coltman, DW, Goodman, SJ, MacLean, I, Pemberton, JM, Williams, JL (1998) Bovine microsatellite loci are highly conserved in red deer (*Cervus elaphus*), sika deer (*Cervus nippon*) and Soay sheep (*Ovis aries*). Anim Genet 29:307–315

Slate, J, Marshall, T, Pemberton, J (2000) A retrospective assessment of the accuracy of the paternity inference program CERVUS. Mol Ecol 9:801–808

Slate, J, Van Stijn, TC, Anderson, RM, McEwan, KM, Maqbool, NJ, Mathias, HC, Bixley, MJ, Stevens, DR, Molenaar, AJ, Beever, JE, Galloway, SM, Tate, ML (2002a) A deer (subfamily Cervinae) genetic linkage map and the evolution of ruminant genomes. Genetics 160:1587–1597

Slate, J, Visscher, PM, MacGregor, S, Stevens, D, Tate, ML, Pemberton, JM (2002b) A genome scan for quantitative trait loci in a wild population of red deer (*Cervus elaphus*). Genetics 162:1863–1873

Slate, J, Phua, SH (2003) Patterns of linkage disequilibrium in mitochondrial DNA of 16 ruminant populations. Mol Ecol 12:597–608

Sorin, AB (2004) Paternity assignment for white-tailed deer (*Odocoileus virginianus*): mating across age classes and multiple paternity. J Mamm 85:356–362

Spraker, TR, Miller, MW, Williams, ES, Getzy, DM, Adrian, WJ, Schoonveld, GG, Spowart, RA, O'Rourke, KI, Miller, JM, Merz, PA (1997) Spongiform encephalopathy in free-ranging mule deer (*Odocoileus hemionus*), white-tailed deer (*Odocoileus virginianus*) and Rocky Mountain elk (*Cervus elaphus nelsoni*) in northcentral Colorado. J Wildl Dis 33:1–6

Strandgaard, H, Simonsen, V (1993) Genetic differentiation in populations of red deer, *Cervus elaphus*, in Denmark. Hereditas 119:171–177

Stratil, A, Glasnak, V, Bobak, P, Cizova, D, Gabrisova, E, Kalab, P (1990) Variation of some serum proteins in red deer, *Cervus elaphus* L. Anim Genet 21:285–293

Stroehlein, H, Herzog, S, Hecht, W, Herzog, A (1993) Biochemical genetic description of German and Swiss populations of red deer. Acta Theriol 38:153–161

Supattapone, S, Muramoto, T, Legname, G, Mehlhorn, I, Cohen, FE, DeArmond, SJ, Prusiner, SB, Scott, MR (2001) Identification of two prion protein regions that modify scrapie incubation time. J Virol 75:1408–1413

Talbot, J, Haigh, J, Plante, Y (1996) A parentage evaluation test in North American elk (wapiti) using microsatellites of ovine and bovine origin. Anim Genet 27:117–119

Tamate, HB, Tsuchiya, T (1995) Mitochondrial DNA polymorphism in subspecies of the Japanese Sika deer, *Cervus nippon*. J Hered 86:211–215

Tate, ML, McEwan, KM (1992) Genetic polymorphism of erythrocyte diaphorase in red deer, *Cervus elaphus* L. Anim Genet 23:449–52

Tate, ML, Mathias, HC, Fennessy, PF, Dodds, KG, Penty, JM, Hill, DF (1995) A new gene mapping resource: interspecies hybrids between Pere David's deer (*Elaphurus davidianus*) and red deer (*Cervus elaphus*). Genetics 139:1383–1391

Tate, ML, Goosen, GJ, Patene, H, Pearse, AJ, McEwan, KM, Fennessy, PF (1997) Genetic analysis of Pere David's x red deer interspecies hybrids. J Hered 88:361–365

Tate, ML, Anderson, RM, McEwan, KM, Goosen, GJ, Pearse, AJ (1998) Genetic analysis of farmed deer hybrids. Acta Vet Hung 46:329–340

Templeton, JW, Sharp, RM, Williams, J, Davis, D, Harmel, D, Armstrong, B, Wardroup, S (1982) Single dominant major gene effect on the expression of antler point number in the white-tailed deer. In: Brown, R. D. (ed) Antler Development in Cervidae. Kingsville, Texas, Caesar Kleberg Wildlife Research Institute

Thevenon, S, Bonnet, A, Claro, F, Maillard, JC (2003) Genetic diversity analysis of captive populations: the Vietnamese sika deer (*Cervus nippon pseuaxis*) in zoological parks. Zoo Biol 22:465–475

Thevenon, S, Thuy, LT, Ly, LV, Maudet, F, Bonnet, A, Jarne, P, Maillard, JC (2004) Microsatellite analysis of genetic diversity of the Vietnamese sika deer (*Cervus nippon pseudaxis*). J Hered 95:11–18

Van Den Bussche, RA, Ross, TG, Hoofer, SR (2002) Genetic variation at a major histocompatibility locus within and among populations of white-tailed deer (*Odocoileus virginianus*). J Mamm 83:31–39

Vankova, D, Bartos, L, Cizova-Schroffelova, D, Nespor, F, Jandurova, O (2001) Mother-offspring bonding in farmed red deer: accuracy of visual observation verified by DNA analysis. Appl Anim Behav Sci 73:157–165

Vial, L, Maudet, C, Luikart, G (2003) Thirty-four polymorphic microsatellites for European roe deer. Mol Ecol Notes 3:523–527

Volmer, K, Hecht, W, Herzog, A, Faltings, V (1995) Genetic investigations of Roe Deer. Z Jagdwiss 41:241–247

Wagener, A, Blottner, S, Goritz, F, Fickel, J (2000) Detection of growth factors in the testis of roe deer (*Capreolus capreolus*). Anim Reprod Sci 64:65–75

Wallis, OC, Bill, LJ, Burt, EJ, Ellis, SA, Wallis, M (2006) Polymorphism of the growth hormone gene of red deer (*Cervus elaphus*). Gen Comp Endocrinol 146:180–185

Wang, M, Schreiber, A (2001) The impact of habitat fragmentation and social structure on the population genetics of roe deer (*Capreolus capreolus L.*) in Central Europe. Heredity 86:703–715

Wang, M, Lang, G, Schreiber, A (2002) Temporal shifts of DNA-microsatellite allele profiles in roe deer (*Capreolus capreolus L.*) within three decades. J Zool Syst Evol Res 40:232–236

Wehner, J, Mueller, HP, Orthwein, L (1991) Investigations on the genetic variability of fallow deer in North Rhine-Westphalia Germany. Z Jagdwiss 37:69–77

Whitehead, GK (1993) The Whitehead Encyclopaedia of Deer. Swan Hill Press, Shrewsbury

Wiehler, J, Tiedemann, R (1998) Phylogeography of the European roe deer *Capreolus capreolus* as revealed by sequence analysis of the mitochondrial control region. Acta Theriol Suppl 5:187–197

Williams, CL, Serfass, TL, Cogan, R, Rhodes, OE (2002) Microsatellite variation in the reintroduced Pennsylvania elk herd. Mol Ecol 11:1299–1310

Williams, ES, Young, S (1982) Spongiform encephalopathy of Rocky Mountain elk. J Wildl Dis 18:465–471

Wilson, GA, Strobeck, C, Wu, L, Coffin, JW (1997) Characterization of microsatellite loci in caribou *Rangifer tarandus*, and their use in other artiodactyls. Mol Ecol 6:697–699

Wilson, PJ, Grewal, S, Rodgers, A, Rempel, R, Saquet, J, Hristienko, H, Burrows, F, Peterson, R, White, BN (2003) Genetic variation and population structure of moose (*Alces alces*) at neutral and functional DNA loci. Can J Zool 81:670–683

Wu, H, Wan, QH, Fang, SG, Zhang, SY (2005) Application of mitochondrial DNA sequence analysis in the forensic identification of Chinese sika deer subspecies. Forensic Sci Int 148:101–105

Wu, HL, Fang, SG (2005) Mitochondrial DNA genetic diversity of black muntjac (*Muntiacus crinifrons*), an endangered species endemic to China. Biochem Genet 43:407–416

Xia, S, Zou, F, Bisong, Y (2006) Six microsatellite loci in forest musk deer, *Moschus berezovskii*. Mol Ecol Notes 6:113–115

Xu, S (2003) Theoretical basis of the Beavis effect. Genetics 165:2259–2268

Yokohama, M, Yamazaki, T, Ishijima, Y (1995) Erythrocyte protein types and restriction endonuclease cleavage of mitochondrial DNA in the Japanese sika deer (*Cervus nippon yesoensis*). J Agric Sci Tokyo 39:303–307

Zachos, FE, Hartl, GB, Apollonio, M, Reutershan, T (2003) On the phylogeographic origin of the Corsican red deer (*Cervus elaphus corsicanus*): evidence from microsatellites and mitochondrial DNA. Mamm Biol 68:284–298

Zachos, FE, Hmwe, SS, Hartl, GB (2006) Biochemical and DNA markers yield strikingly different results regarding variability and differentiation of roe deer (*Capreolus capreolus*, Artiodactyla: Cervidae) populations from northern Germany. J Zool Sci 44:167–174

Zeng, ZB, Liu, J, Stam, LF, Kao, CH, Mercer, JM, Laurie, CC (2000) Genetic architecture of a morphological shape difference between two Drosophila species. Genetics 154:299–310

Zou, F, Yue, B, Xu, L, Zhang, Y (2005) Isolation and characterization of microsatellite loci from forest musk deer (*Moschus berezovskii*). Zoo Sci 22:593–598

CHAPTER 5

Poultry

Michael N. Romanov[1]([✉]), Alexei A. Sazanov[2], Irina Moiseyeva[3], and Aleksandr F. Smirnov[4]

[1] CRES - Conservation and Research for Endangered Species, Zoological Society of San Diego, Arnold and Mabel Beckman Center for Conservation Research, 15600 San Pasqual Valley Road Escondido, CA 92027-7000, USA, mromanov@sandiegozoo.org

[2] All-Russian Institute of Animal Genetics and Breeding, Russian Academy of Agricultural Science, Moskovskoye shosse 55A, St Petersburg, Pushkin 189620, Russia

[3] N.I. Vavilov Institute of General Genetics, Russian Academy of Sciences, Gubkin Street 3, Moscow, GSP-1, 119991, Russia

[4] Biological Research Institute, St Petersburg State University, Oranienbaumskoye shosse 2, St Petersburg, Stary Petergof 198504, Russia

5.1
Introduction

5.1.1
Brief History and Zoological Description

From the time of initial domestication of wild birds, poultry have served humans as a source of food and a subject of cultural use, similar to other livestock species. The role of poultry production in global food provision has been steadily growing since the nineteenth century. Different needs of humans have led to a rise in poultry breeding and use of pure (fancy) breeds, indigenous populations, laboratory lines, and commercial poultry. Being notable for high efficiency and rapid development dynamic, the poultry industry now exceeds other livestock sectors in production growth rate and efficacy. Intensification of commercial poultry production has placed emphasis on selection and improvement of breeds and strains and on the development of novel lines and crosses. This, in turn, has required new genetic and selection approaches and technologies, and utilization of genetic resources adapted to variable and diverse specific environments. The importance of a deeper knowledge of avian biology including heredity, variation, and genomics is paramount. Substantial progress in poultry production can be achieved through advances in several areas, including selection, veterinary, nutrition, avian genetics, and genomics.

Avian species share a common ancestor with humans. The split between synapsids (mammals and their extinct ancestors) and diapsids (reptiles) occurred around 350 million years ago (MYA). Birds are believed to arise from therapod dinosaurs about 150 MYA. The origin of the whole *Galliformes* order is placed in the late Cretaceous at about 90 MYA, while the junglefowl genus, *Gallus*, evolved among the land fowl about 8 MYA (van Tuinen and Dyke 2004). Man began domestication of chickens in Southeast Asia and adjacent areas 8,000–10,000 years ago (Fig. 1). Later, waterfowl species including geese and ducks were domesticated. The turkey and Muscovy duck were domesticated in the New World, and some other birds (e.g., guinea fowl, Japanese quail, and ostrich) were also subject to domestication. Today's poultry breeds and their wild progenitors are separated across globally established poultry meat and egg industries.

There have been four stages of poultry history that have affected genetic diversity, leading to the chickens that exist today (Crawford 1990). The first stage was the process of domestication, involving selection for tameness, changes in body size, and accumulation of morphological and color variants. The second stage was diffusion outward from centers of domestication; genetic drift, migration, and isolation were major genetic forces leading to development of distinctive regional types. The third stage was the "hen craze era" late in the nineteenth century, when nearly all present-day breeds and varieties were created. The fourth stage is in place now, where multinational corporations breed and distribute egg and meat stocks that have remarkable productivity, but are derived from a narrow genetic base (Crawford 1995). During the fourth stage, anthropogenic factors have become more and more important in evolution and development of domestic fowl. Factors

Genome Mapping and Genomics in Animals, Volume 3
Genome Mapping and Genomics in Domestic Animals
N.E. Cockett, C. Kole (Eds.)
© Springer-Verlag Berlin Heidelberg 2009

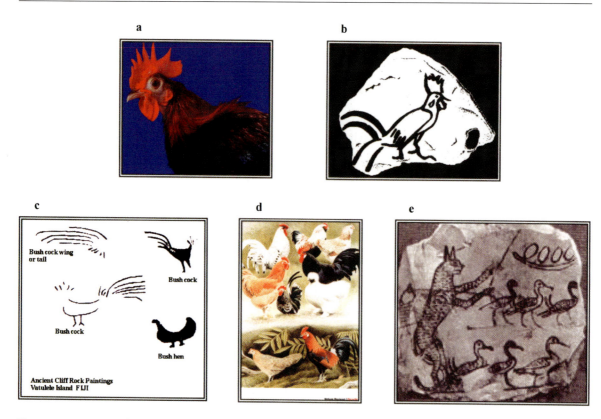

Fig. 1 Domestication of avian species. (**a**) A red junglefowl rooster (*Gallus gallus*), the wild chicken progenitor species (from Pisenti et al. 1999; http://www.grcp.ucdavis.edu/publications/doc20/front.pdf). (**b**) An egg-type cock painted on an ostracon from the tomb of Tutankhamen, Egypt, 1338 BC (Carter and Mace 1923–1933); published by Cassell Plc, currently a division of The Orion Publishing Group (London); attempts at tracing the copyright holder of the image were unsuccessful). (**c**) Ancient Pacific cliff rock paintings, Vatulele Island, Fiji (Ewins 1995; http://www.justpacific.com/fiji/fijianart/cliffart/cliffpaintings.pdf). (**d**) Red junglefowl (*shown at the bottom*) and several domestic chicken breeds (adapted with permission from Macmillan Publishers Ltd: Nature, Andersson 2001, © 2001). (**e**) A typical depiction of domesticated geese in ancient Egypt

such as civilization development, historical events, technological progress (Altukhov 2004), changes in climate, and extreme natural disasters greatly influence poultry genetic variability and may even lead to extinction.

Well-known poultry species include chickens, ducks, geese, turkeys, guinea fowl, quail and pigeons. They play an important role in the world's economy and provide a valuable protein source for people in both developing and developed countries. Birds are often raised as scavengers, i.e., at little cost, in areas where cattle cannot survive, such as those infested by the tsetse fly (*Glossina* spp.). Ostriches, emus, rheas, and cassowaries are all at various stages of domestication for their skins, meat, and other products (Scherf 2000).

Poultry and other avian species raised and kept by man belong to ten orders: *Galliformes, Anseriformes, Columbiformes, Passeriformes, Ciconiiformes, Pelecaniformes, Psittaciformes, Struthioniformes, Rheiformes,* and *Casuariiformes* (Table 1). Representatives of the most useful family of birds are in the *Phasianidae* family, a widely dispersed group of the order *Galliformes*. A more detailed zoological description and origin of the chicken will be narrated below. The turkey will be separately described in Chapter 6 of this volume. Details on other domesticated species of birds (Fig. 2) can be found elsewhere (Barloy 1978; Brothwell and Brothwell 1998; Brown 1929; Crawford 1990, 1995; Darwin 1868; del Hoyo et al. 1992–1996; Dembeck 1965; Hyams 1972; Mason 1984; Petrov 1962; Scherf 2000; Zeuner 1963).

Chapter 5 **Poultry** 77

Fig. 2 Major domesticated avian species other than chicken and turkey. (**a**) A male of the mallard (*Anas platyrhynchos*), the progenitor of domestic ducks. (**b**) A domesticated American anseriform species of Muscovy ducks (*Cairina moschata*). (**c**) A flock of domestic Russian geese (photograph courtesy of Annette Güntherodt, Beberstedt, Germany; © 2004 From Encyclopedia of Animal Science by W.G. Pond and A.W. Bell (ed). Reproduced with permission of Routledge/Taylor & Francis Group, LLC). (**d**) A semidomesticated North American Canada goose (*Branta canadensis*). (**e**) Mute swan (*Cygnus olor*). (**f**) Japanese quail (*Coturnix japonica*). (**g**) Ring-necked pheasant (*Phasianus colchicus*). (**h**) Emu (*Dromaius novaehollandiae*). **a, b**, and **d–h**, USDA Image Gallery (http://www.ars.usda.gov/is/graphics/photos/) and Photography Center (http://www.usda.gov/oc/photo/opchomea.htm)

5.1.2 Chickens

Taxonomy and Wild Ancestors

The chicken is a member of the class *Aves*, subclass *Neornithes*, superorder *Neognathae*, order *Galliformes*, family *Phasianidae*, subfamily *Phasianinae*, genus *Gallus*. Closely related genera are *Meleagris* (turkey), *Pavo* (peafowl), and *Phasianus* (pheasant). Domestic chickens are descendants of junglefowl that now inhabit a wide crescent stretching from Pakistan to Indonesia including India, Indo-China, and South China, as well as the Philippines.

Four junglefowl species are known, including **red junglefowl** (*G. gallus*; Fig. 1a, d, and 3), **gray junglefowl** (*G. sonneratii*), **Ceylon junglefowl** (*G. lafayettei*),

Table 1 Taxonomy of major domesticated, semidomesticated, and caged avian species

Order	Family	Genus	Species	Distribution
Anseriformes	*Anatidae*	*Anser*	*A. anser* (graylag goose)[a]	Eurasia
,,	,,	,,	*A. cygnoides* (swan goose)[a]	Asia
,,	,,	,,	*A. fabalis* (bean goose)[b]	Eurasia
,,	,,	,,	*A. albifrons* (white-fronted goose)[b]	Northern Hemisphere
,,	,,	,,	*A. indicus* (bar-headed goose)[b]	Asia
,,	,,	,,	*A. erythropus* (lesser white-fronted goose)[b]	Eurasia
,,	,,	*Branta*	*B. canadensis* (Canada goose)[c]	North America
,,	,,	*Cygnus*	*C. olor* (mute swan)[c]	Eurasia
,,	,,	*Alopochen*	*A. aegyptiacus* (Egyptian goose)[c]	Africa
,,	,,	*Cairina*	*C. moschata* (Muscovy duck)[a]	Tropical America
,,	,,	*Anas*	*A. platyrhynchos* (mallard)[a]	Northern Hemisphere
Galliformes	*Phasianidae*	*Meleagris*	*M. gallopavo* (turkey)[a]	North America
,,	,,	*Gallus*	*G. gallus* (red junglefowl)[a]	Southeast Asia
,,	,,	,,	*G. sonneratii* (gray junglefowl)	Southwest India
,,	,,	,,	*G. lafayettei* (Ceylon junglefowl)	Ceylon
,,	,,	,,	*G. varius* (green junglefowl)	Java
,,	,,	*Pavo*	*P. cristatus* (Indian peafowl)[a]	India
,,	,,	*Phasianus*	*P. colchicus* (ring-necked pheasant)[c]	Eurasia
,,	,,	,,	*P. versicolor* (green pheasant)[c]	Japan
,,	,,	*Lophura*	*L. nycthemera* (silver pheasant)	Southeast Asia
,,	,,	*Chrysolophus*	*Ch. pictus* (golden pheasant)	China
,,	,,	*Perdix*	*P. perdix* (gray partridge)[c]	Eurasia
,,	,,	*Coturnix*	*C. coturnix* (common quail)[c]	Eurasia, Africa
,,	*Phasianidae*	,,	*C. japonica* (Japanese quail)[a]	Asia
,,	*Numididae*	*Numida*	*N. meleagris* (helmeted guinea fowl)[a]	Africa
,,	*Odontophoridae*	*Colinus*	*C. virginianus* (northern bobwhite)[c]	North America
Passeriformes	*Fringillidae*	*Serinus*	*S. canaria* (island canary)[a]	Madeira, Azores, Canary Islands
,,	*Estrildidae*	*Taeniopygia*	*T. guttata* (zebra finch)	Australia
Columbiformes	*Columbidae*	*Columba*	*C. livia* (rock pigeon)[a]	Eurasia, Africa
,,	,,	*Streptopelia*	*S. roseogrisea* (S. risoria) (African collared dove)[a]	Africa, North America
Pelecaniformes	*Phalacrocoracidae*	*Phalacrocorax*	*P. carbo* (great cormorant)[c]	Asia
Ciconiiformes	*Ardeidae*	*Egretta*	*E. garzetta* (little egret)[c]	Eurasia, Africa, Australia
Psittaciformes	*Psittacidae*	*Psittacula*	*P. krameri* (rose-ringed parakeet)	Africa, Asia
,,	,,	*Melopsittacus*	*M. undulates* (budgerigar)[c]	Australia
Struthioniformes	*Struthionidae*	*Struthio*	*S. camelus* (ostrich)[a]	Africa
Rheiformes	*Rheidae*	*Rhea*	*R. americana* (greater rhea)[c]	South America
,,	,,	*Pterocnemia*	*P. pennata* (lesser rhea)[c]	South America
Casuariiformes	*Dromaiidae*	*Dromaius*	*D. novaehollandiae* (emu)[c]	Australia
,,	*Casuariidae*	*Casuarius*	*C. spp.* (cassowary)[c]	New Guinea, Australia

[a] Domesticated species

[b] Suggested contribution to polyphyletic origin of a domesticated form

[c] Semidomesticated species

and **green junglefowl** (*G. varius*). The red junglefowl, in turn, is subdivided into five subspecies depending on geographic distribution, variation in size of facial wattles and combs, and length and color of the neck hackles in males: Cochin-Chinese (*G. g. gallus*), Burmese (*G. g. spadiceus*), Tonkinese (*G. g. jabouillei*), Indian (*G. g. murghi*), and Javan (*G. g. bankiva*).

The natural habitat of red jungle fowl varies, including most types of forests present in Southeast Asia and in other territories of Asia, field edges, groves, and scrubland. The junglefowl is a highly adaptable species and can thrive in many habitats from sea level to 2,000 meters above sea level. Most junglefowl are found in damp forests, secondary growth, dry scrub, bamboo groves, and small woods. Although not rare, the species is under some hunting pressure (Scherf 2000).

Origin of Domestic Fowl — Monophyletic, Polyphyletic or Intermediate?

Because of varying opinions of zoologists, naturalists, geneticists, and other specialists, there is great interest in exploring the biology of junglefowl species and the origin of the domestic fowl (Moiseyeva et al. 2003). Beginning with Charles Darwin's fundamental work on this subject published in *The Variation of Animals and Plants Under Domestication* (1868), many investigations have been devoted to the specific features of junglefowl. The widely spread species *G. gallus* has been most fully described for discrete morphological and metric quantitative traits (Darwin 1868; Beebe 1918–1922; Delacour 1977; Nishida et al. 1983, 1985a, b; Moiseyeva and Volokhovich 1987; Moiseyeva et al. 1994) and, over last four decades, for biochemical (Baker 1964, 1968; Baker and Manwell 1972; Moiseyeva et al. 1994) and molecular (Siegel et al. 1992; Akishinonomiya et al. 1994, 1996; Romanov and Weigend 2001b; Hillel et al. 2003) markers.

Comparisons of four species of genus *Gallus* and chicken breeds indicate that *G. gallus* (red junglefowl) is the closest to chickens for most traits. This may be evidence of the monophyletic origin of chickens (Darwin 1868; Tegetmeier 1873; Beebe 1918–1922; Ivanov 1924; West and Zhou 1989). Another group of scholars (Dixon 1848; Abozin 1885; Brown 1906; Finsterbusch 1929; Hutt 1949; Smith and Daniel 1975; Plant 1984, 1986) adheres to the theory of polyphyletic origin referring to the fact that some characters known in chickens are absent in *G. gallus*, but present in other

Fig. 3 A lateral view of red junglefowl cock (*G. g. gallus*). An exhibit of the State Darwin Museum, Moscow, Russia. Photograph courtesy of A.A. Nikiforov (Altukhov 2004)

wild species or in extinct progenitor(s). Several investigators support an intermittent point of view, that is, between strong monophyletists and strong polyphyletists, considering *G. gallus* the major ancestor but not excluding small participation of other species in chicken domestication (Baker et al. 1971; Kogan 1979; Crawford 1990; Stevens 1991; Altukhov 2004).

Noteworthy, most genetic studies have relied upon *G. gallus* specimens bred by fancy breeders or in zoos, which may be contaminated with domestic chicken genes. Wild populations in their natural habitats are also quite often mated with village chickens and produce offspring that differ from the pure *G. gallus* (Beebe 1918–1922; Brisbin 1997). This situation leads to an overestimate of kinship between *G. gallus* and chicken breeds as compared with the three other wild species that do not always produce fertile hybrids with domestic chickens.

Based on the comparison of mitochondrial DNA (mtDNA) D-loop sequences, Akishinonomiya et al. (1994, 1996) stated that only one subspecies, *G. g. gallus*, contains all the biodiversity of chickens. Niu et al. (2002) sequenced the first 539 bases of the mtDNA D-loop region in six Chinese local chicken breeds and compared that data to sequences of four juglefowl species in GenBank. The four species of the genus *Gallus* significantly differed from each other, and the Chinese native chickens were closest to the red junglefowl in Thailand and its adjacent regions, suggesting that the Chinese domestic fowl probably originated from the red junglefowl in these regions. It was suggested that the two subspecies of Thailand, *G. g. gallus* and *G. g. spadiceus*, form one subspecies because of their similarity to each other. However, these findings did not provide absolute proof because not all five *G. gallus* subspecies were surveyed and only one or two representatives per taxon were compared. Unfortunately, there was no mention in those studies of the correspondence of the sampled birds to the standards of a species or breeds. In addition, the authors did not acknowledge that local residents often mate different *Gallus* species with domestic chickens. Phylogenetic relationships of species in the genus *Gallus* could not be validated in this study due to insufficient statistical analyses (Nishibori et al. 2005).

On the other hand, supporters of the polyphyletic origin of chickens have not presented any archeological evidence about extinct ancestor(s), at least for heavy Asiatic meat-type breeds. Lack of many traits

in *G. gallus* currently in chickens is not a conclusive argument for the polyphyletic origin. Domestic animal species often have significant breed diversity compared to their ancestral forms. One of the strong arguments in favor of the polyphyletic origin is the observation of fertile progeny produced from a wild fowl species other than *G. gallus* and domestic chickens or red junglefowl (Darwin 1868; Mason 1984; Johnsgard 1999), although hybrid stocks usually terminate in the second and backcross (BC) generations (Crawford 1995).

Remarkably, Darwin (1868) did express an opinion that "we have not such good evidence with fowls as with pigeons, of all the breeds having descended from a single primitive stock." In recent studies, Nishibori et al. (2005) provided evidence at the molecular level for hybridization of species in the genus *Gallus* except *G. varius*. They determined sequences of the whole mtDNA and two genomic sequences (intron) of ornithine carbamoyltransferase and four chicken repeat one elements for the species in the genus *Gallus*. According to the mtDNA sequence-based phylogenetic analyses, two gray junglefowls formed a cluster with red junglefowls and chicken, whereas one gray junglefowl was located in a remote position close to Ceylon junglefowl. The analyses based on the nuclear sequences resulted in alternative clustering of gray junglefowl alleles with those of Ceylon junglefowl and with those of red junglefowls and chicken. Red junglefowl and chicken alleles were also alternatively clustered. These findings could strongly suggest interspecies hybridizations between gray junglefowl and red junglefowl/chicken and between gray junglefowl and Ceylon junglefowl. A question was raised whether these three *Gallus* species are actually subspecies, but examination of more individuals of each species is needed to validate this hypothesis.

Another important quesion is what chicken breeds, i.e., evolutionary branch(es) of chickens, are most closely related to *G. gallus* and, therefore, which domesticated fowl are the most ancient. Studies to address these issues may be complicated because of possible contamination of the wild populations with domestic genes, usage of different markers, different sets of breeds across studies, genetic variation within chicken breeds, crossbreeding used for breed development, and different statistical methods for data analyses. A systems approach for addressing this problem has been proposed (Moiseyeva 1998, 2003).

Time of Origin and Evolutionary Lineages

Darwin (1868) was uncertain about the exact center(s) of chicken origin, saying that "all our breeds are probably the descendants of the Malayan or Indian variety" of the red junglefowl. After Darwin, various authors named different geographical regions as the center of origin, including Burma (Peters 1913), India (Ivanov 1924; Wood-Gush 1959; Zeuner 1963), China (Ho 1977), Southeast Asia (West and Zhou 1989), and Thailand (Akishinonomiya et al. 1994, 1996). The origin of domestic chickens is dated approximately 6000–8000 BC (Ho 1977; Plant 1986; West and Zhou 1989; Crawford 1995). Scherf (2000) proposes that "the first domestication occurred in Southeast Asia some time prior to 6000 BC, before introduction of chickens into China." This was followed by the spread of chickens in the ancient times to the west, north and east including the Mediterranean and, more recently, the Pacific islands (Fig. 1b, c). Around 2500–2100 BC, chickens might have been separately domesticated in the Indus Valley (Pakistan) or diffused over there from Southeast Asia (Scherf 2000). The first domesticated chickens were probably assigned cultural or religious significance or were used for cockfighting, although the possibility of gathering eggs from wild and domestic chickens at the early stage of domestication is possible (Petrov 1941). In ancient beliefs, the rooster symbolized a clock, sun, fire, courage, or fecundity; the hen was related to maternity, housekeeper, and economy; and the egg was associated with development of life. Cockerels and hens were also used in predictions and divinations.

Chickens spread rapidly and their meat and eggs became highly appreciated as an important source of animal protein (Scherf 2000). As early as the times of Plato and Aristotle, chicken varieties were discernible (Moiseyeva and Lisichkina 1996; Scherf 2000). There is strong cumulative evidence that chickens were already present in the Americas at the time of Spanish discovery, and they came from across the Pacific (Carter 1971). However, acceptance of this point of view awaits the discovery of bones from securely dated pre-Columbian sites (Crawford 1995).

Despite the fact that chickens have been subject to domestication for less than 10,000 years, the amount of phenotypic variation accumulated over time is surprising (Jensen 2005). To date, four major evolutionary lineages can be observed among various chicken breeds selected by man (Moiseyeva et al. 2003; Fig. 4): egg-type, or Mediterranean, game, meat-type, and Bantam. Early domesticated chickens were small and shared morphological characteristics with modern egg-type fowl of Mediterranean roots and/or with true Bantams. This hypothesis, developed by Moiseyeva et al. (2003), is in agreement with ancient depictions of domestic chickens, which had the egg-layer morphological type (Brown 1929; Fig. 1b). The game chicken breeds might descend directly from the red junglefowl or from egg-type domestic birds. The meat-type breeds represent the latest chicken lineage and were probably selected from game breeds.

Based on analyses of biological, historical, archeological, etymological, ethnological, and ethnographical evidences, the domestication process of wild forms into chickens might have occurred independently in

Fig. 4 Morphotypological forms of the domestic fowl and possible major evolutionary lineages from the main wild ancestor, *Gallus gallus* (1), to egg-type (2), Bantam-type (3), game (4), and meat-type (5) breeds (Altukhov 2004)

several parts of Asian region and at different times (West and Zhou 1989; Crawford 1995; Moiseyeva 1998; Moiseyeva et al. 2003). Each evolutionary lineage of chickens could contain a different polyphyletic origin, and some types, e.g., Mediterranean and Bantam breeds, might come directly from *G. gallus*. The hypothesis of multiple origins in South and Southeast Asia is supported by molecular analyses of the mtDNA hypervariable segment I for 834 Eurasian domestic chickens, as well as 66 wild red junglefowls from Southeast Asia and China (Liu et al. 2006).

5.1.3
Economic Importance and Nutritional Value

Meat, milk, and eggs produced by domestic animals have long represented important parts of the diets of many people (Pond and Bell 2004). The world poultry industry is a growing part of global agribusiness and also one of the most dynamic components of world agribusiness trade. Over the last four decades, an estimated live poultry population of the world (Table 2)

has increased by 330% (chickens), 440% (ducks), 112% (turkeys), and 724% (geese). On the other hand, the number of fancy breeds and indigenous populations has significantly decreased during the twentieth century.

Poultry meat is defined as meat from chicken, turkey, duck, goose, guinea fowl, and pigeon. Poultry is one of the most consumed meats in the world and is the most consumed meat per capita in the United States (Pond and Bell 2004). The production and consumption of poultry meat, specifically chicken, turkey, duck, and goose, has dramatically increased over the last several decades, with total production of estimated 81.4 million metric tons (MT) in 2005 (Table 3). The bulk of poultry meat is produced from chickens, mostly broilers that are raised for meat production and have been selected for increased meat yield. In the 1950s, it took approximately 11 weeks to raise a 1.6-kilogram (kg) broiler. Currently, a 2.3-kg broiler can be raised in 6–7 weeks depending on feed quality, genetic background of crosses, and local management conditions. Broilers are shipped to the market at various ages and

Table 2 World poultry production: live animal stocks[a]

Stocks (1000)	Year						
	1961	1970	1980	1990	1995	2000	2005
Chickens	3,883,540	5,207,622	7,216,976	10,673,952	12,959,165	14,476,988	16,695,877
Ducks	193,453	256,318	351,979	552,612	797,412	927,973	1,044,736
Geese	36,640	54,578	69,273	131,557	235,103	234,497	301,905
Turkeys	130,745	99,832	200,644	243,042	247,387	268,015	276,821

[a] FAOSTAT (2006)

Table 3 World poultry production: meat[a]

Poultry meat production (MT)	Year						
	1961	1970	1980	1990	1995	2000	2005
Poultry all	8,953,120	15,100,097	25,962,116	41,025,900	54,730,103	69,176,770	81,436,269
Chicken	7,557,158	13,141,695	22,907,219	35,459,934	46,572,934	59,029,981	70,474,502
Turkey	902,220	1,224,183	2,054,235	3,703,989	4,568,167	5,125,341	5,167,560
Duck	335,922	500,841	713,113	1,231,933	2,096,213	3,000,542	3,447,564
Goose	149,717	226,270	282,322	616,534	1,476,930	2,001,974	2,326,683
Pigeon and other birds	8,102	7,108	5,228	13,680	15,859	18,932	19,958

[a] FAOSTAT (2006)

weights and as a variety of products, including whole carcasses weighing around 1 to 4–5 kg broilers subject to meat deboning (Pond and Bell 2004). Modern domestic turkeys have been selected primarily for large body size and rapid growth rate. Commercially, they are usually grown until they reach sexual maturity. For males, this is approximately 20 weeks of age, when they can weigh over 20 kg, compared to a 3-year-old male wild turkey that weighs a mere 9 kg (Pond and Bell 2004).

Eggs produced globally are predominantly from chickens (Table 4), with an increase in egg production of about 50% in the 1990s. While duck eggs lag far behind chicken eggs in importance, they are in high demand in China, several Pacific Rim countries, and Europe. There is commercial production of quail eggs in many countries, although on a smaller scale. In the past, ostrich, pelican, peafowl, swan, and guinea fowl eggs were also valued (Bell and Weaver 2002).

Chicken is the most developed global poultry industry sector, while ducks, geese, and turkeys are raised more regionally (Table 5). The world leader in poultry stocks, except turkeys, is China. Turkeys are

mainly produced in the USA, Europe, and Latin America. Turkey is second after the chicken in economic importance among poultry species in the USA.

Capital investment necessary for an increase in production is roughly US $1 per kilogram for both eggs and broilers. Investment for new facilities in the poultry industry has been US $4 billion annually worldwide. During the 1990s, over US $40 billion was invested in the world chicken industry. It is likely that the world increase in chicken meat and egg production will continue into the twenty-first century, but not at such a rapid pace (Bell and Weaver 2002).

In addition to low production costs, one of the main reasons that poultry meat consumption has increased in the last decade is the nutritional value of the meat (Table 6). The fat in poultry meat is located in the skin and is therefore easily removable compared to other meats, enabling consumers to adopt a more low-fat type of meat in their diets. In addition, the fat in poultry meat is lower in saturated fatty acids and higher in unsaturated fatty acids. This fat deposition can vary among species and is diet-dependent. Therefore, poultry meat can easily be incorporated

Table 4 World poultry production: eggs[a]

Eggs primary production (MT)	Year						
	1961	1970	1980	1990	1995	2000	2005
Chicken eggs	14,408,065	19,538,393	26,215,604	35,243,467	42,854,069	51,678,162	59,433,971
Total	15,133,710	20,412,719	27,414,941	37,524,123	46,891,269	55,797,691	64,576,599

[a] FAOSTAT (2006)

Table 5 World poultry production: live animal stocks (1000) by country in 2005[a]

Country	Chickens	Country	Ducks	Country	Geese	Country	Turkeys
China	4,360,243	China	725,018	China	267,819	USA	88,000
USA	1,950,000	Viet Nam	50,000	Egypt	9,100	France	30,820
Indonesia	1,249,426	Indonesia	34,275	Romania	4,000	Chile	26,500
Brazil	1,100,000	India	33,000	Poland	3,000	Italy	26,000
India	430,000	France	22,406	Madagascar	3,000	Brazil	16,200
Mexico	425,000	Ukraine	22,000	Taiwan	2,819	Germany	10,611
Russia	328,933	Thailand	17,000	Russia	2,750	UK	8,300
Turkey	296,876	Malaysia	16,000	Hungary	2,127	Portugal	7,000
Japan	283,000	Bangladesh	11,700	Israel	1,400	Slovakia	5,800
Iran	280,000	Philippines	10,439	Turkey	1,400	Canada	5,600

[a] Top ten countries in raising each of major poultry species ; FAOSTAT (2006)

into a well-balanced diet to improve health (Pond and Bell 2004).

Eggs are not only palatable, but are also considered to be a healthy food (Bell and Weaver 2002). Its protein value is the highest of all food, it is easy to digest, and its calories and fat content are moderate (Table 7). This image has been tarnished somewhat in the last 20 years due to increased awareness of cholesterol, food safety, and lack of convenience in preparation (Bell and Weaver 2002).

Waterfowl production is traditionally popular in Asia and some countries of Europe and Near East. One of the major features of geese is their capability to forage on grass alone, which is impossible for chickens. Also, geese are willing to eat more than is required. Owing to this peculiarity, they have been used for fattening since very early times and have become too heavy to fly (Scherf 2000). Duck production is not a dominant global sector, especially in most developed countries (Kear 1975), most likely because of a monogamous mating system, the depo-

sition of large amounts of fat below the swim line, a large bone:meat ratio in the carcass, a long incubation period of 28 days, and a breeding season confined to the spring (Scherf 2000). Muscovy duck farming is nowadays very popular in all equatorial countries of Africa and Asia. In Europe and Taiwan, a sterile hybrid, the mulard, is commercially produced by crossing the Muscovy with the common domestic duck (Crawford 1992; Scherf 2000).

Next among poultry farming is the ostrich. The annual world demand for ostrich skins is almost one million, while their world production, largely from South African farms, is less than 250,000 skins a year (Scherf 2000). Small numbers of skins are also supplied to the market from Zimbabwe, Tanzania, and Texas. In Australia, there were more than 35,000 farmed ostriches in 1995 and numbers rapidly increased to 200,000 birds in 2000. In 1995, the local price for ostrich meat in Australia was Aus $40 per kilogram (US$29) and a pair of breeding ostriches sold for Aus$60,000–120,000. Currently, the demand

Table 6 Proximate composition and energy values of raw poultry meat in avian species[a]

Species	Meat type/skin	Moisture %	Protein %	Fat %	Ash %	Energy (kcal/100 g)
Chicken	Light, with skin	68.6	20.3	11.1	0.86	186
Chicken	Light, without skin	74.9	23.2	1.6	0.98	114
Chicken	Dark, with skin	65.4	16.7	18.3	0.76	237
Chicken	Dark, without skin	76.0	20.1	4.3	0.94	125
Turkey	Light, with skin	69.8	21.6	7.4	0.90	159
Turkey	Light, without skin	73.8	23.6	1.6	1.00	115
Turkey	Dark, with skin	71.1	18.9	8.8	0.86	160
Turkey	Dark, without skin	74.5	20.1	4.4	0.93	125
Duck	All, with skin	48.5	11.5	39.3	0.68	404
Duck	All, without skin	73.8	18.3	6.0	1.06	132
Goose	All, with skin	49.7	15.9	33.6	0.87	371
Goose	All, without skin	68.3	22.8	7.1	1.10	161

[a] Pond and Bell (2004)

Table 7 Percentage composition of the chicken egg[a,b]

Component	Water	Protein	Lipid	Carbohydrate	Ash	Calories (kcal/egg)
Albumen	88.0	9.7–10.6	0.03	0.4–0.9	0.5–0.6	19
Yolk	48.2	15.7–16.6	31.8–35.5	0.2–1.0	1.1	65
Whole egg	75.5	12.8–13.4	10.5–11.8	0.3–1.0	0.8–1.0	84

[a] 60 g egg. Shell is not included. Percentages of different components vary in different breeds
[b] After Bell and Weaver (2002)

for ostrich meat is far in excess of supply; world production is only 12,000–15,000 MT as the industry has not yet made a full transition from a breeder market to commercial production. Around 60% of this production is in South Africa (World Ostrich Association, http://www.world-ostrich.org/demand.htm). However, the skin is the most valuable ostrich product (Scherf 2000). High-quality, unprocessed ostrich skins harvested at 14 months of age are worth about US $200 wholesale. The price for a domestic ostrich in South Africa was worth R 150 in 1979 and included 48% for the skin, 40% for the feathers, and 12% for the carcass. In 1994 in Texas, the estimated value of an ostrich was US $900. Processing of ostrich skins is done in South Africa and Germany, while ladies handbags, shoes, briefcases, and wallets are produced from the skins in France and Italy. The greatest demand for these articles is from Japan (Scherf 2000).

The emu farming industry produces meat, skins, and byproducts (oil and feathers) which are sold in Australia and overseas, the key importers being the USA, Japan, France, and Southeast Asia (Scherf 2000). For some farms, tourism is also a source of income. Since 1991, farmed emus have been slaughtered in Australia, and there was an estimate of 85,000 birds available for slaughter in 1995. Emu meat is characterized by a lower fat and cholesterol content, and by a gamey flavor. Emu oil is produced by rendering fat; it is utilized in cosmetics and for muscle and joint pain treatment. An emu was valued as US$450 in Texas in 1994, where there were about 30,000 birds. The outlook for emu farming is very promising, although production and processing costs will need to be decreased (Scherf 2000).

5.1.4
Breeding Objectives

Over the 20th century, modern selective breeding has resulted in spectacular progress in both egg and meat production traits (Burt 2002). By 2002, world egg production was 795 billion/year and broiler meat was at 6.5 million MT/year (Burt 2005). However, these successes have led to an increase in the incidence of undesirable conditions including congenital disorders (e.g., ascites, lameness), reduced fertility, and reduced resistance to infectious disease

in meat-type chickens, and osteoporosis in laying hens. Since genetic progress in egg and meat production could approach its limit within the next 20 years (Burt 2002), the poultry industry priorities would be to reduce losses from undesirable traits, develop new products with higher quality (e.g., increased egg shell strength), and secure greater uniformity and predictability in production. Another growing concern is food safety that requires a reduction in the use of chemicals and antibiotics and to increase genetic resistance to pathogens (Burt 2005).

At present, three major categories of poultry stocks coexist: pure fancy or exhibition breeds, indigenous populations, and commercial poultry. Each of these categories is characterized by specific features that depend on the needs of man. Pure breeds maintained by fanciers may be classified by purpose of use, geographical origin, evolutionary roots, and other criteria. Indigenous flocks are kept locally in primitive conditions and without any selection. Commercial poultry stocks involve in selected egg and meat-type lines and crosses. These three types of poultry stocks also differ in their utility importance. Fancy fowl and indigenous populations represent genetic resources in poultry, i.e., they may be used as the sources of genetic variability for commercial poultry and in creating new breeds and lines. Poultry breeders maintain breed characteristics and economically important traits at the standard level. Commercial poultry is related to commodity output. The main selection task in developing this type of poultry is to increase the productivity and viability of industrial lines and crosses. At the same time, essential efforts are undertaken to lower costs of produce. New genetic and selection approaches and technologies, and genetic resources adapted to various environment conditions are sought by the commercial poultry industry.

Modern poultry breeding industry comprises a limited number of major companies worldwide. These companies maintain the foundation and great grandparent stock to produce commercial meat and egg-type lines (Bell and Weaver 2002). At first pure breeding was used, then, crossbreeding to exploit heterosis was employed between 1930 and 1950, and now crosses of strains and synthetic lines are done routinely but only a few breeds and varieties are used. White Leghorns (WL) dominate white egg production, Rhode Island Red (RIR) and a few others are used for brown eggs, and White Cornish and White Plymouth Rocks for meat (Crawford 1995). However, there is a tendency to utilize more breeds in commercial poultry, especially those breeds adapted to local environments or suitable to meet consumer preferences (Altukhov 2004).

The genetic performance of the birds is a main target of breeding in the poultry industry. Traits for selection, or at least monitoring, by egg-type chicken breeders include age of sexual maturity, rate of lay before and after molt, livability in the growing and laying house, egg weight, body weight (BW), feed conversion, shell color, shell strength, albumen height, egg inclusions (blood and meat spots) and temperament, as well as traits associated with the productivity of the parent (Keeton et al. 2003). Since the early 1980s, the increasing proportion of eggs used in food processing has added such traits as percentage solids and lipids in the egg. Egg production per hen housed remains the single selected trait, although its major component is now considered to be persistency of lay rather than peak rate of lay. Importance of selection for disease resistance varies from one breeding company to another. In meat-type chickens, breeding objectives include broiler growth rate, meat yield traits, livability, hatching egg production, and fertility (Keeton et al. 2003), as well as decreased abdominal and carcass fat, and lower feed conversion. Skeletal problems, such as leg weakness, in commercial broiler, egg laying, and breeder flocks represent another major challenge for poultry breeding and selection (Bennett and Ijpelaar 2003).

5.2
Classic Genetics

5.2.1
Brief History of Poultry Genetics

Analysis of inheritance in the chicken began more than one century ago and led to the development of the classical genetic map (Fig. 5). The first genes assigned to a single chromosome were sex-linked (reviewed by Crawford 1990; Romanov et al. 2004). In about the mid-twentieth century, poultry immunogenetics began. Cytogenetics as a branch of poultry genetics appeared in the early 1960s and became another research avenue in the field of avian hered-

ity. The chicken has been a model for cytogenetic research given that its chromosomal morphology and behavior parallels that of other animal species (Crawford 1990). Chromosome numbers for some avian species are given in Table 8.

Recent progress in molecular biology, cytogenetics, and DNA technologies have resulted in novel tools to address chicken gene mapping and genomics issues. In the 1990s, configurations of chicken molecular and cytogenetic maps were significantly advanced. The application of bacterial artificial chromosome (BAC) libraries, BAC-contig physical maps, expressed sequence tags (EST), and whole-genome sequencing has provided new prospects in chicken genomics (reviewed by Romanov et al. 2004).

The chicken haploid genome has about 1.2×10^9 base pairs of DNA (Stevens 1986; Bloom et al. 1993; Bennett et al. 2003) arranged on 38 pairs of autosomes as well as the Z and W sex chromosomes (Yamashina 1944). Many of the autosomes are small microchromosomes and, unlike the larger macrochromosomes, cannot be identified by size (Ohno 1961; Crawford 1990). This intricacy of the genome composition has impeded mapping of chicken genes and sorting of chicken chromosomes. Current molecular and physical maps for the chicken encompass more than 2,000 genes and markers and other advanced chicken genomic resources also exist (Romanov et al. 2004).

5.2.2
Early Classical Mapping Efforts

After the rediscovery of Mendel's laws, Bateson and Saunders (1902) wrote one of the first articles devoted to hereditary characters or "allelomorphs" (now known as "alleles") in the chicken and some other organisms. This was the first introduction of the domestic fowl as a classical genetic model (see reviews by Pisenti et al. 1999; Romanov et al. 2004). The notion of "linkage" emerged thanks to Sutton (1903) who claimed that "all the allelomorphs represented by one chromosome must be inherited together." Lock (1906) also suggested that linkage might happen if genes lie on the same chromosome. Other geneticists extended these ideas in subsequent decades. Thomas Hunt Morgan demonstrated crossing over, a form of chromosomal recombination between closely linked genes (Morgan 1910, 1911). Morgan received the Nobel Prize in Physiology or

Chapter 5 **Poultry** 87

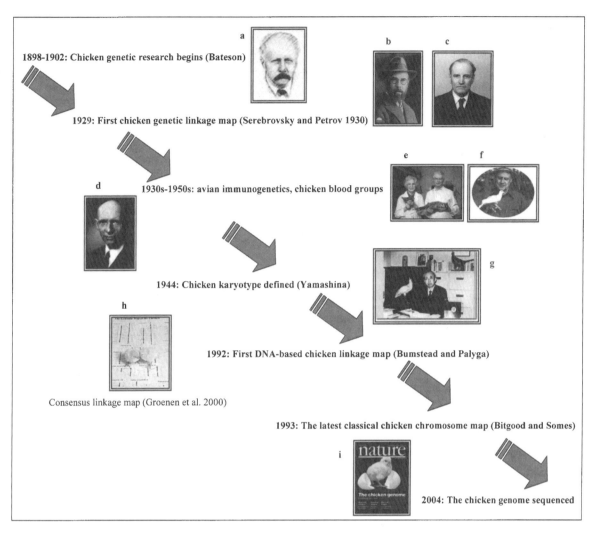

Fig. 5 Outline of poultry genetics history. Around the time of rediscovery of Mendel's laws of inheritance (1900), William Bateson (**a**), the father of modern genetics, conducted with his fellows a series of experiments in the chicken, thus introducing this domestic bird as a classical genetic model. By 1930, the first chicken genetic linkage map was generated by Serebrovsky (**b**) and Petrov (**c**). By the middle of the last century, avian immunogenetics was born and the chicken blood groups were discovered thanks to efforts of L. Cole (**d**), Irwin, McGibbon, E. Briles (**e**), C. Briles, Miller (**f**), and many others. In 1944, Yamashina (**g**) defined the chicken karyotype, as we know it today. With the advent of molecular genetic era, the first DNA-based chicken linkage map was created in UK in 1992. The follow-up development of molecular maps in the USA and the Netherlands led to the generation of the consensus linkage map in 2000 (**h**, USDA Image Gallery, http://www.ars.usda.gov/is/graphics/photos/). The classical chicken chromosome map was last updated in 1993. In 2004, the publication of the draft chicken sequence became a landmark in the history of poultry genetics (**i**, adapted with permission from Macmillan Publishers Ltd: Nature, © 2004)

Medicine in 1933 for postulating the role that chromosomes play in heredity.

The discovery of gene linkage and crossing over became the beginning of classical genetic map development. It was found that the stronger linkage between two genes, the shorter distance between them, and to measure this linkage, the frequency of crossing over was exploited. In honor of Morgan, the map distances were called "centi-Morgan," with 1% of linkage breakage being equal to one centimorgan (1 cM) (reviewed by Romanov et al. 2004).

Sex linkage, as the most obvious variant of genetic linkage, was first reported for the imperfect albinism in canaries by Durham and Marryat (1908)

88 M. N. Romanov et al.

Table 8 Genome size of the selected avian species

Species	Chromosome number (2n)	C value (pg)[a]	Reference
Gallus gallus (red junglefowl)	78	1.25	Crawford (1990), Gregory (2006)
Coturnix japonica (Japanese quail)	78	1.29–1.41	Crawford (1990), Gregory (2006)
C. coturnix (common quail)	78	1.35	Gregory (2006)
Meleagris gallopavo (turkey)	80	1.31–1.68	Crawford (1990), Gregory (2006)
Numida meleagris (helmeted guinea fowl)	78	1.23–1.31	Crawford (1990), Gregory (2006)
Pavo cristatus (Indian peafowl)	76	—	Sasaki et al. (1968)
Phasianus colchicus (ring-necked pheasant)	82	0.97–1.27	Crawford (1990), Gregory (2006)
Chrysolophus pictus (golden pheasant)	82	1.21	Gregory (2006)
Lophura nycthemera (silver pheasant)	80	—	Schmid et al. (2000)
Anas platyrhynchos (mallard)	80	1.24–1.54	Crawford (1990), Gregory (2006)
Cairina moschata (Muscovy duck)	80	1.00–1.34	Crawford (1990), Gregory (2006)
Anser anser (graylag goose)	80	1.08[b]	Crawford (1990), Gregory (2006)
A. cygnoides (swan goose)	80 or 82+	1.08[b]	Crawford (1990), Gregory (2006)
Cygnus olor (mute swan)	80	1.48	Gregory (2006)
Struthio camelus (ostrich)	80	2.16	Gregory (2006)
Rhea americana (greater rhea)	80	—	Gunski and Giannoni (1998)
Dromaius novaehollandiae (emu)	80	1.55–1.63	Gregory (2006)
Casuarius spp. (cassowary)	80	—	Takagi et al. (1972)
Apteryx australis (brown kiwi)	80	—	De Boer (1980)
Columba livia (rock pigeon)	80	1.14–1.65	Gregory (2006)
Streptopelia roseogrisea (African collared dove)	78	—	Schmid et al. (2000)
Serinus canaria (island canary)	80	1.48–1.62	Gregory (2006)
Taeniopygia guttata (zebra finch)	80	1.25	Pigozzi and Solari (1998), Gregory (2006)
Psittacula krameri (rose-ringed parakeet)	68	1.37	Gregory (2006)
Melopsittacus undulates (budgerigar)	58–60	1.02–1.37	Gregory (2006)
Grus grus (common crane)	80	1.54	Gregory (2006)
Ciconia ciconia (white stork)	68	1.58	Takagi and Sasaki (1974), Gregory (2006)
Leptoptilos crumeniferus (Marabou stork)	72	1.34	Gregory (2006)
Gymnogyps californianus (California condor)	80 or 82	1.51	Raudsepp et al. (2002), Gregory (2006)
Pelecanus onocrotalus (great white pelican)	66	1.25	Gregory (2006)
Falco peregrinus (peregrine falcon)	50	1.45	Gregory (2006)
Aquila chrysaetos (golden eagle)	62	1.48	Gregory (2006)

[a] 1 pg = 978 Mb

[b] C value for an unknown goose species (Gregory 2006)

[c] C value for the turtle dove (*Streptopelia turtur*; Gregory 2006)

and for the barring pattern (*BARR***) in Barred Plymouth Rock chickens by Spillman (1908) (reviewed by Romanov et al. 2004). Unlike mammals, the female bird is the heterogametic sex carrying the two different sex chromosomes (now referred to as Z and W), while the male is the homogametic sex (ZZ). There were subsequent assignments of other chicken genes to the sex (Z) chromosome and estimation of linkage between them (Punnett and Bateson 1908; Bateson 1909; Hagedoorn 1909; Bateson and Punnett 1911; Davenport 1911, 1912; Sturtevant 1911, 1912; Morgan and Goodale 1912; Goodale 1917; Haldane 1921; Serebrovsky 1922). Dunn and Jull (1927) found a close linkage between the genes for dominant white (*I*, or *SILV*) and crest (*CR*), and Serebrovsky and Petrov (1928) reported on creeper (*CP*) and rose comb (*R*) that were the first known instances of autosomal linkage. Loci mapped by classical mating are presented in Table 9.

Table 9 Classical chicken loci reviewed by Bitgood and Somes (1993)

Locus symbol[a]	Aliases	Locus/gene name (synonym)	Classical linkage group	Bitgood and Somes (1993) chromosome	Molecular linkage group according to Schmid et al. (2000, 2005) and other considerations[b]
CP	Cp	Creeper	I	Presumably chromosomes 2, 3, or 4	N/A [presumably GGA4 if controlled by FGFR3]
R	—	Comb, rose	I	Presumably chromosomes 2, 3, or 4	N/A [presumably GGA4 due to linking to CP]
U	—	Uropygial	I	Presumably chromosomes 2, 3, or 4	N/A [presumably GGA4 due to linking to R]
LAV	lav	Lavender, plumage color	I	Presumably chromosomes 2, 3, or 4	N/A [presumably GGA4 due to linking to R]
MP	Mp	Ametapodia	I	Presumably chromosomes 2, 3, or 4	N/A [presumably GGA4 due to linking to R]
FR	fr	Fray	II	Presumably chromosomes 2, 3, or 4	N/A [possibly E22C19W28_E50C23 due to linking to SILV]
CR	Cr	Crest, tassel feather length	II	Presumably chromosomes 2, 3, or 4	N/A [possibly E22C19W28_E50C23 due to linking to SILV]
SILV	I, PMEL17, MMP, MMP115	Silver homolog (mouse) [dominant white plumage color; 115-kDa melanosomal matrix protein]	II	Presumably chromosomes 2, 3, or 4	E22C19W28_E50C23
F	f	Frizzle, feather structure	II	Presumably chromosomes 2, 3, or 4	E22C19W28_E50C23
BCDO2	w, W, APOA1	Beta-carotene dioxygenase 2	III	1	GGA24
CPHH	Ea-H, EAH	Erythrocyte alloantigen H (blood group system H)	III	1	N/A [possibly GGA24 due to linking to APOA1]
SE	se	Sleepy-eye	III	1	N/A [possibly GGA1 due to linking to CPJJ]
CPJJ	Ea-J, EAJ	Erythrocyte alloantigen J (blood group system J)	III	1	N/A [possibly GGA1 due to linking to O]

(continued)

Table 9 (Continued)

Locus symbol[a]	Aliases	Locus/gene name (synonym)	Classical linkage group	Bitgood and Somes (1993) chromosome	Molecular linkage group according to Schmid et al. (2000, 2005) and other considerations[b]
P	—	Pea comb	III	1	N/A [possibly GGA1 due to linking to ALVE1]
CHA	cha	Charcoal	III	1	N/A [possibly GGA1 due to linking to P]
DB	Db, ma	Dark brown Columbian-type plumage pattern (marbling)	III	1	N/A [possibly GGA11 due to linking to MC1R]
MC1R	E	Melanocortin 1 receptor (alpha melanocyte stimulating hormone receptor) [extended black plumage pattern]	III	1	GGA11
TAFF	t	Feathering, retarded-tardy feather growth	III	1	N/A [possibly GGA11 due to linking to MC1R]
ML	Ml	Melanotic, plumage pattern	III	1	N/A [possibly GGA11 due to linking to MC1R]
PG	Pg	Patterning gene, penciling, feather pattern	III	1	N/A [possibly GGA11 due to linking to ML]
CPPP	Ea-P, EAP	Erythrocyte alloantigen P (blood group system P)	III	1 (questionable assignment)	GGA3
NA	Na	Naked neck	III	1 (questionable assignment)	GGA3
H	h	Silkiness, feather structure	III	1 (questionable assignment)	N/A [possibly GGA3 due to linking to NA]
FL	Fl	Flightless	III	1 (questionable assignment)	N/A [possibly GGA3 due to linking to H]
CYP19A1	Hf, HF, P450arom, MCW0357, CYP19	Cytochrome P450, family 19, subfamily A, polypeptide 1 (aromatase; henny feathering, feather structure)	III	1	GGA10
GH1	Gh, ROS0118, GH	Growth hormone 1	III	1	GGA27

(continued)

Table 9 (Continued)

Locus symbol[a]	Aliases	Locus/gene name (synonym)	Classical linkage group	Bitgood and Somes (1993) chromosome	Molecular linkage group according to Schmid et al. (2000, 2005) and other considerations[b]
CPDD	*Ea-D, EAD*	Erythrocyte alloantigen D (blood group system D)	III	1	N/A
CPII	*Ea-I, EAI*	Erythrocyte alloantigen I (blood group system I)	III	1	GGA23
ALVE1	*ev1, ev-1,EV1*	Endogenous retrovirus 1	III	1	GGA1
PE	*pe*	Perosis	III	1	N/A [possibly GGA1 due to linking to *SE*]
MB	*Mb*	Muffs and beard, feather length	III	1	N/A [possibly GGA24 due to linking to *CPHH*]
CPCC	*Ea-C, EAC*	Erythrocyte alloantigen C (blood group system C)	III	1	N/A
CPEE	*Ea-E, EAE*	Erythrocyte alloantigen E (blood group system E)	III	1	GGA26
CPAA	*Ea-A, EAA*	Erythrocyte alloantigen A (blood group system A)	III	1	GGA26
PTI(?)[c]	*Pti-?*	Ptilopody, feathered shank	III	1	N/A [possibly GGA24 due to linking to *APOA1*]
ALVE4	*ev4, ev-4, EV4*	Endogenous retrovirus 4	III	1	GGA6
ALVE5	*ev5, ev-5, EV5*	Endogenous retrovirus 5	III	1	N/A [cytogenetically assigned to GGA1; could also be on GGA6 due to linking to *ALVE4*]
ALVE6	*ev6, ev-6, EV6, ALVE6A*	Endogenous retrovirus 6	III	1	GGA1
ALVE13	*ev13, ev-13, EV13*	Endogenous retrovirus 13	III	1	N/A [cytogenetically assigned to GGA1]
ALVE8	*ev8, ev-8, EV8*	Endogenous retrovirus 8	III	1	N/A [cytogenetically assigned to GGA1]
ALVE(?)	*ev(?)*	Avian leukosis virus (ALV) provirus	III	1	N/A

(continued)

Table 9 (Continued)

Locus symbol[a]	Aliases	Locus/gene name (synonym)	Classical linkage group	Bitgood and Somes (1993) chromosome	Molecular linkage group according to Schmid et al. (2000, 2005) and other considerations[b]
HBG2	HGB, HBB, HBB@	Hemoglobin, gamma G (hemoglobin, beta)	III?	1 or 2	GGA1
HBG1 or HBE1	HBE or HBR	Hemoglobin, gamma A or epsilon 1 (globin, epsilon or rho; embryonic beta-like globins)	III?	1 or 2	N/A [assigned to GGA1 or GGA2 by chromosomal fractionation; linked to HBD; assembled on GGA1]
D	—	Cup-, V-type duplex comb	IV	Presumably chromosomes 2, 3, or 4	N/A [probably GGA2 due to linking to LMBR1 and M]
M	—	Spurs, multiple	IV	Presumably chromosomes 2, 3, or 4	N/A [probably GGA2 due to linking to LMBR1]
LMBR1	Po, PO	Limb region 1 homolog (mouse) [polydactyly, duplicate polydactyly]	IV	Presumably chromosomes 2, 3, or 4	GGA2
VLDLR	Ro, RO	Very low-density lipoprotein receptor (restricted ovulator)	V	Z	GGAZ
SH	sh	Shaker	V	Z	N/A
N	n	Naked	V	Z	N/A
PX	px	Paroxysm	V	Z	N/A
LN	ln	Lethal liver necrosis	V	Z	N/A
GHR	dw, DW	Growth hormone receptor (sex-linked dwarfism)	V	Z	GGAZ
WL	wl	Sex-linked wingless	V	Z	N/A
PN	pl	Prenatal lethal	V	Z	N/A
K	ev21	Sex-linked late feathering	V	Z	GGAZ
SLC45A2	S	Solute carrier family 45, member 2 (silver, gold, albinism plumage color)	V	Z	N/A [assembled on GGAZ]
LK	lk	Ladykiller	V	Z	N/A
LI	Li	Light down	V	Z	N/A

(continued)

Chapter 5 **Poultry** 93

Table 9 (Continued)

Locus symbol[a]	Aliases	Locus/gene name (synonym)	Classical linkage group	Bitgood and Somes (1993) chromosome	Molecular linkage group according to Schmid et al. (2000, 2005) and other considerations[b]
ABCA1	*y, Y*	ATP-binding cassette, subfamily A (ABC1), member 1 (sex-linked recessive white skin)	V	Z	N/A [assembled on GGAZ]
BR	*br*	Brown eye	V	Z	N/A
ID	*Id*	Dermal melanin inhibitor	V	Z	GGAZ
BARR	*B*	Barring dilution sex-linked feather pattern	V	Z	N/A
KO	*ko*	Head streak	V	Z	N/A
ALVE7	*ev7, ev-7, EV7*	Endogenous retrovirus 7 (defective ALV provirus)	V	Z	N/A [cytogenetically assigned to GGAZ]
ALVE21	*ev21, ev-21, EV21*	Endogenous retrovirus ev21	V	Z	GGAZ
BA	*ba*	Baldness, congenital	V	Z	N/A
CD	*cd*	Cerebellar degeneration	V	Z	N/A
CHOC	—	Chocolate plumage color, sex-linked	V	Not reviewed	N/A
CHZ	*chz*	Sex-linked chondrodystrophy	V	Z	N/A
CM	*cm*	Sex-linked coloboma	V	Z	N/A
DP4	*dp-4*	Diplopodia-4	V	Z	N/A
GA	*ga*	Gasper	V	Z	N/A
HZ	*H-Z*	Z-linked histoantigen	V	Z	N/A
J	*j*	Jittery	V	Z	N/A
POP	*pop*	Pop-eye	V	Z	N/A
PR	*pr*	Protoporphyrin inhibitor	V	Z	N/A
PW1	*Pw1, Pw₁*	Agglutinogen, pokeweed ("Pw1" agglutinogen)	V	Z	N/A
PW2	*Pw2, Pw₂*	Agglutinogen, pokeweed ("Pw2" agglutinogen)	V	Z	N/A
RG	*rg*	Recessive sex-linked dwarf	V	Z	N/A
SLN	*sln*	Sex-linked nervous disorder	V	Z	N/A

(continued)

Table 9 (Continued)

Locus symbol[a]	Aliases	Locus/gene name (synonym)	Classical linkage group	Bitgood and Somes (1993) chromosome	Molecular linkage group according to Schmid et al. (2000, 2005) and other considerations[b]
ST1	*St1, St$_1$*	Agglutinogen, potato ("St1" agglutinogen)	V	Z	N/A
ST2	*St2, St$_2$*	Agglutinogen, potato ("St2" agglutinogen)	V	Z	N/A
XL	*xl*	Sex-linked lethal	V	Z	N/A
Z	—	Dominant sex-linked dwarf	V	Z	N/A
—	*sex*	Sex-linked lethal Bernier	V	Z	N/A
HW	*H-W*	W-linked histoantigen	VI?	W (questionable assignment)	N/A
PPAT	*Ade-A, ADEA, GPAT*	Phosphoribosyl pyrophosphate amidotransferase (adenine synthesis A)	VII	6	N/A [linked to *ALB* and *PGM2* as shown by somatic cell hybridization; assembled on GGA4]
ALB	*Alb*	Albumin (serum preproalbumin)	VII	6	GGA4
GC	*Gc, VTDB*	Group-specific component (vitamin D-binding protein)	VII	6	GGA4
PGM2	*Pgm-2, RCJMB04_33e1*	Phosphoglucomutase 2	VII	6	N/A [linked to *ALB* and *PPAT* as shown by somatic cell hybridization; assembled on GGA4]
ADEB	*Ade-B*	Adenine synthesis B	VIII	7	N/A
DMD	*dys*	Dystrophin (muscular dystrophy, Duchenne and Becker types)	IX[d]	10	N/A [cytogenetically assigned to GGA10 but assembled on GGA1 sequence]
THRA	*c-erb-A, ERBA1, THRA1*	Thyroid hormone receptor, alpha [erythroblastic leukemia viral (v-erb-a) oncogene homolog, avian]	IX	Presumably chromosomes 10–14	N/A [assigned to a microchromosome by chromosomal fractionation; assembled on UN]

(continued)

Table 9 (Continued)

Locus symbol[a]	Aliases	Locus/gene name (synonym)	Classical linkage group	Bitgood and Somes (1993) chromosome	Molecular linkage group according to Schmid et al. (2000, 2005) and other considerations[b]
ETS1	c-ets, c-ets-1, ETSB, MCW0075, LOC768354	v-ets erythroblastosis virus E26 oncogene homolog 1 (avian)	IX	Presumably chromosomes 9–16	GGA24
FES	c-fps, LOC429374	Feline sarcoma oncogene	IX	Presumably chromosomes 9–16	N/A [assigned to a microchromosome by chromosomal fractionation; assembled on GGA10]
RAF1	c-mil/mht, MIL	v-raf-1 murine leukemia viral oncogene homolog 1	IX	Presumably chromosomes 9–16	N/A [assigned to a microchromosome by chromosomal fractionation; assembled on GGA12 sequence]
SRC	c-src, MCW0050, SDR, PP60C-SCR	v-src sarcoma (Schmidt-Ruppin A-2) viral oncogene homolog (avian)	IX	Presumably chromosomes 10–12	N/A [cytogenetically assigned to a microchromosome 10, 11 or 12; assembled on GGA20 sequence]
HCK	ev3, ev-3, EV3, ALVE3, LOC419280	Hemopoietic cell kinase (endogenous retrovirus 3)	IX	Microchromosome	GGA20
HPRT1	hprt, HPRT	Hypoxanthine phosphoribosyltransferase 1	IX	Microchromosome	GGA4
OVM	Ovm, LOC416236, LOC396462	Ovomucoid	IX	Presumably chromosomes 10–15	GGA13
TF	Tf	Transferrin (ovotransferrin, conalbumin)	IX	Presumably chromosomes 9–12	N/A [cytogenetically assigned to a microchromosome; assembled on GGA9 sequence]
TK1	Tk-F, tk-F, TK	Thymidine kinase 1, soluble (cytosol F)	IX	Microchromosome	N/A [assigned to a microchromosome by somatic cell hybridization; assembled on GGA18 sequence]
HBA1	HBA, HBA@, HBAA	Hemoglobin, alpha 1 (hemoglobin, alpha A)	IX	Presumably chromosomes 10–15	GGA14

(continued)

Table 9 (Continued)

Locus symbol[a]	Aliases	Locus/gene name (synonym)	Classical linkage group	Bitgood and Somes (1993) chromosome	Molecular linkage group according to Schmid et al. (2000, 2005) and other considerations[b]
HBA2	*HBAD, LOC416651*	Hemoglobin, alpha 2 (hemoglobin, alpha D)	IX	Presumably chromosomes 10–15	GGA14
HBZ	—	Hemoglobin, zeta (hemoglobin, pi and pi-prime; embryonic alpha-like globin)	IX	Presumably chromosomes 10–15	N/A [linked to *HBA1* and *HBAD*; assembled on GGA14 sequence]
HLA-B	*B@, MHC, HLA, LOC769497*	Major histocompatibility complex (MHC), class I, B	X	17	GGA16
MHCB	*B-G, Ea-B*	MHC B complex, class IV, B-G region	X	17	GGA16
HLA-G	*B-F, B-FL1, BF1, BF2, MHC1*	MHC B complex, class I, B-F region	X	17	GGA16
HLA-DRB5	*B-L, B-LBL2, B-LBL1, MHC2A, MHC2B, LOC425256*	MHC B complex, class II, B-L region	X	17	GGA16
CPBB	*Ea-B, EAB*	Erythrocyte alloantigen B (blood group system B)	X	17	GGA16
GAT	*Ir-GAT*	Immune response to synthetic polypeptide	X	17	N/A [GGA16 due to linking to *MHC*]
R-Rs-1	*Rs*	Subgroup C Rous sarcoma virus-induced tumor regression	X	17	N/A [GGA16 due to linking to *GAT*]
NOR	—	Nucleolar organizing region	X	17	N/A [GGA16 due to linking to *MHC*]
ACT	*act*	Macrophage activation	X	17	N/A [probably GGA16 due to linking to *MHC*]
ALVE(?)	*ev(?)*	ALV provirus	X	17	N/A [probably GGA16 due to linking to *MHC*]
EGFR	*c-erb-B, LOC396494*	Epidermal growth factor receptor [erythroblastic leukemia viral (v-erb-b) oncogene homolog, avian]	I, II, or IV?	2	GGA2
ALVE2	*ev2, ev-2, EV2*	Endogenous retrovirus 2 (codes for RAV-0)	I, II, or IV?	2	GGA1

Chapter 5 **Poultry** 97

Table 9 (Continued)

Locus symbol[a]	Aliases	Locus/gene name (synonym)	Classical linkage group	Bitgood and Somes (1993) chromosome	Molecular linkage group according to Schmid et al. (2000, 2005) and other considerations[b]
SHL	*shl*	Shankless	I, II, or IV?	2	N/A [assigned to GGA2 based on an X-ray-induced pericentric inversion on 2p]
OV	*Ov, pUN121ov, LOC396058*	Ovalbumin	I, II, or IV?	2 or 3	GGA2
G(3)	*G₃, G3*	Egg white ovoglobulin G(3)	I, II, or IV?	2 or 3	N/A [probably GGA2 due to linking to *OV*]
MYC	*c-myc, CMYCA*	v-myc myelocytomatosis viral oncogene homolog (avian)	I, II, IV, or IX?	Presumably chromosomes 2, 3, or 13–16	GGA2
MYB	*c-myb, ROS0064*	v-myb myeloblastosis viral oncogene homolog (avian)	I, II, IV, or IX?	Presumably chromosomes 2, 3, or 13–16	GGA3
ACTB	*LOC396526, RCJMB04_4h19*	Actin, beta	I, II, IV, or IX?	Presumably chromosomes 2 or 9–12	N/A [cytogenetically assigned to GGA2; assembled on UN sequence]
ALVE14	*ev14, ev-14*	Endogenous retrovirus 14	I, II, or IV?	3	N/A [cytogenetically assigned to GGA3]
CPMM	*Ea-M, EAM*	Erythrocyte alloantigen M (blood group system M)	—	UN	N/A [linked to *CPQQ*]
CPQQ	*Ea-Q, EAQ*	Erythrocyte alloantigen Q (blood group system Q)	—	**UN**	N/A [linked to *CPMM*]
CPOO	*Ea-O, EAO*	Erythrocyte alloantigen O (blood group system O)	—	UN	N/A [linked to *CPSS*]
CPSS	*Ea-S, EAS*	Erythrocyte alloantigen S (blood group system S)	—	UN	N/A [linked to *CPOO*]
ES1	*Es-1*	Serum esterase 1	—	UN	N/A [presumably GGA2 due to linking to *ES2*]
ES2	*Es-2*	Serum esterase 2	—	UN	N/A [presumably GGA2 if controlled by *PON2*; linked to *ES1*]

(continued)

Table 9 (Continued)

Locus symbol[a]	Aliases	Locus/gene name (synonym)	Classical linkage group	Bitgood and Somes (1993) chromosome	Molecular linkage group according to Schmid et al. (2000, 2005) and other considerations[b]
IGY	*IgG-1, IGG*	Immunoglobulin 7S-1 IgG H chain	—	UN	GGA15
IGM	*IgM-1*	Immunoglobulin 17S-1 IgM H chain	—	UN	GGA15
TVA	*Tv-A, tva*	ALV subgroup A receptor	—	UN	GGA28
BTN1A1	*Tv-C, tvc, TVC*	Butyrophilin, subfamily 1, member A1 (ALV subgroup C receptor)	—	UN	GGA28
RHOBTB2	*Tv-B, TVBS3, CAR1, TVB*	Rho-related BTB domain containing 2 (cytopathic ALSV receptor; ALV subgroup B receptor)	—	UN	GGA22
TVE	*Tv-E, SEAR*	ALV subgroup E receptor	—	UN	N/A [GGA22 due to linking to *RHOBTB2*; assembled on GGA22 as *RHOBTB2*]
CHRND	—	Cholinergic receptor, nicotinic, delta	—	UN	N/A [assembled on GGA9; linked to *CHRNG*]
CHRNG	—	Cholinergic receptor, nicotinic, gamma	—	UN	N/A [assembled on GGA9; linked to *CHRND*]
LOC396498	*cryd1, CRYD1, d-cry*	Crystallin, delta 1	—	UN	N/A [assembled on GGA19; linked to *ALS*]
ASL	*cryd2, CRYD2*	Argininosuccinate lyase (crystallin, delta 2)	—	UN	N/A [assigned to GGA19 by RH mapping; assembled on GGA19; linked to *LOC396498*]
BL	*Bl*	Blue plumage color	—	Not reviewed	N/A [possibly GGA3 (if linked to *NA*), GGA1 (if controlled by *KITLG*) or GGA4 (if encoded by *KIT*)]
FM	*Fm*	Fibromelanosis	—	Not reviewed	N/A [possibly GGA11 due to linking to *MC1R* or GGA4 if controlled by *EDNRB2*]

(continued)

Chapter 5 **Poultry** 99

Table 9 (Continued)

Locus symbol[a]	Aliases	Locus/gene name (synonym)	Classical linkage group	Bitgood and Somes (1993) chromosome	Molecular linkage group according to Schmid et al. (2000, 2005) and other considerations[b]
TYR	c	Tyrosinase (oculo-cutaneous albinism IA) [autosomal albinism]	—	Not reviewed	GGA1

N/A, not assigned to the molecular map; UN, unknown chromosome or linkage group.

[a] According to Schmid et al. (2000, 2005). If possible and when there is a homologous human (or other mammalian) gene, the name of the homolog is used (Burt 1999)

[b] N/A, not assigned by molecular linkage mapping. Where applicable, chromosomal location of the genes in the whole-genome sequence assembly is given as found in the NCBI databases (http://www.ncbi.nlm.nih.gov/; accessed August 2006)

[c] There are several ptilopody loci, the one involved here has not been identified. Serebrovsky (1926) suggested at least two dominant and at least two recessive feathered shank genes. Somes (1992) described two loci, *PTI1* and *PTI2*. *PTI1* has two alleles, the Langshan allele (*PTI1'L*) and the Brahma allele (*PTI1'B*). The Brahma allele was shown to be dominant over the Langshan allele. Both the Sultan and Cochin breeds possess two shank-feathering loci, and one of the loci in the Sultan contained the *PTI1'L* allele. The comparable allele in the Cochin breed was hypothesized to be *PTI1'B*. The second locus in both of these breeds appears to be similar, and the symbol *PTI2* is suggested

[d] Discontinued. It was not a classical group, but was convenient for listing loci on microchromosomes (Bitgood and Somes 1990)

5.2.3
First Chicken Map

The first chicken genetic map was constructed by a Russian group led by A.S. Serebrovsky (Serebrovsky and Wassina 1927; Serebrovsky and Petrov 1928, 1930; Petrov 1931; Sungurov 1933). It was also the first linkage map ever developed for any domestic animal and, as such, was a great milestone in the history of genetics (reviewed by Romanov et al. 2004). Serebrovsky's group launched mapping of chicken genes in 1919 at the Central Station for Livestock Genetics, Anikovo. Serebrovsky and Petrov (1930) undertook one of the first attempts to summarize the available information on chicken linkage groups, six years before the article by FB Hutt (1936), but their work was overlooked or not properly credited by others (Moiseyeva et al. 2000; Romanov et al. 2004). The 1930 chicken map comprised four linkage groups with 12 genes and four other unlinked genes (Fig. 6) and was improved in two amendments published by Petrov (1931) and Sungurov (1933). Aggregated together, the map designed by Ser-

ebrovsky, Petrov, and Sungurov, with the acknowledged assistance of Serebrovskaya, Wassina, Rebrina, Kobystina, Ovsyannikova, and Grechka, included 15 chicken genes on six linkage groups: I, or Z chromosome (*ID-BARR-SLC45A2-K*), II (*CP-R*), III (*NA-BL*), IV (*LMBR1-D*), VI (*MC1R-FM*), and IX (*CR-SILV–F*), plus six independent loci (*P, APOA1, MB, TYR, PTI1, and PTI2*), and the recessive ptilopody gene*. The linkage group assignments or independent positions for these loci have been recently confirmed by molecular mapping (Sazanov et al. 1998; Okimoto et al. 1999; Pitel et al. 2000; Smith et al. 2000b, 2001a; Schmid et al. 2000; Kerje et al. 2003; Huang et al. 2006a).

These early linkage mapping efforts were supplemented by Dunn and Jull (1927), Warren (1928, 1933, 1935), Dunn and Landauer (1930), Jull (1930), Landauer (1931), Suttle and Sipe (1932), Hertwig (1933), Hutt (1933), Warren and Hutt (1936), and others. Hutt (1936) prepared the second map that consisted of 18 genes assigned to five linkage groups (Fig. 6). For some reason, Hutt did not provide any appropriate credit to the Serebrovsky and Petrov

*The numbering of true linkage groups (I–IV, VI, and IX) takes into account three independent loci (*P, APOAI*, and *MB*) that were also considered by Serebrovsky, Petrov, and Sungurov as single linkage groups V, VII, and VIII, respectively.

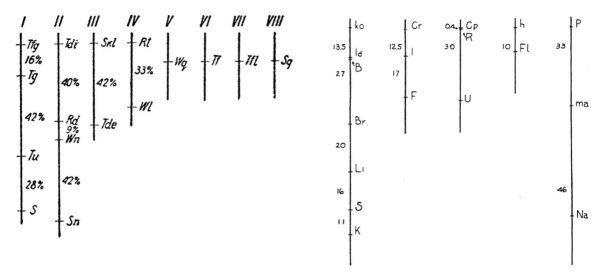

Fig. 6 The Serebrovsky and Petrov (1930) chicken chromosome map (*left*) based on the data obtained at the Central Station for Livestock Genetics, Anikovo by December 1, 1929, and the Hutt (1936) map (*right*). The genes on the Serebrovsky–Petrov map have been designated in accordance with Serebrovsky's own nomenclature (reviewed by Dunn 1928). The currently accepted locus symbols are given below in parentheses:

Linkage group I [GGAZ]: *Tfg (ID)* — *Tg (BARR)* — *Tu (SLC45A2)* — *S (K)*
Linkage group II: *Tdi (SILV)* [E22C19W28_E50C23 (classical group II)] — *Rd (CP)* [GGA4? (classical group I)] — *Wn (R)* [GGA4? (classical group I)] — *Sn (CR)* [E22C19W28_E50C23 (classical group II)]
Linkage group III: *Skl (NA)* [GGA3] — *Tde (BL)* [UN]
Linkage group IV [GGA2 (classical group IV)]: *Rt (LMBR1)* — *Wl (D)*
Chromosome V [GGA1]: *Wq (P)*
Chromosome VI [GGA11]: *Tf (MC1R)*
Chromosome VII [GGA24]: *Tfl (APOA1)*
Chromosome VIII [GGA1 or GGA24]: *Sq (MB)*

(1930) map, although aware of that study and even referring to it (Hutt 1933, 1936, 1949; Warren and Hutt 1936; Hutt and Mueller 1943). Hutt named his 1936 map the "first chromosome map" in subsequent publications (e.g., Hutt and Lamoreux 1940; Hutt 1949). The second map included three more loci (*KO, BR, LI*) on the sex chromosome, one more gene (*U*) linked to the *CP-R* group, one more gene (*ma*, which is now *DB*; Crawford 1990) linked to the *P* gene, and a new linkage group including *H* and *FL* (reviewed by Romanov et al. 2004). On the other hand, *P* and *NA* were located on the same chromosome in the Hutt (1936) map, which is now known not to be true, and did not include two Serebrovsky-Petrov-Sungurov linkage groups (*LMBR1-D* and *MC1R-FM*).

5.2.4
Subsequent Classical Mapping

The importance of genome maps of chicken and other domestic animals for understanding and utilizing the genetic foundations of these species has been acknowledged by both scientists and commodity groups (Romanov et al. 2004). In the 1920s and 1930s, the chromosome topography studies at the Anikovo Station were coordinated with egg production, growth, and other traits of economic values (Serebrovsky and Wassina 1927; Serebrovsky and Petrov 1928). Serebrovsky and Petrov (1930) proposed a "*signal gene*" concept, which is comparable to the modern notions of genetic marker and marker-assisted selection (MAS) (reviewed by

Romanov et al. 2004). The concept suggested that a signal gene does not affect an economic trait by itself. However, if located near a gene for an economic trait, it serves as a landmark for determining the latter. The accuracy of predicting such an association between the signal gene and the economic trait gene depends on the position of the signal gene relative to the trait gene, and the best prediction is achieved with two signal genes closely flanking the economic trait gene. Knowledge of a thorough chromosome map is a prerequisite for using the signal gene approach.

The Serebrovsky and Petrov (1930) and Hutt (1936) maps were further advanced by revisions of Hutt and Lamoreux (1940), Hutt (1949, 1960, 1964), Etches and Hawes (1973), Somes (1973, 1978, 1987), and Crawford (1990). The last update of the map (Bitgood and Somes 1993) listed 140 loci/traits including morphological mutations, biochemical polymorphisms, and chromosomal breakpoints (Romanov et al. 2004). Physical map positions were established for 41 single gene loci on five autosomal linkage groups and the Z chromosome (Tables 9 and 10). Moreover, there were 83 loci/traits assigned to one of the groups or chromosomes but without exact mapping information (including 25 loci/traits placed on micro-

chromosomes), and eight pairs of linked markers are not anchored to a linkage group.

5.3
Molecular Genetics and Whole-Genome Sequence

5.3.1
First-generation Molecular Maps

In the 1990s, three reference mapping populations were developed for the chicken (reviewed by Romanov et al. 2004): the Compton population created at the Institute for Animal Health, UK (Bumstead and Palyga 1992); the East Lansing (EL) population developed at Michigan State University in collaboration with the United States Department of Agriculture (USDA) Avian Disease and Oncology Laboratory and the University of California at Davis, USA (Crittenden et al. 1993; Cheng et al. 1995); and the Wageningen University, Netherlands population (Groenen et al. 1998). The EL population involved 400 back cross (BC) progeny from two highly inbred lines, UCD001 (red junglefowl) and UCD003 (WL). The BC design maximized variation of the DNA markers to be mapped, so that each autosomal

Table 10 Breakdown of the updated classical gene map (Bitgood and Somes 1993) of the chicken

Classical linkage group	I	II	III	IV	V	VI	VII	VIII	X	IX[a]
Chromosome	GGA2, GGA3 or GGA4	GGA2, GGA3 or GGA4	GGA1	GGA2, GGA3 or GGA4	GGAZ	GGAW	GGA6	GGA7	GGA17 (now GGA16)	Other micro-chromosomes
No. of mapped loci	3	4	12	3	17	0	0	0	2	0
No. of assigned loci precisely not mapped[b]	13[c] (7)	See the table footnote c	23 (2) footnote c	See the table	21	1	4	1	8	17 (3)

[a] Discontinued (Crawford 1990)

[b] Numbers of loci assigned to more than one chromosome due to conflicting reports are given in parentheses

[c] Including three loci assigned to GGA2, one to GGA3, seven to more than one chromosome due to conflicting reports, and two to classical group I. If the two unmapped classical group I loci are ignored, the remaining 11 loci might equally belong to classical groups I, II, or IV. The chromosomes containing each of these linkage groups were unknown in 1993, but they were presumably GGA2, GGA3, and GGA4 (Bitgood and Somes 1993)

marker would be biallelic in the BC population. These three chicken linkage maps were integrated into one consensus map by Groenen et al. (2000). By that time, there were 1,965 loci localized on 26 chromosomes and 24 unknown linkage groups (Schmid et al. 2000). Further updates led to a map covering 4,200 cM with 2,261 loci on 53 linkage groups (Schmid et al. 2005). In many cases, smaller subsets of individuals were used to build the afore-mentioned reference maps (e.g., the EL map was based on genotypes for only 52 animals at most marker loci). The Wageningen mapping population included a larger set of animals (reviewed by Romanov et al. 2004).

Another chicken linkage map was developed using the Kobe University, Japan resource family (Lee et al. 2002). The integrated Hiroshima-Tsukuba map was also constructed in Japan using a resource population based on a cross between Japanese Game and White Leghorn chickens, and 301 markers, including 183 new ones, were localized to specific chromosomes either through linkage analysis or by analysis of the chicken draft sequence (Takahashi et al. 2005).

Initially, restriction fragment-length polymorphisms (RFLP) and random amplified polymorphic DNA (RAPD) markers with unknown sequence information were placed on the chicken genetic linkage map. Subsequently, microsatellites (or short tandem repeats), amplified fragment-length polymorphisms (AFLP), single nucleotide polymorphisms (SNP), and other sequence tagged site (STS) markers became the markers of choice. Mapped loci are subdivided in two classes: type I (coding sequences) and type II (anonymous, mainly microsatellites) markers. At present, 31 of the 53 linkage groups have been assigned to a particular chromosome (Masabanda et al. 2004). For the remaining 22 linkage groups, an ExxCxxWxx number is used with capital letters corresponding to the linkage groups of the original three linkage maps: East Lansing, Compton, and Wageningen (Schmid et al. 2005).

To map a mutant, experimental families segregating for one or more mutations are usually constructed (reviewed by Romanov et al. 2004). If no preliminary information of the chromosomal location of a mutant is available, a whole-genome scan using molecular markers is carried out (e.g., Lee et al. 2002). Bulked segregant analysis can essentially lower the cost of this approach (Michelmore et al. 1991; Ruyter-Spira et al. 1997, 1998; Pitel et al. 2000), where genotyping is done on DNA samples pooled according to the phenotype.

Currently, the positions of 62 classical mutants and loci have been determined based on linkage with molecular markers or localization using the whole-genome sequence assembly. There is one instance of discrepancy (the *DMD* locus) between the assembly data and linkage or physical mapping data (Table 9). The 78 classical loci/traits listed in Bitgood and Somes (1993) have to be assigned to the molecular map, not to mention many other reported classical mutations that have not yet been mapped. Only group I on the classical map that involves five mutations has not yet been connected with the molecular map. By analogy with the human and mouse short-limb disorders, the *FGFR3* gene mapped to GGA4 (Suchyta et al. 2001) might contain a causative mutation for the chicken creeper (*CP*), so the position of group I could be expected on this chromosome (Romanov et al. 2004). Mapping of the crest (*CR*) and frizzle (*F*) mutations that flank the dominant white mutation *(SILV)* on linkage group E22C19W28_E50C23 (Ruyter-Spira et al. 1997) would also facilitate the integration of molecular and classical maps (Schmid et al. 2000). A preliminary study by an Indian group (GB Pant University of Agriculture and Technology, http://gbpuat.ac.in/acads/cvsc/gnab.htm) demonstrated that the *F* gene is linked to *ROS0054* and *MCW0188* microsatellite loci on E22C19W28_E50C23. If there is a candidate gene for a mutation, it could be used to map the trait either by fluorescent in situ hybridization (FISH; e.g., Suzuki et al. 1999b) or by identification of a SNP within the gene (e.g., Dunn et al. 1999). Thus, the number of classical genes mapped with molecular markers is expected to increase.

Another approach for development of the genomic maps is the radiation hybrid (RH) panel (Brown et al. 2003). Whole-genome radiation hybrid (WGRH) panels can give higher resolution than conventional recombination analysis, and no polymorphisms are required for RH mapping. A chicken WGRH panel (ChickRH6) was created at Institut National de la Recherche Agronomique (INRA), France, by fusion of irradiated (6,000 rad) chicken embryonic fibroblasts and HPRT-deficient hamster cells (Morisson et al. 2002). The average retention rate of the chicken chromosomes was estimated as 21.9% in 90 clones, although it was lower than 20% for the two largest macrochromosomes.

An enormous collection of chicken ESTs that can be used for marker design and mapping have been generated making it the fourteenth most plentiful organism in the NCBI dbEST database (http://www.ncbi.nlm.nih.gov/dbEST/dbEST_summary.html; as of February 21, 2008). ESTs also serve as a source of information for identification of expressed genes and their function, and for annotating genome sequence and physical maps (Abdrakhmanov et al. 2000; Tirunagaru et al. 2000; Boardman et al. 2002; Brown et al. 2003).

5.3.2
Physical Maps

A clone-based physical map is built using contiguous, overlapping recombinant DNA clone inserts (contigs) that cover all, or almost all, of the genome (reviewed by Romanov et al. 2004). Contigs are often constructed by fingerprinting, which is done by digesting clones, such as BACs, with restriction enzymes. The resulting restriction patterns (or fingerprints) are then analyzed for shared fragments and overlapping clones are assembled into contigs. The integration of the genetic linkage map with the physical map provides the critical bridge between phenotypes, i.e., major mutations and quantitative trait loci (QTL), and their causative gene/allele combinations.

Since the mid-1990s, physical mapping resources have become readily available for major farm animal species including chicken. High-density large-insert libraries have been generated for the chicken and provide 4–10-fold genome coverage (Toye et al. 1997; Zimmer and Verrinder Gibbins 1997; Buitkamp et al. 1998; Crooijmans et al. 2000; Schmid et al. 2000; Kato et al. 2002). Using DNA from the UCD001 red junglefowl female 256 (Fig. 7), three BAC libraries have been produced in a collaboration between Texas A & M University and Michigan State University, USA (Lee et al. 2003). Based on the same UCD001 genome, another BAC library has been made at the Children's Hospital Oakland Research Institute (CHORI), CA, USA (Nefedov et al. 2003). Experimental chicken BAC libraries were also generated in Japan (Hori et al. 2000; Kato et al. 2002) and China (Liu et al. 2003b). These BAC libraries are publicly available (Table 11).

Many of these BAC libraries, including the four UCD001-based libraries and one derived from a WL chicken, have been used in physical mapping and genome sequencing projects (Crooijmans et al. 2000). BAC clones from these libraries were fingerprinted to build a physical contig map covering more than 90% of the chicken genome (Ren et al. 2003; Wallis et al. 2004). In parallel, a BAC-based whole-genome physical map of the chicken genome was integrated with the linkage map by hybridizing probes containing markers to filter-spotted arrays (Lee et al. 2003; Ren et al. 2003; Romanov et al. 2003).

For integrating genetic and physical maps, a high-throughput screening technique is BAC filter hybridization using highly specific OVERGO probes (Romanov et al. 2003), which are overlapping oligo probes derived from specific sequence regions in known genes or markers. The OVERGO probes are synthesized by annealing two oligonucleotides that have an 8-bp overlap, followed by labeling in vitro. Use of OVERGOs facilitates pooling strategies because the melting temperatures for all probes are usually the same.

Typically, a collection of OVERGOs is arranged for three-dimensional screening by plates, rows, and columns. In the case of 6 × 6 × 6 screening scheme, a set of 216 probes is designed, and a pool 36 OVERGOs is used for a single hybridization. Each probe

Fig. 7 A female 256, the inbred red junglefowl line UCD001, served as a DNA source for generating BAC libraries and the draft sequence of the chicken genome. (Photograph courtesy of William S. Payne)

Table 11 Avian genomic BAC libraries

Species/breed (strain)/individual	Average insert size (kb)	No. of clones	Genome coverage	Vector	Cloning site	Country, library code (source's website or e-mail address)
Red junglefowl/inbred line UCD001/female #256	150	38,400	5.2×	pBeloBAC11	*Bam*HI	USA, 031-JF256-BI (http://hbz7.tamu.edu/homelinks/bac_est/bac.htm#animal)
Red junglefowl/inbred line UCD001/female #256	152	38,400	5.3×	pECBAC1	*Eco*RI	USA, 032-JF256-RI (http://hbz7.tamu.edu/homelinks/bac_est/bac.htm#animal)
Red junglefowl/inbred line UCD001/female #256	171	38,400	6.0×	pECBAC1	*Hin*dIII	USA, 033-JF256-H3 (http://hbz7.tamu.edu/homelinks/bac_est/bac.htm#animal)
Red junglefowl/inbred line UCD001/female #256	195	73,700	12.0×	pTARBAC2.1	*Eco*RI	USA, CHORI-261 (http://bacpac.chori.org/chicken261.htm)
Chicken/White Leghorn/a female	130	49,920	5.4×	pECBAC1	*Hin*dIII	USA, The Netherlands, 020-CHK-H3 (http://hbz7.tamu.edu/homelinks/bac_est/bac.htm#animal; http://www.geneservice.co.uk/products/clones/chicken_BAC.jsp)
Chicken/White Leghorn (Julia line)/a female embryo	149	49,152	3.2×	pBAC-Lac	*Hin*dIII	Japan, N/A (s-mizuno@brs.nihon-u.ac.jp)
Chicken/White Silkie/a female	118	138,240	13.34×	pBeloBAC11	*Hin*dIII	China, N/A (ninglbau@public3.bta.net.cn)
Turkey/inbred line, Nicholas Turkey Breeding Farms/a female	190	71,000	11.1×	pTARBAC2.1	*Eco*RI	USA, CHORI-260 (http://bacpac.chori.org/turkey260.htm)
Duck	117.94	84,480	9.84×	pIndig-5	*Hin*dIII	China, N/A (ninglbau@public3.bta.net.cn)
Emu	165	133,632	13.5×	pCCBAC1E	N/A	USA, VMRC16 (http://www.benaroyaresearch.org/investigators/amemiya_chris/libraries.htm)
Zebra finch	134	147,456	15.5×	pCUGIBAC1	*Hin*dIII	USA, TG_Ba (http://www.genome.arizona.edu/orders/direct.html?library=TG_Ba)
California condor/female "Molloko" (Studbook #45)	N/A	89,665	~14×	pTARBAC2.1	*Eco*RI	USA, CHORI-262 (http://bacpac.chori.org/library.php?id=222)

N/A, not available

can be assigned to a number of positive BAC clones common for a particular intersection of plate, row, and column (Romanov et al. 2003).

Screening of the four UCD001 BAC libraries identified 918 genes and markers across all chromosomes and linkage groups, resulting in assignments of nearly 8,000 clones. Most of the OVERGOs were single copy in the chicken genome and the resulting assignments are available online (U.S. Poultry Genome Project http://poultry.mph.msu.edu/resources/Resources.htm#bacdata) and contributed to the alignment of the first-generation BAC-contig map (Ren et al. 2003) to the linkage map (Fig. 8). They also aided in alignment of the second-generation physical map to the linkage map (Wallis et al. 2004), developed in parallel with the whole-genome sequence, and resulted in the assignment of BAC contigs to specific chicken chromosomes (Fig. 9). The second-generation physical map was made at 20-fold coverage and contained 260 contigs of 180,000+ overlapping clones. It covers about 91% of the chicken genome and has been used for determining chicken BACs aligned to positions in other sequenced genomes (Wallis et al. 2004).

Additionally, the physical map has been integrated with the cytogenetic map. Many BACs positive for genes have been hybridized by FISH to several chicken chromosomes (e.g., Sazanov et al. 2004a, b; Fig. 10), and a detailed analysis of microchromosome 17 using FISH has been conducted (Romanov et al. 2005). The GGA17 map orientation was demonstrated to be different and reversed from that currently proposed for the linkage map and draft sequence.

5.3.3
Whole-Genome Sequence

Over the previous 100 years of chicken genetics, efforts have been aimed at genetic mapping in order to identify, characterize, and locate genes associated

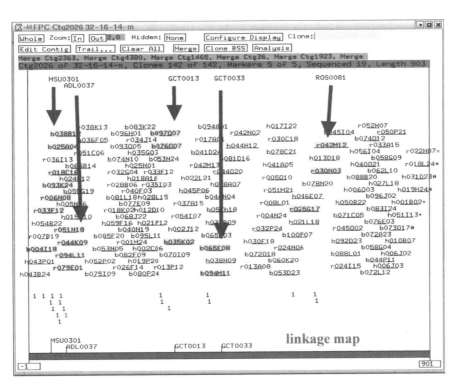

Fig. 8 First-generation BAC physical map of the chicken genome (after Ren et al. 2003). Example of a BAC contig anchored to the GGA1 genetic map. This contig consists of 142 clones from three source BAC libraries (prefixed with "h," "b," and "r"), contains 903 unique fingerprint bands, and is estimated to span 4.01 Mb. The contig was anchored to the region around 361 cM of the GGA1 genetic map using five DNA markers, *MSU0301, ADL0037, GCT0013, GCT0033,* and *ROS0081* (Groenen et al. 2000) as shown with the arrows. The highlighted clones indicate the positive clones identified by DNA marker hybridization

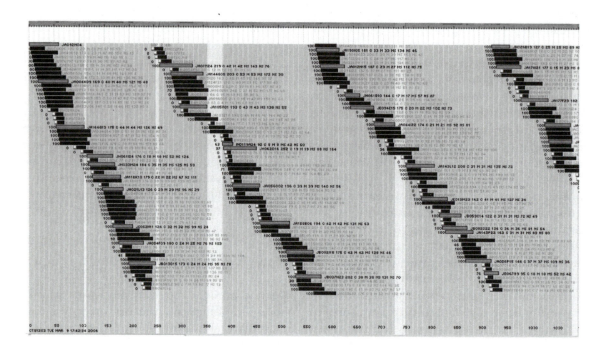

Fig. 9 Chicken BAC tiling set from the fingerprint map for, List 003, Ctg 1203 (Martin Krzywinski, Genome Sciences Centre, Vancouver, Canada). The estimated minimum tiling path set consisted of 9,210 BAC clones with an average clone overlap of 77 kb (Wallis et al. 2004)

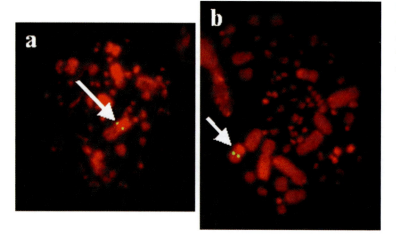

Fig. 10 FISH of the chicken BAC clones. (a) Clone b071F17 (*KITLG*, GGA1; Sazanov et al. 2004a). (b) Clone b027G23 (*CTSL*, GGAZ; Sazanov et al. 2004b; *arrows* indicate sites of specific hybridization)

with productivity and health of the species (Romanov et al. 2004). To elucidate genomic architecture underlying productivity and disease resistance traits, further progress in chicken gene discovery and, eventually, the complete genome sequence will be required. The whole-genome sequence is the ultimate physical map and the basis for a high-resolution linkage map (Dodgson 2003).

In February 2002, a "white paper" for sequencing the chicken genome was submitted to the US National Human Genome Research Institute (NHGRI). The proposal stated that because of its evolutionary distance from mammals (around 310 MYA), the chicken would make a significant contribution to comparative genomics at the sequence level. Due to a notable level of conservation in gene order between mammalian and chicken genomes, the chicken genome is also a perfect model for studying the evolution of gene order and arrangement (Burt et al. 1999; Groenen et al. 2000; Waddington et al. 2000; Suchyta et al. 2001). The NHGRI added the chicken to the list of high-priority genomes for sequencing, making it the first sequenced bird genome and also the first sequenced agricultural species (Jensen 2005). The project objectives were to provide the assembly of a 6-fold whole-genome shotgun coverage of the UCD001 genome and ordered the resulting sequence scaffolds by alignment to BAC, fosmid, and plasmid-paired end reads in a comprehensive contig map at the Washington University Genome Sequencing Center (WUGSC, St. Louis, MO, USA) (reviewed by Romanov et al. 2004). This strategy led to a high-quality assembly thanks to the relatively small size of the chicken genome (1/3 that of a mammal) and low repetitive DNA content (only 11% compared with 40–50% in mammals) (Burt 2005).

The approximately 1 Gigabase sequence published in *Nature* by the International Chicken Genome Sequencing Consortium (2004) is based on DNA from the inbred red junglefowl female 256. At a later stage, gaps in the genome sequence should be finished and errors eliminated in the contig assembly. In particular, a substantial number of clone contigs have unknown or ambiguous chromosome assignment (Aerts et al. 2005). As a contribution to the assembly, Aerts et al. (2005) mapped 86 SNP markers derived from 86 clones on the genetic map and, thus, anchored 56 clone contigs and 13 individual BACs that correspond to a total of 57,145 clones. Another problem is a poor assembly of the sex chromosome sequences, which currently contains only 30% of the Z [expected 100 Megabases (Mb)] and 2% of the W (expected 30 Mb) chromosomes, due to single copies of these chromosome in the female used for sequencing as well as high repeat content in the W chromosome. Owing to overlaps with an additional set of BACs sequenced to high quality, sequence coverage on the autosomes was 98% (Burt 2005). The sequence was only obtained for 30 chromosomes, although the final goal is to have linkage and sequence maps for all 39 chromosomes in the chicken genome (Schmid et al. 2005).

The draft sequence is also being annotated in terms of aligning and characterizing genes and other genome elements. Overlaps with cDNA clones suggested 5–10% of genes were missing from the final assembly because of gene duplications (e.g., MHC

region) and GC-rich sequences (Burt 2005). The sequence annotation will eventually contain an estimate of 20,000–23,000 chicken genes (International Chicken Genome Sequencing Consortium 2004).

Advanced bioinformatics resources involving genome browsers, genetic maps, marker and gene expression databases, and other related poultry genetics and genomics information are available on the World Wide Web (see the selected list at the end of this chapter).

5.3.4
Chicken Genome and Sequence Features

Birds are characterized by the greatest conservatism of genome size among vertebrate animals, with the diploid nuclear DNA content per nucleus ranging between 2.5 and 3.0 picograms (1 pg = 978 Mb) (as reviewed by Romanov et al. 2004). Haploid DNA content (C value) for various avian species, including the chicken, is presented in Table 8. The average avian haploid genome is 1.45 pg; flightless birds have larger genome sizes, with the largest one being 2.16 pg in the ostrich. The chicken genome is at 2.8-fold less than the average mammalian genome (Gregory 2006).

The size of the avian genome positively correlates with red blood cell and nucleus sizes and negatively with metabolic rate. There is no correlation with developmental rate or longevity, and no cases of polyploidy in birds are known. Comparatively low DNA content could be because of the "necessity of flight," i.e., as a response to selection for high metabolism/flight, or due to high evolutionary conservatism of this parameter, taking into consideration monophyletic origin of the class *Aves* (Kadi et al. 1993; Gregory 2006).

The avian karyotype is characterized by a remarkably large number and heterogeneity of chromosomes. The avian karyotype contains several macrochromosomes (3–8 m) and numerous microchromosomes (0.3–3 m) (Schmid et al. 2000). The chicken karyotype is thought to represent an ancestral type of avian karyotype (Rodionov 1997; Derjusheva et al. 2004).

In the past, the number of macrochromosomes varied in the literature between 6 and 10 pairs, including the Z and W in the heterogametic female (Schmid et al. 2005). The International Chicken Genome

Sequencing Consortium (2004) designated three chromosome size groups: large macrochromosomes (GGA1–5), intermediate chromosomes (GGA6–10), and 28 microchromosomes (GGA11–38). Masabanda et al. (2004) and Schmid et al. (2005) proposed a new, definitive classification system. In accordance with this classification, group A is composed of chromosomes 1–10, Z, and W (cytogenetically distinguishable macrochromosomes tractable in a flow karyotype). Group B is composed of chromosomes 11–16 (large microchromosomes up to and including the nucleolar organizing region chromosome). Group C is composed of chromosomes 17–32 (small microchromosomes most of which associated with known linkage groups) and group D chromosomes 33–38 (smallest microchromosomes not yet associated with known linkage groups).

Chicken microchromosomes constitute about 23% of the genome and possess not less than 50% of the avian genes (Smith et al. 2000b; Schmid et al. 2000). Furthermore, there are many indirect indications of the increased gene density on the microchromosomes (e.g., Andreozzi et al. 2001; Habermann et al. 2001). The recombination frequencies in macro- and microchromosomes are one crossover per 30 and 12 megabases (Mb), respectively, which is two and five times less than in mammals (Rodionov 1996). Detailed elaboration of the structural and functional organization of chicken microchromosomes would be useful for both enlightening minimally required elements of eukaryotic chromosomes and studying the evolution of vertebrate karyotypes (Romanov et al. 2004).

As a result of the comparative genomics analysis, 80 or more regions of evolutionary conservation have been suggested on the aligned human and chicken chromosomes (Burt et al. 1999; Burt 2002). This level of conserved grouping of orthologous genes, also called conserved synteny ("gene loci in different organisms located on a chromosomal region of common evolutionary ancestry"; Passarge et al. 1999), was even higher than that between the human and the mouse (Burt et al. 1999).

As a straightforward approach for direct physical mapping, FISH of chicken chromosomes is normally used. The intrachromosomal localization has been identified for around 250 type I markers (reviewed by Romanov et al. 2004). Moreover, many cytogenetically assigned large-insert clones that include coding sequences can be employed for com-

parative genome anchoring. There are a number of FISH techniques for determining hybridization signals that are used for chicken genome mapping. Using large-insert clones of genomic libraries like BACs as DNA probes for FISH achieves almost 100% efficiency of hybridization, making it one of the most perspective approaches (Buitkamp et al. 1998; Smith et al. 2000b; Sazanov et al. 2002). Sets of chromosome-specific clones and whole chromosome paints represent powerful tools for microchromosome detection and ordering (Zimmer et al. 1997; Fillon et al. 1998; Guillier-Gensik et al. 1999). To improve the resolution of FISH technique, lampbrush chromosomes, in addition to mitotic ones, can be effectively used (Mizuno and Macgregor 1998; Rodionov et al. 2002). Using the confocal microscopy, spatial distribution of the chromosome paints in the chicken nucleus can be examined to better understand micro- and macrochromosome localization features during interphase (Habermann et al. 2001). Chromosome microisolation and microcloning followed by isolation and mapping of microsatellite markers is another approach for increasing map density (Ambady et al. 2002). The combination of molecular and cytological approaches was demonstrated in a study of the W chromosome by Itoh and Mizuno (2002).

Ultimately, the 6.6-fold coverage draft genome sequence was generated and its analysis revealed the following major features (International Chicken Genome Sequencing Consortium 2004):

- The chicken genome is characterized by a substantial decrease in interspersed repeat content, pseudogenes and segmental duplications, and in intron size. This reduction accounts for the nearly 3-fold difference in size between the chicken and mammalian genomes.
- There are long blocks of conserved synteny that contain chicken–human aligned segments (Fig. 11).
- When comparing macro- vs. microchromosomes, there is a negative correlation between the size of chicken chromosomes and recombination rate (Fig. 12), G + C and CpG content, and gene density, but there is a positive correlation between chromosome size and repeat density.
- Genes in both chicken microchromosomes and in subtelomeric regions of macrochromosomes show higher synonymous substitution rates.

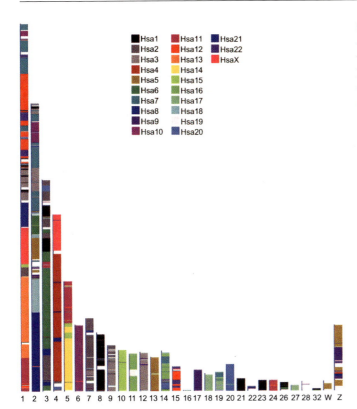

Fig. 11 Maps of conserved synteny between chicken chromosomes and human chromosomes (reprinted with permission from Macmillan Publishers Ltd: Nature, International Chicken Genome Sequencing Consortium 2004,© 2004): chicken compared to human (*top*), and human compared to chicken (*bottom*)

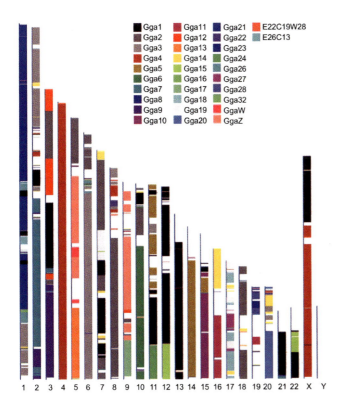

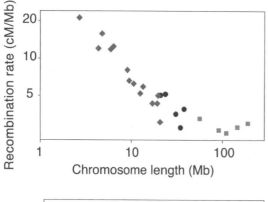

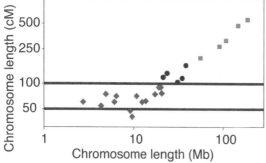

Fig. 12 Relationships between chicken chromosome characteristics for chromosomes 1–28: comparison of recombination rate and sequence length (*top*), and comparison of genetic and sequence length (*bottom*). Both plots exclude chromosomes 16, 22, 23, 25, which have insufficient genetic markers or sequence. *Upright squares*: macrochromosomes; *circles*: intermediate chromosomes; *diagonal squares*: microchromosomes (adapted with permission from Macmillan Publishers Ltd: Nature, International Chicken Genome Sequencing Consortium 2004, © 2004)

- Unlike other vertebrate genomes, the chicken genome has had no active insertions of short interspersed nucleotide elements (SINE) over the last 50 MYR.
- At least 70 Mb of the chicken–human aligned sequences seem to be functional in both species.
- Alignment of the chicken–human noncoding sequences often led to their localization far from genes and in clusters that are likely to be under selection for unknown functions.

In a parallel article, the International Chicken Polymorphism Map Consortium (2004) described 2.8 million SNPs that represented the first chicken genome-wide genetic variation map. This set of SNPs was designed by comparing the sequences of three domestic chicken breeds (a broiler, a layer, and a Chinese Silkie) and the red junglefowl. At least 90% of the variant sites were true SNPs, and at least 70% were common SNPs showing segregation in many domestic breeds. For almost every possible comparison between domestic breeds and junglefowl, average nucleotide diversity was about five SNPs per 1 kilobase (kb), which contradicts with previous views of domestic animals as highly inbred in comparison with their wild progenitors. Most of the chicken SNPs seem to have arisen prior to domestication, and little evidence of selective sweeps for adaptive alleles was found on length scales greater than 100 kb.

The chicken genome sequence and genetic polymorphisms are expected to benefit agriculture and medicine, shed new light on animal domestication, and provide an ideal model for studies in development and evolution as well as comparative research in 9,600 extant avian species (Burt 2005).

5.3.5
Genetics and Molecular Mapping in Other Birds

Until recently, genetic studies and gene mapping in the other poultry, semidomesticated and caged species (Table 1) have been carried out at a significantly slower pace despite the fact that duck, Muscovy duck, canary, pigeon and budgerigar, along with chicken, mouse, rat, rabbit, and three fish species, were among the first vertebrates in which sex-linked and autosomal-linked genes were found (Durham and Marryat 1908; Spillman 1908; Hutt 1936; Table 12). The limited classical linkage maps of the Japanese quail and turkey Z chromosome involved only two and three morphological loci, respectively (Crawford 1990; Minvielle et al. 2000). There were also four known classical Z-linked loci in the turkey, five in the ring-necked pheasant, three in the guinea fowl, three in the peafowl, five in the pigeon, one in the African collared dove, four in the domestic duck, two in the Muscovy duck, one in the mute swan and, presumably, up to five in the goose (Table 12). Additionally, four classical autosomal linkage relationships have been reported in the Japanese quail, two in each of the turkey and dove, and one in each of the duck and pigeon. Several cases of sex and autosomal linkage have been discovered in major caged birds.

Chapter 5 **Poultry** 111

Table 12 Linkage in other avian species raised by man

Loci linked (locus alleles) <aliases>	Trait name (synonym)	Chromosome (Z, W, 1) or autosomal linkage (AL)	Reference
Anseranser (goose)			
b	Diluted feet	Z	Staško (1970)
*b*¹	Buff celler	Z	Hollander (1990)
G	Gray	Z (linked to *Sd*)	Crawford (1990)
Sd	Dilution	Z (linked to *G*)	Crawford (1990)
*Sp*²	Solid pattern	Z	Crawford (1990)
Cygnusolor (muteswan)			
r	Polish	Z	Lancaster (1977)
Cairina moschata (Muscory duck)			
ALDOB	Aldolase B, fructose-bisphosphate	Z	Nanda and Schmid (2002)
*ch*³	Chocolate	Z	Sokolovskaya (1935); Hollander (1970); Crawford (1990)
—	Crest	Z	Sokolovskaya (1935)
HBA1	Hemoglobin, alpha 1 (globin, alpha A)	AL (linked to *HBA2* and *HBAZ*)	Niessing et al. (1982); Erbil and Niessing (1984)
HBA2	Hemoglobin, alpha 2 (globin, alpha D)	AL (linked to *HBA1* and *HBAZ*)	Niessing et al. (1982); Erbil and Niessing (1984)
HBAZ	Hemoglobin, zeta (embryonic alpha-globin pi-prime)	AL (linked to *HBA1* and *HBA2*)	Niessing et al. (1982); Erbil and Niessing (1984)
HBG1	Hemoglobin, gamma A (globin, epsilon)	AL (linked to *HBG2*)	Lin and Paddock (1984)
HBG2	Hemoglobin, delta (globin, beta)	AL (linked to *HBG1*)	Lin and Paddock (1984)
Anas platyrhynchos (domestic duck) Buff dilution			
bu		Z	Crawford (1990)
*d*³	Brown dilution	Z	Crawford (1990)
IFNA1	Interferon, alpha 1	Z	Nanda et al. (1998)
IFNB1	Interferon, beta 1, fibroblast	Z	Nanda et al. (1998)
WPKCI-8 <Wpkci, HINTW, Wpkci-7, ASW>	W chromosome-specific histidine triad nucleotide-binding protein 1	W	Hori et al. (2000)
ASL <CRYD2>	Argininosuccinate lyase (crystallin, delta 2)	AL (linked to *CRYD1*)	Li et al. (1995)
CRYD1<d-cry>	Crystallin, delta 1	AL (linked to *ASL*)	Li et al. (1995)
E	Black	AL (linked to *S*)	Crawford (1990)
HLA-B	MHC, class I, B	AL (linked to *TAP2*)	Mesa et al. (2004)
S	Bib	AL (linked to *E*)	Crawford (1990)
TAP2	Transporter 2, ATP-binding cassette, subfamily B (MDR/TAP)	AL (linked to *HLA-B*)	Mesa et al. (2004)

(continued)

Table 12 (Continued)

Loci linked (locus alleles) <aliases>	Trait name (synonym)	Chromosome (Z, W, 1) or autosomal linkage (AL)	Reference
		Meleagris gallopavo (turkey) Achondroplasia	
ach		Z	Crawford (1990)
bo	Bobber	Z	Crawford (1990)
e	Brown	Z	Crawford (1990)
K	Late feathering	Z	Crawford (1990)
n (n, n^{al})	Narragansett, imperfect albinism	Z	Crawford (1990)
tt	Tetanic torticollar spasm	Z	Savage et al. (1993)
vi	Vibrator	Z	Crawford (1990)
WPKCI-8 <Wpkci, HINTW, Wpkci-7, ASW>	W chromosome-specific histidine triad nucleotide-binding protein 1	W	Hori et al. (2000)
D	Slate	AL (linked to *ga*)	Crawford (1990)
ga	Glaucoma	AL (linked to *D*)	Crawford (1990)
ha	Hairy	AL (linked to r)	Crawford (1990)
r	Red	AL (linked to *ha*)	Crawford (1990)
		Pavo cristatus (Indian peafowl)	
ca <d>	Cameo (silver-dun)	Z	Somes and Burger (1988); Hollander (1990)
—	Purple	Z	Legg[4]
—	Peach	Z	Legg[4]
WPKCI-8 <Wpkci, HINTW, Wpkci-7, ASW>	W chromosome-specific histidine triad nucleotide-binding protein 1	W	Hori et al. (2000)
a	*Phasianus colchicus* (ring-necked pheasant) Incomplete albinism	Z	Crawford (1990)
Ba	Barring	Z	Crawford (1990)
di	Dilute	Z	Crawford (1990)
DMRT1	Doublesex and mab-3-related transcription factor 1	Z	Nanda et al. (2000)
id	Dermal melanin	Z	Crawford (1990)
s	Gold	Z	Crawford (1990)
		P. versicolor (green pheasant)	
WPKCI-8 <Wpkci, HINTW, Wpkci-7, ASW>	W chromosome-specific histidine triad nucleotide-binding protein 1	W	Hori et al. (2000)
		Chrysolophus pictus (golden pheasant)	
ALDOB	Aldolase B, fructose-bisphosphate	Z	Nanda and Schmid (2002)

(continued)

Chapter 5 **Poultry** 113

Table 12 (Continued)

Loci linked (locus alleles) <aliases>	Trait name (synonym)	Chromosome (Z, W, 1) or autosomal linkage (AL)	Reference
DMRT1	Doublesex and mab-3-related transcription factor 1	Z	Nanda et al. (2000)

Coturnix coturnix (common quail)

DMRT1	Doublesex and mab-3-related transcription factor 1	Z	Nanda et al. (2000)

C. japonica (Japanese quail)

ACO1 <IREBP>	Aconitase 1, soluble	Z	Saitoh et al. (1993)
ALDOB	Aldolase B, fructose-bisphosphate	Z	Suzuki et al. (1999a), Nanda and Schmid (2002)
SLC45A2 <AL (al, al^C, al^D, al^reb)>	Solute carrier family 45, member 2 (imperfect albino, cinnamon, dark-eyed dilute, red-eyed brown)	Z	Crawford (1990); Minvielle et al. (2000); Gunnarsson et al. (2007)
BR (br, ro)	Brown, roux	Z	Crawford (1990); Minvielle et al. (2000)
EMB <ZOV3>	Embigin homolog (mouse)	Z	Saitoh et al. (1993)
GHR	Growth hormone receptor	Z	Suzuki et al. (1999a)
MUSK	Muscle, skeletal, receptor tyrosine kinase	Z	Suzuki et al. (1999a)
PRLR	Prolactin receptor	Z	Suzuki et al. (1999a)
WPKCI-8 <Wpkci, HINTW, Wpkci-7, ASW>	W chromosome-specific histidine triad nucleotide-binding protein 1	W	Hori et al. (2000)
Bh	Black at hatch	1	Niwa et al. (2003)
ALB <Alb>	Albumin	AL (linked to s and *GC*)	Crawford (1990); Shibata and Abe (1996)
GC	Group-specific component (vitamin D-binding protein)	AL (linked to *ALB*)	Shibata and Abe (1996)
EDNRB2 <S>	Endothelin receptor B subtype 2 gene (panda)	AL (linked to *ALB*)	Crawford (1990); Miwa et al. (2006, 2007)
E	Extended brown	AL (linked to *GPI*)	Crawford (1990)
GPI <Pgi>	Glucose phosphate isomerase	AL (linked to *E*)	Crawford (1990)
HIST1H1A <H1.a>	Erythrocyte histone H1.a	AL (linked to *HIST1H1B* and *HIST1H1Z*)	Palyga (1998)
HIST1H1B <H1.b>	Erythrocyte histone H1.b	AL (linked to *HIST1H1A* and *HIST1H1Z*)	Palyga (1998)
HIST1H1Z <H1.z>	Erythrocyte histone H1.z	AL (linked to *HIST1H1A* and *HIST1H1B*)	Palyga (1998)
wb	White-breasted	AL (linked to *y*)	Crawford (1990)
y	Yellow	AL (linked to *wb*)	Crawford (1990)

(continued)

Table 12 (Continued)

Loci linked (locus alleles) <aliases>	Trait name (synonym)	Chromosome (Z, W, 1) or autosomal linkage (AL)	Reference
Numida meleagris (guinea fowl)			
ACO1 <*Acon*[B]>	Aconitase 1, soluble (cytoplasmic aconitase)	Z	Crawford (1990)
is	Brown, dundotte	Z	Hollander (1990)
k	Sex-linked feathering	Z	Crawford (1990)
Serinus canaria (island canary)			
cin	Cinnamon (brown)	Z	Mason[5]
ino (ino, ino[ag])	Sex-linked imperfect albinism, agate	Z	Durham and Marryat (1908); Onsman[6]
—	Hearing and song in the Belgian Waterslager canary	Z	Wright et al. (2004)
Taeniopygia guttata (zebra finch)			
ACO1 <*IREBP*>	Aconitase 1, soluble	Z	Lacson and Morizot (1988), Itoh et al. (2006)
ATP5A1	ATP synthase, H+ transporting, mitochondrial F1 complex, alpha subunit 1, cardiac muscle	Z	Itoh et al. (2006)
B	Brown (fawn)	Z	Miller (1992)
C	Chestnut flanked white	Z	Miller (1992)
CHD1	Chromodomain helicase DNA-binding protein 1	Z	Itoh et al. (2006)
DMRT1	Doublesex and mab-3-related transcription factor 1	Z	Itoh et al. (2006)
GHR	Growth hormone receptor	Z	Itoh et al. (2006)
HINT1	Histidine triad nucleotide-binding protein 1	Z	Itoh et al. (2006)
HSD17B4	Hydroxysteroid (17-beta) dehydrogenase 4	Z	Itoh et al. (2006)
NIPBL	Nipped-B homolog (Drosophila)	Z	Itoh et al. (2006)
NR2F1	Nuclear receptor subfamily 2, group F, member 1	Z	Itoh et al. (2006)
NTRK2	Neurotrophic tyrosine kinase, receptor, type 2	Z	Chen et al. (2005), Itoh et al. (2006)
PAM	Peptidylglycine alpha-amidating monooxygenase	Z	Itoh et al. (2006)
S	Silver	Z	Miller (1992)
SMAD2	SMAD family member 2	Z	Itoh et al. (2006)
SPIN1	Spindlin 1	Z	Itoh et al. (2006)
UBE2R2	Ubiquitin-conjugating enzyme E2R 2	Z	Itoh et al. (2006)

(continued)

Table 12 (Continued)

Loci linked (locus alleles) <aliases>	Trait name (synonym)	Chromosome (Z, W, 1) or autosomal linkage (AL)	Reference
—	Light back	Z	Miller (1992)
WPKCI-8 <Wpkci, HINTW, Wpkci-7, ASW>	W chromosome-specific histidine triad nucleotide-binding protein 1	W	O'Neill et al. (2000)
	*Columba livia* (pigeon)		
ACO1 <IREBP>	Aconitase 1, soluble	Z	Saitoh et al. (1993)
b (B^A, b)	Ash-red, brown	Z	Hollander (1990)
d (D^P, d)	Pale, dilute	Z	Hollander (1990)
EMB <ZOV3>	Embigin homolog (mouse)	Z	Saitoh et al. (1993)
R (r, r^RU)	Reduced, rubella	Z	Hollander (1990); Huntley[7]
St (St, St^H, St^Q, St^F, St^Sa, St^Fr, St^C)	Almond, hickory, qualmond, faded, sandy, frosty, chalky	Z	Hollander (1990); Huntley[7]
Wl	Web lethal	Z	Hollander and Miller (1982); Hollander (1990)
WPKCI-8 <Wpkci, HINTW, Wpkci-7, ASW>	W chromosome-specific histidine triad nucleotide-binding protein 1	W	Hori et al. (2000)
C (C^T, C^D, C, C^L, c)	T-pattern, dark checker, checker, light checker, barless	AL (linked to o and S)	Miller and Hollander (1978); Huntley[7]
o	Opal	AL (linked to C and S)	Miller and Hollander (1978); Huntley[7]
S	Spread pattern	AL (linked to C and o)	Miller and Hollander (1978); Huntley[7]
	*Streptopelia roseogrisea (S. risoria)* (African collared dove)		
d (d, d^B, d^w)	Dark, blond (fawn), white	Z	Cole (1930)
ALB <H-R>	*S. tranquebarica humilis*-specific albumin	AL (linked to hu-y)	Miller and Weber (1969)
hu-8	*S. tranquebarica humilis*-specific erythrocyte alloantigen hu-8	AL (linked to L)	Miller (1964)
hu-y	*S. tranquebarica humilis*-specific erythrocyte alloantigen hu-y	AL (linked to ALB)	Miller and Weber (1969)
L	Silky	AL (linked to hu-8)	Miller (1964)
	*Psittacula krameri* (rose-ringed parakeet)		
cin	Cinnamon	Z	Rašek[8]
ino (ino, ino^pd, ino^py)	Ino, pallid (lime), pearly	Z	Onsman[6]; Rašek8
op	Opaline	Z	Rašek[8]
bl (bl, bl^tq, bl^aq)	Blue, turquoise(parblue), aqua(parblue)	AL (linked to D)	Rašek[8]
D	Dark	AL (linked to bl)	Rašek[8]

Table 12 (Continued)

Loci linked (locus alleles) <aliases>	Trait name (synonym)	Chromosome (Z, W, 1) or autosomal linkage (AL)	Reference
	Melopsittacus undulates (budgerigar)		
ACO1 <IREBP>	Aconitase 1, soluble	Z	Saitoh et al. (1993)
cin	Cinnamon	Z (closely linked to *ino* and *sl*)	Mason[5]; Onsman[6]
EMB <ZOV3>	Embigin homolog (mouse)	Z	Saitoh et al. (1993)
ino (ino^cb, ino^l, ino^pl, ino)	Clearbody, lime, platinum, lutino (sex-linked imperfect albinism)	Z (closely linked to *cin* and *sl*)	Mason[5]; Onsman[6]
op	Opaline	Z	Mason[5]; Onsman[6]
sl	Slate	Z (closely linked to *cin* and *ino*)	Mason[5]; Onsman[6]
bl (bl^1, bl^2, bl^tq, bl^gf; or *bl^1, bl^2, bl^yf2, bl^gf) <s, b>[9]*	Blue (blue 1, blue 2, turquoise, blue goldenface, blue yellowface 2)	AL (linked to *D*)	Onsman[6]; Hesford[10]
D	Dark	AL (linked to *bl*)	Mason[5]; Olszewski[11]
ACO1 <IREBP>	*Struthio camelus (ostrich)* Aconitase 1, soluble	Z	Ogawa et al. (1998); Tsuda et al. (2007)
ATP5A1	ATP synthase, H+ transporting, mitochondrial F1 complex, alpha subunit 1, cardiac muscle	Z, W	Tsuda et al. (2007)
CHD1	Chromodomain helicase DNA-binding protein 1	Z, W	Tsuda et al. (2007)
EMB <ZOV3>	Embigin homolog (mouse)	Z, W	Ogawa et al. (1998)
GHR	Growth hormone receptor	Z, W	Tsuda et al. (2007)
HINT1	Histidine triad nucleotide-binding protein 1	Z	Tsuda et al. (2007)
NTRK2	Neurotrophic tyrosine kinase, receptor, type 2	Z, W	Tsuda et al. (2007)
RPS6	Ribosomal protein S6	Z, W	Tsuda et al. (2007)
SPIN1	Spindlin 1	Z, W	Tsuda et al. (2007)
TMOD1	Tropomodulin 1	Z, W	Tsuda et al. (2007)
WPKCI-8 <Wpkci, HINTW, Wpkci-7, ASW>	W chromosome-specific histidine triad nucleotide-binding protein 1	W	O'Neill et al. (2000); Tsuda et al. (2007)
	Dromaius novaehollandiae (emu)		
ACO1 <IREBP>	Aconitase 1, soluble	Z, W	Ogawa et al. (1998)
DMRT1	Doublesex and mab-3-related transcription factor 1	Z	Shetty et al. (2002)
EMB <ZOV3>	Embigin homolog (mouse)	Z, W	Ogawa et al. (1998)
WPKCI-8 <Wpkci, HINTW, Wpkci-7, ASW>	W chromosome-specific histidine triad nucleotide-binding protein 1	W	O'Neill et al. (2000)

(continued)

Table 12 (Continued)

Loci linked (locus alleles) <aliases>	Trait name (synonym)	Chromosome (Z, W, 1) or autosomal linkage (AL)	Reference
Casuarius casuarius (southern cassowary)			
ACO1 <IREBP>	Aconitase 1, soluble	Z	Nishida-Umehara et al. (1999)
EMB <ZOV3>	Embigin homolog (mouse)	Z, W	Nishida-Umehara et al. (1999)

[1] This locus appears to be the same as *g*

[2] Some authors (e.g., Lancaster, 1977) consider this gene to be autosomal

[3] Sokolovskaya (1935) and Hollander (1970) showed a homology between the loci ch in the Muscovy duck and d in the domestic duck

[4] Legg B. http://www.leggspeafowl.com/peafowlcolors.htm

[5] Mason AD. http://www.cabinsoftware.biz/Genetics_Tutorial/Part1.htm

[6] Onsman I. http://www.euronet.nl/users/hnl/; http://www.euronet.nl/users/dwjgh/

[7] Huntley RR. http://www.angelfire.com/ga/huntleyloft/

[8] Rašek M. http://www.gencalc.com/

[9] A universal nomenclature is required for this locus

[10] Hesford C. http://ourworld.compuserve.com/homepages/clivehesford/parblu04.html

[11] Olszewski A. http://www.petcraft.com/docs/avgen.shtml

Over the last decade, molecular tools and genetic maps have been developed in other avian species, and comparative avian genome studies have been boosted by creating the chicken genomic resources. From the field of comparative cytogenetics including ZOO-FISH studies, comparative mapping of BAC clones, and comparative chromosome G-banding, it is known that chicken chromosome 2 has split into chromosomes 3 and 6 of turkey and pheasants, while chicken chromosome 4 is a fusion between chromosome 4 and a microchromosome in many other birds (Shetty et al. 1999; Schmid et al. 2000, 2005; Raudsepp et al. 2002; Guttenbach et al. 2003; Derjusheva et al. 2004; Itoh and Arnold 2005). In the guinea fowl, chromosome 4 is the result of a centric fusion of chicken chromosome 9 with the q arm of chicken chromosome 4 (Shibusawa et al. 2002). Guinea fowl chromosome 5 represents the fusion of chicken chromosomes 6 and 7. A pericentric inversion in guinea fowl chromosome 7 corresponds to chicken chromosome 8. Chicken chromosome-specific paints from macrochromosomes 1–9 and Z hybridized to metaphases of the Japanese quail and red-legged partridge revealed no interchromosomal rearrangements (Schmid et al. 2000; Shibusawa et al. 2002; Kasai et al. 2003). Comparative FISH mapping of selected chicken BAC clones specific for macrochromosomes (GGA1–8, GGAZ) suggested strong conservation between sequences of the chicken, quail, turkey, and duck (Schmid et al. 2005) that represent two early evolutionary avian lineages split nearly 90 MYA. Several intrachromosomal rearrangements, fusions, or fissions were detected in four species. Evolution of karyotypes in birds seems to have proceeded slower in time than in mammals, which have more radical karyotype rearrangements. Avian karyotypes could have evolved via many fusion/fission and/or inversions instead of reciprocal translocations (Burt et al. 1999; Burt 2002; Schmid et al. 2005).

Comparative investigations also contributed to chicken genome mapping and cross-species application of molecular tools in chicken, turkey, guinea fowl, Japanese quail, duck, and pigeon (Pimentel-Smith et al. 2000; Smith et al. 2000a, 2001b; Reed et al. 2003; Schmid et al. 2005). Using FISH mapping and direct sequencing of genomic regions, several loci have been assigned to the sex chromosomes and autosomes in the turkey, peafowl, pheasants, quails, ducks, pigeon, ostrich, emu, cassowary, budgerigar, zebra finch (Table 12), and some other birds.

Comparative mapping using BAC contigs can provide a critical component of the genomic research in other birds. Large-insert BAC libraries are also available for several other avian species (Table 11). For instance, a zebra finch BAC library (Clayton 2004) with ~16-fold coverage was made at the Arizona

Genome Institute, an emu BAC library (13.5×) at the US Department of Energy Joint Genome Institute (Kellner et al. 2005), and a California condor (*Gymnogyps californianus*, family *Cathartidae*, order *Ciconiiformes*) BAC library with ~14-fold coverage by CHORI (Nefedov et al. 2003; Romanov et al. 2006). Genomic cosmid libraries have been constructed for the Japanese quail, pigeon, goose, emu and two passerines, red-winged blackbird (*Agelaius phoeniceus*) and brown-headed cowbird (*Molothrus ater*) (Kameda and Goodridge 1991; Edwards et al. 1998; Longmire et al. 1999; Shiina et al. 1999; Roots and Baker 2002; Takahashi et al. 2003), and a fosmid library for the domestic duck (Moon and Magor 2004). Large-insert contig physical maps of other avian genomes, aligned with the chicken sequence, would be valuable resources. Furthermore, these comparative maps would aid in the analysis and application of the chicken whole-genome sequence.

Thomas et al. (2002) demonstrated that so-called universal OVERGO probes, or Uprobes, can be used to identify orthologous BACs in a variety of mammals (primates, cat, dog, cow, pig) and, more recently, between vertebrate orders (Kellner et al. 2005). OVERGOs are designed from regions of high sequence conservation and then used to probe unsequenced genomes. Romanov and Dodgson (2006) analyzed cross-species hybridizations using OVER-GOs that were derived from chicken genomic and zebra finch EST sequences and probed to turkey and zebra finch BAC libraries. OVERGOs within coding sequences were more effective than those within untranslated region (UTR), intron or flanking sequences. In general, interspecies hybridization was more successful between chicken and turkey than for more distant evolutionary comparisons (chicken-zebra finch or zebra finch-turkey). This strategy can be used to align BAC contig maps of other avians along the chicken genome sequence and to construct interspecific comparative maps.

Molecular markers and tools have been generated for the Japanese quail (e.g., Pang et al. 1999; Kayang et al. 2002), duck and goose (Maak et al. 2003; Huang et al. 2005, 2006b), pheasant (Baratti et al. 2001), peafowl (Hanotte et al. 1991; Hale et al. 2004), pigeon (Traxler et al. 2000), ostrich (Tang et al. 2003), emu (Taylor et al. 1999), budgerigar (Kamara et al. 2007), and other avian species. These tools should facilitate linkage map construction, which has lately become reality for the

duck (Huang et al. 2006b). Its preliminary linkage map was developed by segregation analysis of microsatellite markers using an inbred Peking duck resource population that consisted of 12 full-sib families with a total of 224 F_2 individuals. As a result, 115 loci were placed into 19 linkage groups and 34 markers were unlinked. The total length of the preliminary sex-averaged linkage map for duck is 1,387.6 cM, as in other species. Integration of the genetic and cytogenetic map of the duck genome was done by FISH using chicken BAC clones, and 11 of 19 linkage groups were assigned to ten duck chromosomes (Huang et al. 2006b). The construction of a duck BAC library (Yuan et al. 2006) will pave the way for genome research in this poultry species.

A considerable breakthrough in genetic mapping of the Japanese quail genome has been achieved by designing molecular maps using AFLP and microsatellite markers. The first genetic linkage map contained 258 AFLP markers assigned to 39 autosomal linkage groups plus the Z and W sex chromosomes (Roussot et al. 2003). The first-generation microsatellite linkage map of this species included 58 markers resolved into 12 autosomal linkage groups and Z chromosome (Kayang et al. 2004). On the second-generation genetic linkage map, 1,660 AFLP and eight microsatellite markers, phenotype of a genetic disease (neurofilament-deficient mutant) and sex phenotype were assigned to 44 multipoint linkage groups, the W chromosome and 21 two-point linkage groups (Kikuchi et al. 2005). Six more microsatellite loci derived from ESTs (Mannen et al. 2005) and nine EST markers derived from cDNA-AFLP fragments (Sasazaki et al 2006a) were added to this map.

A subsequent contribution to the Japanese quail molecular linkage map and its enrichment with classical markers, such as plumage colors and blood proteins, was done by Miwa et al. (2005). These authors constructed maps for 14 autosomal linkage groups and the Z chromosome and the maps contained 69 microsatellite markers and five classical markers: yellow (*Y*), black at hatch (*Bh*), hemoglobin (*Hb-1*), transferrin (*Tf*), and prealbumin-1 (*Pa-1*). The study confirmed an earlier observation from FISH studies that the *Bh* locus was mapped on the long arm of chromosome 1 (CJA1) using the flanking sequence of *Bh* as a probe (Niwa et al. 2003). Miwa et al. (2006) mapped five other microsatellite markers and the panda (*s*) character to chromosome 4 (CJA4), suggesting the endothelin receptor B subtype 2 gene

(*EDNRB2*) as a candidate for the *s* locus that was confirmed in a follow-up study by Miwa et al. (2007).

Two microsatellite and the AFLP quail genetic maps were integrated and amended with the alignment of the quail linkage groups on the chicken genome sequence assembly and with interspecific FISH. Kayang et al. (2006) obtained a total of 14 autosomal and Z chromosome-specific linkage groups with 92 loci and aligned them with the AFLP map. The total map distance was 904.3 cM with an average spacing of 9.7 cM between loci. After aligning the quail linkage groups and the chicken sequence, marker order for nine macrochromosomes and 14 microchromosomes was found to be very similar between the two species. No interchromosomal rearrangements were detected for all 23 chromosomes, suggesting conservation of the aligned syntenic segments (Kayang et al. 2006).

In a separate effort, Sasazaki et al. (2006b) developed another integrated map for quail that comprised 1,995 markers, including 1,933 AFLP, three phenotypic loci (*Quv, LWC*, and sex) and 59 genes/ESTs, assigned to 66 linkage groups (including the W chromosome). The total linkage map length was 3,199 cM and an average marker interval of 5.0 cM. There were similar positions of the genes and their orders in the quail and chicken except within a known inversion on quail chromosome 2 (CJA2; Shibusawa et al. 2001). On the other hand, low map resolution did not allow detection of three other inversions previously found in CJA1, CJA4, and CJA8.

Another well-known laboratory bird is the zebra finch, referred to as "the mouse, or Drosophila of the avian world" (Arnold and Clayton 2004). Zebra finch is an Australian songbird that is a widely studied behavioral model, especially for mechanisms of learning and control of the male song, adult neurogenesis, and steroid synthesis in brain. There is a remarkable sexual dimorphism in brain regions controlling song. Interest in the genetic regulation of zebra finch behaviors has led to the generation of a BAC library, two EST projects, and a cDNA microarray (Arnold and Clayton 2004; Wade et al. 2004; Luo et al. 2006). Comparative cytogenetic analysis in the zebra finch using chicken chromosome paints suggested a very few chromosomal rearrangements since the evolutionary divergence of these two species, and a high conserved synteny of chicken genes and zebra finch orthologs (Itoh and Arnold 2005). Two major intrac-

hromosomal rearrangements were detected that split chicken chromosome 1 into two macrochromosomes in zebra finches, and chicken chromosome 4 into a zebra finch macrochromosome and a microchromosome. Later on, zebra finch BAC end sequences and the whole BAC sequence were aligned with the chicken sequence, and a high degree of conserved synteny between two genomes was verified (Luo et al. 2006). BACs assigned by Romanov and Dodgson (2006) to zebra finch genes using cross-species hybridization are available online (US Poultry Genome Project, http://poultry.mph.msu.edu/resources/Resources.htm#bacdatafinch). The zebra finch will be the third avian species for which the BAC-contig physical map and sequence of the whole genome are available (Clayton et al. 2005).

A partial linkage map has been built for another passerine bird, great reed warbler (*Acrocephalus arundinaceus*; Hansson et al. 2005), as well as a comparative chicken-passerine microsatellite map (Dawson et al. 2006). The first linkage map for a passerine species included 43 microsatellite markers on 11 autosomal linkage groups and seven loci on the Z chromosome (Hansson et al. 2005). A predicted passerine map (Dawson et al. 2006) was based on the sequence similarity between 550 passerine microsatellites and the draft chicken genome sequence, and was also aligned with the Hansson et al. (2005) great reed warbler linkage map. A SNP-based Z chromosome map for 23 genes was created by Backström et al. (2006) using a natural population of collared flycatchers (*Ficedula albicollis*) and chicken genome sequence; conserved synteny with gene order rearrangements on the avian Z chromosome was demonstrated.

To initiate genomic studies for the California condor and take advantage of progress in chicken genomics, Raudsepp et al. (2002) attempted a broad cytogenetic analysis in this endangered species. As a result, a chromosome number of 80 was established (with a likelihood of an extra pair of microchromosomes), and information on the centromeres, telomeres, and nucleolar organizing regions was obtained. By hybridizing individual chicken chromosome-specific paints for 1–9 and Z and W on condor metaphase spreads, condor and chicken macrochromosomes were compared. Good correspondence of the chicken macrochromosomes with a single condor macrochromosome was observed, except for chromosomes 4 and Z. GGA4 was homologous to condor chromosomes 4 and 9, supporting the idea that the latter are ancestral avian chromosomes. The GGAZ paint hybridized to both Z and W in the condor, suggesting incomplete differentiation of the condor sex chromosomes during evolution, contrary to data for sex chromosomes in all other nonratites studied (Raudsepp et al. 2002).

Additionally, a first-generation comparative chicken-condor physical map was developed using a condor BAC library and OVERGO hybridization approach (Romanov et al. 2006). The OVERGOs were designed using chicken (164 probes) and New World vulture (8 probes) sequences. After screening a 2.8 × subset of the total library, 236 BAC-gene assignments were identified, with an average success rate of 2.5 positive BAC clones per probe. A preliminary comparative chicken-condor BAC-based map contained 93 genes. Alignment of selected condor BAC sequences with orthologous chicken sequences showed a high conserved synteny between the two avian genomes. This study has created indispensable resources for seeking candidate loci for chondrodystrophy in condors and assisting genetic management of this disease (Romanov et al. 2006).

Currently, the active genome mapping and sequencing projects in birds include chicken, turkey, duck, Japanese quail, zebra finch, brown kiwi, and California condor (NCBI Entrez Genome Project database, http://www.ncbi.nlm.nih.gov/entrez/query.fcgi?db=genomeprj). The majority of 1,075,888 nucleotides and 73,649 protein avian sequences deposited in GenBank (as of May 7, 2007) belong to chicken, turkey, zebra finch, duck, and condor; while chicken, pigeon, common and Japanese quail, and duck remain key avian models for biomedical research as assessed by the number of PubMed accessions (Table 13). Details on turkey genetics and genome mapping are given in Chapter 6 of this volume.

5.4
QTL and Functional Genomics

5.4.1
QTL Analysis

The identification of genes underlying the expression of economically important traits is a main research focus in agricultural genomics. Most of these traits are characterized by a wide variation in the expression of genes at certain loci called QTL (Cheng et al. 1995), which are polymorphic loci associated with variation in a phenotypic trait like egg production, body weight (BW), and so on. Characterization of the chromosomal regions carrying QTL can be applied in MAS to improve breeding efficiency (Grisart et al. 2002). Molecular linkage map, in combination with powerful statistical methods, facilitates the genetic dissection of complex traits, and the chicken is ideally suited for this task due to a relatively short life cycle and large number of progenies (Vallejo et al. 1998). Two major approaches are employed to understand genomic architecture underlying economically important traits: QTL mapping and, more recently, functional genomics.

QTL studies in chickens were started in the middle of 1990s using minisatellite markers and a technique known as genomic fingerprinting. In one of these studies, crosses of two genetically distinct lines of layer-type chickens and a single-trait animal model were used to identify genetic markers linked to QTL (Lamont et al. 1996). Analysis of associations of individual DNA fingerprint (DFP) bands of sires and their progeny phenotypic performance revealed QTL linked to specific traits of growth, reproduction, and egg quality QTL.

QTL have been identified for a variety of traits in chickens including growth (Groenen et al. 1997; van Kaam et al. 1998, 1999a; Tatsuda and Fujinaka 2001), feed efficiency (van Kaam et al. 1999a), carcass traits (van Kaam et al. 1999b), resistance to Marek's disease (MD; Vallejo et al. 1998; Xu and Goodridge 1998; Yonash et al. 1999; Lipkin et al. 2002), fatness (Ikeobi et al. 2002), and egg quality (Tuiskula-Haavisto et al. 2002; Wardecka et al. 2002), based on high-resolution genetic maps. Mapping information from QTL studies has enabled the further localization of 45 microsatellites on the consensus map resulting in a total number of 2,306 markers (Schmid et al. 2005).

5.4.2
QTL: Growth, Meat Quality, and Productivity

To elucidate QTL that affect growth, genome-wide scans with microsatellite markers has been employed. For example, van Kaam et al. (1999a) performed a whole-genome scan for QTL affecting growth and

Table 13 Number of accessions in the NCBI GenBank (Nucleotide and Protein) and PubMed databases for avian species with more than 300 deposited nuclear and mitochondrial nucleotide sequences[a]

Species/Order	Nucleotide/Protein	PubMed
Gallus gallus (chicken)/*Galliformes*	923,090/31,335	81,156
Taeniopygia guttata (zebra finch)/*Passeriformes*	67,726/263	275
Meleagris gallopavo (turkey)/*Galliformes*	18,052/411	299
Anas platyrhynchos (mallard)/*Anseriformes*	4,072/549	787
Gymnogyps californianus (California condor)/*Ciconiiformes*	970/4	8
Motacilla flava (yellow wagtail)/*Passeriformes*	890/91	4
Pygoscelis adeliae (Adelie penguin)/*Sphenisciformes*	671/17	55
Ficedula hypoleuca (European pied flycatcher)/*Passeriformes*	575/271	72
Ficedula albicollis (collared flycatcher)/*Passeriformes*	568/286	41
Coturnix japonica (Japanese quail)/*Galliformes*	538/425	4,093
Motacilla alba (white wagtail)/*Passeriformes*	498/259	5
Parus major (great tit)/*Passeriformes*	498/153	159
Luscinia svecica (bluethroat)/*Passeriformes*	422/200	10
Parus montanus (willow tit)/*Passeriformes*	416/168	18
Parus caeruleus (blue tit)/*Passeriformes*	411/84	74
Anas strepera (gadwall)/*Anseriformes*	409/47	9
Carpodacus erythrinus (common rosefinch)/*Passeriformes*	387/190	4
Columba livia (domestic pigeon)/*Columbiformes*	380/188	9,258
Strix aluco (tawny owl)/*Strigiformes*	328/175	42
Coturnix coturnix (common quail)/*Galliformes*	314/195	4,208
Dendrocopos major (great spotted woodpecker)/*Piciformes*	300/144	2

[a] As of May 7, 2007

feed efficiency in chickens and detected four QTL on GGA1, GGA2, GGA4, and GGA23 that exceeded the significance thresholds. The same research group has carried out a whole-genome scan in chicken for QTL affecting carcass traits (van Kaam et al. 1999b). Two QTL were shown to be located on GGA1 and GGA2. These results were confirmed and refined using Bayesian analysis (van Kaam et al. 2002).

Tatsuda and Fujinaka (2002) detected QTL affecting BW closely aligned with those reported using a reference population derived from a cross of a Satsumadori (slow-growing, light-weight Japanese native breed used as a meat chicken) sire and a White Plymouth Rock (early maturing, heavy weight broiler) dam. Two QTL affecting BW at 13 and 16 weeks were mapped at 220 cM on GGA1 and at 60 cM on GGA2. The closest QTL markers were *LEI0071* on GGA1 and *LMU0013* and *MCW0184* on GGA2.

QTL for BW at 3, 6, and 9 weeks of age were investigated by Sewalem et al. (2002) using a broiler × layer cross. A QTL on GGA13 influenced BW at all three ages and QTL significant at the genome-wide level that affected BW at two ages were found on chromosomes 1, 2, 4, 7, and 8.

Identification of QTL for meat quality and production in a commercial population of broilers was done by de Koning et al. (2004). Using genotypes for 52 microsatellite loci spanning regions of nine chicken chromosomes and a half-sib analyses with a multiple QTL model, linkage between these nine regions and growth, carcass and feed intake traits was established.

QTL affecting fatness in the chicken were investigated and mapped by Ikeobi et al. (2002) in an F_2 population developed by crossing a broiler line with a layer line. Using within-family regression analyses of 102 microsatellite loci in 27 linkage groups, the QTL for abdominal fat weight were identified on chromosomes 3, 7, 15, and 28; abdominal fat weight adjusted for carcass weight on chromosomes 1, 5, 7, and 28; skin and subcutaneous fat on chromosomes 3, 7, and 13; skin fat weight adjusted for carcass weight on chromosomes 3 and 28; and skin fat weight adjusted for abdominal fat weight on chromosomes 5, 7. and 15. Significant positive and negative QTL alleles were detected in both lines. Several QTL affecting fatness in broilers were detected by Jennen et al. (2004) using two genetically different outcross broiler dam lines, originating from the White Plymouth Rock breed.

Genetic architecture of growth and body composition was investigated in reference chicken populations obtained by crossing one modern broiler male from a commercial broiler breeder male line with females from two unrelated highly inbred lines (Deeb and Lamont 2003). Traditionally selected phenotypic traits in broilers were suggested to be controlled by a large number of genes with small epistatic effects, while fitness-related traits could be determined by a lower number of genes with major effects.

After simultaneous mapping epistatic QTL in a chicken F_2 intercross, clusters of QTL pairs with similar genetic effects on growth were found by Carlborg et al. (2004). The authors used simultaneous mapping of interacting QTL pairs to study growth traits. This approach improved the number of detected QTL by 30%. The genetic variance of growth was significantly influenced by epistasis, the largest impact being on early growth (before 6 weeks of age). Because early growth was shown to be associated with a discrete set of interacting loci involved in early growth, these results provided further insight into different genetic regulations in early and late growth in chicken found in other studies.

5.4.3
QTL: Egg Quality and Productivity

Genome-wide scans for egg quality and productivity QTL have been done using reference populations, while a line cross between two egg layer lines was used in the study by Tuiskula-Haavisto et al. (2002). The authors determined 14 genome-wide significant and six suggestive QTL located on chromosomes 2, 3,

4, 5, 8, and Z. The most interesting area was found on GGA4, with QTL for BW, egg weight, and feed intake. A related investigation was done by Wardecka et al. (2002) to determine influence of genotypes of the Rhode Island Red (RIR) and Green-legged Partrigenous (GLP) breeds on egg production and quality traits based on analysis of 23 microsatellite markers. Significant effects were demonstrated for 16 traits.

Marker loci detected by QTL mapping can serve as multiple entry points into the physical BAC-contig map and sequence of the chicken genome. For example, two QTL from the aforementioned study were selected for FISH mapping using microsatellite-specific large-insert clones (Sazanov et al. 2005). This strategy helps to specify genes that might underlie QTL and is known as QTL positional cloning.

Genetic mapping of QTL affecting egg characters, egg production, and BW in F_2 White Leghorn (WL) × RIR intercross chickens was done by Sasaki et al. (2004) using 123 microsatellite markers. The authors assigned 96 markers to 25 autosomal linkage groups and 13 markers to the Z chromosome, including eight previously unmapped markers. Significant QTL were discovered for BW on chromosomes 4 and 27, egg weight on GGA4, the short length of egg on GGA4, and redness of egg shell color on GGA11. A significant QTL on GGAZ was found for age at first egg. Overall, 6–19% of the phenotypic variance in the F_2 population may be explained by these QTL.

5.4.4
QTL: Disease Resistance

Immune response and disease resistance can be improved by selection. Because these quantitative traits have low to moderate heritability, they may respond more efficiently to marker-assisted selection than to conventional selection (Yonash et al. 2001).

As an alternative to vaccination control, increased genetic resistance to Marek's disease (MD) represents an attractive solution for lowering disease outbreaks. Genetic mapping of QTL affecting susceptibility to MD virus-induced tumors was performed by Vallejo et al. (1998) and was the first to report the mapping of non-major histocompatibility complex (MHC) QTL involved in MD susceptibility in chickens. Two significant and two suggestive MD QTL were detected on four chromosomal regions. These loci explained 11–23% of the phenotypic MD variation, or 32–68% of the genetic variance.

Another QTL for MD explaining 7.2% of the total disease variation was revealed on GGA4 by Xu et al. (1998) using a heterogeneous residual variance model, which is considered to be computationally much faster than the mixture model approach. These and other studies that used the same F_2 cross between two experimental lines (Bumstead 1998; Yonash et al. 1999; Liu et al. 2001, 2003a) identified QTL on chromosomes 1, 2, 4, 7, 8, 12, and 17 that affect MD resistance. QTL associated with MD resistance (defined as survival time following challenge) were tested in a cross between lines of commercial layer chickens (McElroy et al. 2005). In this study, genotyping was performed using 81 microsatellites selected based on prior results with selective DNA pooling, and several markers associated with MD survival were identified. One of these markers corresponds to a QTL identified on GGA2 near the region identified for MD susceptibility by Vallejo et al. (1998) and Yonash et al. (1999), which is around 90 cM on the consensus map.

A very important issue for poultry production and food safety is the contamination with *Salmonella enteritidis* (SE). Kaiser et al. (2002) identified genetic markers of antibody (Ab) response to SE vaccine in broiler chicks and confirmed this linkage in broiler-cross offspring. Interactions of microsatellite marker alleles with dam line and sex were also detected.

Several QTL for immune response to sheep red blood cells (SRBC) were detected by Siwek et al. (2003a) in laying hens using 170 microsatellite markers, and F_2 resource populations originated from a cross of two divergently selected lines for high and low primary Ab response to SRBC. A half-sib model and a line-cross model, both based on the regression interval method, were used to identify QTL. The QTL involved in the primary Ab response to keyhole lympet hemocyanin and *Mycobacterium butyricum* were detected in two independent populations of laying hens (Siwek et al. 2003b). The genetic regulation of Ab responses to two different T- cell dependent antigens was suggested to differ.

QTL affecting the immune response were investigated using a linkage disequilibrium approach with microsatellites in hybrids of highly inbred males of two MHC-congenic Fayoumi chicken lines and highly inbred G-B1 Leghorn hens (Zhou et al. 2003). The QTL that affect Ab kinetics were localized on chromosomes 3, 5, 6, and Z.

A genome wide scan using 119 microsatellite loci allowed Zhu et al. (2003) to map QTL associated with disease resistance to avian coccidiosis to GGA1.

QTL associated with immune response to SRBC, Newcastle disease virus, and *E. coli* and with survival were investigated by Yonash et al. (2001). Three markers were shown to have significant association with these traits.

Besides its own economic importance, the chicken can be considered as a model object for human diseases, e.g., for genetic susceptibility to form-deprivation myopia (Guggenheim et al. 2002; Dodgson and Romanov 2004; Jensen 2005).

5.4.5
QTL: Behavior

Several QTL affecting feather pecking (FP) behavior (which is a major problem in large group housing systems) and stress response in laying hens were detected by Buitenhuis et al. (2003) Using genotypes at 180 microsatellite loci, one significant QTL for severe FP was detected on GGA2, and suggestive QTL for gentle FP on GGA1, GGA2, and GGA10.

A genome-wide scan using 104 microsatellite markers was performed to identify QTL affecting foraging behavior and social motivation QTL in F_2 progeny from a WL × red junglefowl intercross (Schutz et al. 2002). Significant QTL were found for preference of free food without social stimuli and low contra-freeloading on GGA27 and GGA7, respectively. Interestingly, the location of the QTL coincided with known QTL for growth rate and BW.

QTL studies in the chicken have rapidly expanded, and a specialized chicken QTL database has been created (NAGRP, http://www.animalgenome.org/QTLdb/chicken.html). With the availability of dense genetic linkage maps, QTL studies are becoming more feasible in other poultry species (e.g., Minvielle et al. 2005; Beaumont et al. 2005; Huang et al. 2007).

5.4.6
Toward Functional Genomics of Poultry

The chicken has been an attractive model organism in the field of fundamental biology and medicine for at least 100 years, for instance with the discovery of B cells and tumor viruses (Brown et al. 2003; Romanov et al. 2004). The avian embryo is an ideal system for studies of vertebrate development (e.g., limb bud) because of the ease of access and manipulation using incubated eggs (Stern 2004, 2005). Avian functional genomics is a new promising research area due to increased genetic resources and tools including EST programs, DNA microarrays, electroporation of chicken embryos, use of RNAi to knock down gene expression, and transgenic technologies (Brown et al. 2003; Stern 2004; Burt 2005).

DNA microarrays have become a powerful tool for determining functional genes in several organisms including human, rodents, fruit fly, chicken, etc. An international, US-French consortium for systems-wide chicken gene expression profiling was established in 2000 (Cogburn et al. 2003). The goal of the project was to provide genomic resources (ESTs and DNA microarrays), examine global gene expression in target tissues of chickens, and facilitate discovery of functional genes. Another chicken functional genomics initiative led by the UK consortium resulted in the collection of 339,314 ESTs from 64 cDNA libraries derived from 21 tissues of adult hens and chicken embryos (Boardman et al. 2002). These DNA sequences were organized in 85,486 contigs corresponding to 89% of estimated total number of chicken genes. Around 180,000 of these ESTs represented novel coding sequences in the chicken, while 38% of them were orthologous to sequences in other species. Later on, the merging of the UK (300,000) and US (30,000) EST collections took place (reviewed by Romanov et al. 2004). Currently, there are 599,330 ESTs deposited in the NCBI dbEST database (as of May 7, 2007).

Affymetrix, Inc. has developed the first commercially available GeneChip Chicken Genome Array (Affymetrix, http://www.affymetrix.com/products/arrays/specific/chicken.affx). This array includes 25-mer oligonucleotide probes for identifying 32,773 transcripts corresponding to over 28,000 chicken genes, as well as 689 probe sets for 684 transcripts from 17 avian viruses. Other chicken whole-genome long oligo arrays include: (1) the Operon Biotechnologies, Inc. 70-mer array with 21,120 features designed by ARK-Genomics (Roslin, UK) and manufactured by the University of Arizona Genomics Research Lab (GRL, Tucson, AZ, USA); (2) the Chicken Consortium cDNA array with 11,136 × 2 features also

produced by GRL; and (3) the NimbleGen Systems, Inc. Chicken ChIP- chip that tiles every 100 bp across nonrepetitive regions. A new 44,000-element long oligonucleotide chicken array was also made by Agilent Corp. A collaborative team from the Roslin Institute (Edinburgh, UK), University of Delaware (Newark, DE, USA), GSF Institute of Molecular Radiation Biology (Neuherberg, Germany) and the Fred Hutchinson Cancer Research Center (Seattle, WA, USA) produced a publicly available microarray containing ~13,000 chicken ESTs (Burnside et al. 2005). Additional chicken microarrays include four University of Delaware custom arrays (UD_Liver_3.2K, UD 7.4K Metabolic/Somatic Systems, Chicken Neuroendocrine System 5K, and the DEL- MAR 14K Integrated Systems), three ARK-Genomics arrays (an 1153 clone chicken embryo array, a 5,000 cDNA chicken immune array, and a 4,800 clone chicken neuroendocrine array) (US Poultry Genome Project, http://poultry.mph.msu.edu/about/Poultry%20Coord%20report%20for%2006.pdf; Smith et al. 2006).

These and other chicken microarray resources have been used for analyzing gene expression profiles in connection with immune responses to infectious diseases (e.g., Bliss et al. 2005; Smith et al. 2006; van Hemert et al. 2007), growth traits, lipid metabolism and fatness (Cogburn et al. 2003; Bourneuf et al. 2006; Wang et al. 2006), differentially expressed transcripts in shell glands (Yang et al. 2006), and embryonic development (Afrakhte and Schultheiss 2004; Ellestad et al. 2006) in experimental and commercial strains.

5.5
Other Molecular Applications

5.5.1
Biodiversity Studies

Genetic resources refer to races or populations with unique genetic characteristics. Agricultural resources need to be conserved for genetic adaptation to changes in agricultural production conditions and consumer preferences as well as for preservation of native (sometimes called local or "heritage") breeds. Thus, genetic resources in agricultural production systems require further identification, evaluation, and proper utilization for the welfare of humanity and nature. Although often underestimated, the need to conserve and utilize genetic resources as a safeguard against an unpredictable future is evident (FAO 1997–2004; El Bassam 1998; Romanov and Weigend 2001a, b).

There is a growing loss of genetic diversity in all agriculturally used species, and poultry genetic resources are one of the most vulnerable (Scherf 2000; Weigend and Romanov 2001, 2002). The current market-oriented breeding strategies in poultry concentrate on a few specialized breeds that may cause a significant erosion of local breeds, leading to the loss of valuable genetic variability and unique characteristics of these breeds (Weigend et al. 1995). For instance, few decades ago there were more than 50 chicken breeds in North America, while only two for meat production are left, the others being mainly lost (Scherf 2000).

On the whole, in the world poultry market a limited number of breeding companies dominate and use a similar gene pool. The economic importance of single-purpose high-performance breeds is distorting the perception of the value of multipurpose breeds that are adapted to local conditions, from the point of view of the broader gene pool. According to Sørensen (1997), there is concern by the general public and the poultry industry that "cage-adapted populations of laying hens seem to have lost some of their abilities when returned to the old floor/free range systems." Due to a growing concentration of all components of the poultry production, less than ten world's breeding companies are the source of most egg laying hens. Until recently, these companies have little interest in improving genetic material for the West-European region, with noticeable consumer preferences for eggs produced in noncaged systems (Sørensen 1997). This situation could get worse under global epidemic challenges like avian influenza.

On the other hand, genetic studies (e.g., Dunnington et al. 1994; International Chicken Polymorphism Map Consortium 2004) showed that the pure lines of broiler and layer stocks in USA and other world regions still contain a considerable reservoir of genetic variation as estimated by DNA fingerprint (DFP) analysis and other molecular tools.

The evaluation of genetic diversity within and between both native and commercial chicken populations has been undertaken so far using the following molecular markers and techniques:

- RFLP (e.g., Wakana et al. 1986; Akishinonomiya et al. 1994; Wang et al. 1994),
- DFP (e.g., Dawe et al. 1988; Siegel et al. 1992; Haberfeld et al. 1992; Wimmers et al. 1992; Dunnington et al. 1994; Yamashita et al. 1994; Plotsky et al. 1995; Meng et al. 1996; Semyenova et al. 1996),
- RAPD (e.g., Plotsky et al. 1995; Romanov and Weigend 2001a),
- AFLP (e.g., Lee et al. 2000),
- microsatellites (e.g., Romanov and Weigend 2001b; Hillel et al. 2003),
- sequencing and SNP (e.g., Akishinonomiya et al. 1994; Schmid et al. 2005).

Multiple applications of molecular markers for biodiversity studies in poultry, mostly, in chickens, have been reviewed and listed elsewhere (e.g., Weigend and Romanov 2001, 2002; Soller et al. 2006; Michigan State University, http://www.msu.edu/ ~romanoff/biodiversity/studiesdb.htm).

5.5.2
Molecular Sexing

Sex identification methods in domestic and wild birds have been developed to distinguish between males and females, when no or weak sexual dimorphism is evident either at hatch, maturity, or in ovo. The usefulness of DNA sex determination has been demonstrated for evolutionary studies, ecological and conservation problems, and management of endangered species in the wild and captivity (e.g., Millar et al. 1996; Ellegren and Fridolfsson 1997; Kahn and Quinn 1999; Bermudez-Humaran et al. 2002).

Traditionally, sex linkage of external traits (autosexing; e.g., Spillman 1908; Staško 1970; Lancaster 1977; Romanov and Bondarenko 1988), vent sexing, surgical gonad examination and later, analyses of karyotype and the amount of DNA per cell (e.g., Wang and Shoffner 1974; Nakamura et al. 1990) have been applied for this purpose. With the advance of sophisticated molecular tools, it is now possible to obtain gender-specific DNA fingerprints. Because in birds the homogametic sex is the male with two Z chromosomes, and the heterogametic sex is the female with one Z and one W chromosomes, the molecular sexing techniques in avian species are principally based on targeting the repetitive (e.g., Kagami et al. 1990; D'Costa and Petitte 1998; Cassar et al. 1998; Trefil et al. 1999), nonrepetitive (Ogawa et al. 1997), or coding regions in the W chromosome that are absent or different from their homologs in the Z chromosome.

PCR-amplified molecular markers make it possible to discriminate bird sexes based on Z- and W-chromosome-specific homologous sequences. Most known avian sexing markers are derivatives of two conserved gene homologs, *CHD1Z* and *CHD1W*, which encodes chromodomain helicase DNA-binding protein 1 that plays an important role in gene regulation (Ellegren 1996; Griffiths et al. 1996, 1998; Griffiths and Korn 1997; Kahn et al. 1998; Fridolfsson and Ellegren 1999).

5.6
Conclusions

In conclusion, contemporary avian genetics addresses biological questions at the genome-wide level. In the course of the last century, an enormous wealth of information has been accumulated regarding genetics, physiology, and biochemistry of poultry species. The chicken exemplifies both an important agricultural species and a model organism for studying the evolution of vertebrate genomes and developmental mechanisms. The success of the chicken genome project has been preceded by decades of genetic linkage mapping.

As a prominent experimental model in the last century for various fundamental and applied biologic disciplines, the chicken will keep its significance in the twenty-first century. The chicken genome sequence annotated with gene functions will pave the way for improving traits of economic importance and value in poultry (Romanov et al. 2004). The recent draft of the chicken genome sequence can also be used as a reference in comparative mapping, making up for the lack of knowledge in genetics and genomics of other domestic and wild birds and addressing global questions in biology of avian and vertebrate genomes.

References

Abdrakhmanov I, Lodygin D, Geroth P, Arakawa H, Law A, Plachy J, Korn B, Buerstedde JM (2000) A large database of chicken bursal ESTs as a resource for the analysis of vertebrate gene function. Genome Res 10:2062–2069

Abozin II (1885) Chicken breeding: detailed description of various chicken breeds, with the recommendations of care of them, breed improvement by crossbreeding and selection of breeders, pt 2. Moscow, Russia

Aerts JA, Veenendaal T, van der Poel JJ, Crooijmans RPMA, Groenen MAM (2005) Chromosomal assignment of chicken clone contigs by extending the consensus linkage map. Anim Genet 36:216–222

Afrakhte M, Schultheiss TM (2004) Construction and analysis of a subtracted library and microarray of cDNAs expressed specifically in chicken heart progenitor cells. Dev Dyn 230:290–298

Akishinonomiya F, Miyake T, Sumi S, Takada M, Ohno S, Kondo N (1994) One subspecies of the red junglefowl (*Gallus gallus gallus*) suffices as the matriarchic ancestor of all domestic breeds. Proc Natl Acad Sci USA 91:12505–12509

Akishinonomiya F, Miyake T, Takada M, Shingu R, Endo T, Gojobori T, Kondo N, Ohno S (1996) Monophyletic origin and unique dispersal patterns of domestic fowls. Proc Natl Acad Sci USA 93:6792–6795

Altukhov Y (ed) (2004) Dynamics of population gene pools under anthropogenic pressures. Nauka, Moscow, Russia

Ambady S, Cheng HH, Ponce De Leon FA (2002) Development and mapping of microsatellite markers derived from chicken chromosome-specific libraries. Poult Sci 81:1644–1646

Andersson L (2001) Genetic dissection of phenotypic diversity in farm animals. Nat Rev Genet 2:130–138

Andreozzi L, Federico C, Motta S, Saccone S, Sazanova AL, Sazanov AA, Smirnov AF, Galkina SA, Lukina NA, Rodionov AV, Carels N, Bernardi G (2001) Compositional mapping of chicken chromosomes and identification of the gene-richest regions. Chrom Res 9:521–532

Arnold AP, Clayton D (2004) Proposal for construction of a BAC library of the genome of the zebra finch (*Taeniopygia guttata*). National Human Genome Research Institute, USA. http://www.genome.gov/Pages/Research/Sequencing/BACLibrary/zebraFinch.pdf (accessed August 20, 2008)

Backstrom N, Brandstrom M, Gustafsson L, Qvarnstrom A, Cheng H, Ellegren H (2006) Genetic mapping in a natural population of collared flycatchers (*Ficedula albicollis*): conserved synteny but gene order rearrangements on the avian Z chromosome. Genetics 174:377–386

Baker CMA (1964) Molecular genetics of avian proteins. III. The egg proteins of an isolated population of Jungle Fowl, *Gallus gallus L.* Comp Biochem Physiol 12:389–403

Baker CMA (1968) Molecular genetics of avian proteins. IX. Interspecific and intraspecific variation of egg white proteins of genus Gallus. Genetics 58:211–226

Baker CMA, Manwell C (1972) Molecular genetics of avian proteins. XI. Egg proteins of *Gallus gallus, G. sonnerati* and hybrids. Anim Blood Groups Biochem Genet 3:101–107

Baker CMA, Manwell C, Jayaprakash N, Francis N (1971) Molecular genetics of avian proteins. X. Egg white protein polymorphism of indigenous Indian chickens. Comp Biochem Physiol B Comp Biochem 40:147–153

Baratti M, Alberti A, Groenen M, Veenendaal T, Fulgheri FD (2001) Polymorphic microsatellites developed by cross-species amplifications in common pheasant breeds. Anim Genet 32:222–225

Barloy JJ (1978) Man and Animals. 100 Centuries of Friendship. Gordon and Cremonesi, London, UK

Bateson W (1909) Mendel's Principles of Heredity. Cambridge University Press, Cambridge, UK

Bateson W, Punnett RC (1911) The inheritance of peculiar pigmentation of the Silky fowl. J Genet 1:185–203

Bateson W, Saunders ER (1902) Experimental studies in the physiology of heredity. Rep Evol Comm R Soc I:1–160

Beaumont C, Roussot O, Feve K, Vignoles F, Leroux S, Pitel F, Faure JM, Mills AD, Guémené D, Sellier N, Mignon- Grasteau S, Le Roy P, Vignal A (2005) A genome scan with AFLP™ markers to detect fearfulness-related QTL in Japanese quail. Anim Genet 36:401–407

Beebe W (1918–1922) A Monograph of Pheasants, vols I-IV. H.F. and G. Witherby, London, UK

Bell DD, Weaver WD Jr (2002) Commercial Chicken Meat and Egg Production, 5th edn. Kluwer, Norwell, USA

Bennett MD, Leitch IJ, Price HJ, Johnston JS (2003) Comparisons with *Caenorhabditis* (~100 Mb) and Drosophila (~175 Mb) using flow cytometry show genome size in Arabidopsis to be ~157 Mb and thus ~25% larger than the Arabidopsis genome initiative estimate of ~125 Mb. Ann Bot (Lond) 91:547–557

Bennett RM, Ijpelaar ACE (2003) Economics of Livestock Diseases. Department of Agricultural and Food Economics, University of Reading, Reading, UK. http://www.apd.rdg.ac.uk/AgEcon/livestockdisease/index.htm (accessed August 20, 2008)

Bermudez-Humaran LG, Garcia-Garcia A, Leal-Garza CH, Riojas-Valdes VM, Jaramillo-Rangel G, Montes-de-Oca-Luna R (2002) Molecular sexing of monomorphic endangered Ara birds. J Exp Zoo 292:677–680

Bitgood JJ, Somes RG Jr (1993) Gene map of the chicken (*Gallus gallus* or *G. domesticus*). In: O'Brien S (ed) Genetic Maps, 6th edn. Cold Spring Harbor Lab Press, Cold Spring Harbor, USA, pp 4332–4342

Bliss TW, Dohms JE, Emara MG, Keeler CL Jr (2005) Gene expression profiling of avian macrophage activation. Vet Immunol Immunopathol 105:289–299

Bloom SE, Delany ME, Muscarella DE (1993) Constant and variable features of avian chromosomes. In: Etches RJ, Verrinder Gibbins AM (eds) Manipulation of the Avian Genome. CRC Press, Boca Raton, USA, pp 39–59

Boardman PE, Sanz-Ezquerro J, Overton IM, Burt DW, Bosch E, Fong WT, Tickle C, Brown WR, Wilson SA, Hubbard SJ (2002) A comprehensive collection of chicken cDNAs. Curr Biol 12:1965–1969

Bourneuf E, Herault F, Chicault C, Carre W, Assaf S, Monnier A, Mottier S, Lagarrigue S, Douaire M, Mosser J, Diot C (2006) Microarray analysis of differential gene expression in the liver of lean and fat chickens. Gene 372:162–170

Brisbin IL (1997) Concerns for the genetic integrity and conservation status of the red junglefowl. SPPA Bull 2:1–2

Brothwell D, Brothwell P (1998) Food in Antiquity: A Survey of the Diet of Early Peoples. expanded edn. Johns Hopkins University Press, Baltimore, USA

Brown E (1906) Races of Domestic Poultry. Edward Arnold, London, UK

Brown E (1929) Poultry Breeding and Production, vols I and II. Ernst Benn Ltd, London, UK

Brown WR, Hubbard SJ, Tickle C, Wilson SA (2003) The chicken as a model for large-scale analysis of vertebrate gene function. Nat Rev Genet 4:87–98

Buitenhuis AJ, Rodenburg TB, Siwek M, Cornelissen SJB, Nieuwland MGB, Crooijmans RPMA, Groenen MAM, Koene P, Bovenhuis H, van der Poel JJ (2003) Identification of quantitative trait loci for receiving pecks in young and adult laying hens. Poult Sci 82:1661–1667

Buitkamp J, Ewald D, Schalkwyk L, Weiher M, Masabanda J, Sazanov A, Lehrach H, Fries R (1998) Construction and characterisation of a gridded chicken cosmid library with four-fold genomic coverage. Anim Genet 29:295–301

Bumstead N (1998) Genomic mapping of resistance to Marek's disease. Avian Pathol 27:S78–S81

Bumstead N, Palyga J (1992) A preliminary linkage map of the chicken genome. Genomics 13:690–697

Burnside J, Neiman P, Tang J, Basom R, Talbot R, Aronszajn M, Burt D, Delrow J (2005) Development of a cDNA array for chicken gene expression analysis. BMC Genomics 6:13

Burt DW (1999) Chick. In: Wood R (ed) Genetic Nomenclature Guide. Elsevier, West Sussex, UK, Trends Genet 15(Nov Suppl):S34–S36

Burt DW (2002) Origin and evolution of avian microchromosomes. Cytogenet Genome Res 96:97–112

Burt DW (2005) Chicken genome: current status and future opportunities. Genom Res 15:1692–1698

Burt DW, Pourquié O (2003) Chicken genome – science nuggets to come soon. Science 300:1669

Burt DW, Bruley C, Dunn IC, Jones CT, Ramage A, Law AS, Morrice DR, Paton IR, Smith J, Windsor D, Sazanov A, Fries R, Waddington D (1999) The dynamics of chromosome evolution in birds and mammals. Nature 402:411–413

Carlborg O, Hocking PM, Burt DW, Haley CS (2004) Simultaneous mapping of epistatic QTL in chickens reveals clusters of QTL pairs with similar genetic effects on growth. Genet Res 83:197–209

Carter GF (1971) Pre-Columbian chickens in America. In: Riley CL, Kelly JC, Pennington CW, Rands RL (eds) Man

Across the Sea. Problems of Pre-Columbian Contacts. University of Texas Press, Austin, USA, pp 178–218

Carter H, Mace AC (1923–1933) The Tomb of Tut-ankh-amen Discovered by the Late Earl of Carnarvon and Howard Carter, 3 vols. Cassell, London, UK

Cassar G, Mohammed M, John TM, Gazdzinski P, Etches RJ (1998) Differentiating between parthenogenetic and "positive development" embryos in turkeys by molecular sexing. Poult Sci 77:1463–1468

Chen X, Agate RJ, Itoh Y, Arnold AP (2005) Sexually dimorphic expression of trkB, a Z-linked gene, in early posthatch zebra finch brain. Proc Natl Acad Sci USA 102:7730–7735

Cheng HH, Levin I, Vallejo RL, Khatib H, Dodgson JB, Crittenden LB, Hillel J (1995) Development of a genetic map of the chicken with markers of high utility. Poult Sci 74:1855–1874

Clayton DF (2004) Songbird genomics: methods, mechanisms, opportunities, and pitfalls. Ann NY Acad Sci 1016:45–60

Clayton D, Arnold A, Warren W, Dodgson J (2005) Proposal for construction of a physical map of the genome of the zebra finch (*Taeniopygia guttata*). University of California, Los Angeles, USA. http://www.physci.ucla.edu/html/images/Zebra_finch_genome_white_paper.pdf (accessed August 20, 2008)

Cogburn LA, Wang X, Carre W, Rejto L, Porter TE, Aggrey SE, Simon J (2003) Systems-wide chicken DNA microarrays, gene expression profiling, and discovery of functional genes. Poult Sci 82:939–951

Cole LJ (1930) A triple allelomorph in doves and its interspecific transfer. Anat Rec 47:389 (Aviculture 2:27–30)

Crawford RD (1990) Poultry Breeding and Genetics. Elsevier, Amsterdam, The Netherlands

Crawford RD (1992) A global review of the genetic resources of poultry. In: Management of Global Animal Genetic Resources. FAO Animal Production and Health Paper, No 104, Rome, Italy, pp 205–214

Crawford RD (1995) Origin, history, and distribution of commercial poultry. In: Hunton P (ed) Poultry Production. Elsevier, Amsterdam, The Netherlands, pp 1–20

Crittenden LB, Provencher L, Santangelo L, Levin I, Abplanalp H, Briles RW, Briles WE, Dodgson JB (1993) Characterization of a Red Jungle Fowl by White Leghorn backcross reference population for molecular mapping of the chicken genome. Poult Sci 72:334–348

Crooijmans RPMA, Vrebalov J, Dijkhof RJM, van der Poel JJ, Groenen MAM (2000) Two-dimensional screening of the Wageningen chicken BAC library. Mamm Genom 11:360–363

Darwin C (1868) The Variation of Animals and Plants under Domestication. John Murray, London, UK, pp 273–335

Davenport CB (1911) Another case of sex-limited heredity in poultry. Proc Soc Exp Biol Med 9:19–20

Davenport CB (1912) Sex-limited inheritance in poultry. J Exp Zool 13:1–26

Dawe Y, Kuhnlein U, Zadworny D, Gavora J (1988) DNA fingerprinting: a tool for assessing parentship, strain relationship and genetic variability in poultry. Proc 18th Worlds Poult Congr, Nagoya, Japan, September 4–9, pp 507–508

Dawson DA, Burke T, Hansson B, Pandhal J, Hale MC, Hinten GN, Slate J (2006) A predicted microsatellite map of the passerine genome based on chicken-passerine sequence similarity. Mol Ecol 15:1299–1320

D'Costa S, Petitte JN (1998) Sex identification of turkey embryos using a multiplex polymerase chain reaction. Poult Sci 77:718–721

de Boer LEM (1980) Do the chromosomes of the kiwi provide evidence for a monophyletic origin of the ratites? Nature 287:84–85

Deeb N, Lamont SJ (2003) Use of a novel outbred by inbred F_1 cross to detect genetic markers for growth. Anim Genet 34:2051–212

de Koning D-J, Haley CS, Windsor D, Hocking PM, Griffin H, Morris A, Vincent J, Burt DW (2004) Segregation of QTL for production traits in commercial meat-type chickens. Genet Res 83:211–220

Delacour J (1977) The Pheasants of the World, 2nd edn. Spur, Hindhead, Surrey, UK pp 119–136

del Hoyo J, Elliott A, Sargatal J (eds) (1992–1996) Handbook of the Birds of the World, vols 1–3. Lynx Edicions, Barcelona, Spain

Dembeck H (1965) Animals and Men. The American Museum of Natural History. The Natural History Press, Garden City USA

Derjusheva S, Kurganova A, Habermann F, Gaginskaya E (2004) High chromosome conservation detected by comparative chromosome painting in chicken, pigeon and passerine birds. Chrom Res 12:715–723

Dixon ES (1848) Ornamental and Domestic Poultry: Their History and Management, 1st edn. Gardener's Chronicle, London, UK

Dodgson JB (2003) Chicken genome sequence: a centennial gift to poultry genetics. Cytogenet Genome Res 102:291–296

Dodgson JB, Romanov MN (2004) Use of chicken models for the analysis of human disease. In: Dracopoli NC, Haines JL, Korf BR, Moir DT, Morton CC, Seidman CE, Seidman JG, Smith DR (eds) Current Protocols in Human Genetics. Wiley, Hoboken, USA, Unit 15.5, pp 15.5.1–15.5.11

Dunn IC, Sharp PJ, Paton IR, Burt DW (1999) Mapping of the gene responsible for henny feathering (CYP19/aromatase) to chicken chromosome E29C09W09. Proc Poult Genet Symp, Mariensee, Germany, October 6–8, 1999, p 114

Dunn LC (1928) The genetics of the domestic fowl. J Hered 19:511–519

Dunn LC (1929) The genetics of the domestic fowl: Memoirs of the Anikowo Genetical Station, 1926. II. The genetics of leg feathering. J Hered 20:111–118

Dunn LC, Jull MA (1927) On the inheritance of some characteristics of the Silky fowl. J Genet 19:27–63

Dunn LC, Landauer W (1930) Further data on a case of autosomal linkage in the domestic fowl. J Genet 22:95–101

Dunnington EA, Stallard LC, Hillel J, Siegel PB (1994) Genetic diversity among commercial chicken populations estimated from DNA fingerprints. Poult Sci 73:1218–1225

Durham FM, Marryat DCE (1908) Note on the inheritance of sex in canaries. Rep Evol Comm R Soc IV:57–60

Edwards SV, Gasper J, March M (1998) Genomics and polymorphism of *Agph-DAB1*, an *Mhc* class II B gene in red-winged blackbirds (*Agelaius phoeniceus*). Mol Biol Evol 15:236–250

El Bassam N (1998) Sustainable development in agriculture – global key issues. Landbauforsch Völkenrode 48:1–11

Ellegren H (1996) First gene on the avian W chromosome (CHD) provides a tag for universal sexing of non-ratite birds. Proc Roy Soc Lond B Biol Sci 263:1635–1641

Ellegren H, Fridolfsson AK (1997) Male-driven evolution of DNA sequences in birds. Nat Genet 17:182–184

Ellestad LE, Carre W, Muchow M, Jenkins SA, Wang X, Cogburn LA, Porter TE (2006) Gene expression profiling during cellular differentiation in the embryonic pituitary gland using cDNA microarrays. Physiol Genom 25:414–425

Erbil C, Niessing J (1984) Chromosomal arrangement of the duck α-globin genes and primary structure of the embryonic α-globin gene π'. Gene 32:161–170

Etches RJ, Hawes RO (1973) A summary of linkage relationships and a revised linkage map of the chicken. Can J Genet Cytol 15:553–570

Ewins R (1995) Proto-Polynesian art? The cliff paintings of Vatulele, Fiji. J Polyn Soc 103:23–74

FAO (1997–2004) Secondary Guidelines for Development of Na-tional Farm Animal Genetic Resources Management Plans. Measurement of Domestic Animal Diversity (MoDAD): Recommended Microsatellite Markers. New Microsatellite Marker Sets — Recommendation of joint ISAG/FAO Standing Committee (to be presented at ISAG 2004). Initiative for Domestic Animal Diversity, FAO, Rome, Italy

FAOSTAT (2006) FAOSTAT Database Collections. Food and Agriculture Organization of the United Nations, Rome, Italy. http://faostat.fao.org/ (accessed August 25, 2006)

Fillon V, Morisson M, Zoorob R, Auffray C, Douaire M, Gellin J, Vignal A (1998) Identification of 16 chicken microchromosomes by molecular markers using two-colour fluorescence in situ hybridization (FISH). Chrom Res 6:307–313

Fridolfsson AK, Ellegren H (1999) A simple and universal method for molecular sexing of non-ratite birds. J Avian Biol 30:116–121

Finsterbusch CA (1929) Cock Fighting all over the World. Grit and Steel, Gaffney, UK

Goodale HD (1917) Crossing-over in the sex chromosome of the male fowl. Science 46:213

Gregory TR (2006) Animal Genome Size Database. University of Guelph, Canada. http://www.genomesize.com/ (accessed August 25, 2006)

Griffiths R, Korn RM (1997) A *CHD1* gene is Z chromosome linked in the chicken *Gallus domesticus*. Gene 197:225–229

Griffiths R, Daan S, Dijkstra C (1996) Sex identification in birds using two CHD genes. Proc Roy Soc Lond B Biol Sci 263:1251–1256

Griffiths R, Double MC, Orr K, Dawson RJ (1998) A DNA test to sex most birds. Mol Ecol 7:1071–1075

Grisart B, Coppieters W, Farnir F, Karim L, Ford C, Berzi P, Cambisano N, Mni M, Reid S, Simon P, Spelman R, Georges M, Snell R (2002) Positional candidate cloning of a QTL in dairy cattle: identification of a missense mutation in the bovine *DGAT1* gene with major effect on milk yield and composition. Genom Res 12:222–231

Groenen MAM, Crooijmans RPMA, Veenendaal A, van Kaam JBCHM, Vereijken ALJ, van Arendonk JAM, van der Poel JJ (1997) QTL mapping in chicken using a three generation full sib family structure of an extreme broiler × broiler cross. Anim Biotechnol 8:41–46

Groenen MAM, Crooijmans RPMA, Veenendaal A, Cheng HH, Siwek M, van der Poel JJ (1998) A comprehensive microsatellite linkage map of the chicken genome. Genomics 49:265–274

Groenen MAM, Cheng HH, Bumstead N, Benkel BF, Briles WE, Burke T, Burt DW, Crittenden LB, Dodgson J, Hillel J, Lamont S, Ponce de Leon FA, Soller M, Takahashi H, Vignal A (2000) A consensus linkage map of the chicken genome. Genom Res 10:137–147

Guggenheim JA, Erichsen JT, Hocking PM, Wright NF, Black R (2002) Similar genetic susceptibility to form-deprivation myopia in three strains of chicken. Vision Res 42:2747–2756

Guillier-Gensik Z, Bernheim A, Coullin P (1999) Generation of whole-chromosome painting probes specific to each chicken macrochromosomes. Cytogenet Cell Genet 87:282–285

Gunnarsson U, Hellström AR, Tixier-Boichard M, Minvielle F, Bed'hom B, Ito S, Jensen P, Rattink A, Vereijken A, Andersson L (2007) Mutations in *SLC45A2* cause plumage color variation in chicken and Japanese quail. Genetics 175:867–877

Gunski RJ, Giannoni ML (1998) Nucleolar organizer regions and a new chromosome number for *Rhea americana* (Aves: Rheiformes). Genet Mol Biol 21:207–210

Guttenbach M, Nanda I, Feichtinger W, Masabanda JS, Griffin DK, Schmid M (2003) Comparative chromosome painting of chicken autosomal paints 1–9 in nine different bird species. Cytogenet Genome Res 103:173–184

Haberfeld A, Dunnington EA, Siegel PB (1992) Genetic distances estimated from DNA fingerprints in crosses of White Plymouth Rock chickens. Anim Genet 23:165–173

Habermann F, Cremer M, Walter J, Kreth G, von Hase J, Bauer K, Wienberg J, Cremer C, Cremer T, Solovei I (2001)

Arrangement of macro- and microchromosomes in chicken cells. Chrom Res 9:569–584

Hagedoorn AL (1909) Mendelian inheritance of sex. Wilhelm Roux' Arch Entwicklungsmech Organ 28:1–34

Hale ML, Petrie M, Wolff K (2004) Polymorphic microsatellite loci in peafowl (*Pavo cristatus*). Mol Ecol Notes 4:528–530

Haldane JBS (1921) Linkage in poultry. Science 54:663

Hanotte O, Burke T, Armour JA, Jeffreys AJ (1991) Hypervariable minisatellite DNA sequences in the Indian peafowl *Pavo cristatus*. Genomics 9:587–597

Hansson B, Åkesson M, Slate J, Pemberton JM (2005) Linkage mapping reveals sex-dimorphic map distances in a passerine bird. Proc Biol Sci 272:2289–2298

Hertwig P (1933) Geschlechtsgebundene und autosomale Koppelungen bei Hühnern. Verh Dtsch Zool Ges 35:112–118

Hillel J, Groenen MAM, Tixier-Boichard M, Korol AB, David L, Kirzhner VM, Burke T, Barre-Dirie A, Crooijmans RPMA, Elo K, Feldman MW, Freidlin PJ, Mäki-Tanila A, Oortwijn M, Thomson P, Vignal A, Wimmers K, Weigend S (2003) Biodiversity of 52 chicken populations assessed by microsatellite typing of DNA pools. Genet Sel Evol 35:533–557

Ho PT (1977) The indigenous origin of Chinese agriculture. In: Reed CA (ed) Origins of Agriculture. Mouton, The Hague, The Netherlands, pp 413–484

Hollander WF (1970) Sex-linked chocolate coloration in the Muscovy Duck. Poult Sci 49:594–596

Hollander WF (1990) ABC's of Poultry Genetics. Stromberg, Pine River, USA

Hollander WF, Miller WJ (1982) A new sex-linked mutation, "web-lethal" from Racing Homers. Am Racing Pigeon News 98:50–51

Hori T, Asakawa S, Itoh Y, Shimizu N, Mizuno S (2000) *Wpkci*, encoding an altered form of *PKCI*, is conserved widely on the avian W chromosome and expressed in early female embryos: implication of its role in female sex determination. Mol Biol Cell 11:3645–3660

Huang Y, Tu J, Cheng X, Tang B, Hu X, Liu Z, Feng J, Lou Y, Lin L, Xu K, Zhao Y, Li N (2005) Characterization of 35 novel microsatellite DNA markers from the duck (*Anas platyrhynchos*) genome and cross-amplification in other birds. Genet Sel Evol 37:455–472

Huang YQ, Deng XM, Du ZQ, Qiu X, Du X, Chen W, Morisson M, Leroux S, Ponce de Leon FA, Da Y, Li N (2006a) Single nucleotide polymorphisms in the chicken *Lmbr1* gene are associated with chicken polydactyly. Gene 374:10–18

Huang Y, Zhao Y, Haley CS, Hu S, Hao J, Wu C, Li N (2006b) A genetic and cytogenetic map for the duck (*Anas platyrhynchos*). Genetics 173:287–296

Huang Y, Haley CS, Wu F, Hu S, Hao J, Wu C, Li N (2007) Genetic mapping of quantitative trait loci affecting carcass and meat quality traits in Beijing ducks (*Anas platyrhynchos*). Anim Genet 38:114–119.

Hutt FB (1933) Genetics of the fowl. II. A four-gene autosomal linkage group. Genetics 18:82–94

Hutt FB (1936) Genetics of the fowl. VI. A tentative chromosome map. In: Ag V (ed) Neue Forschungen in Tierzucht und Abstammungslehre (Festschrift zum 60. Geburtstag von Prof. Dr. J. Ulrich Duerst). Verbandsdruckerei, Bern, Switzerland, pp 105–112

Hutt FB (1949) Genetics of the Fowl. McGraw-Hill, New York, USA

Hutt FB (1960) New loci in the sex chromosome of the fowl. Heredity 15:97–110

Hutt FB (1964) Animal Genetics. Ronald, New York, USA

Hutt FB, Lamoreux WF (1940) Genetics of the fowl. 11. A linkage map for six chromosomes. J Hered 31:231–235

Hyams E (1972) Animals in the Service of Man: 10000 Years of Domestication. J.M. Dent and Sons, London, UK

Ikeobi CO, Woolliams JA, Morrice DR, Law A, Windsor D, Burt DW, Hocking PM (2002) Quantitative trait loci affecting fatness in the chicken. Anim Genet 33:428–435

International Chicken Genome Sequencing Consortium (2004) Sequence and comparative analysis of the chicken genome provide unique perspectives on vertebrate evolution. Nature 432:695–716

International Chicken Polymorphism Map Consortium (2004) A genetic variation map for chicken with 2.8 million single-nucleotide polymorphisms. Nature 432:717–722

Itoh Y, Arnold AP (2005) Chromosomal polymorphism and comparative painting analysis in the zebra finch. Chromosome Res 13:47–56

Itoh Y, Mizuno S (2002) Molecular and cytological characterization of *Ssp*I-family repetitive sequence on the chicken W chromosome. Chromosome Res 10:499–511

Itoh Y, Kampf K, Arnold AP (2006) Comparison of the chicken and zebra finch Z chromosomes shows evolutionary rearrangements. Chromosome Res 14:805–815

Ivanov MF (1924) Poultry Breeds. Ekonomicheskaya zhizn', Moscow, USSR

Jennen DGJ, Vereijken ALJ, Bovenhuis H, Crooijmans RPMA, Veenendaal A, van der Poel JJ, Groenen MAM (2004) Detection and localization of quantitative trait loci affecting fatness in broilers. Poult Sci 83:295–301

Jensen P (2005) Genomics: the chicken genome sequence. Heredity 94:567–568

Johnsgard PA (1999) The Pheasants of the World; Biology and Natural History, 2nd edn. Smithsonian Institution Press, Washington, DC, USA

Jull MA (1930) The association of comb and crest characters in the domestic fowl. J Hered 21:21–28

Kadi F, Mouchiroud D, Sabeur G, Bernardi G (1993) The compositional patterns of the avian genomes and their evolutionary implications. J Mol Evol 37:544–551

Kagami H, Nakamura H, Tomita T (1990) Sex identification in chickens by means of the presence of the W chromosome specific repetitive DNA units. Jap Poult Sci 27:379–384

Kahn NW, Quinn TW (1999) Male-driven evolution among Eoaves? A test of the replicative division hypothesis in a heterogametic female (ZW) system. J Mol Evol 49:750–759

Kahn NW, St John J, Quinn TW (1998) Chromosome-specific intron size differences in the avian CHD gene provide an efficient method for sex identification in birds. Auk 115:1074–1078

Kaiser MG, Deeb N, Lamont SJ (2002) Microsatellite markers linked to *Salmonella enterica* serovar *enteritidis* vaccine response in young F$_1$ broiler-cross chicks. Poult Sci 81:193–201

Kamara D, Geng T, Xu J, Guynn S, Hopwood K, Smith EJ (2007) Isolation and characterization of microsatellite markers from the budgerigar, *Melopsittacus undulatus*. Mol Ecol Notes 7:507–509

Kameda K, Goodridge AG (1991) Isolation and partial characterization of the gene for goose fatty acid synthase. J Biol Chem 266:419–426

Kasai F, Garcia C, Arruga MV, Ferguson-Smith MA (2003) Chromosome homology between chicken (*Gallus gallus domesticus*) and the red-legged partridge (*Alectoris rufa*); evidence of the occurrence of a neocentromere during evolution. Cytogenet Genome Res 102:326–330

Kato J, Hattori T, Ohba S, Tamaki Y, Yamada N, Taguchi T, Ogihara J, Ohya K, Itoh Y, Hori T, Asakawa S, Shimizu N, Mizuno S (2002) Efficient selection of genomic clones from a female chicken bacterial artificial chromosome library by four-dimensional polymerase chain reactions. Poult Sci 81:1501–1508

Kayang BB, Inoue-Murayama M, Hoshi T, Matsuo K, Takahashi H, Minezawa M, Mizutani M, Ito S (2002) Microsatellite loci in Japanese quail and cross-species amplification in chicken and guinea fowl. Genet Sel Evol 34:233–253

Kayang BB, Vignal A, Inoue-Murayama M, Miwa M, Monvoisin JL, Ito S, Minvielle F (2004) A first-generation microsatellite linkage map of the Japanese quail. Anim Genet 35:195–200

Kayang BB, Fillon V, Inoue-Murayama M, Miwa M, Leroux S, Feve K, Monvoisin JL, Pitel F, Vignoles M, Mouilhayrat C, Beaumont C, Ito S, Minvielle F, Vignal A (2006) Integrated maps in quail (*Coturnix japonica*) confirm the high degree of synteny conservation with chicken (*Gallus gallus*) despite 35 million years of divergence. BMC Genomics 7:101

Kear J (1975) How wildfowl could improve our domestic breeds. In: Waterfowl Yearbook. Buyer's Guide 1975–1976, pp 37–41

Keeton GW, Muir WM, Aggrey SE (eds) (2003) Poultry Genetics, Breeding and Biotechnology. CABI, Oxon, UK

Kellner WA, Sullivan RT, Carlson BH, Thomas JW (2005) Uprobe: a genome-wide universal probe resource for comparative physical mapping in vertebrates. Genome Res 15:166–173

Kerje S, Lind J, Schütz K, Jensen P, Andersson L (2003) Melanocortin 1-receptor (*MC1R*) mutations are associated with plumage colour in chicken. Anim Genet 34:264–274

Kikuchi S, Fujima D, Sasazaki S, Tsuji S, Mizutani M, Fujiwara A, Mannen H (2005) Construction of a genetic linkage map of Japanese quail (*Coturnix japonica*) based on AFLP and microsatellite markers. Anim Genet 36:227–231

Kogan ZM (1979) Exterior and Interior Characters in Chickens (Genetics and Economical Importance). Nauka, Novosibirsk, USSR

Lacson JM, Morizot DC (1988) Confirmation of avian sex-chromosome linkage of liver cytosolic aconitase (ACO1). Cytogenet Cell Genet 48:244–245

Lamont SJ, Lakshmanan N, Plotsky Y, Kaiser MG, Kuhn M, Arthur JA, Beck NJ, O'Sullivan NP (1996) Genetic markers linked to quantitative traits in poultry. Anim Genet 27:1–8

Lancaster FM (1977) Sex-linkage and autosexing in waterfowl. Bull Nat Inst Poult Husbandry, Newport, UK, No 1

Landauer W (1931) The linkage relationships of the autosomal genes for Creeper and Rose comb in the fowl. Anat Rec 51:123

Lee EJ, Mannen H, Mizutani M, Tsuji S (2000) Genetic analysis of chicken lines by amplified fragment length polymorphism (AFLP). Anim Sci J 71:231–238

Lee EJ, Yoshizawa K, Mannen H, Kikuchi H, Kikuchi T, Mizutani M, Tsuji S (2002) Localization of the muscular dystrophy *AM* locus using a chicken linkage map constructed with the Kobe University resource family. Anim Genet 33:42–48

Lee MK, Ren CW, Yan B, Cox B, Zhang HB, Romanov MN, Sizemore FG, Suchyta SP, Peters E, Dodgson JB (2003) Construction and characterization of three complementary BAC libraries for analysis of the chicken genome. Anim Genet 34:151–152

Li X, Wistow GJ, Piatigorsky J (1995) Linkage and expression of the argininosuccinate lyase/delta-crystallin genes of the duck: insertion of a CR1 element in the intergenic spacer. Biochim Biophys Acta 1261:25–34

Lin FK, Paddock GV (1984) Characterization of duck genome fragments containing beta and epsilon globin genes. Gene 31:59–64

Lipkin E, Fulton J, Cheng H, Yonash N, Soller M (2002) Quantitative trait locus mapping in chickens by selective DNA pooling with dinucleotide microsatellite markers by using purified DNA and fresh or frozen red blood cells as applied to marker-assisted selection. Poult Sci 81: 283–292

Liu HC, Kung HJ, Fulton JE, Morgan RW, Cheng HH (2001) Growth hormone interacts with the Marek's disease virus SORF2 protein and is associated with disease resistance in chicken. Proc Natl Acad Sci USA 98:9203–9208

Liu HC, Niikura M, Fulton J, Cheng HH (2003a) Identification of chicken stem lymphocyte antigen 6 complex, locus E (*LY6E*, alias *SCA2*) as a putative Marek's disease resistance gene via a virus-host protein interaction screen. Cytogenet Genome Res 102:304–308

Liu W, Liu Z, Hu X, Zhang Y, Yuan J, Zhao R, Li Z, Xu W, Gao Y, Deng X, Li N (2003b) Construction and characterization of a novel 13.34-fold chicken bacterial artificial chromosome library. Anim Biotechnol 14:145–153

Liu YP, Wu GS, Yao YG, Miao YW, Luikart G, Baig M, Beja-Pereira A, Ding ZL, Palanichamy MG, Zhang YP (2006) Multiple maternal origins of chickens: out of the Asian jungles. Mol Phylogenet Evol 38:12–19

Lock RH (1906) Recent progress in the study of variation, heredity, and evolution. E.P. Dutton, New York, USA

Longmire JL, Hahn DC, Roach JL (1999) Low abundance of microsatellite repeats in the genome of the brown-headed cowbird (*Molothrus ater*). J Hered 90:574–578

Luo M, Yu Y, Kim HR, Kudrna D, Itoh Y, Agate RJ, Melamed E, Goicoechea JL, Talag J, Mueller C, Wang W, Currie J, Sisneros NB, Wing RA, Arnold AP (2006) Utilization of a zebra finch BAC library to determine the structure of an avian androgen receptor genomic region. Genomics 87:181–190. Erratum in: Genomics 87:678–679

Maak S, Wimmers K, Weigend S, Neumann K (2003) Isolation and characterization of 18 microsatellites in the Peking duck (*Anas platyrhynchos*) and their application in other waterfowl species. Mol Ecol Notes 3:224–227

Mannen H, Murata K, Kikuchi S, Fujima D, Sasazaki S, Fujiwara A, Tsuji S (2005) Development and mapping of microsatellite markers derived from cDNA in Japanese quail (*Coturnix japonica*). J Poult Sci 42:263–271

Masabanda JS, Burt DW, O'Brien PCM, Vignal A, Fillon V, Walsh PS, Cox H, Tempest HG, Smith J, Habermann F, Schmid M, Matsuda Y, Ferguson-Smith MA, Crooijmans RPMA, Groenen MAM, Griffin DK (2004) Molecular cytogenetic definition of the chicken genome: the first complete avian karyotype. Genetics 166:1367–1373

Mason IL (ed) (1984) Evolution of Domesticated Animals. Longmann, New York, USA

McElroy JP, Dekkers JC, Fulton JE, O'Sullivan NP, Soller M, Lipkin E, Zhang W, Koehler KJ, Lamont SJ, Cheng HH (2005) Microsatellite markers associated with resistance to Marek's disease in commercial layer chickens. Poult Sci 84:1678–1688

Meng A, Gong G, Chen D, Zhang H, Qi S, Tang H, Gao Z (1996) DNA fingerprint variability within and among parental lines and its correlation with performance of F_1 laying hens. Theor Appl Genet 92:769–776

Mesa CM, Thulien KJ, Moon DA, Veniamin SM, Magor KE (2004) The dominant MHC class I gene is adjacent to the polymorphic TAP2 gene in the duck, Anas platyrhynchos. Immunogenetics 56:192–203

Michelmore RW, Paran I, Kesseli RV (1991) Identification of markers linked to disease-resistance genes by bulked segregant analysis: a rapid method to detect markers in specific genomic regions by using segregating populations. Proc Natl Acad Sci USA 88:9828–9832

Millar CD, Lambert DM, Anderson S, Halverson JL (1996) Molecular sexing of the communally breeding pukeko: an important ecological tool. Mol Ecol 5:289–293

Miller WJ (1964) First linkage of a species antigen in the genus Streptopelia. Science 143:1179–1180

Miller WJ (1992) Color mutants in zebra finches, *Poephila guttata*. Friends'N Feathers, Mid-America Cage Bird Society, Des Moines, USA, Oct Issue, pp 3–7

Miller WJ, Hollander WF (1978) The quest for linkages. Pigeon Sci Genet Newsl 8:13–14

Miller WJ, Webber JL (1969) A new species-antigen in doves and its linkage with the species-albumin type. Genetics (Suppl):s40–s41

Minvielle F, Ito S, Inoue-Murayama M, Mizutani M, Wakasugi N (2000) Genetic analyses of plumage color mutations on the Z chromosome of Japanese quail. J Hered 91: 499–501

Minvielle F, Kayang BB, Inoue-Murayama M, Miwa M, Vignal A, Gourichon D, Neau A, Monvoisin JL, Ito S (2005) Microsatellite mapping of QTL affecting growth, feed consumption, egg production, tonic immobility and body temperature of Japanese quail. BMC Genomics 6:87

Miwa M, Inoue-Murayama M, Kayang BB, Minvielle F, Monvoisin JL, Takahashi H, Ito S (2005) Mapping of plumage colour and blood protein loci on the microsatellite linkage map of the Japanese quail. Anim Genet 36:396–400

Miwa M, Inoue-Murayama M, Kobayashi N, Kayang BB, Mizutani M, Takahashi H, Ito S (2006) Mapping of panda plumage color locus on the microsatellite linkage map of the Japanese quail. BMC Genet 7:2

Miwa M, Inoue-Murayama M, Aoki H, Kunisada T, Hiragaki T, Mizutani M, Takahashi H, Ito S (2007) *Endothelin receptor B2* (*EDNRB2*) is associated with the *panda* plumage colour mutation in Japanese quail. Anim Genet 38:103–108

Mizuno S, Macgregor H (1998) The ZW lampbrush chromosomes of birds: a unique opportunity to look at the molecular cytogenetics of sex chromosomes. Cytogenet Cell Genet 80:149–157

Moiseyeva IG (1998) Ancient evidence for the origin and distribution of domestic fowl. Proc 10th Eur Conf "The Poultry Industry Towards the 21st Century", Jerusalem, Israel, June 21–26, 1998, vol I, pp 244–245

Moiseyeva IG, Lisichkina MG (1996) Origin and evolution of the domestic fowl. Priroda 5:88–96

Moiseyeva IG, Volokhovich VA (1987) Variation of qualitative traits of chicken exterior. In: Selection and Technological Processes in Poultry Industry. Stiintsa, Chisinau, USSR, pp 70–74

Moiseyeva IG, Semyenova SK, Bannikova LV, Filippova ND (1994) Genetic structure and origin of an old Russian Orloff chicken breed. Genetika 30:681–694

Moiseyeva I, Romanov M, Pigaryev N (2000) Obituary: sergey petrov. Worlds Poult Sci J 56:437–438

Moiseyeva IG, Romanov MN, Nikiforov AA, Sevastyanova AA, Semyenova SK (2003) Evolutionary relationships of Red Jungle Fowl and chicken breeds. Genet Sel Evol 35:403–423

Moon DA, Magor KE (2004) Construction and characterization of a fosmid library for comparative analysis of the duck genome. Anim Genet 35:417–418

Morgan TH (1910) The method of inheritance of two sex limited characters in the same animal. Proc Soc Exp Biol Med 8:17–19

Morgan TH (1911) An attempt to analyze the constitution of the chromosomes on the basis of sex-limited inheritance in Drosophila. J Exp Zool 11:365–412

Morgan TH, Goodale HD (1912) Sex-linked inheritance in poultry. Ann N Y Acad Sci 22:113–133

Morisson M, Lemiere A, Bosc S, Galan M, Plisson-Petit F, Pinton P, Delcros C, Feve K, Pitel F, Fillon V, Yerle M, Vignal A (2002) ChickRH: a chicken whole-genome radiation hybrid panel. Genet Sel Evol 34:521–533

Nakamura D, Tiersch TR, Douglass M, Chandler RW (1990) Rapid identification of sex in birds by flow cytometry. Cytogenet Cell Genet 53:201–205

Nanda I, Schmid M (2002) Conservation of avian Z chromosomes as revealed by comparative mapping of the Z-linked aldolase B gene. Cytogenet Genome Res 96:176–178

Nanda I, Sick C, Munster U, Kaspers B, Schartl M, Staeheli P, Schmid M (1998) Sex chromosome linkage of chicken and duck type I interferon genes: further evidence of evolutionary conservation of the Z chromosome in birds. Chromosoma 107:204–210

Nanda I, Zend-Ajusch E, Shan Z, Grutzner F, Schartl M, Burt DW, Koehler M, Fowler VM, Goodwin G, Schneider WJ, Mizuno S, Dechant G, Haaf T, Schmid M (2000) Conserved synteny between the chicken Z sex chromosome and human chromosome 9 includes the male regulatory gene *DMRT1*: a comparative (re)view on avian sex determination. Cytogenet Cell Genet 89:67–78

Nefedov M, Zhu B, Thorsen J, Shu CL, Cao Q, Osoegawa K, de Jong P (2003) New chicken, turkey, salmon, bovine, porcine and sheep genomic BAC libraries to complement world wide effort to map farm animals genomes. Proc Plant Anim Genome XI Int Conf, San Diego, USA, January 11–15, 2003, p 96, Abstr P87

Niessing J, Erbil C, Neubauer V (1982) The isolation and partial characterization of linked α^A- and α^D-globin genes from a duck DNA recombinant library. Gene 18:187–191

Nishibori M, Shimogiri T, Hayashi T, Yasue H (2005) Molecular evidence for hybridization of species in the genus *Gallus* except for *Gallus varius*. Anim Genet 36:367–375

Nishida T, Hayashi Y, Hashiguchi T, Mansjoer SS (1983) Ecological and morphological studies on the jungle fowl in Indonesia. Rep Soc Res Native Livest 10:155–170

Nishida T, Hayashi Y, Fujioka T, Tsugiyama I, Mochizuki K (1985a) Osteometrical studies on the phylogenetic-relationships of Japanese native fowls. Jap J Vet Sci 47:25–37

Nishida T, Hayashi Y, Hashiguchi T (1985b) Somatometrical studies on the morphological relationships of Japanese native fowls. Jap J Zootech Sci 56:645–657

Nishida-Umehara C, Fujiwara A, Ogawa A, Mizuno S, Abe S, Yoshida MC (1999) Differentiation of Z and W chromosomes revealed by replication banding and FISH mapping of sex-chromosome-linked DNA markers in the cassowary (Aves, Ratitae). Chromosome Res 7:635–640

Niu D, Fu Y, Luo J, Ruan H, Yu XP, Chen G, Zhang YP (2002) The origin and genetic diversity of Chinese native chicken breeds. Biochem Genet 40:163–174

Niwa T, Shibusawa M, Matsuda Y, Terashima A, Nakamura A, Shiojiri N (2003) The *Bh* (black at hatch) gene that causes abnormal feather pigmentation maps to chromosome 1 of the Japanese quail. Pigment Cell Res 16:656–661

Ogawa A, Solovei I, Hutchison N, Saitoh Y, Ikeda JE, Macgregor H, Mizuno S (1997) Molecular characterization and cytological mapping of a non-repetitive DNA sequence region from the W chromosome of chicken and its use as a universal probe for sexing carinatae birds. Chromosome Res 5:93–101

Ogawa A, Murata K, Mizuno S (1998) The location of Z- and W-linked marker genes and sequence on the homomorphic sex chromosomes of the ostrich and the emu. Proc Natl Acad Sci USA 95:4415–4418

Ohno S (1961) Sex chromosomes and microchromosomes of *Gallus domesticus*. Chromosoma 11:484–498

Okimoto R, Stie JT, Takeuchi S, Payne WS, Salter DW (1999) Mapping the melanocortin 1-receptor (MC1-R) gene and association of MC1-R polymorphisms with *E* locus phenotypes. Poult Sci 78(Suppl):60

O'Neill M, Binder M, Smith C, Andrews J, Reed K, Smith M, Millar C, Lambert D, Sinclair A (2000) *ASW*: a gene with conserved avian W-linkage and female specific expression in chick embryonic gonad. Dev Genes Evol 210:243–249

Palyga J (1998) Genes for polymorphic H1 histones are linked in the Japanese quail genome. Biochem Genet 36:93–103

Pang SW, Ritland C, Carlson JE, Cheng KM (1999) Japanese quail microsatellite loci amplified with chicken-specific primers. Anim Genet 30:195–199

Passarge E, Horsthemke B, Farber RA (1999) Incorrect use of the term synteny. Nat Genet 23:387

Peters JP (1913) The cock. J Am Orient Soc 33:363–396

Petrov SG (1931) Plan of the chromosomes of the domestic fowl. Zh Eksp Biol 7:71–76

Petrov SG (1941) Origin of the domestic fowl. DSc (Biol) Thesis, Moscow, USSR

Petrov SG (1962) Origin and evolution of domestic fowl. In: Penionzhkevich EE (ed) Poultry Science and Practice. Israel Program Sci Transl U S Dept Commer, Springfield, MS, USA, vol 1 (translated 1968)

Pigozzi MI, Solari AJ (1998) Germ cell restriction and regular transmission of an accessory chromosome that mimics a sex body in the zebra finch, *Taeniopygia guttata*. Chromosome Res 6:105–113

Pimentel-Smith GE, Shi L, Drummond P, Tu Z, Smith EJ (2000) Amplification of sequence tagged sites in five avian species using heterologous oligonucleotides. Genetica 110:219–226

Pisenti JM, Delany ME, Taylor RL, Jr, Abbott UK, Abplanalp H, Arthur JA, Bakst MR, Baxter-Jones C, Bitgood JJ, Bradley F, Cheng KM, Dietert RR, Dodgson JB, Donoghue A, Emsley AE, Etches R, Frahm RR, Gerrits RJ, Goetinck PF, Grunder AA, Harry DE, Lamont SJ, Martin GR, McGuire PE, Moberg GP, Pierro LJ, Qualset CO, Qureshi M, Schultz F, Wilson BW (1999) Avian genetic resources at risk: an assessment and proposal for conservation of genetic stocks in the USA and Canada. Rep No 20, Univ Calif, Div Agric Nat Resour, Genet Resour Conserv Program, Davis, CA, USA

Pitel F, Berge R, Coquerelle G, Crooijmans RPMA, Groenen MAM, Vignal A, Tixier-Boichard M (2000) Mapping the Naked Neck (*NA*) and Polydactyly (*PO*) mutants of the chicken with microsatellite molecular markers. Genet Sel Evol 32:73–86

Plant WJ (1984) The Origin, Evolution, History and Distribution of the Domestic Fowl, Pt 2. Chicken Bone Recoveries. Privately published, 54 Bonar Street, Maitland 2320, N.S.W., Australia

Plant WJ (1986) The Origin, Evolution, History and Distribution of the Domestic Fowl, Pt 3. The Gallus Species. Jungle Fowls. Privately published, 54 Bonar Street, Maitland 2320, N.S.W., Australia

Plotsky Y, Kaiser MG, Lamont SJ (1995) Genetic characterization of highly inbred chicken lines by two DNA methods: DNA fingerprinting and polymerase chain reaction using arbitrary primers. Anim Genet 26:163–170

Pond WG, Bell AW (eds) (2004) Encyclopedia of Animal Science. Marcel Dekker, New York, USA

Punnett RC, Bateson W (1908) The heredity of sex. Science 27:785–787

Raudsepp T, Houck ML, O'Brien PC, Ferguson-Smith MA, Ryder OA, Chowdhary BP (2002) Cytogenetic analysis of California condor (*Gymnogyps californianus*) chromosomes: comparison with chicken (*Gallus gallus*) macrochromosomes. Cytogenet Genome Res 98:54–60

Reed KM, Chaves LD, Garbe JR, Da Y, Harry DE (2003) Allelic variation and genetic linkage of avian microsatellites in a new turkey population for genetic mapping. Cytogenet Genome Res 102:331–339

Ren CW, Lee MK, Yan B, Ding K, Cox B, Romanov MN, Price JA, Dodgson JB, Zhang HB (2003) A BAC-based physical map of the chicken genome. Genome Res 13:2754–2758

Rodionov AV (1996) Micro versus macro: a review of structure and function of avian micro- and macrochromosomes. Genetika 32:597–608

Rodionov AV (1997) Evolution of avian chromosomes and linkage groups. Rus J Genet 33:605–617

Rodionov AV, Lukina NA, Galkina SA, Solovei I, Saccone S (2002) Crossing over in chicken oogenesis: cytological and chiasma-based genetic maps of chicken lampbrush chromosome 1. J Hered 93:125–129

Romanov MN, Bondarenko YV (1988) Improvement of a colour-sexing cross of chickens. Nauchno-tekhnicheskiy byulleten, Ukr Poult Res Inst, Kharkiv, USSR, No 24:8–10

Romanov MN, Dodgson JB (2006) Cross-species overgo hybridization and comparative physical mapping within avian genomes. Anim Genet 37:397–399

Romanov MN, Weigend S (2001a) Using RAPD markers for assessment of genetic diversity in chickens. Arch Geflügelkd 65:145–148

Romanov MN, Weigend S (2001b) Analysis of genetic relationships between various populations of domestic and jungle fowl using microsatellite markers. Poult Sci 80:1057–1063

Romanov MN, Price JA, Dodgson JB (2003) Integration of animal linkage and BAC contig maps using overgo hybridization. Cytogenet Genome Res 102:277–281

Romanov MN, Sazanov AA, Smirnov AF (2004) First century of chicken gene study and mapping — a look back and forward. Worlds Poult Sci J 60:19–41

Romanov MN, Daniels LM, Dodgson JB, Delany ME (2005) Integration of the cytogenetic and physical maps of chicken chromosome 17. Chromosome Res 13:215–222

Romanov MN, Koriabine M, Nefedov M, de Jong PJ, Ryder OA (2006) Construction of a California condor BAC library and first-generation chicken-condor comparative physical map as an endangered species conservation genomics resource. Genomics 88:711–718

Roots EH, Baker RJ (2002) Distribution and characterization of microsatellites in the emu (*Dromaius novaehollandiae*) genome. J Hered 93:100–106

Roussot O, Feve K, Plisson-Petit F, Pitel F, Faure JM, Beaumont C, Vignal A (2003) AFLP linkage map of the Japanese quail *Coturnix japonica*. Genet Sel Evol 35:559–72

Ruyter-Spira CP, Gu ZL, van der Poel JJ, Groenen MAM (1997) Bulked segregant analysis using microsatellites: mapping of the dominant white locus in the chicken. Poult Sci 76:386–391

Ruyter-Spira CP, de Groof AJC, van der Poel JJ, Herbergs J, Masabanda J, Fries R, Groenen MAM (1998) The HMGI-C gene is a likely candidate for the autosomal dwarf locus in the chicken. J Hered 89:295–300

Saitoh Y, Ogawa A, Hori T, Kunita R, Mizuno S (1993) Identification and localization of two genes on the chicken Z chromosome: implication of evolutionary conservation of the Z chromosome among avian species. Chromosome Res 1:239–251

Sasaki M, Ikeuchi T, Makino S (1968) A feather pulp culture technique for avian chromosomes, with notes on the chromosomes of the peafowl and the ostrich. Experientia 24:1292–1293

Sasaki O, Odawara S, Takahashi H, Nirasawa K, Oyamada Y, Yamamoto R, Ishii K, Nagamine Y, Takeda H, Kobayashi E, Furukawa T (2004) Genetic mapping of quantitative trait loci affecting body weight, egg character and egg production in F_2 intercross chickens. Anim Genet 35:188–194

Sasazaki S, Hinenoya T, Fujima D, Kikuchi S, Fujiwara A, Mannen H (2006a) Mapping of EST markers with cDNA-AFLP method in Japanese quail (*Coturnix japonica*). Anim Sci J 77:42–46

Sasazaki S, Hinenoya T, Lin B, Fujiwara A, Mannen H (2006b) A comparative map of macrochromosomes between chicken and Japanese quail based on orthologous genes. Anim Genet 37:316–320

Sazanov A, Masabanda J, Ewald D, Takeuchi S, Tixier-Boichard M, Buitkamp J, Fries R (1998) Evolutionarily conserved telomeric location of *BBC1* and *MC1R* on a microchromosome questions the identity of *MC1R* and a pigmentation locus on chromosome 1 in chicken. Chromosome Res 6:651–654

Sazanov AA, Trukhina AV, Smirnov AF, Jaszczak K (2002) Two chicken genes *APOA1* and *ETS1* are physically assigned to the same microchromosome. Anim Genet 33:321–322

Sazanov AA, Sazanova AL, Tzareva VA, Kozyreva AA, Smirnov AF, Romanov MN, Price JA, Dodgson JB (2004a) Refined localization of the chicken KITLG, MGP and TYR genes on GGA1 by FISH mapping using BACs. Anim Genet 35:148–150

Sazanov AA, Sazanova AL, Stekolnikova VA, Kozyreva AA, Smirnov AF, Romanov MN, Dodgson JB (2004b) Chromosomal localization of *CTSL*: expanding of the region of evolutionary conservatism between GGAZ and HSA9. Anim Genet 35:260

Sazanov AA, Romanov MN, Ward&ecedil;cka B, Sazanova AL, Korczak M, Stekol'nikova VA, Kozyreva AA, Smirnov AF, Jaszczak K, Dodgson JB (2005) Chromosomal localization of fifteen large insert BAC clones containing three microsatellites on chicken chromosome 4 (GGA4) which refine its centromere position. Anim Genet 36:161–163

Savage TF, Harper JA, Engel HN, Jr (1993) Inheritance of tetanic torticollar spasms in turkeys. Poult Sci 72:1212–1217

Scherf BD (ed) (2000) World watch list for domestic animal diversity, 3rd edn. Food and Agriculture Organization of the United Nations, Rome, Italy http://dad.fao.org/

Schmid M, Nanda I, Guttenbach M, Steinlein C, Hoehn M, Schartl M, Haaf T, Weigend S, Fries R, Buerstedde J-M, Wimmers K, Burt DW, Smith J, A'Hara S, Law A, Griffin DK, Bumstead N, Kaufman J, Thomson PA, Burke T, Groenen MAM, Crooijmans RPMA, Vignal A, Fillon V, Morisson M, Pitel F, Tixier-Boichard M, Ladjali-Mohammedi K, Hillel J, Mäki-Tanila A, Cheng HH, Delany ME, Burnside J, Mizuno S (2000) First report on chicken genes and chromosomes 2000. Cytogenet Cell Genet 90:169–218

Schmid M, Nanda I, Hoehn H, Schartl M, Haaf T, Buerstedde J-M, Arakawa H, Caldwell RB, Weigend S, Burt DW, Smith J, Griffin DK, Masabanda JS, Groenen MAM, Crooijmans RPMA, Vignal A, Fillon V, Morisson M, Pitel F, Vignoles M, Garrigues A, Gellin J, Rodionov AV, Galkina SA, Lukina NA, Ben-Ari G, Blum S, Hillel J, Twito T, Lavi U, David L, Feldman MW, Delany ME, Conley CA, Fowler VM, Hedges SB, Godbout R, Katyal S, Smith C, Hudson Q, Sinclair A, Mizuno S (2005) Second report on chicken genes and chromosomes 2005. Cytogenet Genome Res 109:415–479

Schütz K, Kerje S, Carlborg O, Jacobsson L, Andersson L, Jensen P (2002) QTL analysis of a red junglefowl × White Leghorn intercross reveals trade-off in resource allocation between behavior and production traits. Behav Genet 32:423–433

Semyenova SK, Filenko AL, Vasilyev VA, Prosnyak MI, Sevastyanova AA, Ryskov AP (1996) Differentiation of chicken breeds of different origin by polymorphic DNA markers. Genetika 32:795–803

Serebrovsky AS (1922) Crossing-over involving three sex-linked genes in chickens. Am Nat 56:571–572

Serebrovsky AS (1926) Studies on genetics of domestic fowl. In: Koltzoff NK (ed) Genetics of the domestic fowl: memoirs of Anikowo Genetical Station near Moscow. Commissariat Agric, Novaia Derevnia, Moscow, USSR, pp 3–74. (Abstracted in: Dunn, 1929)

Serebrovsky AS, Petrov SG (1928) A case of close autosomal linkage in the fowl. J Hered 19:306–306

Serebrovsky AS, Petrov SG (1930) On the composition of the plan of the chromosomes of the domestic hen. Zh Eksp Biol 6:157–180

Serebrovsky AS, Wassina ET (1927) On the topography of the sex-chromosome in fowls. J Genet 17:211–216

Sewalem A, Morrice DM, Law A, Windsor D, Haley CS, Ikeobi CO, Burt DW, Hocking PM (2002) Mapping of quantitative trait loci for body weight at three, six, and nine weeks of age in a broiler layer cross. Poult Sci 81:1775–1781

Shetty S, Griffin DK, Graves JA (1999) Comparative painting reveals strong chromosome homology over 80 million years of bird evolution. Chromosome Res 7:289–295

Shetty S, Kirby P, Zarkower D, Graves JA (2002) DMRT1 in a ratite bird: evidence for a role in sex determination and discovery of a putative regulatory element. Cytogenet Genome Res 99:245–251

Shibusawa M, Minai S, Nishida-Umehara C, Suzuki T, Mano T, Yamada K, Namikawa T, Matsuda Y (2001) A comparative cytogenetic study of chromosome homology between chicken and Japanese quail. Cytogenet Cell Genet 95: 103–109

Shibusawa M, Nishida-Umehara C, Masabanda J, Griffin DK, Isobe T, Matsuda Y (2002) Chromosome rearrangements between chicken and guinea fowl defined by comparative chromosome painting and FISH mapping of DNA clones. Cytogenet Genome Res 98:225–230

Shibata T, Abe T (1996) Linkage between the loci for serum albumin and vitamin D binding protein (GC) in the Japanese quail. Anim Genet 27:195–197

Shiina T, Shimizu C, Oka A, Teraoka Y, Imanishi T, Gojobori T, Hanzawa K, Watanabe S, Inoko H (1999) Gene organization of the quail major histocompatibility complex (MhcCoja) class I gene region. Immunogenetics 49:384–394

Sibley CG (1996) Birds of the world. Thayer Birding Software, Version 2.0, Dec 1996 (a computerized book on two compressed 3.5 inch diskettes). Thayer Birding Software, Naples, USA. http://www.thayerbirding.com/Default.aspx? TabId=581; Sibley's Sequence, http://www.scricciolo.com/classificazione/sequence.htm (accessed August 20, 2008)

Siegel PB, Haberfeld A, Mukherjee TK, Stallard LC, Marks HL, Anthony NB, Dunnington EA (1992) Jungle fowl-domestic fowl relationships: a use of DNA fingerprinting. Worlds Poult Sci J 48:147–155

Siwek M, Cornelissen SJB, Nieuwland MGB, Buitenhuis AJ, Bovenhuis H, Crooijmans RPMA, Groenen MAM, de Vries-Reilingh G, Parmentier HK, van der Poel JJ (2003a) Detection of QTL for immune response to sheep red blood cells in laying hens. Anim Genet 34:422–428

Siwek M, Buitenhuis AJ, Cornelissen SJB, Nieuwland MGB, Bovenhuis H, Crooijmans RPMA, Groenen MAM, de Vries-Reilingh G, Parmentier HK, van der Poel JJ (2003b) Detection of different quantitative trait loci for antibody responses to keyhole lympet hemocyanin and *Mycobacterium butyricum* in two unrelated populations of laying hens. Poult Sci 82:1845–1852

Smith E, Shi L, Drummond P, Rodriguez L, Hamilton R, Powell E, Nahashon S, Ramlal S, Smith G, Foster J (2000a) Development and characterization of expressed sequence tags for the turkey (*Meleagris gallopavo*) genome and comparative sequence analysis with other birds. Anim Genet 31:62–67

Smith EJ, Shi L, Drummond P, Rodriguez L, Hamilton R, Ramlal S, Smith G, Pierce K, Foster J (2001a) Expressed sequence tags for the chicken genome from a normalized 10-day-old White Leghorn whole embryo cDNA library: 1. DNA sequence characterization and linkage analysis. J Hered 92:1–8

Smith EJ, Shi L, Prevost L, Drummond P, Ramlal S, Smith G, Pierce K, Foster J (2001b) Expressed sequence tags for the chicken genome from a normalized, ten-day-old white leghorn whole embryo cDNA library. 2. Comparative DNA sequence analysis of guinea fowl, quail, and turkey genomes. Poult Sci 80:1263–1272

Smith J, Bruley CK, Paton IR, Dunn I, Jones CT, Windsor D, Morrice DR, Law AS, Masabanda J, Sazanov A, Waddington D, Fries R, Burt DW (2000b) Differences in gene density on chicken macrochromosomes and microchromosomes. Anim Genet 31:96–103

Smith J, Speed D, Hocking PM, Talbot RT, Degen WG, Schijns VE, Glass EJ, Burt DW (2006) Development of a chicken 5 K microarray targeted towards immune function. BMC Genomics 7:49

Smith P, Daniel C (1975) The Chicken Book. Little, Brown, Toronto, Quebec, Canada

Sokolovskaya II (1935) Sex-linked characters in hybrids between the Muscovy duck (*Cairina moschata*) and Khaki duck (*Anas platyrincha*). In: Nurinov AA (ed) Hybridization and Acclimatization of Farm Animals in Askania Nova. VASKhNIL, Moscow Leningrad, USSR, Issue 4, vol II, pp 144–156

Soller M, Weigend S, Romanov MN, Dekkers JCM, Lamont SJ (2006) Strategies to assess structural variation in the chicken genome and its associations with biodiversity and biological performance. Poult Sci 85:2061–2078

Somes RG Jr (1973) Linkage relationships in domestic fowl. J Hered 64:217–221

Somes RG Jr (1978) New linkage groups and revised chromosome map of the domestic fowl. J Hered 69:401–403

Somes RG Jr (1987) Linked loci of the chicken — *Gallus gallus* (*G. domesticus*). In: O'Brien S (ed) Genetic Maps, 4th edn. Cold Spring Harbor Laboratory Press, Cold Spring Harbor, USA, pp 422–429

Somes RG Jr (1992) Identifying the ptilopody (feathered shank) loci of the chicken. J Hered 83:230–234

Somes RG Jr, Burger RE (1988) A sex-linked mutation in the Indian blue peafowl (*Pavo cristatus*). Poultry Sci 66(Suppl 1):158

Sørensen P (1997) The population of laying hens loses important genes: a case history. Anim Genet Resour Inf 22:71–78

Spillman WJ (1908) Spurious allelomorphism: results of some recent investigations. Am Nat 42:610–615

Staško J (1970) K autosexingu u husi chovanych na Slovensky. Vedecke Prace—Hydinarstvo, No 9:5–13

Stern CD (2004) The chick embryo—Past, present and future as a model system in developmental biology. Mech Dev 121:1011–1013

Stern CD (2005) The chick: a great model system becomes even greater. Dev Cell 8:9–17

Stevens L (1986) Gene structure and organisation in the domestic fowl (*Gallus domesticus*). Worlds Poult Sci J 42:232–242

Stevens L (1991) Genetics and evolution of the domestic fowl. Cambridge University Press, Cambridge, UK

Sturtevant AH (1911) Another sex-limited character in fowls. Science 33:337–338

Sturtevant AH (1912) An experiment dealing with sex-linkage in fowls. J Exp Zool 12:499–518

Suchyta SP, Cheng HH, Burnside J, Dodgson JB (2001) Comparative mapping of chicken anchor loci orthologous to genes on human chromosomes 1, 4 and 9. Anim Genet 32:12–18

Sungurov AN (1933) On the plan of the fowl chromosomes. Biol Zh 2:196–201

Suttle AD, Sipe GR (1932) Linkage of genes for crest and frizzle. J Hered 23:135–142

Sutton WS (1903) The chromosomes in heredity. Biol Bull 4:231–251

Suzuki T, Kansaku N, Kurosaki T, Shimada K, Zadworny D, Koide M, Mano T, Namikawa T, Matsuda Y (1999a) Comparative FISH mapping on Z chromosomes of chicken and Japanese quail. Cytogenet Cell Genet 87:22–26

Suzuki T, Kurosaki T, Shimada K, Kansaku N, Kuhnlein U, Zadworny D, Agata K, Hashimoto A, Koide M, Koike M, Takata

M, Kuroiwa A, Minai S, Namikawa T, Matsuda Y (1999b) Cytogenetic mapping of 31 functional genes on chicken chromosomes by direct R-banding FISH. Cytogenet Cell Genet 87:32–40

Takagi N, Itoh M, Sasaki M (1972) Chromosome studies in four species of Ratitae (Aves). Chromosoma 36:281–291

Takagi N, Sasaki M (1974) A phylogenetic study of bird karyotypes. Chromosoma 46:91–120

Takahashi H, Tsudzuki M, Sasaki O, Niikura J, Inoue-Murayama M, Minezawa M (2005) A chicken linkage map based on microsatellite markers genotyped on a Japanese Large Game and White Leghorn cross. Anim Genet 36:463–467

Takahashi R, Akahane K, Arai K (2003) Nucleotide sequences of pigeon feather keratin genes. DNA Seq 14:205–210

Tang B, Huang YH, Lin L, Hu XX, Feng JD, Yao P, Zhang L, Li N (2003) Isolation and characterization of 70 novel microsatellite markers from ostrich (*Struthio camelus*) genome. Genome 46:833–840

Tatsuda K, Fujinaka K (2001) Genetic mapping of QTL affecting body weight in chickens using a F$_2$ family. Br Poult Sci 42:333–337

Taylor EL, Vercoe P, Cockrem J, Groth D, Wetherall JD, Martin GB (1999) Isolation and characterization of microsatellite loci in the emu, *Dromaius novaehollandiae*, and cross-species amplification within Ratitae. Mol Ecol 8:1963–1964

Tegetmeier WB (1873) The Poultry Book: Comprising the Breeding and Management af Profitable and Ornamental Poultry. G. Routledge, London, UK

Thomas JW, Prasad AB, Summers TJ, Lee-Lin SQ, Maduro VV, Idol JR, Ryan JF, Thomas PJ, McDowell JC, Green ED (2002) Parallel construction of orthologous sequence-ready clone contig maps in multiple species. Genome Res 12:1277–1285

Tirunagaru VG, Sofer L, Cui J, Burnside J (2000) An expressed sequence tag database of T-cell-enriched activated chicken splenocytes: sequence analysis of 5251 clones. Genomics 66:144–1451

Toye AA, Schalkwyk L, Lehrach H, Bumstead N (1997) A yeast artificial chromosome (YAC) library containing 10 haploid chicken genome equivalents. Mamm Genome 8:274–276

Traxler B, Brem G, Muller M, Achmann R (2000) Polymorphic DNA microsatellites in the domestic pigeon, *Columba livia* var. *domestica*. Mol Ecol 9:366–368

Trefil P, Bruno MM, Mikus T, Thoraval P (1999) Sexing of chicken feather follicle, blastodermal and blood cells. Folia Biol (Praha) 45:253–256

Tsuda Y, Nishida-Umehara C, Ishijima J, Yamada K, Matsuda Y (2007) Comparison of the Z and W sex chromosomal architectures in elegant crested tinamou (*Eudromia elegans*) and ostrich (*Struthio camelus*) and the process of sex chromosome differentiation in palaeognathous birds. Chromosoma 116:159–173

Tuiskula-Haavisto M, Honkatukia M, Vikki J, de Koning D-J, Schulman NF, Mäki-Tanila A (2002) Mapping of quantitative trait loci affecting quality and production traits in eggs layers. Poult Sci 81:919–927

Vallejo RL, Bacon LD, Liu HC, Witter RL, Groenen MAM, Hillel J, Cheng HH (1998) Genetic mapping of quantitative trait loci affecting susceptibility to Marek's disease virus induced tumors in F$_2$ intercross chickens. Genetics 148:349–360

van Hemert S, Hoekman AJW, Smits MA, Rebel JMJ (2007) Immunological and gene expression responses to a *Salmonella* infection in the chicken intestine. Vet Res 38:51–63

van Kaam JBCHM, van Arendonk JAM, Groenen MAM, Bovenhuis H, Vereijken ALJ, Crooijmans RPMA, van der Poel JJ, Veenendaal A (1998) Whole genome scan in chickens for quantitative trait loci affecting body weight in chickens using a three generation design. Livest Prod Sci 54:133–150

van Kaam JBCHM, Groenen MAM, Bovenhuis H, Veenendaal A, Vereijken ALJ, Van Arendonk JAM (1999a) Whole genome scan in chickens for quantitative trait loci affecting growth and feed efficiency. Poult Sci 78:15–23

van Kaam JBCHM, Groenen MAM, Bovenhuis H, Veenendaal A, Vereijken ALJ, van Arendonk JAM (1999b) Whole genome scan in chickens for quantitative trait loci affecting carcass traits. Poult Sci 78:1091–1099

van Kaam JBCHM, Bink MCAM, Bovenhuis H, Quaas RL (2002) Scaling to account for heterogeneous variances in a Bayesian analysis of broiler quantitative trait loci. J Anim Sci 80:45–56

van Tuinen M, Dyke GJ (2004) Calibration of galliform molecular clocks using multiple fossils and genetic partitions. Mol Phylogenet Evol 30:74–86

Waddington D, Springbett AJ, Burt DW (2000) A chromosome-based model for estimating the number of conserved segments between pairs of species from comparative genetic maps. Genetics 154:323–332

Wade J, Peabody C, Coussens P, Tempelman RJ, Clayton DF, Liu L, Arnold AP, Agate R (2004) A cDNA microarray from the telencephalon of juvenile male and female zebra finches. J Neurosci Methods 138:199–206

Wakana S, Watanabe T, Hayashi Y, Tomita T (1986) A variant in the restriction endonuclease cleavage pattern of mitochondrial DNA in the domestic fowl, *Gallus gallus domesticus*. Anim Genet 17:159–168

Wallis JW, Aerts J, Groenen M, Crooijmans R, Layman D, Graves T, Scheer D, Kremitzki C, Higgenbotham J, Gaige T, Mead K, Walker J, Albracht D, Davito J, Yang S-P, Leong S, Chinwalla A, Hillier L, Sekhon M, Wylie K, Dodgson J, Romanov MN, Cheng H, de Jong PJ, Zhang H, McPherson JD, Krzywinski M, Schein J, Mardis E, Wilson R, Warren WC (2004) A physical map of the chicken genome. Nature 432:761–764

Wang H, Li H, Wang Q, Wang Y, Han H, Shi H (2006) Microarray analysis of adipose tissue gene expression profiles between two chicken breeds. J Biosci 31:565–573

Wang N, Shoffner RN (1974) Trypsin G- and C-banding for interchange analysis and sex identification in the chicken. Chromosoma 47:61–69

Wang W, Lan H, Liu AH, Shi LM (1994) Variation of mitochondrial DNA among domestic fowl and red jungle fowl. Zool Res 15:55–60

Wardęcka B, Olszewski R, Jaszczak K, Zęba C, Pierzchała M, Wicirińska K (2002) Relationship between microsatellite marker alleles on chromosome 1–5 originating from the Rhode Island Red and Green-legged Partrigenous breeds and egg production and quality traits in F_2 mapping population. J Appl Genet 43:319–329

Warren DC (1928) Sex-linked characters of poultry. Genetics 13:421–433

Warren DC (1933) Nine independently inherited autosomal factors in the domestic fowl. Genetics 18:68–81

Warren DC (1935) A new linkage group in the fowl (*Gallus domesticus*). Am Nat 69:82

Warren DC, Hutt FB (1936) Linkage relations of crest, dominant white and frizzling in the fowl. Am Nat 70:379–394

Weigend S, Romanov MN (2001) Current strategies for the assessment and evaluation of genetic diversity in chicken resources. Worlds Poult Sci J 57:275–288

Weigend S, Romanov MN (2002) The World Watch List for Domestic Animal Diversity in the context of conservation and utilisation of poultry biodiversity. Worlds Poult Sci J 58:519–538

Weigend S, Vef E, Wesch G, Meckenstock E, Seibold R, Ellendorff F (1995) Concept for conserving genetic resources in poultry in Germany. Arch Geflügelkd 59:327–334

West B, Zhou BX (1989) Did chicken go North? New evidence for domestication. Worlds Poult Sci J 45:205–218

Wimmers K, Valle-Zarate A, Mathur PK, Horst P, Wittig B (1992) Oligonucleotide fingerprinting in chickens. Proc 19th Worlds Poult Congr, Amsterdam, The Netherlands, September 19–14, 1992, vol 1, pp 539–540

Wood-Gush DGM (1959) A history of the domestic chicken from antiquity to the 19th century. Poult Sci 38:321–326

Wright TF, Brittan-Powell EF, Dooling RJ, Mundinger PC (2004) Sex-linked inheritance of hearing and song in the Belgian Waterslager canary. Proc Biol Sci 271(Suppl 6): S409–S412

Xu G, Goodridge AG (1998) A CT repeat in the promoter of the chicken malic enzyme gene is essential for function at an alternative transcription start site. Arch Biochem Biophys 358:83–91

Xu S, Yonash N, Vallejo RL, Cheng HH (1998) Mapping quantitative trait loci for binary traits using a heterogeneous residual variance model: an application to Marek's disease susceptibility in chickens. Genetica 104:171–178

Yamashina Y (1944) Karyotype studies in birds. I. Comparative morphology of chromosomes in seventeen races of domestic fowl. Cytologia (Tokyo) 13:270–296

Yamashita H, Okamoto S, Maeda Y, Hashiguchi T (1994) Genetic relationships among domestic and jungle fowls revealed by DNA fingerprinting analysis. Jap Poult Sci 31:335–344

Yang KT, Lin CY, Liou JS, Fan YH, Chiou SH, Huang CW, Wu CP, Lin EC, Chen CF, Lee YP, Lee WC, Ding ST, Cheng WT, Huang MC (2006) Differentially expressed transcripts in shell glands from low and high egg production strains of chickens using cDNA microarrays. Anim Reprod Sci 101:113–124.

Yonash N, Bacon LD, Witter RL, Cheng HH (1999) High resolution mapping and identification of new quantitative trait loci (QTL) affecting susceptibility to Marek's disease. Anim Genet 30:126–135

Yonash N, Cheng HH, Hillel J, Heller DE, Cahaner A (2001) DNA microsatellites linked to quantitative trait loci affecting antibody response and survival rate in meat-type chickens. Poult Sci 80:22–28

Yuan X, Zhang M, Ruan W, Song C, Ren L, Guo Y, Hu X, Li N (2006) Construction and characterization of a duck bacterial artificial chromosome library. Anim Genet 37:599–600

Zeuner FE (1963) A History of Domesticated Animals. Hutchinson, London, UK

Zhou H, Li H, Lamont SJ (2003) Genetic markers associated with antibody response kinetics in adult chickens. Poult Sci 82:699–708

Zhu JJ, Lillehoj HS, Allen PC, Van Tassell CP, Sonstegard TS, Cheng HH, Pollock D, Sadjadi M, Min W, Emara MG (2003) Mapping quantitative trait loci associated with resistance to coccidiosis and growth. Poult Sci 82:9–16

Zimmer R, Verrinder Gibbins AM (1997) Construction and characterization of a large-fragment chicken bacterial artificial chromosome library. Genomics 42:217–226

Zimmer R, King WA, Verrinder Gibbins AM (1997) Generation of chicken Z-chromosome painting probes by microdissection for screening large-insert genomic libraries. Cytogenet Cell Genet 78:124–130

World Wide Web Resources on Poultry Genetics and Genomics (accessed August 20, 2008)

A Brief Review of Avian Genetics, by Darrel K. Styles, Texas A & M University, USA. http://www.oldworldaviaries.com/text/styles/genetics.html

AceBrowser, Wageningen University, The Netherlands [chicken and pig genome mapping databases]. https://acedb.asg.wur.nl/

Apteryx australis (Brown Kiwi) Genome Project, NCBI, USA. http://www.ncbi.nlm.nih.gov/sites/entrez?db=genomeprj&cmd=Retrieve&dopt=Overview&list_uids=12704

ArkDB, Roslin Institute, UK [chicken, turkey and Japanese quail genome mapping databases]. http://www.thearkdb.org/

ARK-Genomics Microarrays: Chickens, Roslin Institute, UK. http://www.ark-genomics.org/microarrays/bySpecies/chicken

Avian Genetic Resource Laboratory, University of British Columbia, Canada. http://www.landfood.ubc.ca/avian_research/avian_genetic_pgm.htm

Avian Genetic Resources, USDA-CSREES, USA. http://www.csrees.usda.gov/nea/animals/sri/an_breeding_sri_avian.html

Avian Genetic Resources at Risk, by Pisenti et al. (1999), University of California, Davis, USA. http://www.grcp.ucdavis.edu/publications/indexa.htm

Avian Genetic Resources Task Force, University of California, Davis, USA. http://www.grcp.ucdavis.edu/projects/projdet.htm#grtf

Avian Genetics, by Alan D. Mason, Cabin Software, USA. http://www.cabinsoftware.biz/Genetics_Tutorial/Part1.htm

Avian Genetics, by Anthony Olszewski, PETCRAFT, USA. http://www.petcraft.com/docs/avgen.shtml

Avian Molecular Genetics and Genome Projects, by Terry Burke, Sheffield University, UK. http://www.shef.ac.uk/misc/groups/molecol/projects.html

AVIANDIV (Development of Strategy and Application of Molecular Tools to Assess Biodiversity in Chicken Genetic Resources) Project, Institute for Animal Breeding Mariensee, Germany, and European Commission (1998–2000). http://w3.tzv.fal.de/aviandiv/index.html; http://rosenberglab.bioinformatics.med.umich.edu/aviandiv.html

AvianNET (Avian Information Network), Roslin Institute, UK. http://www.chicken-genome.org/

BAC-based, Integrated Physical and Genetic Map of the Chicken, Texas A & M University, USA. http://hbz7.tamu.edu/homelinks/phymap/chicken/chick_home.htm

Basic Peafowl Genetics, by Brad Legg, Legg's Peafowl Farm, Kansas City, MO, USA. http://www.leggspeafowl.com/basicgenetics.htm

Biodiversity in Poultry, by M.N. Romanov, Michigan State University, USA. http://www.msu.edu/~romanoff/biodiversity/bdmain.htm

Biotechnology and Biological Sciences Research Council (BBSRC) ChickEST Database, University of Manchester Institute of Science and Technology, UK. http://www.chick.manchester.ac.uk

Bird Library Resources, BACPAC Resources Center, CHORI, USA. http://bacpac.chori.org/libraries.php?o=1_12_14_40#beg

Birds, Animal Genome Size Database, by T. Ryan Gregory, University of Guelph, Canada. http://www.genomesize.com/search.php?search=type&value=Birds&display=300

Chick EST Database, University of Delaware, USA. http://www.chickest.udel.edu/

Chicken (*Gallus gallus*) Genome Browser Gateway, University of California, Santa Cruz, USA. http://genome.ucsc.edu/cgi-bin/hgGateway

Chicken 13k cDNA Microarray, Fred Hutchinson Cancer Research Center, USA. http://www.fhcrc.org/science/shared_resources/genomics/SpottedArrays/expression_unlabeled.html

Chicken B-cell Line DT40 and Bursal Transcript Database, Institute for Molecular Radiobiology, Germany. http://pheasant.gsf.de/DEPARTMENT/dt40.html

Chicken Classical Plumage Color Genes, by F.P. Jeffrey (1979), Bantam Chickens, USA. http://www.geocities.com/Heartland/Plains/4175/genes.html

Chicken Color Genetics and Silky Chickens, by W.F. Hollander, Ames, IA, USA. http://www.ringneckdove.com/Wilmer's%20WebPage/SILKYCHI.htm

Chicken Cosmid Library, RZPD German Resource Center for Genome Research, Germany. http://www.imagenes-bio.de/libraryinfo?LibNum=0125

Chicken Development, by Mark Hill, University of New South Wales, Australia. http://embryology.med.unsw.edu.au/OtherEmb/Chicken.htm

Chicken Full-length cDNA Database, Department of Zoology and BIOSUPPORT, University of Hong Kong, Hong Kong. http://bioinfo.hku.hk/chicken/

Chicken Genome Array, Affymetrix, USA. http://www.affymetrix.com/products/arrays/specific/

Chicken Genome Mapping, Wageningen University, The Netherlands. http://137.224.73.223/abg-org/hs/research/molecular/intro.html

Chicken Genome Mapping and Biodiversity, by Terry Burke, Sheffield University, UK. http://www.shef.ac.uk/misc/groups/molecol/chicken.html

Chicken Genome Project, NCBI, USA. http://www.ncbi.nlm.nih.gov/sites/entrez?db=genomeprj&cmd=Retrieve&dopt=Overview&list_uids=10804

Chicken Genome Resources, NCBI, USA. http://www.ncbi.nlm.nih.gov/genome/guide/chicken/

Chicken IMAGE (improvement of chicken IMmunity and resistance to disease based upon Analysis of Gene Expression) Project, INRA, France, and European Commission (2000–2003). http://www.vjf.cnrs.fr/image/chicken/; http://ec.europa.eu/research/quality-of-life/ka5/en/01591.html; http://ec.europa.eu/research/quality-of-life/wonderslife/project06_en.html

Chicken IPI (International Protein Index), European Molecular Biology Laboratory (EMBL) — European Bioinformatics Institute (EBI) and Wellcome Trust Sanger Institute, UK. http://www.ebi.ac.uk/IPI/IPIchicken.html

Chicken Map Viewer [*Gallus gallus* (Chicken) Genome View], NCBI, USA. http://www.ncbi.nlm.nih.gov/mapview/map_search.cgi?taxid=9031

Chicken Public EST Contig Browser, Wellcome Trust Sanger Institute, UK. http://public-contigbrowser.sigenae.org:9090/Gallus_gallus/

Chicken Quantitative Trait Loci database (ChickenQTLdb), National Animal Genome Research Program (NAGRP)

Bioinformatics Coordination and U.S. Livestock Genome Research Projects, USDA, USA. http://www.animalgenome.org/QTLdb/chicken.html

Chicken Retroviral Vector Project, Harvard Medical School, USA. http://genepath.med.harvard.edu/~cepko/vector/index.html

Chicken SAGE (Serial Analysis of Gene Expression) Website, Claude Bernard University, France. http://www.cgmc.univ-lyon1.fr/Gandrillon/chicken_SAGE.php

Chicken Sequencing White Paper, WUGSC/NHGRI, USA. http://genome.wustl.edu/ancillary/data/whitepapers/Gallus_gallus_WP.pdf; http://www.genome.gov/Pages/Research/Sequencing/SeqProposals/Chicken_Genome.pdf

ChickFPC, Wageningen University, The Netherlands [chicken FPC (fingerprinted contigs) database]. http://www.bioinformatics.nl/gbrowse/cgi-bin/gbrowse/ChickFPC/

ChickGO, AgBase, Mississippi State University, USA [GO (gene ontology) functional annotations for chicken gene products]. http://www.agbase.msstate.edu/information/ChickGO.html

ChickMaP Project Home Page (1995–1997 version), Roslin Institute, UK. [Chicken Genome Mapping Project, European Commission (1995–1998)] http://www.projects.roslin.ac.uk/chickmap/ChickMappingInfo.html

ChickMaP Project Home Page (1998–2000 version), Roslin Institute, UK http://www.projects.roslin.ac.uk/chickmap/ChickMapHomePage.html

ChickRH (RH mapping on INRA chicken radiation hybrid panel), INRA, France. http://www-lgc.toulouse.inra.fr/internet/index.php/Tools/Chickrh.html

ChickVD (Chicken Variation Database), Beijing Genomics Institute, China. http://chicken.genomics.org.cn/index.jsp

Colour Genetics of Parrots and Budgerigars, by Clive Hesford, UK. http://www.birdhobbyist.com/parrotcolour/

Dromaius novaehollandiae (Emu) BAC Library, Genome Project Solutions, Inc., USA. http://genomeprojectsolutions.com/Dromaius.html

Ensembl Chicken Genome Server, EMBL—EBI and Wellcome Trust Sanger Institute, UK. http://www.ensembl.org/Gallus_gallus/index.html

ESTIMA (Songbird EST Project), W.M. Keck Center for Comparative and Functional Genomics, University of Illinois, Urbana-Champaign, USA. http://titan.biotec.uiuc.edu/cgi-bin/ESTWebsite/estima_start?seqSet=songbird

G. gallus Genomic Sequencing Project, Wellcome Trust Sanger Institute, UK. http://www.sanger.ac.uk/Projects/G_gallus/

Gallus gallus (chicken), KEGG: Kyoto Encyclopedia of Genes and Genomes, Japan. http://www.genome.jp/kegg-bin/show_organism?org=gga

Gallus gallus Genome, WUGSC, USA. http://www.genome.wustl.edu/genome.cgi?GENOME=Gallus%20gallus

Gallus gallus Trace Archive, NCBI, USA [archival data storage of chicken sequence traces]. http://www.ncbi.nlm.nih.gov/Traces/trace.cgi?cmd=retrieve&s=search&m=obtain&retrieve=Search&val=SPECIES_CODE='GALLUS+GALLUS'

Gallus gallus Trace Server, Ensembl, EMBL—EBI and Wellcome Trust Sanger Institute, UK [repository of chicken sequence traces]. ftp://ftp.ensembl.org/pub/traces/gallus_gallus

Gallus Genome GBrowser, University of Delaware, USA [chicken genome sequence aligned with California condor, turkey and zebra finch sequence datasets]. http://birdbase.net/cgi-bin/

GallusKB, University of Delaware, USA [chicken gene (ESTs and mRNA) and pathway database]. http://www.udel.edu/dnasequence/UDSGC/Links.html

GEISHA: *Gallus gallus* (chicken) EST and In Situ Hybridization Analysis database, University of Arizona, USA. http://geisha2.biosci.arizona.edu/data

Gene Mapping in Passerines, Sheffield University, UK. http://www.shef.ac.uk/misc/groups/molecol/Genemapping Passerines.html

Genetic Calculator, by Martin Rašek, Czech Republic. [mutation gallery for 43 caged avian species]. http://www.gen-calc.com/

Genetics of the Fowl, by F.B. Hutt (1949), Cornell University Library Digital Collections, USA [online book]. http://chla.library.cornell.edu/cgi/t/text/text-idx?c=chla;idno=2837819

Genome Studies for Conservation of the California Condor, Zoological Society of San Diego, CA, USA. http://cres.sandiegozoo.org/projects/gr_condor_genome.html

Meleagris gallopavo (Common Turkey) Genome Project, NCBI, USA. http://www.ncbi.nlm.nih.gov/entrez/query.fcgi?db=genomeprj&cmd=Retrieve&dopt=Overview&list_uids=10805

Meleagris gallopavo Trace Archive, NCBI, USA [archival data storage of turkey sequence traces]. http://www.ncbi.nlm.nih.gov/Traces/trace.cgi?cmd=retrieve&s=search&m=obtain&retrieve=Search&val=SPECIES_CODE='MELEAGRIS+GALLOPAVO'

Meleagris gallopavo Trace Server, Ensembl, EMBL—EBI and Wellcome Trust Sanger Institute, UK [repository of turkey sequence traces]. ftp://ftp.ensembl.org/pub/traces/meleagris_gallopavo

MUTAVI Research and Advice Group, by Inte Onsman, The Netherlands [genetics in budgerigars and other birds]. http://www.euronet.nl/users/hnl/

NAGRP Chicken Genome Community Upcoming Events, USA. http://www.animalgenome.org/share/mtg_chic.html

NC1008 [formerly NC-168]: Advanced Technologies for the Genetic Improvement of Poultry, Multistate Research Project, NCRA, USA. http://lgu.umd.edu/lgu_v2/home-pages/home.cfm?trackID=2874

OMIA — Online Mendelian Inheritance in Animals, by Frank Nicholas, University of Sydney, Australia and NCBI, USA [including budgerigar, chicken, duck, emu, goose, guinea fowl, Indian peafowl, pigeon, quail, rhea, turkey]. http://omia.angis.org.au/; http://www.ncbi.nlm.nih.gov/entrez/query.fcgi?db=omia

Peafowl Genetics, by Valerie Farris, 3 Peas Bird Farm, Cook, NE, USA. http://www.birdfarm.bravepages.com/blugenex.html

Pigeon Genetics for Fanciers, by Frank Mosca, Montclair, CA, USA. http://www.angelfire.com/ga3/pigeongenetics/

Population Genetics and Molecular Evolution of Plumage Genes, by Terry Burke, Sheffield University, UK. http://www.shef.ac.uk/misc/groups/molecol/mohsen.html

Poultry and Avian Research Resources, University of California, Davis, USA. http://animalscience.ucdavis.edu/Avian-Resources/index.htm

Poultry Genetics for the Non-professional, by the Sellers family, Brookings, SD, USA. http://www.pekinbantams.com/genetics.asp

Proposal for a Chicken Genome Project, by Paul Goetinck et al., National Institutes of Health, USA. http://www.nih.gov/science/models/nmm/appd.html

Psittacine Genetics Reference Bank, MUTAVI Research and Advice Group, by Inte Onsman, The Netherlands, and Dirk van den Abeele, Belgium. http://www.euronet.nl/users/dwjgh/

Ron Huntley's Rare Color Homing Pigeons, by Ronald R. Huntley, Warner Robins, GA, USA [Basic Pigeon Genetic Information]. http://www.angelfire.com/ga/huntleyloft/Page1.html

Single Gene Traits in Turkey, by Thomas F. Savage and Elzbieta I. Zakrzewska, Oregon State University, USA. http://oregonstate.edu/instruct/ans-tgenes/tc.html

Songbird Brain Transcriptome Database, Jarvis Laboratory of Duke University, Duke Bioinformatics, USA, and The Genomics Group of RIKEN, Japan. http://songbirdtranscriptome.net/

Songbird Genomics (Songbird Genome Sequencing Project), Songbird Genomics Organization, USA. (International Consortium). http://www.songbirdgenome.org/index.html

Songbird Neurogenomics Initiative, University of Illinois, Urbana-Champaign, USA. http://titan.biotec.uiuc.edu/songbird/

Summa Gallicana (Chicken Genetics), by Elio Corti, Valenza, Italy. http://www.summagallicana.it/

Taeniopygia guttata (Zebra Finch) Genome Project, NCBI, USA. http://www.ncbi.nlm.nih.gov/entrez/query.fcgi?db=genomeprj&cmd=Retrieve&dopt=Overview&list_uids=12898

Taeniopygia guttata Trace Archive, NCBI, USA. [archival data storage of zebra finch sequence traces]. http://www.ncbi.nlm.nih.gov/Traces/trace.cgi?cmd=retrieve&s=search&m=obtain&retrieve=Search&val=SPECIES_CODE='TAENIOPYGIA+GUTTATA'

Taeniopygia guttata Trace Server, Ensembl, EMBL—EBI and Wellcome Trust Sanger Institute, UK. [repository of zebra finch sequence traces] ftp://ftp.ensembl.org/pub/traces/taeniopygia_guttata

Taxonomy Browser (*Gallus gallus*), NCBI, USA http://www.ncbi.nlm.nih.gov/Taxonomy/Browser/wwwtax.cgi?id=9031

Texas A & M BAC Libraries: Animals, Texas A & M University, USA [including red junglefowl and chicken BAC libraries]. http://hbz7.tamu.edu/homelinks/bac_est/bac.htm#animal

Trans-NIH Gallus Initiative, Model Organisms for Biomedical Research, NIH, USA. http://www.nih.gov/science/models/gallus/

Turkey Genome Mapping Project, University of Minnesota, USA. http://www.tc.umn.edu/~reedx054/Turkeygenome.htm

USDA-NAGRP ChickBASE (U.S. Node), Iowa State University, USA. [chicken genome mapping]. http://www.animalgenome.org/chickmap/

USDA-NAGRP Poultry Genome Project, Michigan State University, USA http://poultry.mph.msu.edu/

Wageningen Chicken Genomic BAC Library, Wageningen University, The Netherlands and Geneservice Ltd, UK. http://137.224.73.223/abg-org/hs/proj/project.php?projid=25; http://www.geneservice.co.uk/products/clones/chicken_BAC.jsp

William Bateson: Two Levels of Genetic Information [including the paper by Bateson and Saunders (1902)], by Donald Forsdyke, Queen's University at Kingston, Canada. http://post.queensu.ca/~forsdyke/bateson1.htm

Wilmer Jay Miller's Web Site, Ames, IA, USA [genetics of avian caged and poultry species]. http://www.ringneckdove.com/

Zebra Finch BAC Library, Arizona Genomics Institute, University of Arizona, USA. http://www.genome.arizona.edu/orders/direct.html?library=TG__Ba

CHAPTER 6

Turkey

Kent M. Reed

Department of Veterinary and Biomedical Sciences, College of Veterinary Medicine,
University of Minnesota,
St. Paul, MN 55108 USA, reedx054@umn.edu

6.1
Introduction

6.1.1
Origin and Domestication

The turkey (*Meleagris gllopavo*) is the only major agricultural animal species native to the New World. The limited fossil record indicates that the species evolved in the Americas from a pheasant-like ancestor. Evidence for several related species dates to the upper Pliocene and remains, identifiable as the modern species, are described from Pleistocene sites (12–20 thousand years bp) in both the eastern and southwestern United States.

The indigenous people of Mexico and the southwestern United States first domesticated the turkey. The earliest evidence of domestic turkeys was found in Tehuacan and dates between 200 BC and 700 AD. Archeological sites such as Tularosa Cave, New Mexico, support a second and perhaps independent domestication by the Mogollon and Anasazi. Turkeys were important for both food and fiber (feathers and bone) and are depicted in the artwork of both Mayan and Mimbres cultures. Clearly, domestication of turkeys spread throughout Central America prior to the discovery of Mexico by the Spanish explorer Juan de Grijalva in 1518. More detailed accounts of the history and domestication of the wild turkey in the Americas can be found in Schorger (1966) and Aldrich (1967).

Turkeys were first brought to Europe from the Americas (Mexico) in the 1500s during the Spanish conquest. Accounts of these early introductions are difficult to verify in part because of confusion in common names used for the Asiatic peafowl and African guinea fowl, which were also traded at the time. The exact origin of the common name "turkey" is unclear.

Some attribute the name to the confusion with the peafowl brought to Europe via Asia Minor or "Turkey." Others suggest origin in the "turk turk" call made by the bird or Anglicanization of terms applied to the species in other languages.

Regardless of its derivation, the uncertainty on the origin and taxonomic position of the turkey is reinforced in its scientific naming. The scientific name of the turkey (*Meleagris gallopavo*) was designated by Linnaeus in 1758 based on description of domestic birds imported to Europe. *Meleagris* is derived from "meleagrides," a term rooted in Ancient Greek mythology and commonly used by both Greek and Roman cultures to refer to the African guinea fowl. The species name *gallopavo* is further indicative of then taxonomic uncertainties, being a combination of terms applied to other species (*gallus* = chicken and *pavo* = peafowl).

After their first introduction to Spain, turkeys were rapidly spread to other parts of the Old World and were well established in parts of Europe by 1530. Archbishop Cramner first authentically reported the species in England in 1541, and fixed prices appeared in the London markets by the mid-1550s. Early selective breeding led to distinct turkey breeds such as the Royal Palm and Spanish Black. In an interesting twist, English settlers brought domestic turkeys back to North America in the early 1600s. By then, these domestic birds were both smaller and distinctly colored from the wild turkey that inhabited Eastern North America. In America, domestic turkeys were propagated and even occasionally interbred with eastern wild turkeys. Wild turkeys, on the other hand, were exploited for food and nearly extirpated from the eastern US. Recent conservation and reintroduction efforts have dramatically expanded the range and abundance of the wild turkey in the United States.

Genome Mapping and Genomics in Animals, Volume 3
Genome Mapping and Genomics in Domestic Animals
N.E. Cockett, C. Kole (Eds.)
© Springer-Verlag Berlin Heidelberg 2009

The foundation for the broad-breasted commercial birds of today is credited to Jesse Throssel, an English gamekeeper who selected turkeys for meat quality, early maturity, and hatchability. Throssel immigrated to Canada in 1926 and imported turkeys of his own stock in 1927 (Crawford 1990 and references therein). Determined efforts to advance domestic turkey stocks through genetic selection for improved performance began in the 1930s. At that time, there was a strong consumer demand for turkeys that weren't too large, so breeders at the USDA Bureau of Animal Industry set out to create a smaller turkey (Beltsville White) to meet that demand. After World War II, the availability of reliable refrigeration and the advent of commercial breeding companies such as Orlopp, Nicholas, Hybrid, and BUT ushered in the era of intensive turkey farming in the US as well as Europe. By the 1950s small birds like the Beltsville Whites were being raised and sold for slaughter in almost every American state, but their popularity was short lived with the development of the new broad-breasted bronze rapid-growing turkeys.

The most visible change during the past 50 years in the domestic turkey was the conversion of most commercial birds from bronze to white plumage. Consumers objected to the dark pinfeathers and black pigment retained in the follicles after processing (Brant 1998). In the 1950s, poultry processors began to seek broad-breasted turkeys with less visible pinfeathers. By the early 1960s breeders selecting for the recessive white-feathered allele (*c*) (Robertson et al. 1943) at a single autosomal gene completed the conversion from bronze to white. Several pigmented varieties such as Bourbon Red, Jersey Buff, Slate, Black Spanish, and Narragansett are still raised as "heritage" or standard varieties within a growing specialty-food market niche.

6.1.2
Taxonomy and Zoological Description

The turkey is a member of the order Galliformes and phylogenetically placed in the family Phasianidae that includes chicken, pheasants, partridge, grouse, quail, cailles, and faisans. Turkeys occupy a monogeneric subfamily Meleagridinae containing two species: the wild turkey (*M. gallopavo*) which inhabits mixed forests, open woodlands, and savannahs of North and Central America; and the Mexican Ocellated turkey (*M. ocellata*; Cuvier 1820). Six subspecies distinguished by coloration, size, and distribution are recognized in the wild turkey. The type originally brought to Europe was most certainly the Mexican wild turkey (*M. g. gallopavo*). Current status of this subspecies is uncertain. The remaining five subspecies are found in North America and include the eastern wild turkey (*M. g. silvestris*) that inhabits roughly the eastern half of the United States; the Florida wild turkey (*M. g. osceola*) found only on the peninsula of Florida; the Rio Grande wild turkey (*M. g. intermedia*) native to the central plains states, the southern Great Plains, western Texas, and northeastern Mexico; Merriam's wild turkey (*M. g. merriami*) found in the ponderosa pine, western mountain regions of the United States; and Gould's wild turkey (*M. g. mexicana*) found in portions of Arizona and New Mexico, as well as northern Mexico. Current subspecies designations are generally supported by molecular phylogenetics with the exception of the eastern (*M. g. silvestris*) and Florida (*M. g. osceola*) subspecies, which in recent analyses consistently formed a single taxonomic group (Mock et al. 2002). Wild turkeys have dark, iridescent plumage with black flight feathers with brown stripes and white bars. Males have a red wattle, a caruncle, and a blackish breast tuft. Their heads may be red, blue, or white depending on the season. Although weight varies considerably, wild male turkeys (toms) typically weigh 6.8–11 kilograms (kg) and hens usually weigh 3.6–5.4 kg.

The second species in the genus, the ocellated turkey (*M. ocellata*), is also a large bird around 70–90 centimeters (cm) long and 3 kg (female) to 4 kg (male) in weight. The species can be distinguished from its North American cousin by the bronze-green iridescent color mixture of the body feathers. This species has, in the past, been placed in a separate genus *Agriocharis*. Ocellated turkeys are known by several common names including *pavo*, *pavo ocelado*, or in Mayan as *ucutz il chican*. They exist only in a 50,000 square mile area comprising the Yucatan Peninsula.

6.1.3
Modern Breeding Objectives

The turkey industry, as we know it today, is the result of several technological developments since the late

1950s. Most importantly, these include breeding for increased performance coupled with improvements in animal husbandry and disease control. Modern broad-breasted turkeys have been selected for rapid growth, efficient feed conversion, and an increased proportion of lean white breast meat. In addition, better knowledge of the nutritional requirements of turkeys and the use of growth promoters have led to diets specifically designed to exploit the genetic potential for rapid growth. The development of environmentally controlled housing and improved hygiene coupled with better disease control methods has enabled producers to increase production densities. The turkey of today has been selected to handle a wide range of environmental conditions and perform well commercial settings.

Since 1970, the average live weight at 18 weeks of age for commercial toms has increased from 16.9 to 32.6 pounds (lb) (Table 1). At the same time feed efficiencies and growth rate have steadily increased such that only 132 days are required to produce a 35 lb bird. Data for commercial hens show a similar trend. Genetic selection for rapid muscle growth and muscling in turkey has altered many characteristics of the bird including muscle composition and meat quality. Intense selection for increased production has also resulted in increased incidence of metabolic disorders (Julian 2005), growth-induced myopathies (Velleman et al. 2003), skeletal deformities such as tibiodischondroplasia (TD; Julian 1998), and meat-quality defects (Sosnicki and Wilson 1991). One concern in the turkey industry is the prevalence of the meat-quality condition termed pale, soft, exudative muscle (PSE), which may contain a genetic component (Le Bihan-Duval et al. 2003). In response, breeders have altered selection strategies to include depth of body (greater cardiovascular efficiency)

and stronger, more stable legs and joints (selection against TD).

6.1.4
Economic Importance

During the past 45 years turkey production has seen a dramatic increase. In 1961 (earliest year for which Food and Agriculture Organization statistics are available), worldwide turkey production was estimated at just less than 1 million metric ton (Table 2). At that time the United States (US) accounted for over 75% of world production. By 2005, turkey production had increased roughly five-fold with an estimated worldwide production of just over 5 million metric tons. Although turkey production in the United States increased almost four-fold during this period, its percent of the worldwide production dropped to 47.3% while countries such as France, Germany, and Brazil had increases in their contribution to the worldwide market. Turkey farming remains a minor component of overall world poultry production. Although over 5 million metric tons of turkey meat was produced in 2005, worldwide chicken production was estimated at over 70 million metric tons. In the US alone, the combined value of production from chicken (broilers and eggs) and turkey was $28.9 billion with 11% attributable to turkeys (2004 statistics, US Poultry and Egg Association).

Since 2000, overall turkey production has reached a plateau but regional patterns of import/export continue to evolve (Windhorst 2005). For example, in 1970 the United States accounted for 87.1% of worldwide turkey meat exports. By 2003, the majority of exports (26.2%) were from France with the US and Brazil contributing 23.1% and 12.5%, respectively.

Table 1 Average live weights, feed efficiencies, and market maturity rates of commercial Large White toms and hens[a]

Year	Commercial toms			Commercial hens		
	LW (18 wk)	Feed/Gain (0–18 wk)	Days to 35 lbs	LW (14 wk)	Feed/Gain (0–14 wk)	Days to 16 lbs
1970	16.9	3.10	235+	9.0	3.10	151
1980	22.4	2.70	185	12.2	2.30	134
1990	27.1	2.80	156	14.2	2.44	107
2000	31.9	2.62	143	15.5	2.36	100
2004	32.6	2.53	132	15.9	2.17	98

[a] Data from Ferket (2004)

Table 2 Worldwide turkey meat production for the top 25 producing countries in 2005[a,b]

Country	Year										
	1961		1970		1980		1990		2000		2005	
USA	678,207	0.752	784,044	0.640	1,075,000	0.523	2,047,500	0.553	2,419,000	0.473	2,461,000	0.475
France	41,700	0.046	56,456	0.046	203,000	0.099	439,000	0.119	738,000	0.144	633,000	0.122
Germany	7,000	0.008	16,000	0.013	44,000	0.021	127,500	0.034	295,500	0.058	380,000	0.073
Italy	14,000	0.016	65,000	0.053	225,000	0.110	279,100	0.075	327,000	0.064	300,000	0.058
United Kingdom	18,125	0.020	69,000	0.056	122,077	0.059	170,517	0.046	255,000	0.050	221,000	0.043
Brazil	3,800	0.004	7,000	0.006	19,000	0.009	53,085	0.014	137,000	0.027	220,000	0.042
Canada	65,241	0.072	102,197	0.083	100,454	0.049	128,964	0.035	152,594	0.030	146,000	0.028
Israel	4,000	0.004	12,000	0.010	39,000	0.019	57,000	0.015	137,429	0.027	129,000	0.025
Hungary	2,500	0.003	8,000	0.007	13,500	0.007	46,200	0.012	98,000	0.019	92,000	0.018
Chile	–	0.000	–	0.000	–	0.000	–	0.000	62,000	0.012	85,800	0.017
Czechoslovakia (Czech Republic/Slovakia)	2,195	0.002	3,293	0.003	15,999	0.008	22,540	0.006	(CR) 12,000	0.002	12,000	0.002
									(S) 63,800	0.012	72,000	0.014
Portugal	0	0.000	0	0.000	13,000	0.006	30,000	0.008	43,600	0.009	45,000	0.009
The Netherlands	0	0.000	0	0.000	14,854	0.007	30,000	0.008	54,700	0.011	40,000	0.008
Argentina	2,856	0.003	4,845	0.004	12,750	0.006	42,024	0.011	34,680	0.007	35,190	0.007
Ireland	5,000	0.006	4,800	0.004	7,200	0.004	26,600	0.007	34,000	0.007	33,000	0.006
Australia	1,700	0.002	5,943	0.005	11,000	0.005	23,360	0.006	25,200	0.005	28,700	0.006
Mexico	13,860	0.015	18,150	0.015	33,891	0.016	24,595	0.007	17,688	0.003	27,242	0.005
Poland	3,000	0.003	9,000	0.007	14,000	0.007	22,000	0.006	17,500	0.003	26,000	0.005
Austria	186	0.000	453	0.000	1,892	0.001	11,772	0.003	23,794	0.005	25,000	0.005
Tunisia	–	0.000	–	0.000	–	0.000	4,650	0.001	25,750	0.005	25,000	0.005
Spain	2,400	0.003	3,000	0.002	14,000	0.007	29,000	0.008	22,000	0.004	21,000	0.004
Islamic Rep of Iran	2,640	0.003	3,180	0.003	4,380	0.002	10,500	0.003	16,800	0.003	15,000	0.003
Algeria	150	0.000	150	0.000	150	0.000	260	0.000	4,020	0.001	14,912	0.003
Turkey	3,360	0.004	4,660	0.004	6,500	0.003	9,000	0.002	11,800	0.002	13,600	0.003
Egypt	2,900	0.003	3,275	0.003	3,730	0.002	6,750	0.002	11,750	0.002	10,500	0.002
World	902,220	0.970	1,224,183	0.964	2,054,235	0.971	3,703,989	0.983	5,119,544	0.985	5,185,043	0.986

[a] For each year the total meat production (Metric tons) and the proportion of world production are given. Total worldwide production and proportion attributed to the included countries are given at the bottom

[b] Data from FAO-STAT (accessed May 2006)

Germany remains the leading EU importer of turkey meat but has been recently surpassed by Mexico, Russia, and China in overall imports. New challenges such as shifts in the global economy and unforeseen events including disease outbreaks will continue to significantly impact global patterns of turkey production. For example, global poultry meat prices declined sharply in 2006 with renewed outbreaks and concern over the spread of the avian influenza virus.

6.1.5
Karyotype and Genome

The karyotype of the turkey ($2n = 80$) includes nine pairs of distinct macrochromosomes, the Z/W sex chromosomes, and a graded series of 30 pairs of microchromosomes (Fig. 1). The macrochromosomes are in the same size range as typical mammalian chromosomes, but the microchromosomes are considerably smaller (Bloom et al. 1993). Macrochromosomes include a single large metacentric chromosome (MGA1), five acrocentric chromosomes (MGA2, 4, 6, 8, and 9) and three chromosomes with distinguishable short arms (MGA3, 5, and 7) (Schmid et al. 2005). The metacentric Z chromosome is the fourth largest chromosome and the metacentric W chromosome is approximately the size of chromosome 6 (Krishan and Shoffner 1966). The repetitive ribosomal RNAs (nucleolus organizer region, NOR) are located on a microchromosomal pair (Chaves et al. 2007). Standardized cytogenetic banding patterns have been described for the nine macrochromosomes that correspond to the eight macrochromosomes found in the chicken (Schmid et al. 2000).

The genome of the turkey is thought to be approximately the same size as the chicken genome (1.05 gigabases or Gbp for $1n$ sequence, 2.33 picograms or pg DNA in $2n$ nucleated red blood cells) or about one-third the size of the human genome. The repetitive component of the turkey genome has not been extensively characterized, but is likely similar to the 10–15% reported for the chicken (Wicker et al. 2005). A female-specific 0.4 kilobase (kb) PstI repeat and two repetitive elements enriched on microchromosomes including a 41 base pair (bp) tandem repeat (Matzke et al. 1992) and a 0.5 kb MspI repeat (unpublished GenBank accessions) have been described. Ultralong telomere repeat arrays $(TTAGGG)_n$ reported for the chicken and other avian species are also found in the turkey (Delany et al. 2000; Chaves et al. 2007). These appear to be enriched on the W-chromosome and numerous microchromosomes (Fig. 2). The reduced amount of repetitive DNA in the chicken relative to mammalian genomes is also reflected in a lower frequency of microsatellite repeats (Primmer et al. 1997). In the chicken, the most common dinucleotide motifs are $(CA)_n$ and $(GA)_n$, and this is also the case in the turkey (Dranchak et al. 2003). A significant number of $(GGA)_n$ trinucleotide motifs are also present in the turkey genome.

6.1.6
Classical Mapping Efforts and Limitations

As with most agricultural species, early genetic studies in the turkey were for the most part limited to examinations of the inheritance of genes responsible for novel phenotypic variants. Savage (1990) compiled

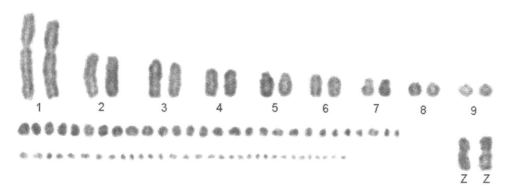

Fig. 1 Karyotype of the turkey (*Meleagris gallopavo*). Figure courtesy of Valérie Fillon, INRA

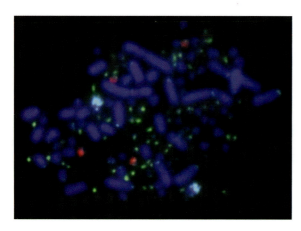

Fig. 2 Dual-color FISH of turkey metaphase chromosomes (tetraploid female nucleus) hybridized with an rDNA BAC clone (*red*) and a telomere probe (*green*). Note the enhanced telomere signal on several microchromosomes and the intense staining of the W chromosomes

a summary of mutations for which inheritance had been determined in experimental crosses. The majority of these phenotypic traits were autosomally inherited with either codominant or recessive alleles responsible for the observed variants (Table 3). Epistatic effects were observed for some feathering traits indicating involvement of multiple genes in the observed phenotype and variable expression was also noted. For example sex-influenced expression was observed for the autosomally inherited metabolic trait Bowed hocks (*bh*).

The extent to which traits are genetically mapped (i.e., linked) is limited mainly to those traits for which sex linkage has been observed (Table 3). This is due primarily to the lack of multitrait genetic crosses. Sex-linked traits include feather pigmentation mutants (*e*, *n*, and *n^{al}*), feathering traits (*K*), and neurological mutations (*bo*, *tt*, and *vi*). Two exceptions are the autosomal linkage observed between glaucoma (*ga*) and slate (*D*) (Bitgood and Somes 1990) and the feather structure mutant Hairy (Smyth 1954). This latter trait results from mutation in a gene responsible for feather structure, and affected birds have abnormal barbule structure and number. Genetic crosses determined that this feather mutant was the product of an autosomal recessive allele (*ha*). Interestingly, birds with red plumage (*r*) were used in the crosses to determine inheritance of Hairy, and suggestive linkage between these two traits was found.

6.2 Construction of Genetic Maps

6.2.1 Genetic Markers

Traditionally several marker types have been used for genetic linkage mapping, including both coding sequences (expressed sequence tags, ESTs or Type I) and highly polymorphic markers such as microsatellites (Type II). For the turkey, the earliest gene sequences were deposited in GenBank in 1992 with the first microsatellite sequences accessioned in 1995. More recently, other anonymous marker types have been developed including amplified fragment-length polymorphisms (AFLPs; Paxton et al. 2005), sequence characterized amplified regions (SCARs), random amplified polymorphic DNAs (RAPDs; Smith et al. 1996), and single nucleotide polymorphisms (SNPs; Smith et al. 2000; Reed et al. 2006b). Each of these marker classes has its own strengths and weaknesses and represents new technological advances in both marker development and genotyping. However, because of their "transportability" between laboratories, populations, and even species, microsatellites have been the marker of choice for linkage map construction.

Cross-species amplification of microsatellite loci has been of tremendous utility in the construction of genetic maps of agricultural species. Given the advanced state of the chicken genome map and the potential for comparative studies, several investigations of chicken microsatellites were undertaken in the turkey. Early studies with different sets of markers reported mixed results (Levin et al. 1995; Liu et al. 1996; Hanotte et al. 1997). In one of the first studies, Levin et al. (1995) examined 48 chicken markers with 92% amplification success on turkey genomic DNA. Significant polymorphism was observed in a subset of these markers leading to the conclusion that a major portion of chicken microsatellite markers would be useful for genetic mapping in the turkey. A second study (Liu et al. 1996) examined 88 markers

Chapter 6 **Turkey** 149

Table 3 Genetic mutations in the turkey[a]

Trait	Inheritance
Feather pigmentation	
Black (*B*)	Autosomal codominant
Black-winged Bronze (*b¹*)	Autosomal codominant
Brown (*e*)	Sex-linked recessive
Faded bronze (*fb*)	Autosomal
Gray (*c^g*)	Autosomal codominant
Narragansett (*n*)	Sex-linked codominant
Imperfect albinism (*n^{al}*)	Sex-linked recessive
Nonbarring (*1*)	Autosomal codominant
Palm (*p*)	Autosomal recessive
Red (*r*)	Autosomal recessive
Slate (*D*)	Autosomal dominant
Slate (*s1*)	Autosomal recessive
Spotting (*sp*)	Autosomal recessive
White (*c*)	Autosomal recessive
Feather structure	
Hairy (*ha*)	Autosomal recessive
Feathering	
Late feathering (*K*)	Sex-linked dominant
Naked (*n*)	Autosomal recessive
Skin	
Melanization (*m*)	Autosomal recessive
Metabolic	
Binucleated erythrocytes (*bn*)	Autosomal recessive
Bowed hocks (*bh*)	Autosomal recessive (sex influenced)
Pendulous crop	Autosomal recessive / polygenic
Hereditary glaucoma (*ga*)	Autosomal recessive
Muscle	
Degenerative myopathy	Polygenic
Muscular dystrophy (*dy*)	Autosomal recessive
Neurological	
Bobber (*bo*)	Sex-linked recessive
Titanic torticollar spasm (*tt*)	Sex-linked recessive[b]
Congenital loco (*lo*)	Autosomal recessive
Vibrator (*vi*)	Sex-linked recessive
Calcifide tissues	
Achrondroplasia (*ach*)	Sex linked
Crooked-neck dwarf (*cn*)	Autosomal recessive
Shortening of bones (*s*)	Autosomal recessive/codominant
Embryonic lethal	
Chrondrodystrophy (*ch*)	Autosomal recessive
Chrondrodystrophy-m (*ch-m*)	Autosomal recessive
Hemimelia (*hm*)	Autosomal recessive
Knobby (*kn*)	Autosomal recessive
Micromelia-like (*mm*)	Autosomal recessive
Ring (*rl*)	Autosomal recessive

(Continued)

Table 3 (Continued)

Trait	Inheritance
Short spined (*sh*)	Autosomal recessive
Swollen plumules (*sdp*)	Autosomal recessive
Biochemical	
Albumin	Autosomal codominant
Adenosine deaminase	Autosomal codominant

[a] Summarized from Buss 1989 and Savage (1990)
[b] Savage et al. (1993)

and came to the opposite conclusion. Although positive amplification of turkey genomic DNA was achieved for 69% of the markers, low allelic variation and heterozygosity in their test population led to the conclusion that chicken microsatellites would be of little utility in construction of a turkey genetic map.

With support from both government and industry, a large collection of chicken microsatellite primer sets were made available in 1996 through the US Poultry Genome Coordinator. An intensive effort was then made to screen these markers for use in turkey genetic mapping. In a preliminary screen, positive amplification was obtained for 55% of 141 primer pairs (Reed et al. 1999). This led to an expanded survey of 520 primer sets (Reed et al. 2000a), of which 280 (54%) of the chicken primer pairs produced amplification products in the turkey. Examination of genetic polymorphism at a subset of 57 markers suggested that 20% of the available chicken microsatellite markers would be useful for mapping the turkey genome. Although this percentage was low given the total number of available markers, the potential resource for comparative mapping was significant in that chicken markers could serve as anchor points for linkage group alignments. A total of nearly 800 chicken microsatellites have now been examined in the turkey (Reed et al. 2003a) with roughly 10% assigned to the turkey genetic linkage map (see below).

Much of the early efforts in marker development for the turkey grew from an industry desire for efficient parentage testing in addition to genetic mapping. In fact, the majority of nonmapping marker applications have been for parentage testing (Donoghue et al. 1999) and population genetic evaluations of commercial (Szöke et al. 2004; Smith et al. 2005) and wild turkeys (Latch et al. 2002; Mock et al. 2002).

One difficulty in the development of microsatellite markers is the overall lower frequency of these repeats in avian genomes as compared to mammals (Primmer et al. 1997). Microsatellites for the turkey were first identified by screening enriched genomic libraries (Huang et al. 1999). This first report included 38 markers (TUM, Tuskegee University) from libraries enriched for TG, GAT, and CCT repeats. A second group of 32 microsatellite markers was developed at the Wageningen University (M. A. M. Groenen, unpublished) and later tested for amplification of pheasant DNA (Baratti et al. 2001). Seven additional microsatellite markers (WT) developed from turkey sequences in GenBank were used to examine genetic variation in wild turkey populations (Latch et al. 2002). At the University of Minnesota, 20 loci containing CA and GT repeats (MNT) were initially isolated from a small-insert (*Mbo*I fragments) genomic library (Reed et al. 2000a, b, 2002). Later, Dranchak et al. (2003) identified 42 microsatellite markers (MNT) associated with expressed sequences by screening a complementary DNA (cDNA) library for CA, GA, AGG, and AAAC repeats.

Large-scale marker development projects for the turkey were initiated in the late 1990s. Screening of two additional small-insert genomic libraries for CA, GA, AGG, ACC, AGC, GGAA, and AAAC repeats resulted in a large group of 322 markers (MNT; Reed et al. 2003b; Knutson et al. 2004). During the same time, researchers at the Roslin Institute created a microsatellite-enriched genomic library and screened for GT-containing clones. A total of 277 primer sets (RHT; Burt et al. 2003) were developed and tested on turkey DNA bringing the total number of reported turkey microsatellite markers to over 700.

6.2.2
Primary Genetic Linkage Maps

The first genetic linkage map of the turkey genome was initiated by Nicholas Turkey Breeding Farms and later published by Harry et al. (2003). This initial effort was designed to both create a linkage map and obtain DNA sequences for expressed genes. Construction of this map relied on the inheritance of restriction fragment-length polymorphisms (RFLPs) detected with cDNA probes. A cDNA library was constructed from RNA isolated from a 24-day turkey embryo. Clones were chosen at random for use as hybridization probes to genomic DNA digested with one of five restriction enzymes and immobilized on nylon membranes. Marker genotypes were scored across the Nte (Nicholas turkey EST) population.

The Nte population was derived from a single F_1 sire (an intercross between two Nicholas commercial lines A and B) backcrossed to two dams from line B. Whole blood was collected from grandparents, parents, and euthanized poults (1 day posthatch). A total of 84 offspring was included in the analysis, 44 poults from one dam and 40 from the other. In total 138 cDNA/RFLP loci were scored in the Nte families. Following linkage analysis, 113 loci were arranged into 22 linkage groups and an additional 25 loci remained unlinked (Fig. 3). The average distance between linked markers was 6 centimorgans (cM) with the largest linkage group (17 loci) measuring 131 cM. The total map distance contained within the 22 linkage groups was 651 cM. With 20 cM ascribed to each unlinked marker (a conservative value), total coverage was approximately 1,150 cM. Although the number of linkage groups (22) totaled roughly half the number of chromosomes, the cDNA/RFLP map provided an important initial framework for genome mapping in the turkey.

Concurrent with the cDNA/RFLP mapping of Harry et al. (2003), Dave Burt at the Roslin Institute and colleagues developed a preliminary microsatellite map of the turkey genome (Burt et al. 2003). The Roslin mapping effort used multiple backcross reference populations developed at both the Roslin Institute and at Brigham Young University. The Roslin (ROS) families were derived from crosses between the Nebraska Spot and Large White turkeys (British United Turkeys T5). F_1 progeny were first derived from mating Nebraska Spot sires to BUT dams, and reciprocal backcross families were then created either from Nebraska Spot parental F_0 sires mated to one of their F_1 daughters, or F_1 males mated to one of their parental BUT F_0 dams. Only the two largest backcross families (ROS-5 and ROS-12) were used for mapping. These families contained 93 and 75 backcross progeny, respectively. The Brigham Young University (BYU) backcross families were derived from crosses between wild turkeys and partially inbred Orlopp line C turkeys. Mating wild turkey sires to Orlopp line C dams generated F_1 progeny. Reciprocal backcross families were created either from wild parental F_0 sires mated to F_1 daughters or F_1 males mated to Orlopp parental F_0 dams. Four backcross families (BYU-A, -B, -I and -II) were selected. These included 34, 39, 51, and 37 backcross progeny, respectively.

Over 300 markers (including 277 new microsatellites developed from enriched libraries) were evaluated in the Roslin study and 111 were found to be polymorphic in the mapping families (Burt et al. 2003). A total of 20 linkage groups (including the Z-chromosome) containing 74 markers were established with an additional 37 markers remaining unassigned. The number of markers per linkage group ranged from two to six with the largest being 76.3 cM. The total genetic distance contained in the 20 linkage groups was 700 cM which represents an estimated 25% coverage of the turkey genome, based on the length of the chicken genetic map (3,800 cM). Again, with 20 cM ascribed to each unlinked marker, total coverage of the map extended to approximately 70% of the genome.

6.2.3
Second-generation Linkage Map

Recently, a medium-density genetic linkage map for the turkey was constructed (Reed et al. 2005a) based on segregation analysis of markers in a reciprocal backcross population, the University of Minnesota/Nicholas Turkey Breeding Farms mapping population (UMN/NTBF; Reed et al. 2003a). The UMN/NTBF mapping population was of F_2 reciprocal design and was initiated by crossing two divergent production lines (high breast meat yield x high fecund) to create the F_1 generation. A single male and 16 females were retained as breeders from within each of five initial yield x fecund families. The five F_1 males were then each mated to 12 unrelated F_1 females using a modified

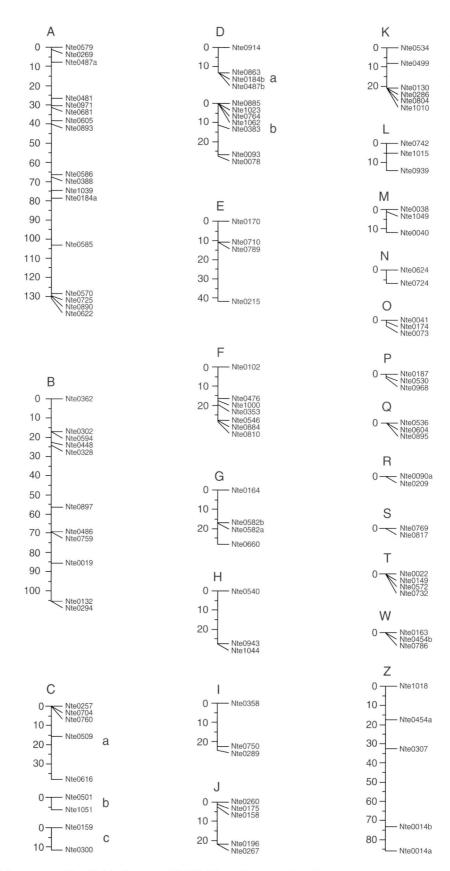

Fig. 3 First linkage groups identified in the turkey. Modified from Harry et al. (2003)

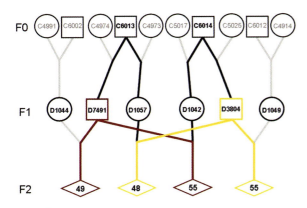

Fig. 4 Pedigree of the UMN/NTBF turkey genetic mapping population

factorial design. F_2 poults form each of the 60 F_2 families hatched in early 2000 and produced over 2,000 total F_2 offspring. Families of four dams from two of the five F_1 sires were chosen based on total number of offspring per family, distribution of offspring across the F_1 dams, and relatedness of the F_1 parents, for use in map construction (Fig. 4). These four families included 207 F_2 offspring and were related to each other through two grandsires in the F_0 generation.

The UMN mapping effort included 515 genetic markers with 507 microsatellite-based markers developed for the turkey, chicken, and quail and 8 SNPs corresponding to markers on the cDNA/RFLP map of Harry et al. (2003). A total of 370 markers (microsatellites and SNPs) were informative in the UMN/NTBF mapping population and 354 had significant linkage by two-point analysis (likelihood odds ratio or LOD > 3.0). Fine mapping arranged 314 of the 354 markers into 29 linkage groups. Linkage groups ranged from 0.3 to 413.3 cM with an average marker interval of 6.43 cM. The largest linkage group included 47 loci and 7 linkage groups were each defined by only two loci. Forty markers could not be fine mapped, but were aligned within linkage groups by two-point scores. Only 16 of the 515 markers remained unlinked and the total genetic distance encompassed by the linkage groups was 2,011 cM (sex averaged).

6.2.4
Integrative Mapping

With the independent construction of three separate turkey maps, integration of maps and their associated markers is important to support future map-based genetic association studies and quantitative trait loci (QTL) scans. Integration of genetic maps requires a set of common markers be genotyped across a single mapping population. For example, the marker set of Reed et al. (2005a) was essentially independent of the Roslin marker set (Burt et al. 2003). Thus, a significant number of additional genetic markers could be added to the genetic map with additional genotyping and combined analysis.

In order to integrate the two microsatellite-based maps, markers from the Roslin study were genotyped on the UMN/NTBF mapping families (Reed et al., 2006a). A total of 279 primer sets was tested and 240 markers were subsequently screened for polymorphism. Eighty-nine of the 240 markers (37%) were genetically informative in at least one of the four UMN/NTBF dam families and were used for genotyping F_2 offspring. Significant genetic linkage (LOD > 3.0) within the UMN/NTBF families was found for 84 markers from the Roslin study. Of these 84 markers, 49 were genetically linked in both the Roslin and UMN studies allowing for alignment of 18 of the 20 Roslin linkage groups with the UMN map (Table 4). In two cases, evidence for union of linkage groups from the Roslin map was found. Markers on Roslin linkage groups (8, 9, 11, 13, and 14) were each significantly linked with markers from UMN linkage group M1 (turkey chromosome 1, MGA1) and markers on Roslin linkage groups (3 and 17) were significantly linked with markers on UMN linkage group M5 (MGA5). The remaining two linkage groups were assigned to UMN linkage groups based on comparative sequence analysis (see later).

With integration of the Roslin and UMN maps, work continues to integrate the cDNA/RFLP markers from the Nicholas Turkey EST (Nte) map (Harry et al. 2003) with the UMN microsatellite-based map. One study (Reed et al. 2006a–c) focused on development of SNP-based markers directly from the Nte marker sequences. In this study, 48 primer sets were designed and tested with putative SNPs detected in 20 (61%) of the amplified gene fragments. Ten SNP markers were genotyped in the UMN/NTBF families by PCR/RFLP for segregation analysis. This study found evidence for union of linkage groups B + S and K + M from the cDNA/RFLP map, and identified a new linkage group for the turkey. Tentative alignment for some the 22 Nte linkage groups with the UMN turkey genome map was indicated, but additional work is needed to confirm homologies of the SNP and RFLP markers.

Table 4 Correspondence between the Roslin genetic linkage map (Burt et al. 2003) and the UMN map (Reed et al. 2005a)

Roslin linkage group[a]	UMN linkage group (chromosome)	
1*	M5	(MGA5)
2	M3	(MGA3)
3	M5	(MGA5)
4	M16	
5	M6	(MGA6)
6	M24	
7	M8	
8	M1	(MGA1)
9	M1	(MGA1)
10	M2	(MGA2)
11	M1	(MGA1)
12	M4	(MGA4)
13	M1	(MGA1)
14	Z	(MGAZ)
15	M23	
16	M26	(MGA1)
17	M5	(MGA5)
18	M1	(MGA1)
19	M11	
Z*	Z	(MGAZ)

[a] Asterisks denote homologies determined by sequence alignments

Final integration of the three initial turkey maps (Nte, Roslin and UMN) (Reed et al., 2007b) was recently completed at the University of Minnesota. To this end, the majority of the informative microsatellites and SNPs from the Roslin and UMN mapping efforts have been genotyped in the Nte families. In addition, 58 markers that were noninformative in the Roslin ($n = 15$) or UMN ($n = 43$) studies have also been genotyped in the Nte families and will be included in the next map build. The next-generation integrated genetic linkage map for the turkey includes over 700 markers (Reed et al., 2007b).

6.2.5
Comparative Mapping

Comparative mapping approaches are particularly useful for "data-poor" species because they facilitate extrapolating information from "data-rich" species. Because of its lesser rank as an agricultural commodity, support for turkey genetics is less significant

and the turkey genome was not a priority species for full genome sequencing. As such, the extent to which turkey genetic markers and maps can be aligned with the chicken genome sequence will determine the degree to which turkey researchers and breeders can use the wealth of chicken genetic resources.

Attempts at comparative map alignments that predated the release of the chicken whole-genome sequence were limited in scope. For example, as part of the mapping effort of Harry et al. (2003) sequence data were retrospectively obtained from stored cDNA clones used to create the cDNA/RFLP map. Sequences averaging 505 bases in length were obtained for 114 of the 138 clones used to create the Nte map. Searches of GenBank (BLASTn and tBLASTx) at that time found 26 (23%) of the cDNA sequences highly similar to sequences from chicken, and 58 (51%) similar to sequences from other species. These sequence comparisons allowed for a first attempt at comparative mapping and very tentative alignments of some turkey linkage groups to the human and chicken genetic maps. As part of the Roslin mapping study (Burt et al. 2003), marker sequences were also used to search against sequence databases to identify homologs (again, analysis was done prior to release of the chicken whole-genome sequence). Significant matches to chicken, human, or mouse sequences were observed for six markers including *RHT0202* with significant similarity to a sequence on chicken chromosome 5 (GGA5). Comparative alignments deduced from these initial maps could not sufficiently test the extent of marker synteny between the species.

One objective in constructing a genetic linkage map for the turkey was to align the map with the genetic map of the chicken, thereby producing a comparative map. Initial work on cross-species amplification of chicken microsatellites, described previously, was intended to provide anchor loci on a comparative map. To this end, linkage relationships among chicken markers were elucidated in the turkey and compared to those observed in the chicken (Reed et al. 2003a). Ultimately these efforts resulted in the inclusion of 61 chicken markers in the UMN linkage map (Reed et al. 2005a) and confirmatory anchor points for the major turkey linkage groups.

The ability to make direct comparison between the turkey and chicken genome maps has dramatically changed since the release of the chicken whole-genome sequence (International Chicken Genome

Sequencing Consortium 2004). Construction of the UMN second-generation turkey genetic linkage map was completed concurrent with the release of the draft assembly of the chicken sequence (WASHUC 1, Mar 2004). Sequences of all turkey markers included in the UMN mapping effort were queried against the chicken whole-genome sequence by BLASTn searches, and significant matches (341 of 515 comparisons) were used to construct a physical map by assigning turkey markers to positions on the chicken whole-genome assembly. This physical map was then aligned with the turkey genetic map to construct a turkey/chicken comparative map. Most of the homologous syntenic groups were identified, and all but one turkey linkage group aligned with the physical map of the chicken genome. By comparing lengths of the genetic and physical maps, it was estimated that the UMN genetic map encompassed 90% of the turkey genome and that 1 Mbp corresponds to 2.38 cM on the UMN turkey map.

6.2.6
Other Comparative Studies

In the past, genetic markers that could not be assigned to linkage maps because they lacked informativeness remained unmapped and thus were unlikely to be included in future studies. Since completion of the turkey/chicken comparative map, subsequent studies used in silico approaches to tentatively assign turkey DNA sequences to the comparative map based on alignments with the chicken whole-genome sequence. Reed (2005) examined 278 unpublished turkey genomic sequences and found 236 significant alignments with the chicken whole-genome sequence. Similarly, Reed et al. (2005c) compared 877 turkey ESTs and found 788 significant sequence similarities. Of these, 530 sequences were assignable to the 34 designated chicken chromosomes in the WASHUC 1 genome sequence build and 65 sequences corresponded to unassigned contigs.

Sequence comparisons were also used to align the markers from the Roslin map (Burt et al. 2003) with the chicken whole-genome sequence (Reed et al., 2006a). Significant matches were found for 260 sequences including 175 of the 189 markers that could not be genetically mapped in the UMN/NTBF families. Based on BLASTn sequence alignments,

165 of these noninformative sequences were assignable to designated chicken chromosomes. Using the UMN turkey/chicken comparative map the position of these markers in the turkey genome can be predicted.

As a result of combined in silico mapping over 1,700 turkey sequences have now been comparatively assigned to positions in the chicken genome. Using these data, each of the turkey genetic maps (Nte, Roslin, and UMN) can be aligned with the chicken genome sequence. Comparative alignments provide starting points for map integration and enhance incorporation of new markers into the turkey genetic map because markers can be tentatively assigned to linkage groups based on position in the chicken genome sequence prior to initial linkage analysis. For integration of the three turkey maps, linkage groups were placed relative to each other using the chicken whole-genome sequence as a backbone. For example, each mapping effort had identified linkage groups corresponding to the p arm of chicken chromosome 2 (GGA2p, Fig. 5).

Usefulness of the chicken whole-genome sequence and comparative alignments in turkey genetic mapping was further demonstrated by Chaves et al. (2006). In this study, partial turkey microsatellite sequences were aligned with the chicken genome sequence to obtain the additional flanking sequence needed for development of PCR primers. A total of 78 new primer sets were tested and 65 (83%) amplified turkey genomic DNA. This provided an additional set of turkey microsatellites and demonstrated the ability to use the chicken genome sequence to target areas of the turkey genome (and potentially other avian species) for genomic resource development.

In addition to in silico mapping via whole-genome sequences, radiation hybrid cell panels allow comparative positional mapping of nonsegregating markers. RH panels are currently not available for the turkey. However, the National Institute for Agronomic Research (INRA) ChickRH6 whole-genome radiation hybrid panel (Morisson et al. 2002) has been used to comparatively map turkey microsatellite markers. Reed et al. (2005b) examined microsatellites that were monomorphic and/or noninformative in the UMN/ NTBF and/or Nte mapping families. Of the markers tested for use on the panel, 54% amplified an appropriate sized fragment from chicken genomic DNA and 41 markers were ultimately genotyped on the RH

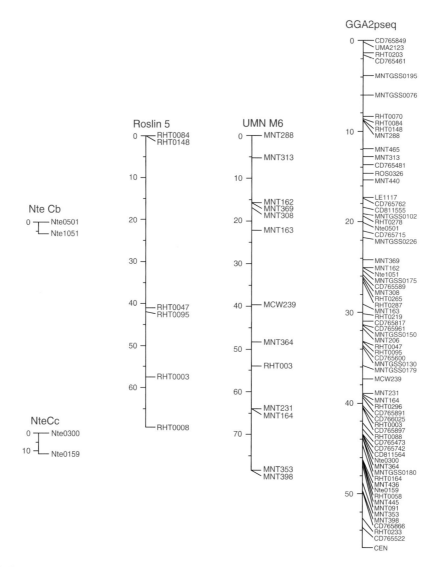

Fig. 5 Turkey linkage groups (Nte, Harry et al. 2003; Roslin, Burt et al. 2003; and UMN, Reed et al. 2005a) corresponding to the p arm of chicken chromosome 2 (GGA2p/MGA4). Linkage groups are aligned with a physical map (units = Mbp) of marker sequences in the chicken genome sequence (GGA2p$_{seq}$)

panel. Thirty-three of the 41 markers typed on the RH panel had significant linkage. Over 50 turkey markers have now been typed on the chicken RH panel.

When combined with sequence analysis, the chicken RH panel provides confirmatory positioning of these markers in the chicken genome and thus on the comparative map. The INRA chicken RH panel has also been used to provide additional resolution for the turkey/chicken comparative map. For example, one turkey linkage group (M22; Reed et al. 2005a) could not be aligned with the chicken genome because it did not contain anchor loci nor had significant sequence matches to contigs assigned to a chicken chromosome. Markers comprising M22 were evaluated on the INRA ChickRH6 whole-genome radiation hybrid panel (Reed and Mendoza 2006). Significant linkage was found with markers on the framework map for chicken linkage group E26C13, thereby providing alignment of turkey linkage group M22 with the chicken map.

These types of comparative studies increase the value of the turkey genetic maps and associated markers and expand their use in genetic and evolutionary studies. One of the more interesting comparisons examined the rate of nucleotide divergence between the turkey and chicken genomes (Axelsson et al. 2005). This study examined patterns of nucleotide substitution in chicken-turkey sequence alignments for both introns and exons. Significant differences were found in substitution rate when alignments were classified by chromosome type (macro versus micro) with higher rates observed for the microchromosomes. This suggests that mutation rates vary significantly in different regions of the avian genome. The increasing availability of DNA sequences from the turkey will allow for broader applications of chicken genetics and the chicken genome sequence in turkey biology.

6.2.7
Physical Mapping

Few studies have reported on physical mapping of the turkey genome. Cross-species experiments hybridizing chicken bacterial artificial chromosome (BAC) clones to turkey metaphase chromosomes (Robertson et al. 2005) and chromosome painting experiments (Fig. 6; Shibusawa et al. 2004) have confirmed the chromosomal homologies and rearrangements predicted between the species based on banded karytoypes. Two recent studies used florescent in situ hybridization (FISH) to localize turkey BAC clones. Reed et al. (2006c) mapped BAC clones from the CHORI-260 library (see later) containing microsatellite markers from the three turkey linkage groups (M1, M18, and M26) each with homology to regions of GGA1. FISH analysis assigned all three of these linkage groups to a single turkey chromosome (MGA1). In a second study, Chaves et al. (2007) examined the physical locations of the MHC-B and MHC-Y loci in relation to the ribosomal DNA cluster NOR) in the turkey. The NOR of the chicken occupies most of the distal end of a microchromosome (GGA16; Bloom and Bacon 1985) that also contains the major histocompatibility (MHC) region (Fillon et al. 1996; Miller et al. 1996). FISH with BAC clones clearly positioned the turkey rDNA locus distal to the MHC region (MHC-Y locus on the q-arm and the MHC-B on the p-arm; Chaves et al., 2007).

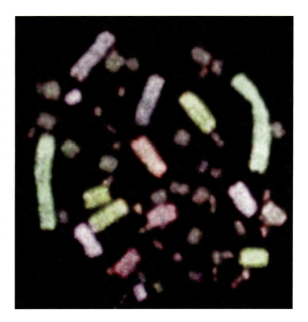

Fig. 6 Turkey metaphase chromosomes painted with fluorescent probes complementary to chicken chromosomes 1–5. Chicken chromosome 2 is represented as two chromosomal pairs in the turkey due to a fission/fusion event. Image courtesy of D. Griffin (Kent UK) and J. Wienberg (Munich, Germany)

6.2.8
The Next-Generation Physical Maps

Among the newer genetic resources available for the turkey is a large-insert BAC library. The CHORI-260 BAC library was constructed by Michael Nefedov in Pieter de Jong's laboratory at BACPAC Resources, Children's Hospital Oakland Research Institute with funding from the USDA (http://bacpac.chori.org/turkey260.htm). Genomic DNA for the creation of the library was isolated from blood collected at Nicholas Turkey Breeding Farms by David Harry. The female bird used to create the library was selected from an inbred sub-line derived from commercial stock. The BAC library (*Eco*RI inserts cloned into the pTAR-BAC2.1 vector) contains in excess of 11× genome redundancy (~71,000 clones) with inserts averaging 190 kb. The library has been arrayed into 384-well microtiter dishes and gridded onto four 22 × 22 cm nylon high-density filters (each membrane represents over 18,000 distinct turkey BAC clones, stamped in duplicate) for screening by probe hybridization.

Extensive analysis of the turkey BAC library has already produced meaningful comparative results.

Romanov and Dodgson (2006) used the CHORI-260 filter sets to perform cross-species hybridizations for comparative mapping. Overgo probes (415 in total) designed from chicken genomic sequence and zebra finch ESTs were used to probe the high-density filters. Successful hybridization was obtained for 81% of the overgos and as a result, 336 markers or genes were assigned to 3,772 turkey BACs with 11.2 positive clones identified for each successful probe. These include genes on almost every defined chicken chromosome.

Work is currently underway at Michigan State University and Texas A&M University to further characterize the CHORI-260 library and in doing so place turkey genes on a detailed physical map aligned with the chicken whole-genome sequence. To accomplish this, approximately 40,000 BACs (6.3× genome coverage) from the CHORI-260 turkey BAC library will be fingerprinted, and 20,000 BAC end sequences will be generated for assembling a BAC contig map. A detailed comparative turkey-chicken genome map will then be generated based on alignment of the BAC contig map, the turkey linkage map, and the chicken whole-genome sequence. The extensive characterization of the turkey BAC library will create a wealth of new sequence data for the species and significantly increase the value of the BAC library resource.

Expanded sequencing of individual BACs in the turkey has been limited to two comparative studies. The National Institute of Health (NIH) Intramural Sequencing Center (NISC) Comparative Sequencing Initiative involves the generation of genomic sequence from 66 vertebrate species (including the turkey), for use in developing and refining computational tools. Working from BAC contig maps, minimal sets of overlapping BACs are sequenced to create sets of orthologous sequences for the same large genomic region from multiple vertebrates. A total of 31 gene regions in the turkey genome have been targeted for sequencing using BAC contigs constructed from the CHORI-260 library. Over 77,000 single-pass shotgun sequence reads have been generated for 28 turkey BACs and are available online through the National Center for Biotechnology Information (NCBI) (http://www.nisc.nih.gov/). A second comparative sequence study underway in the turkey is of the major histocompatibility complex. BACs containing sequences homologous to the MHC-B and MHC-Y loci have been sequenced for comparative analysis with the chicken and other species (Chaves et al., unpubl). Sequences derived from this study are being used to examine variation in the MHC region in and between commercial and wild turkeys.

6.3
Advanced Works, Functional Genomics

6.3.1
ESTs, Microarrays, and SAGE

Expressed sequence tags (ESTs) represent the transcribed portion of the genome and are an effective way to identify genes for interspecies comparisons and expression studies including microarray analysis. Large-scale sequencing of turkey ESTs has lagged behind such efforts in the chicken. As of July 2006, Genbank records only included 857 ESTs from two sources. Twenty-two sequences were derived from a pituitary cDNA library (Smith et al. 2000) and 835 sequences were derived from an embryonic cDNA library (24-day turkey embryo) developed by Nicholas Turkey Breeding Farms (Harry et al. 2003; Chaves et al. 2005). In characterizing the embryonic library, Chaves et al. (2005) sequenced and analyzed 694 ESTs. A total of 437 unique sequences were identified including 76 assembled contigs and 361 singletons. Sequences were compared to those in GenBank with the majority showing significant comparative matches to sequences reported from the chicken. This study provided the first catalog of expressed genes during embryonic hyperplasia.

Even with the limited number of gene sequences, detailed study of gene expression within defined physiological contexts is possible in the turkey. For example, during incubation developing poultry embryos must switch from lipid-based energy metabolism to a carbohydrate-based metabolism dependant on glycogen reserves. To characterize this process in the turkey, de Oliveira et al. (2006) used a 90-gene microarray to examine the time course of gene expression associated with both lipid and carbohydrate metabolism at 20, 22, 24, 26, and 28 days of incubation. Results from this study found a shift from lipid to carbohydrate metabolism initiated at day 22, and that storage and usage of carbohydrates seemed to gradually increase until day 26 corresponding to pipping. As a greater number of gene sequences become available, functional genomic approaches such as this will increasingly be used to study specific elements of growth and development in the turkey.

As discussed in Section 6.1.3, genetic selection for muscle growth and increased muscling in turkey has altered many characteristics of the bird including muscle composition and meat quality. A catalog of genes involved in muscle growth and development and elucidating patterns of gene expression will be important in understanding these characteristics. The first skeletal muscle genes to be extensively characterized in the turkey were myosin light chain genes (Chaves et al. 2003). Myosin light chain proteins associate with the motor protein myosin and are believed to play a role in the regulation of its actin-based ATPase activity. cDNA clones corresponding to the skeletal muscle essential light chain (*MELC1F*) and to the smooth muscle isoform of myosin light chain (*MLC*) were isolated and sequenced. The skeletal *MELC* gene was genetically mapped by SNP genotyping to turkey chromosome 7 (MGA7). Additional sequencing of turkey ESTs identified several other muscle-associated genes (Chaves et al. 2005) and four of these (*α-Tropomyosin, β-Actin, Myosin binding protein-H, Troponin-T$_{skeletal}$*) are genetically mapped in the turkey.

A large-scale EST project in the turkey involving researchers at the Michigan State University, the Ohio State University, and the University of Minnesota is currently underway. The goal of this project is to characterize the suite of genes expressed in skeletal muscle and study gene expression profiles associated with muscle growth and development. To this end, a turkey skeletal muscle microarray representing 10,000 nonredundant ESTs is being developed. To construct the array, three turkey skeletal muscle cDNA libraries corresponding to three developmental stages were examined. Developmental stages included in the study are 18-day embryo, 1-day posthatch poult, and 16-week posthatch adult turkey. These represent the critical time periods (hyperplasia, hypertrophy, and market age) in turkey skeletal muscle development, respectively. The project is expected to reveal differential expression of genes associated with muscle characteristics and postmortem meat quality.

A second approach to assess the transcribed portion of the genome is through serial analysis of gene expression (SAGE). The SAGE methodology is based on the assumption that a short sequence tag (10–14 bp) contains sufficient information to uniquely identify a gene transcript. Sequence tags are linked together as long serial molecules that are cloned and sequenced. The number of times a particular tag is observed in the sequenced clones is an indication of the expression level of the corresponding transcript. Thus, SAGE allows for both qualitative (which genes are expressed) and quantitative (relative copy number of expressed genes) analysis of the entire transcriptome at distinct physiological states within tissues. The SAGE approach was recently used to investigate the biology of artificial insemination and sperm storage. Commercial turkey breeding relies extensively on artificial insemination. Even with use of extenders, turkey sperm lose viability within 8–18 h when stored as liquid semen. In contrast, following a single insemination, sperm are maintained within the hen's sperm storage tubules for 45 or more days. Long et al. (2003) used SAGE to compare gene expression in sequences recovered from turkey hens after artificial insemination with extended semen or extender alone to identify and characterize differentially expressed genes that may underlie the physiology underling prolonged sperm storage. This study recovered 95,325 10 bp tags that represented 27,430 unique genes with approximately 1% of the putative unique genes being differentially expressed (P < 0.05) between treatments. One of the genes that were upregulated and more extensively studied was *avidin*, a protein that binds and sequesters biotin within the egg.

6.3.2
Candidate Gene Mapping

With the lack of detailed QTL studies, candidate gene mapping studies in the turkey have been limited primarily to genes involved in metabolic pathways established in other species. One example is round heart (rh) syndrome, which is a spontaneous cardiomyopathy in avian species. The condition was first identified in turkeys in 1962 and occurs as both a spontaneous (idiopathic) and inducible (furazolidone) condition. Researchers at Virginia Tech University have examined the cardiac-associated candidate genes *troponin T* (*cTnT*) and *phospholamban* (*PLN*) for association with toxin-induced round heart. Although no significant differences were observed in the expression of *cTnT* or *PLN* in affected versus unaffected birds,

12 SNPs were detected in *cTnT* and *PLN* sequences (Lin 2006). Linkage analysis showed that *cTnT* was unlinked on the UMN genetic map (*cTnT* is on microchromosome 32 in the chicken and a homologous linkage group has not yet been identified in turkey). Recently, Paxton et al. (2005) identified several AFLPs statistically associated with either phenotypically normal turkeys or birds with spontaneous occurrence of round heart. These markers were later mapped as SNPs at the University of Minnesota (Reed et al., 2007c) to identify candidate genes for future analysis.

A second example of use of metabolic pathways is the study of the mycotoxin aflatoxin B_1 (AFB$_1$), which is a potent hepatotoxin and hepatocarcinogen in humans and animals (Coulombe 1993). In the early 1960s, "turkey X disease" was responsible for widespread deaths of turkeys and other poultry throughout Europe. AFB$_1$ discovered in poultry feed contaminated with Brazilian peanut meal was determined to be the etiological agent (Asao et al. 1965). AFB$_1$ is a natural toxin produced by the ubiquitous fungi *Aspergillus flavus* and *A. parasiticus*. It is widespread in foods and feeds, and is always present to some extent in poultry feeds. In poultry (chickens, turkeys, ducks, quail, and others), AFB$_1$ causes severe economic losses from reductions in growth rate, feed efficiency, hatchability, and increased susceptibility to disease (Coulombe et al. 2005). Modern commercial turkeys are substantially more susceptible than other poultry and are the most susceptible animal to AFB$_1$ examined to date (Klein et al. 2000).

Hypersensitivity of commercial turkeys to AFB$_1$ is largely explained by the unfortunate combination of extremely efficient *cytochrome P450-(CYP)* mediated bioactivation of AFB$_1$ to the AFB$_1$–8,9-epoxide (AFBO) and a deficiency of *glutathione S-transferase* (*GST*) isoforms able to detoxify AFBO (Coulombe et al. 2005). One of the *CYP* genes involved in AFB$_1$ bioactivation (*CYP1A5*) was the first functional gene to be cloned, sequenced, and expressed from turkey (Yip and Coulombe 2006). Using comparative alignment of the turkey *CYP1A5* cDNA with the chicken genome sequence, introns in the *CYP1A5* gene were examined for sequence polymorphism. Identified SNPs were used to map the gene to turkey linkage group M16 in the UMN/NTBF population (Reed et al., 2007a). Work is currently underway to further characterize the *CYP* and *GST* genes in the turkey using

comparative approaches. The long-term goal is to determine if resistance traits are present in wild turkeys and other progenitor strains that possess greater genetic diversity than commercial turkeys. If so, it may be possible to restore or enhance protective traits in domestic breeds by introgression of wild turkey genes into domestic lines.

6.4
Conclusion

The past few years have brought dramatic changes in the availability of genomic resources for the turkey, from just a handful of genetic markers and sequences to over hundreds of genetic markers and thousands of DNA sequences. Three genetic linkage maps have been constructed, and these were integrated into a single medium-density map aligned with the chicken whole-genome sequence. The turkey linkage maps are already providing the framework for functional genomic studies such as round heart and AFB$_1$ toxicity. Refinement of the genetic, physical, and comparative maps will enhance future studies determining the location of quantitative trait loci with significant effects on production traits and disease susceptibility.

Acknowledgment. The author thanks Dr. Michael Romanov for his insightful comments on this work.

References

Aldrich JW (1967) Historical background. In: Hewitt OH (ed) The Wild Turkey and Its Management. The Wildlife Society, Washington, DC, USA, pp 3–16

Asao T, Buechi G, Abdel-Kader MM, Chang SB, Wick EL, Wogan GN (1965) The structures of aflatoxins B and G. J Am Chem Soc 87:882–886

Axelsson E, Webster MT, Smith NG, Burt DW, Ellegren H (2005) Comparison of the chicken and turkey genomes reveals a higher rate of nucleotide divergence on microchromosomes than macrochromosomes. Genome Res 15:120–125

Baratti M, Groenen M, Veenendaal T, Fulgheri FD (2001) Polymorphic microsatellites developed by cross-species amplifications in common pheasant breeds. Anim Genet 32:222–225

Bitgood JJ, Somes Jr RG (1990) Linkage relationships and gene mapping. In: Crawford RD (ed) Poultry Breeding and Genetics. Elsevier, Amsterdam, The Netherlands, pp 469–495

Bloom SE, Bacon LD (1985) Linkage of the major histocompatibility (B) complex and the nucleolar organizer in the chicken. Assignment to a microchromosome. J Hered 76:146–154

Bloom SE, Delany ME, Muscarella DE (1993) Constant and variable features of avian chromosomes. In: Etches RL, Verrinder Gibbins AM (eds) Manipulation of the Avian Genome. CRC Press, Boca Raton, USA, pp 39–59

Brant AW (1998) A brief history of the turkey. World's Poult Sci J 54:365–373

Burt DW, Morrice DR, Sewalem A, Smith J, Paton IR, Smith EJ, Bentley J, Hocking PM (2003) Preliminary linkage map of the turkey (*Meleagris gallopavo*) based on microsatellite markers. Anim Genet 34:399–409

Buss EG (1989) Genetics of turkeys: origin and development. In: Nixey C, Grey TC (eds) Recent Advances in Turkey Science. Poult Sci Symp No 21. Butterworth and Co, London, UK, pp 11–30

Chaves LD, Ostroski BJ, Reed KM (2003) Myosin light chain genes in the turkey (*Meleagris gallopavo*). Cytogenet Genome Res 102:340–346

Chaves LD, Rowe JA, Reed KM (2005) Survey of a cDNA library from the turkey (*Meleagris gallopavo*). Genome 48:12–17

Chaves LD, Knutson TP, Kreuth SB, Reed KM (2006) Using the chicken genome sequence in the development and mapping of genetic markers in the turkey (*Meleagris gallopavo*). Anim Genet 37:130–138

Chaves LD, Kreuth SB, Reed KM (2007) Characterization of the turkey MHC chromosome through genetic and physical mapping. Cytogenet Genome Res 117:213–220

Coulombe RA, Jr (1993) Biological action of mycotoxins. J Dairy Sci 76:880–891

Coulombe RA, Guarisco JA, Klein PJ, Hall JO (2005) Chemoprevention of aflatoxicosis in poultry by dietary butylated hydroxytoluene. Anim Feed Sci Technol 121:217–225

Crawford RD (1990) Poultry genetic resources: evolution, diversity, and conservation. In: Crawford RD (ed) Poultry Breeding and Genetics, Elsevier Health Sciences, Edinburgh, Scotland, pp 43–60

de Oliveira J, Ferket P, Ashwell C, Uni Z (2006) Assessing liver energy metabolism of late term turkey embryos using microarrays. Metabolism and Nutrition. Abstr no 49. Annu Meet, Poult Sci Assoc, Edmonton, Canada, July 16–19, p 35

Delany ME, Krupkin AB, Miller MM (2000) Organization of telomere sequences in birds: evidence for arrays of extreme length and for in vivo shortening. Cytogenet Cell Genet 90:139–145

Donoghue AM, Sonstegard TS, King LM, Smith EJ, Burt DW (1999) Turkey sperm mobility influences paternity in the context of competitive fertilization. Biol Reprod 61: 422–427

Dranchak P, Chaves LD, Rowe JA, Reed KM (2003) Turkey microsatellite loci from an embryonic cDNA library. Poult Sci. 82:526–531

Ferket PR (2004) Tom weights up seven percent. WATT Poultry USA, July, pp 32–42

Fillon V, Zoorob R, Yerle M, Auffray C, Vignal A (1996) Mapping of the genetically independent chicken major histocompatibility complexes B@ and RFP-Y@ to the same microchromosome by two-color fluorescent in situ hybridization. Cytogenet Cell Genet 75:7–9

Hanotte O, Puch A, Maucher C, Dawson D, Burke T (1997) Nine novel chicken microsatellite loci and their utility in other Galliformes. Anim Genet 28:308–322

Harry DE, Marini PJ, Zaitlin D, Reed KM (2003) A first generation map of the turkey genome. Genome 46:914–924

Huang HB, Song YQ, Hsei M, Zahorchak R, Chiu J, Teuscher C, Smith EJ (1999) Development and characterization of genetic mapping resources for the turkey (*Meleagris gallopavo*). J Hered 90:240–242

International Chicken Genome Sequencing Consortium (2004) Sequence and comparative analysis of the chicken genome provide unique perspectives on vertebrate evolution. Nature 432:695–716

Julian RJ (1998) Rapid growth problems: ascites and skeletal deformities in broilers. Poult Sci 77:1773–1780

Julian RJ (2005) Production and growth related disorders and other metabolic diseases of poultry-a review. Vet J 169:319–320

Klein PJ, Buckner R, Kelly J, Coulombe RA, Jr (2000) Biochemical basis for the extreme sensitivity of turkeys to aflatoxin B_1. Toxicol Appl Pharmacol 165:45–52

Knutson TP, Chaves LD, Hall MK, Reed KM (2004) One hundred fifty-four genetic markers for the turkey (*Meleagris gallopavo*). Genome 47:1015–1028

Krishan A, Shoffner RN (1966) Sex chromosomes in the domestic fowl (*Gallus domesticus*), turkey (*Meleagris gallopavo*), and the Chinese pheasant (*Phasianus colchicus*). Cytogenetics 5:53–63

Latch EK, Smith EJ, Rhodes OE, Jr (2002) Isolation and characterization of microsatellite loci in wild and domestic turkeys (*Meleagris gallopavo*). Mol Ecol Notes 2:176–178

Le Bihan-Duval E, Berri C, Baeza E, Sante V, Astruc T, Remignon H, Le Pottier G, Bentley J, Beaumont C, Fernandez X (2003) Genetic parameters of meat technological quality traits in a grand-parental commercial line of turkey. Genet Sel Evol 35:623–635

Levin I, Cheng HH, Baxter-Jones C, Hillel J (1995) Turkey microsatellite DNA loci amplified by chicken-specific primers. Anim Genet 26:107–110

Lin K (2006) Candidate gene expression and SNP analyses of toxin-induced dilated cardiomyopathy in the turkey (*Meleagris gallopavo*). MS Thesis. Dept Animal Poult Sci, Virg Polytech Univ, Blacksburg, VA

Liu Z, Crooijmans RPMA, van der Poel JJ, Groenen MAM (1996) Use of chicken microsatellite markers in turkey: a pessimistic view. Anim Genet 27:191–193

Long EL, Sonstegard TS, Long JA, Van Tassell CP, Zuelke KA (2003) Serial analysis of gene expression in turkey sperm storage tubules in the presence and absence of resident sperm. Biol Reprod 69:469–474

Matzke AJ, Varga F, Gruendler P, Unfried I, Berger H, Mayr B, Matzke MA (1992) Characterization of a new repetitive sequence that is enriched on microchromosomes of turkey. Chromosoma 102:9–14

Miller MM, Goto RM, Taylor RL Jr, Zoorob R, Auffray C, Briles RW, Briles WE, Bloom SE (1996) Assignment of Rfp-Y to the chicken major histocompatibility complex/NOR microchromosome and evidence for high-frequency recombination associated with the nucleolar organizer region. Proc Natl Acad Sci USA 93:3958–3962

Mock KE, Theimer TC, Rhodes OE Jr, Greenberg DL, Keim P (2002) Genetic variation across the historical range of the wild turkey (*Meleagris gallopavo*). Mol Ecol 11:643–657

Morisson M, Lemiere A, Bosc S, Galan M, Plisson-Petit F, Pinton P, Delcros C, Feve K, Pitel F, Fillon V, Yerle M, Vignal A (2002) ChickRH6: a chicken whole-genome radiation hybrid panel. Genet Sel Evol 34:521–533

Paxton CN, Pierpont ME, Kooyman DL (2005) Identification of AFLP markers associated with round heart syndrome in turkeys. Intl J Poult Sci 4:133–137

Primmer CR, Raudsepp T, Chowdhary BP, Moller AP, Ellegren H (1997) Low frequency of microsatellites in the avian genome. Genome Res 7:471–482

Reed KM (2005) In silico comparative mapping of 278 genomic survey sequences (GSS) from the turkey (*Meleagris gallopavo*). Anim Genet 36:438–443

Reed KM, Mendoza KM (2006) Assignment of turkey linkage group M22 on the comparative turkey/chicken map. Anim Genet 37:84

Reed KM, Mendoza KM, Beattie CW (1999) Utility of chicken-specific microsatellite primers for mapping the turkey genome. Anim Biotechnol 10:137–141

Reed KM, Mendoza KM, Beattie CW (2000a) Comparative analysis of microsatellite loci in chicken and turkey. Genome 43:796–802

Reed KM, Roberts MC, Murtaugh J, Beattie CW, Alexander LJ (2000b) Eight new dinucleotide microsatellite loci in turkey (*Meleagris gallopavo*). Animal Genet 31:140

Reed KM, Chaves LD, Rowe JA (2002) Twelve new turkey microsatellite loci. Poult Sci 81:1789–1791

Reed KM, Chaves LD, Garbe JR, Da Y, Harry DE (2003a) Allelic variation and genetic linkage of avian microsatellites in a new turkey population for genetic mapping. Cytogenet Genom Res 102:331–339

Reed KM, Chaves LD, Hall MK, Knutson TP, Rowe JA, Torgerson AJ (2003b) Microsatellite loci for genetic mapping in the turkey (*Meleagris gallopavo*). Anim Biotechnol 14:119–131

Reed KM, Chaves LD, Hall MK, Knutson TP, Harry DE (2005a) A comparative genetic map of the turkey genome. Cytogenet Genom Res 111:118–127

Reed KM, Holm J, Morisson M, Leroux S, Vignal A (2005b) Assignment of non-informative turkey genetic markers through comparative approaches. Cytogenet Genom Res 109:527–532

Reed KM, Knutson TP, Krueth SB, Sullivan LR, Chaves LD (2005c) In silico mapping of ESTs from the turkey (*Meleagris gallopavo*). Anim Biotechnol 16:81–102

Reed KM, Chaves LD, Hall MK, Knutson TP, Kreuth SB, Ashwell CM, Burt DW (2006a) Integration of microsatellite-based genetic maps for the turkey (*Meleagris gallopavo*). Genome 49:1308–1318

Reed KM, Hall MK, Chaves LD, Knutson TP (2006b) Single nucleotide polymorphisms for integrative mapping in the turkey (*Meleagris gallopavo*). Anim Biotechnol 17:73–80

Reed KM, Sullivan LR, Foster LK, Chaves LD, Ponce de León FA (2006c) Assignment of linkage groups to turkey chromosome 1 (MGA1). Cytogenet Genom Res17:73–80

Reed KM, Mendoza KM, Coulombe RA, Jr (2007a) Structure and genetic mapping of the Cytochrome P4501A5 gene in the turkey (*Meleagris gallopavo*). Cytogenet Genom Res 116:104–109

Reed KM, Mendoza KM, Chaves LD (2007b) An integrated and comparative map of the turkey genome. Cytogenet Genome Res 119:113–126

Reed KM, Mendoza KM, Hu GR, Sullivan LR, Grace M, Chaves LD, Kooyman DL (2007c) Genomic analysis of genetic markers associated with inherited cardiomyopathy (round heart disease) in the turkey (*Meleagris gallopavo*). Animal Genet 38:211–217

Robertson WRB, Bohren RR, Warren DC (1943) The inheritance of plumage color in the turkey. J Hered 34:246–256

Robertson LB, Tempest HG, Patel AP, Vignal A, Fillon V, Crooijmans RPMA, Groenen MAM, Hillier LW, Morrice DR, Speed D, Bentley J, Masabanda JS, Burt DW, Griffin DK (2005) Production of a comparative physical genome map of the turkey (*Melagris gallopavo*). Abstr. Chicken Genom Dev Workshop, Cold Spring Harbor, NY, USA, p 39

Romanov MN, Dodgson JB (2006) Cross-species overgo hybridization and comparative physical mapping within avian genomes, Anim Genet 37:397–399

Savage TF (1990) Mutations and major variants in turkeys. In: Crawford RD (ed) Poultry Breeding and Genetics. Elsevier Health Sciences, Edinburgh, Scotland, pp 317–332

Savage TF, Harper JA, Engel HN, Jr (1993) Inheritance of tetanic torticollar spasms in turkeys. Poult Sci 72:1212–1217

Schmid M, Nanda I, Guttenbach M, Steinlein C, Hoehn M, Schartl M, Haaf T, Weigend S, Fries R, Buerstedde J-M, Wimmers K, Burt DW, Smith J, A'Hara S, Law A, Griffin DK,

Bumstead N, Kaufman J, Thomson PA, Burke T, Groenen MAM, Crooijmans RPMA, Vignal A, Fillon V, Morisson M, Pitel F, Tixier-Boichard M, Ladjali-Mohammedi K, Hillel J, Mäki-Tanila A, Cheng HH, Delany ME, Burnside J, Mizuno S (2000) First report on chicken genes and chromosomes 2000. Cytogenet Cell Genet 90:169–218

Schmid M, Nanda I, Hoehn H, Schartl M, Haaf T, Buerstedde J-M, Arakawa H, Caldwell RB, Weigend S, Burt DW, Smith J, Griffin DK, Masabanda JS, Groenen MAM, Crooijmans RPMA, Vignal A, Fillon V, Morisson M, Pitel F, Vignoles M, Garrigues A, Gellin J, Rodionov AV, Galkina SA, Lukina NA, Ben-Ari G, Blum S, Hillel J, Twito T, Lavi U, David L, Feldman MW, Delany ME, Conley CA, Fowler VM, Hedges SB, Godbout R, Katyal S, Smith C, Hudson Q, Sinclair A, Mizuno S (2005) Second report on chicken genes and chromosomes 2005. Cytogenet Genom Res 109:415–479

Schorger AW (1966) The Wild Turkey: Its History and Domestication. University of Oklahoma Press, Norman, Oklahoma, USA

Shibusawa M, Nishibori M, Nishida-Umehara C, Tsudzuki M, Masabanda J, Griffin DK, Matsuda Y (2004) Karyotypic evolution in the Galliformes: an examination of the process of karyotypic evolution by comparison of the molecular cytogenetic findings with the molecular phylogeny. Cytogenet Genom Res 106:111–119

Smith E, Shi L, Drummond P, Rodriguez L, Hamilton R, Powell E, Nahashon S, Ramlal S, Smith G, Foster J (2000) Development and characterization of expressed sequence tags for the turkey (*Meleagris gallopavo*) genome and comparative sequence analysis with other birds. Anim Genet 31:62–67

Smith EJ, Geng T, Long E, Pierson FW, Sponenberg DP, Larson C, Gogal R (2005) Molecular relatedness of five domestic turkey strains. Biochem Genet 43:35–47

Smith EJ, Jones CP, Bartlett J, Nestor KE (1996) Use of randomly amplified polymorphic DNA markers for the genetic analysis of relatedness and diversity in chickens and turkeys. Poult Sci 75:579–784

Smyth JR, Jr (1954) Hairy, a gene causing abnormal plumage in the turkey. J Hered 45:197–200

Sosnicki AJ, Wilson BW (1991) Pathology of turkey skeletal muscle: implications for the poultry industry. Food Struct 10:317–326

Szöke S, Komlósi, Korom E, Ispány M, Mihók S (2004) A statistical analysis of population variability in bronze turkey considering gene conservation. Archiv Tierzucht 4:377–385

Velleman SG, Anderson JW, Coy CS, Nestor KE (2003) Effect of selection for growth rate on muscle damage during turkey breast muscle development. Poult Sci 82:1069–1074

Wicker T, Robertson JS, Schulze SR, Feltus FA, Magrini V, Morrison JA, Mardis ER, Wilson RK, Peterson DG, Paterson AH, Ivarie R (2005) The repetitive landscape of the chicken genome. Genome Res 15:126–136

Windhorst H–W (2005) Changing regional patterns of turkey production and turkey meat trade. In: Hafez HM (ed), Turkey Production: Prospects on Future Development, Mensch & Buch Verlag, Berlin, pp. 25–41

Yip SM, Coulombe RA, Jr (2006) Molecular cloning and expression of a novel cytochrome P450 from turkey liver with aflatoxin B_1 oxidizing activity. Chem Res Toxicol 19:30–37

CHAPTER 7

Rabbit

Claire Rogel-Gaillard[1](✉), Nuno Ferrand[2], and Helene Hayes[3]

[1] INRA CEA, UMR314 Laboratoire de Radiobiologie et Etude du Génome, F-78350 Jouy-en-Josas, France, claire.rogel-gaillard@jouy.inra.fr
[2] CIBIO, Centro de Investigação em Biodiversidade e Recursos Genético, Campus Agrário de Vairão, 4485-661 Vairão, and Departamento de Zoologia e Antropologia, Faculdade de Ciências, Universidade do Porto, Praça Gomes Teixeira, 4099-002 Porto, Portugal
[3] INRA, UR339 Unité de Génétique biochimique et Cytogénétique, F-78350 Jouy-en-Josas, France

7.1
Introduction

7.1.1
History

Because both wild and domestic European rabbits belong to the species *Oryctolagus cuniculus*, the designation "domestic rabbit" refers to rabbits confined by man in cages or other enclosures. Until about 7000 BC, man hunted wild European rabbits as a source of meat. The use of rabbits as food progressively declined as bigger animals, i.e., wild boars and deer were hunted. About 1000 BC, Phoenician sailors discovered rabbits upon reaching the coasts of Spain. Impressed by the large numbers of these small burrowing mammals, which resembled the hyraxes of their homeland, they named this new country "I-Sapham-Im," i.e., country of saphan or hyraxes. The Romans latinized this name into Hispania, hence Spain.

Rabbits are closely associated with Spain. The oldest fossil of the genus *Oryctolagus* dating back to the Miocene (24 millions–5 millions BCE) has been found near Granada. Today, the genus *Oryctolagus* contains only one species, *O. cuniculus*, and the oldest fossil known for *O. cuniculus* dating back to the middle Pleistocene (800,000–130,000 BCE) comes from Baza in Andalusia. From there, rabbits spread to France, Western Europe, and during the Neolithic period, to Northwestern Africa. At present, wild rabbits are still restricted to Western Europe from the south of Sweden to Spain and North Africa, and are absent from Italy and Switzerland (except in the Neuchâtel region). Rabbits were introduced into Great

Britain in the eleventh century and only recently into Australia (1859), New Zealand and the Kerguelen Islands (1874), and Chile (around 1910) (Rougeot 1981; Niederberger 1989). The introduction of rabbits into China probably dates back to the period of the Roman Empire, around 30 BC when regular communications and trade had been developed between Europe and Asia, especially via the Silk Road. To date, no fossils of domesticated or wild rabbits have been discovered in China, and analyses of the mitochondrial DNA control region sequence from so-called Chinese and other introduced domestic rabbit breeds in China support the hypothesis that present-day Chinese rabbits were derived from European rabbits (Long et al. 2003).

Rabbit is the only domesticated mammal originating from Western Europe. In addition, the history of its wild populations is well documented through both archeological and genetic studies. Thus, genetic diversity analyses have been numerous for this species (Hardy et al. 1994, 1995; Mougel et al. 1997; Branco et al. 2000, 2002; Queney et al. 2000, 2001, 2002).

7.1.2
Taxonomic Position

The scientific name of the European rabbit, *Oryctolagus cuniculus* (Linné 1758), originates from the Greek "oruktês" (burrower) and "lagôs" (hare) for *Oryctolagus* and "cuniculus," which is Latin for rabbit, itself derived from the Iberian language and transcribed as "ko(n)niklos" in Greek. This small mammal belongs to the order Lagomorpha, which includes hares and pikas, the family *Leporidae* (Gray 1821) and the

---

Genome Mapping and Genomics in Animals, Volume 3
Genome Mapping and Genomics in Domestic Animals
N.E. Cockett, C. Kole (Eds.)
© Springer-Verlag Berlin Heidelberg 2009

subfamily *Leporinae* (Trouessart 1880). Traditionally, lagomorphs are thought to be phylogenetically close to rodents because they are the only small mammals having evergrowing incisors adapted for gnawing, two pairs of upper incisors in lagomorphs and one pair in all rodents. Based on morphological data, some authors place Lagomorpha and Rodentia together in the Glires clade (Simpson 1945; Liu and Miyamoto 1999), but the monophyly of this group is still a matter of debate among scientists (Misawa and Janke 2003; Douzery and Huchon 2004).

7.1.3
Physical Characteristics

Rabbits are small mammals with the two sexes looking alike. They have large ears, a short tail, and strong hind legs adapted for running and leaping. Wild rabbits weigh between 1.2 and 2.5 kg and generally have gray and brown fur with a characteristic white patch on the underside of the tail. They are gregarious animals and prefer cultivated, open, or brushy areas. They live in extensive and complex underground burrows or warrens, which they either take over from other animals or dig themselves. In appropriate conditions, they form large colonies containing six to ten adults. They are well known for their prolificacy; they start breeding at 3–4 months of age and produce five to six young per litter. Domestic rabbits are year-round breeders, while wild rabbits are anestrous during autumn and winter (Boyd et al. 1987). They can live up to ten to thirteen years but are delicate animals susceptible to many bacterial (pasteurellosis, enterotoximia, tularemia, colibacillosis, salmonellosis, treponematosis), viral (myxomatosis, viral hemorrhagic disease), and parasitic (coccidiosis, taenia) diseases. These pathogens cause high mortality rates and severe reduction in productivity. For example, an epidemic outbreak of the very contagious hemorrhagic disease can lead to 90% mortality and quickly exterminate the entire rabbit population on a farm. Therefore, preventing the occurrence and spread of diseases is critical in intensive and large-scale rabbit farms and requires strict breeding conditions, good hygiene, ventilation and lighting, and regular attention.

In regions, such as Australia and New Zealand, where no natural predators exist to control population numbers, rabbits are major economical animal pests. They cause considerable environmental damage and decreased agricultural production because of loss of vegetation and land degradation. Controlling rabbit populations costs governments and land-owners several hundred million dollars per year, part of it being spent on research for new control methods.

7.1.4
Breeds

The European rabbit, *Oryctolagus cuniculus*, is the ancestor of all domestic rabbit breeds, which vary greatly in size, fur type, and color. Domestication of rabbits probably began with the early Romans who raised these animals for food, in large walled colonies called "leporaria" but without selective breeding. It was not until the Middle Ages that true domestication with selective breeding took place in monasteries, with emphasis on meat and fur. By the sixteenth century, several breeds of various sizes and colors were recorded (Lebas et al. 1997). In the following centuries, with the progress in selection practices and genetics, numerous rabbit breeds were created for meat, fur, wool, and even for exhibitions and as pets. Today, about 60–90 pure breeds are described, all originating from the European wild rabbit. They are referenced and described via official standards controlled in France by the Fédération Française de Cuniculiculture (http://www.ffc.asso.fr/), in Germany by the Zentralverband Deutscher Rasse-Kaninchenzüchter (http://www.kaninchenzucht.de/rassen/), in Great Britain by the British Rabbit Council (http://www.thebrc.org/breeds.htm), and in the USA by the American Rabbit Breeders Association (http://www.arba.net/index.htm), among others.

Lebas et al. (1997) compiled a publication for the Food and Agriculture Organization (FAO) entitled "The Rabbit – Husbandry, Health and Production." This publication is a reference work covering all the aspects in the field, which we recommend reading and from which we have taken most of the data presented here. These authors distinguish four types of rabbit breeds:

- Primitive, primary, or geographic breeds directly derived from the wild rabbit.
- Breeds obtained through artificial selection from the aforementioned, such as the Fauve de Bourgogne

(Burgundy Fawn) or the New Zealand White or Red.

- Synthetic breeds obtained by planned crosses of several breeds, such as the Blanc du Bouscat or the Californian.
- Mendelian breeds obtained by fixation of a new character of simple genetic determination, via mutation, such as the Castorrex, the Angora, the Satin, or the Japanese.

Rabbit breeds can also be classified according to adult weight:

- Heavy breeds characterized by an adult weight exceeding 5 kg, such as the Flemish Giant (7–9 kg) and the French Giant Papillon (6–7 kg).
- Average breeds, which are the most numerous, with an adult weight ranging from 3.5 to 4.5 kg. They have been used as the basic stock for intensive rabbit production for meat in western Europe, such as the Fauve de Bourgogne and the Large Chinchilla; and in the United States, such as the New Zealand White and the Californian.
- Lightweight breeds with an adult weight around 2.5–3 kg, such as the Small Himalayan, the Small Chinchilla, the Dutch, and the French Havana.
- Small or dwarf breeds, which do not exceed 1 kg and are essentially pet animals, such as the Polish rabbit.

In addition, private breeders and research laboratories select and maintain inbred rabbit strains to study their biological and breeding characteristics and for specific breeding programs. A strain is a closed group of animals that is maintained by inbreeding for several generations so that particular characteristics can be retained. For example, the National Institute of Agronomic Research (INRA) in Toulouse conducts selection programs for meat on strains such as INRA2066 and INRA1077 and for fur traits on strains such as Orylag® and Laghmere® (H de Rochambeau and D Allain, personal communication).

7.1.5
Domestication, Phylogeny, and Genetic Diversity

As mentioned earlier, the European rabbit (*Oryctolagus cuniculus*) emerged in the Iberian Peninsula in the middle Pleistocene period (Lopez-Martinez 2006). Fossil evidence also suggests that the history of the rabbit lineage leading to *Oryctolagus* is found within Iberia and Southern Europe. While extinct species in Spain, Southern France, and Italy include *O. laynensis* and *O. lacosti,* no transitional forms between those extinct species and modern rabbit populations have been found. However, the fossil record for rabbits is very incomplete and its interpretation problematic, making it impossible to establish the phylogenetic relationships of *Oryctolagus* within *Leporidae*. As an alternative to the paleontological approach, molecular phylogeny studies may provide useful insights into the history of taxa. Recently, Matthee et al. (2004) proposed a molecular supermatrix, based on five nuclear and two mitochondrial DNA (mtDNA) fragments, to phylogenetically resolve multiple genera included in *Leporidae*. According to their results, *Oryctolagus* could be related to *Caprolagus, Bunolagus,* and *Pentalagus,* but these relationships were equivocal. It is possible that these four genera diverged approximately ten million years ago and this fact, in combination with the rapid radiation of all *Leporidae*, makes it difficult to determine their phylogenetic relationships.

Genetic Structure of Wild Rabbit Populations

In spite of being recognized as an important model species for biomedical research (Weisbroth et al. 1974), knowledge about the genetic diversity and structure of wild and domestic rabbit populations has remained limited. The first relevant data came from the analysis of the so-called immunoglobulin allotypes (Mage et al. 1973). The study of the *IgKC1* and *IgVH1* loci revealed a remarkable contrast between the high levels of diversity found in rabbits from the Iberian Peninsula and the homogeneity of other European populations, including very similar domestic breeds (Cazenave et al. 1987; van der Loo 1987; van der Loo et al. 1987, 1991). However, a coherent picture of the evolutionary history of the rabbit started to emerge only a few years later, with the work of Biju-Duval et al. (1991) based on the analysis of mtDNA. These authors described two highly divergent mtDNA lineages within the Iberian Peninsula and also suggested that these two lineages shared an ancestral form of mtDNA approximately two million years ago. More recently, Branco et al. (2000, 2002) conducted comprehensive studies focused on the geographical distribution of the two mtDNA lineages and the mechanisms that determined these patterns. These studies have shown that one lineage was essentially circumscribed to the Iberian Southwest, while the other occurred mostly in the Northeast. Based on these results, the authors proposed that two rabbit populations evolved in allopatry in two refugia during the Pleistocene Ice Ages. After the climatic amelioration that followed the last glacial maximum, both populations experienced a significant expansion, leading to the establishment of a contact zone in the central Iberian Peninsula. Other detailed studies of French populations have indicated that only the mtDNA lineage occurring in the Iberian Northeast is present today and, in addition, revealed a very clear subset of the haplotype diversity previously described (Monnerot et al. 1994; Hardy et al. 1995).

An extensive analysis of protein polymorphism in Iberian wild rabbit populations has provided data to support mtDNA results (reviewed by Ferrand and Branco 2006). First, high levels of polymorphism have been demonstrated, with approximately one-third of the loci exhibiting some variation, a remarkable finding among mammals (Nei 1987). Second, two main population groups have been found, corresponding essentially to those defined by the divergent mtDNA clades. Third, French populations show a subset of the genetic diversity described in the Iberian Peninsula. Taken together, mtDNA and protein polymorphism data are concordant with a historical scenario of two differentiated populations that correspond to the subspecies *O. c. algirus* in the Southwest and *O. c. cuniculus* in the Northeast. These subspecies could result from independent evolution in the Iberian Peninsula since at least the middle Pleistocene. The recent use of highly variable microsatellites in the analysis of rabbit populations has greatly improved our understanding of the recent geographical expansion of the species (Queney et al. 2001). While essentially no differences have been observed between the two divergent subspecies, possibly due to extensive homoplasy in these markers, microsatellite allele distributions in France strongly suggest that populations arrived postglacially and followed two main routes to the central-northern parts of the country. Importantly, the fact that microsatellite loci also show a clear subset of the diversity described in the Iberian Peninsula is

a clear indication that rabbits arrived recently in the South of France. If this migration had occurred earlier, the high mutation rate of these markers would have reconstituted genetic diversity to previous levels.

More recent investigations have focused on the sex chromosomes. First, Geraldes et al. (2005) described evidence for the occurrence of two very divergent lineages based on the Y-chromosome of Iberian rabbits, which was geographically concordant with analyses of mtDNA data (Geraldes and Ferrand 2006). This means that the historical processes that sculpted genetic variation in both maternal and paternal lineages are probably very similar. Second, Geraldes et al. (2006) studied four independent X-linked markers across the contact zone of the two subspecies and showed that two divergent lineages are observed at each locus. In addition, using coalescent simulations, the authors suggested that the patterns described at the X-chromosome are the result of allopatric evolution of the two subspecies during the Pleistocene, followed by secondary contact and asymmetric introgression. Specifically, they found that centromeric regions of the X-chromosome are geographically highly structured, while telomeric regions are apparently completely admixed.

Taken together, available genomic data suggest that wild rabbit populations in Iberia and Southern France exhibit a remarkable contrast between most of their genomic DNA. Indeed, they exhibit a molecular signature of a deep split and yet they are completely admixed. Other populations that are quite rare bear the same molecular deep split, but they are almost parapatric. These notable aspects of the genetic structure of wild rabbit populations will be of great relevance in understanding the process of domestication, as explained in the next section.

Process of Domestication

Rabbits are particularly interesting because they represent the only mammalian species that has been domesticated in Western Europe. However, a comprehensive study addressing the process of rabbit domestication is still lacking. For example, it would be interesting to know if the two rabbit subspecies have contributed to rabbit domestication, and if genetic diversity has been significantly lost during the process. It would be equally interesting to know which genes have been involved in the notable changes that now separate domestic breeds from wild populations and how many are involved. In spite of being a recent event, as compared to other animals (e.g., cattle, sheep, and goats), conflicting interpretations about the history of rabbit domestication are found in the literature. For example, Clutton-Brock (1987, 1992) asserts that rabbits have been domesticated in the Iberian Peninsula during the Roman occupation, while Zeuner (1963) suggests that this process did not take place until at least the 15th century.

Initial mtDNA studies have revealed that domestic rabbits possess only one of the two divergent lineages found in wild populations (Biju-Duval et al. 1991). This is lineage B, typically present in the northeastern part of Iberia and the south of France. In addition, subsequent studies have shown that domestic breeds exhibit only a small subset of the intra-haplotypic diversity of lineage B, namely, B1 and B3–4 (Monnerot et al. 1994), and that these types occur in natural wild rabbit populations in southern France (Hardy et al. 1995). From a mitochondrial perspective, available data suggest that a single subspecies (*O. c. cuniculus*) might have contributed to the foundation of domestic stocks, and that a domestication process within Iberia, as previously proposed by Clutton-Brock (1987, 1992), is difficult to support. Very similar results have been obtained with the male Y-chromosome. In fact, only one of the two deeply differentiated lineages described by Geraldes et al. (2005) on the basis of limited sequencing of the *SRY* region has been detected in several domestic breeds. This lineage is the one that Geraldes and Ferrand (2006) have shown to be exclusively present in northeastern Spain and southern France. Although mtDNA and the Y-chromosome are independent, uniparental loci can frequently reveal the differential contribution of sexes to the evolutionary history of a species, in this case, they tell the same story.

Multiloci analyses using both allozymes and microsatellites are in clear agreement with the previous scenario. Ferrand and Branco (2006) compared the protein data published by Arana et al. (1989) and Peterka and Hartl (1992) to those obtained in a representative population of each subspecies. While the wild rabbit sample belonging to *O. c. algirus* has always been very distinct, Portuguese domestic rabbits are hardly separated from several other domestic breeds (e.g., Californian and New Zealand), but also from many wild rabbit populations from northern Spain, France, England, Austria, and Australia. Based on a more detailed analysis of 20 protein polymorphisms, Ferrand and Branco (2006) showed that the genetic diversity of the relatively unselected Portuguese domestic rabbits corresponds to a subset of that observed in French wild rabbits (Fig. 1a). In another study, Queney et al. (2002) used six microsatellite loci to characterize patterns of genetic variability in 25 rabbit breeds and compared their results with previously studied wild rabbits from the Iberian Peninsula and France (Queney et al. 2001). Notably, population relationships are essentially the same, with all domestic breeds clustering with wild rabbit populations from northern France (Fig. 1b). Together, these results represent very strong nuclear evidence supporting the derivation of all domestic breeds from the subspecies *O. c. cuniculus*. In addition, they are also in agreement with a scenario of a domestication process that may have occurred in southern France during the Middle Ages (Callou 2003).

The nuclear data described by Queney et al. (2002) and Ferrand and Branco (2006) have also permitted, for the first time, an estimation of how much diversity was lost during the bottleneck that led to rabbit domestication. Taking the unselected stock of Portuguese domestic rabbits as representative of the genetic diversity in present-day breeds and two French wild populations analyzed for the same set of 20 protein loci, heterozygosity estimates were 0.170 and 0.201, respectively (Ferrand and Branco 2006). These estimates translate into a value of 84.6% of genetic diversity retained in domestic rabbits. When the same procedure was applied to the microsatellite data reported in Queney et al. (2002), heterozygosity was estimated at 0.489 in a set of 25 domestic breeds, and 0.596 in a set of 13 French wild populations. These estimates suggest that 82% of the diversity as measured by microsatellite markers has been retained in domestic animals. The two independent estimates are remarkably similar and suggest that loss of genetic diversity during rabbit domestication was limited.

Similar studies in other domesticated species are still rare and prevent a more complete analysis of these findings. In fact, most of these studies have been based solely on the analysis of mtDNA (Bruford

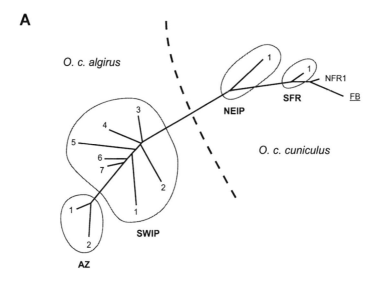
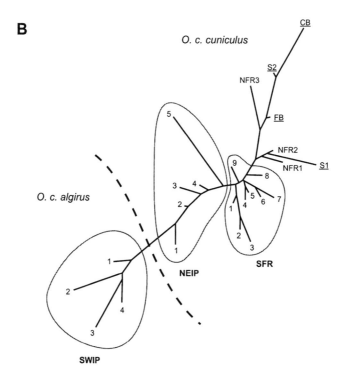

Fig. 1 Unrooted Neighbor-Joining networks relating wild and domestic rabbit populations from the Iberian Peninsula and France. (**a**) Network based on 20 protein loci and Nei's standard genetic distance (modified from Ferrand and Branco 2006). (**b**) Network based on six microsatellite loci and Nei's standard genetic distance (modified from Queney et al. 2002). Wild rabbit populations from southwestern and northeastern Iberian Peninsula (SWIP and NEIP, respectively), the Azores (AZ), and Southern France (SF) are highlighted. The dashed line represents the separation between subspecies *O. c. algirus* and *O. c. cuniculus*. Wild rabbit populations from northern France are indicated (NFR). Domestic breeds are underlined (FB, fancy breeds; S, strains; CB, commercial breeds)

et al. 2003) and, in addition, these studies include species for which wild ancestors are now extinct (e.g., cattle and horse). However, among plants, the domestication of maize has been addressed using multiple types of molecular markers (see for review Doebley 2004). In this case, analyses of genetic polymorphism in both teosinte and maize populations suggest that about 30% of the original diversity was lost during the domestication process. Thus, it seems that the first pictures to emerge from comprehensive genetic studies of rabbit domestication involving multiple molecular markers and both wild and domesticated populations indicate the conservation of a considerable fraction of the original diversity. Recently, Vilà et al. (2005) analyzed patterns of molecular diversity in *MHC* loci and have suggested that the genetic variability of many mammalian domesticated species could have been increased through the occurrence of backcrosses with their wild ancestors. The low estimates of genetic diversity loss during rabbit domestication (15–18%) can also be explained by a similar process and are compatible with a scenario of strong selective pressures over a limited number of genes responsible for morphological, physiological, and behavioral modifications and the maintenance of relatively high levels of gene flow with wild rabbit populations.

7.1.6
Economic Importance

High prolificacy and high efficiency of fodder to meat conversion make the rabbit one of the major economically interesting species. Indeed, meat rabbit breeds have average litter sizes exceeding nine young, a 31–32 day gestation period, and sexual maturity at four months. Thus, one female rabbit can have up to 50 young per year. The metabolic system of rabbit can convert up to 20% of the proteins contained in the fodder into high-quality meat proteins, whereas this value is only 16–18% in pig and 8–12% in cattle. In addition, rabbits are ready for slaughter at about 10–12 weeks of age. Thus, rabbit meat production can be a profitable venture, especially in countries without cereal surpluses. Other rabbit breeds are reared for fur or wool or as companion pets or as laboratory animals where they are extensively used in medical research and toxicology studies.

Rabbit Meat Production

In 2004, the world production of rabbit meat was $1,121 \times 10^3$ tons carcass equivalent (tce) per year. The largest producers representing about 78% of the world production are China (460×10^3 tce), Italy (222×10^3 tce), Spain (111×10^3 tce), and France (85×10^3 tce). The rabbit meat industry predominately uses the Californian and New Zealand White breeds because of their prolificacy. Rabbits produce white, delicate, and tender meat that has good nutritional value because it contains more protein, less cholesterol, less fat, and fewer calories than pork, beef, lamb, or chicken. Traditionally, the countries where rabbit meat is most consumed are France, Italy, and Malta in Europe and the central-western province of Sichuan in China.

Rabbit Fur and Wool Production

Commercial use of rabbits also involves the production of pelt and shorn hair or wool. The world production of rabbit pelts is estimated at 10^3 millions, which far exceeds that of any other fur species. Today, in the rabbit meat industry, pelts are only minor byproducts because intensive production techniques require slaughtering the animals at 10–12 weeks when their coat is not suitable for the fur trade. Rabbit pelts, which are recovered from the meat industry, are sorted into two classes, i.e., good quality for ordinary fur garment industry and poor quality for manufacturing textile, felt, glue, or fertilizers.

In addition to this secondary source of pelts, some rabbit breeds are raised specifically for the production of high-quality fur and wool, usually by small-scale producers or intensive factory farms. About 40 rabbit pelts are needed to make a coat. Although no official statistics are available, the rabbit fur industry (coats, trims, accessories) is thought to represent several million dollars. The most popular breeds for fur production are the Rex and the American Chinchilla. The Rex rabbit is a medium-sized animal weighing 3–3.5 kg with characteristic silky, dense, and homogeneous fur. This trait is due to a recessive mutation causing the degeneration of the central and lateral primary hair follicles and the reduced size of the guard hairs. This degeneration is more or less complete among strains, which explains the variability of the observed phenotypes. Nevertheless, in the Rex pelt, guard hairs and other hair types are approximately the same length, which provide high-quality fur for the fashion industry. Rex rabbits are raised in intensive production systems and are slaughtered at the age of five to six months during the winter when they are fully developed and their coat is sufficiently mature. Special mention is made of the unique Orylag® strain, selected by the INRA during the 1970s and 1980s, which provides one of the most precious furs. The Orylag® strain is described in more detail in the next section.

Rabbit angora fiber production is the third largest animal fiber industry in the world (7,000–8,000 tons per year) after sheep wool and goat mohair, which are 1.5 million and 12,000 tons per year, respectively. China, with 10 million rabbits bred per year, dominates the world angora fiber trade at approximately 98% of the total production. China (and on a much smaller scale, Chile and Argentina) has taken over the position in the world market traditionally held by France as the main source of angora fiber. Angora fiber is mainly produced from the French and German Angora breeds. The origin of the Angora breed is not well known, but has been documented since Roman times. Angora rabbits carry a recessive mutation responsible for unusually long and very fine hair mainly used to manufacture high-quality knitwear. The animals are not slaughtered to recover the wool, but their hair is either pulled at a precise stage of their development for the French fiber or sheared every three months with hand scissors or electric shears. A rabbit can yield on average 250–500 g at each shearing, i.e., up to 1,500–2,000 g per year, over eight to ten years. Furthermore, the Laghmere® strain produces albino animals carrying both the absence of guard hairs and long fine hair; these animals have very long and soft fur suitable for dyeing (D Allain and H de Rochambeau, personal communication).

Orylag® Luxury Fur Production

In the late 1970s, J-L Vrillon became interested in the selection of an improved rabbit strain which would carry the perfect fur that is ideal for the fur industry and ethically acceptable. He started with the Rex rabbit breed, itself originating from a spontaneous mutation in a litter of wild gray rabbits in France in the 1920s and which is characterized by its thick and velvet-like coat as described earlier. However, the fur of the Rex rabbit contains guard hairs thicker than the down hairs, so the aim of J-L Vrillon was to develop a strain, which produced no guard hairs and thus an

even softer fur. After 15 years of inbreeding selection, the "perfect" rabbit strain was produced, combining a precious fur (commercialized under the brand name "Orylag®") and high-quality meat (commercialized under the brand name "Rex du Poitou®") (Thebault et al. 2000).

The first step in producing this strain required the improvement of zootechnical standards of the Rex breed, which were very poor. Rex rabbits were first crossed with the INRA1077 strain, which is highly prolific and has excellent quality meat. Then, through a process of crossing and selecting, the quality of the fur was improved using three criteria:

- Absence of guard hairs. (The Orylag® strain does not possess any guard hair but only very fine down hairs about $15\,\mu$ in diameter)
- Number of hairs per centimeter $(cm)^2$ in order to obtain a very dense and warm fur
- Length of hairs. (A length of 20 millimeter (mm) is optimal for a light and high-quality fur)

However, because guard hairs and number of hairs per cm^2 are positively correlated, simultaneous selection of the absence of guard hairs and higher hair density was a difficult and long process. Finally, a density of 8,000–10,000 hairs per cm^2 was reached. The resulting Orylag® - Rex du Poitou® strain currently represents the top of the rabbit trade and only the Cooperative of Orylag Breeders (Coopérative des éleveurs d'Orylag®), totaling about 20 breeders in the Poitou-Charentes region of France, can produce and commercialize these rabbits under license. In 2005, 100,000 Orylag® pelts were sold, a small number compared to the 35 million mink pelts sold annually, but production is increasing continuously. Indeed, more designers and furriers are choosing Orylag® as a luxury fur, paying 15–30 € per pelt, comparable to that of mink. However, an understanding of the genetics behind the characteristics of the Orylag® strain is lacking. The need for further research will be discussed in Sect 7.2.7.

Production of Rabbits for Laboratory Stock and Pets

A small but not negligible contribution to the rabbit breeding industry is the supply of animals for research and as pets. As detailed below, rabbits are often used as biomedical laboratory animals because they are easy to handle and breed within laboratory facilities and their physiology is fairly close to humans. However, because of the growing emphasis on animal welfare and the development and validation of nonanimal testing methods, the number of rabbits used in research is gradually decreasing. In the early 1990s, there were 600,000 laboratory rabbits in the USA and 113,770 in France, while by the early 2000s, these numbers had fallen to 244,000 laboratory animals in the USA and 54,000 in France. Researchers strive to reduce the number of animals needed for experiments, to replace them with nonanimal models when possible, and to avoid any unnecessary pain or stress, laboratory animals still remain irreplaceable in many research domains.

High-quality breeding conditions are essential for the production of laboratory rabbits. Depending on the demands, they can be raised under standard conditions, but most often laboratories choose specific pathogen-free or barrier-specific rabbits. In addition, research laboratories may have specific requirements for breed, sex, size, and age, although the albino New Zealand White and the Dutch breeds are most commonly used.

Rabbits are also raised as pets and for exhibitions. They make excellent pets and offer a range of phenotypes such as color, size, fur length, and ear carriage and length. The most popular pet breeds are the smaller ones, such as the Dutch, Holland Lop, Mini Lop, Mini Rex, Netherland Dwarf, and Polish breeds. The pet care industry is a major segment of the world economy (over $38 million in the USA alone), but it is mainly concentrated on cats and dogs. Although statistics are not available for rabbits, in 2004–2005 there were 18 million small pet mammals including rabbits but excluding cats and dogs in the USA and 2.3 million in France (first and second worldwide, respectively).

Rabbits as Laboratory Animals

Laboratory animals have played an essential role in many major biomedical advances, including health and fundamental research, for more than a hundred years. They are used to develop new or improved techniques for diagnosis and to elaborate protocols to treat diseases affecting humans. They are also used to test the toxicity of new products and to learn more about biological mechanisms that cannot be studied in nonanimal models, i.e., computer models, tissue, and cell cultures.

An animal is a valid biomedical model if it satisfies part or all the following criteria:

- Small size
- Short gestation period
- High prolificacy
- High quality and controlled breeding practices

Rabbit possesses all these characteristics and thus provides a very valuable animal model and research tool. In addition:

- The physiology of cells, sera, tissues, and organs of rabbits has been well studied (Manning et al. 1994) and, in many aspects, is more similar to humans than mice and rats.
- Its size is sufficient to collect adequate quantities of tissue (for example, blood and milk) for experiments throughout its life span and also to conduct surgical operations, which would be difficult in mice or rats.
- Its immune response to antigens is excellent.
- Its small size makes it more economical and easier to raise as compared to larger animals, such as pigs, sheep, and primates.
- Its short generation time and large litter size reduces the length, and therefore the cost of research programs involving animals.
- The environment (diet, temperature, lighting) in which the rabbits are bred can be easily controlled.
- Many inbred rabbit strains and mutants with natural genetic deficiencies are available, and it is possible to develop genetically modified rabbits for clinical studies.

The rabbit as laboratory animal has played a significant role in many areas of biomedical research including parasitology, physiology, embryology, reproduction, growth and development, metabolism, microbiology, immunity, hematopoiesis, metabolism, cancer, and cardiovascular diseases (Fox 1984). A few examples are presented below to illustrate the wide range of possible applications for the use of rabbits as laboratory animals.

Rabbits as a Tool for Toxicity Testing

Since the late 1930s, rabbits have been extensively used to test the safety of new medical treatments and cosmetic products, because their small size requires only minor amounts of the product under test while being sufficiently large to visualize the effects on various organs, for example the fetus. Unlike mice, rabbits are hypersensitive to teratogenic agents and their cellular responses are much more similar to humans. Rabbit skin and eye irritation tests provide reliable means of evaluating the toxicity of chemical agents alone or applied together with ultraviolet light, in order to observe possible interactions and to develop appropriate procedures for handling these agents.

Rabbits for the Production of Proteins

For over 40 years, rabbits have been well known for their ability to produce large amounts of polyclonal antibodies to specific antigens. Adequate amounts of blood can be collected from the ear marginal vein for small-scale analyses or by heart puncture for large-scale purification. Efficient protocols for the purification of rabbit immunoglobulins are available (Stills 1994).

Many systems permit the production of recombinant proteins including bacteria, yeast, insect, plant, and mammalian cells. However, most of them cannot carry out the post-translational modifications of proteins necessary for normal biological activity such as glycosylations, formation of disulfide bonds for protein folding and site-specific cleavages. Mammalian cells can fulfill these activities, but they are delicate and expensive to maintain and have low yields. Transgenic animals offer a good alternative and in particular, rabbits with a size intermediate to that of mice and pigs or cattle can produce recombinant proteins in adequate quantities. Rabbit recombinant proteins are usually recovered from the milk because a lactating doe can produce up to 250 milliliter (ml) per day (Houdebine 2002) and, depending on the expression level of the transgene, the concentration of the desired protein contained in the milk can range from 1 to 10 gram (g) per liter. The French company BioProtein Technologies (website at http://www.bioprotein.com) has specialized in the production of human therapeutic proteins and vaccines in the milk of transgenic rabbits (monoclonal antibodies, plasma proteins, hormones, peptides, and multivalent VLP-based vaccines). Human recombinant therapeutic proteins for medical treatments, such as alpha-antitrypsin, erythroprotein, and human growth factor (Fan and Watanabe, 2003) as well as acid alpha-glucosidase (van den Hout et al. 2000 and 2001), have also been produced by transgenic rabbits.

Cloning Genetically Identical Rabbits

In the absence of rabbit embryonic stem cells, cloning by somatic nuclear transfer from cumulus cells (Chesné et al. 2002) and from fibroblasts (Li et al. 2006) provides genetically identical rabbits in large numbers for experimental purposes. For example, rabbits that are knock-outs for specific genes can be used to investigate the mechanisms of diseases, and cloned transgenic rabbits can be used for large-scale production of recombinant proteins.

Rabbits as Models for Medical Research

Rabbits have been very important in the study of human diseases as listed in the reviews by Fox (1984) and Fan and Watanabe (2003), including:

- Cardiovascular diseases such as hypertension and atherosclerosis, which are the major causes of mortality in Western countries. As early as 1908, Ignatowski mentions the use of rabbits in the study of atherosclerosis. Rabbits are valuable models to investigate atherosclerosis because they quickly develop hypercholesterolemia and atheromatic lesions when fed a high cholesterol diet. In addition, the generation of transgenic and knock-out rabbits with gene modifications known to be associated with hyperlipidemia and atherosclerosis offers approaches to understand their pathogenesis.
- Infectious diseases caused by bacterial (*Treponema pallidum, Mycobacterium tuberculosis,* enterohemorrhagic *E. coli*), fungal (*Candida, Aspergillus* sp., *Coccidioides immitis*), and viral (papillomavirus, herpes simplex virus, human immunodeficiency virus) pathogens. For instance, when domestic rabbits are experimentally infected with cottontail rabbit papillomavirus, they develop benign skin lesions (warts), which become malignant in 25% of cases (Kreider and Bartlett 1981; Breitburd et al. 1997). These skin tumors, which resemble those observed in humans, can either regress spontaneously or deteriorate into carcinoma and form metastases in other organs. Thus, the rabbit is the most suitable model to investigate the mechanisms of progression or regression of this tumor and to test vaccines against papillomavirus infection.
- Cancers, such as renal tumors caused by ethylnitrosourea in rabbits when given transplacentally, which are very similar to human Wilm's tumor

(Sharpe and Franco 1995). In addition, transgenic rabbits expressing specific oncogenes have been developed to serve as models for oncology studies and for comparisons of the therapeutic efficacy of various antitumor treatments.

- Respiratory diseases such as asthma.
- Eye conditions, in particular entropion and glaucoma.

Rabbits as Models for Developmental Biology

Rabbits are valid experimental models for studies of gene function and regulation during embryo implantation and development, which cannot be undertaken in humans for ethical reasons. The advantages of the rabbit compared to other animal models are that:

- Rabbit and human placentas are similar. In particular, their trophoblast is in direct contact with maternal blood unlike other animal species such as sheep, pig, dog, and cat.
- The kinetics of transcriptional activation of the rabbit embryonic genome is closer to that of humans than mice (Christians et al. 1994) and is not associated with epigenetic modifications as prevalent as in mice (Shi et al. 2004).
- In rabbits, extraembryonic tissues undergo massive growth at the blastocyst stage, which rapidly increases surface contact between the embryo and its environment. Thus, rabbits can be used to study metabolic disorders due to effects of the environment on embryo development.
- During gastrulation, the rabbit has a flat embryonic disc, which serves as an early morphological landmark to investigate the presumptive area of the formation of the primitive streak and mesoderm (Viebahn et al. 2002).

7.2
Molecular Genetics

7.2.1
Cytogenetics

The Rabbit Karyotype

Although the rabbit is a domesticated species and can be used in a wide range of domains, there are relatively few studies concerning its genome and the genetics of its phenotypes and diseases. As early as 1926, TS Painter established that the rabbit karyotype contains $2n = 44$ chromosomes. Later, chromosomal analyses using cultured cells (Nichols et al. 1965; Ray and Williams 1966; Issa et al. 1968) confirmed the pioneering observations made by Painter. However, it was only with the advent of chromosome banding in the late 1980s that precise identification of all 22 rabbit chromosome pairs and their classification was achieved. The banding patterns of rabbit chromosomes have been described in detail by several groups: G-banded metaphase chromosomes (Echard 1973; Chan et al. 1977; Ford et al. 1980; Committee for Standardized Karyotype of *Oryctolagus cuniculus* 1981), G-, Q-, and R-banded metaphase chromosomes (Hageltorn and Gustavsson 1979), G-banded late prophase chromosomes (Yerle et al. 1987), R-banded prometaphase chromosomes (Poulsen et al. 1988; Rønne et al. 1993, Rønne 1995), and G- and R-banded midmetaphase chromosomes (Hayes et al. 2002). All these studies show that the rabbit autosomes fall into four groups according to the position of the centromere: twelve metacentric, five submetacentric, six submetacentric with a shorter centromeric index and four acrocentric autosomes, plus a fifth group corresponding to the sex chromosomes. The X-chromosome is one of the larger submetacentric chromosomes, while the Y-chromosome is also submetacentric but the smallest of the karyotype. Classification of rabbit chromosomes into five groups has been confirmed by measurements of the relative DNA content, the centromeric index and the relative DNA content of the short and long arms of each rabbit metaphase chromosome carried out by cytophotometry on chromic gallocyanin stained chromosomes after prior Q-banding (Ashworth et al. 1979). Finally, it has been shown that chromosomes 13, 16, and 20 (telomeric p arm) and 21 (telomeric q arm) bear the nucleolar organizer regions (Martin-DeLeon 1980; Zijlstra et al. 2002).

The first international chromosome nomenclature system for the rabbit karyotype providing chromosome and band numbering was published in 1981 (Committee for Standardized Karyotype of *Oryctolagus cuniculus* 1981) and was based on G-band patterns. In spite of several publications on R-banded chromosomes, no definite correlation between G- and R-banded rabbit chromosomes was available until 2002. With the development of fluorescent in situ hybridization (FISH) to determine the cytogenetic position of genes on R-banded metaphase spreads it became necessary to establish an R-banded rabbit karyotype nomenclature in agreement with the 1981 G-banded standard nomenclature. In 2002, a collaborative French/Dutch team established unambiguous correlations between G- and R-banded rabbit chromosomes by combining classical karyotyping and FISH localization of chromosome specific markers on both G- and R-banded chromosomes (Hayes et al. 2002). This work led to the preparation of R-banded rabbit idiograms shown in Fig. 2.

To date, no spontaneous chromosome abnormalities (structural or numerical) have been reported in rabbits, which may be explained by the absence of systematic cytogenetic surveys in rabbit populations.

Comparative Cytogenetics among Leporidae Species

Within the Leporidae family, which includes 56–60 recognized species depending on classifications, the chromosome numbers vary from $2n = 38$ to 52. Several studies have compared the rabbit karyotype with the other lagomorph species (Gustavsson 1972; Stock 1976; Schroder and van der Loo 1979; Robinson 1980; Robinson et al. 1983). Using available banded karyotypes for seven of the eleven Leporidae genera, Robinson et al. (1981) reconstituted a presumed ancestral karyotype with $2n = 48$ chromosomes conserved with those of the brown hare (*Lepus europaeus*). Rabbit karyotype ($2n = 44$) differs from that of brown hare ($2n = 48$) by the presence of two centric fusions. This study and others (Robinson and Matthee 2005) show that within Leporidae, karyotypes have either retained the presumed *lepus*-like ancestral state or have undergone various chromosomal rearrangements leading to chromosome patterns unique to each lineage. For example, the refined analysis carried out by Robinson et al. (2002) using whole chromosome paints of the rabbit 21 autosomes and X-chromosome as probes hybridized to chromosome spreads of six Leporidae species, shows that at least 18 fusions and six fissions are necessary to reconstitute the presumed *lepus*-like ancestral karyotype from that of the six analyzed species. The riverine rabbit (*Bunolagus monticularis*, $2n = 44$) has the most derived karyotype, differing from the presumed ancestral state by seven fusions and five fissions, followed by Smith's red rockhare (*Pronolagus rupestris*, $2n = 42$) with four fusions and one fission.

Rabbit and Human Karyotypes Are Very Similar

Dutrillaux et al. (1980) compared the banded karyotype of rabbit with that of several primates and humans. Surprisingly, this analysis revealed that the banding patterns are very similar among the karyotypes of these species and that only around 50 chromosome rearrangements differentiate human and rabbit karyotypes. These observations suggest a high level of chromosome conservation since the divergence of the ancestral forms of humans and rabbits, which occurred about 65 million years ago. Korstanje et al. (1999) confirmed this extensive conservation between the human and rabbit genomes by heterologous chromosome painting analyses, which are described in Sect. 7.2.7.

7.2.2
Genetic Molecular Markers

A report of the European program referred to as RESGEN (RESGEN CT95-060, regulation 1467/94, 1996–2000) was prepared by Bolet et al. (1999) and included an evaluation of rabbit genetic resources that were available for biodiversity and zootechnical studies. The following year, a very complete inventory reviewing available genetic markers for rabbits was published (Bolet et al. 2000). This large-scale study included restriction fragment-length polymorphism (RFLP) and sequence analysis of mtDNA, genotypes for 28 microsatellites, RFLP-based haplotyping of the major histocompatibility complex (MHC), and protein polymorphism analyses of milk caseins, red blood cell proteins, and immunoglobulins. A 2006 update for molecular genetic markers available for rabbits is presented later, including DNA markers targeting specific chromosomal loci and whole-genome maps developed with amplified fragment-length polymorphism (AFLP) and microsatellite markers.

Mitochondrial DNA Markers

mtDNA is maternally inherited and thus suitable for animal assignment to maternal lineages. Rabbit mtDNA has been sequenced and is 17,145 bp in length (Gissi et al. 1998). Molecular markers have been designed from mtDNA to evaluate genetic diversity in wild and domesticated rabbit breeds both in Europe (Bolet et al. 2000; Queney et al. 2002) and in China (Long et al. 2003), to study rabbit

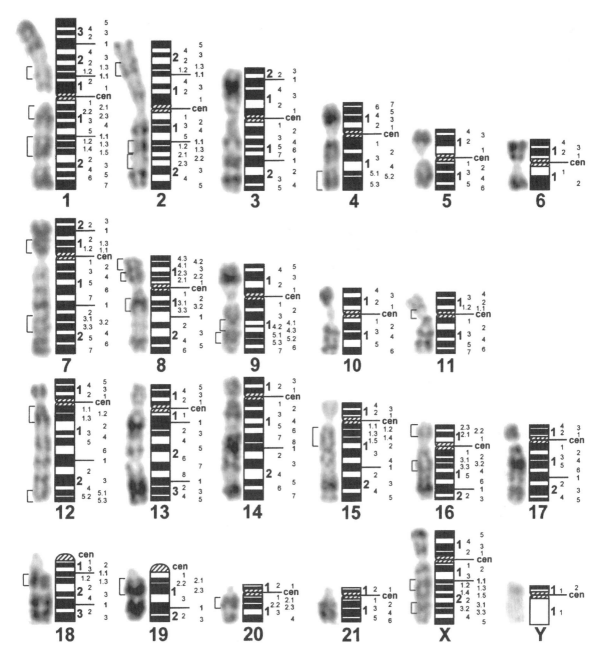

Fig. 2 Rabbit R-banded chromosomes with corresponding standard idiograms (reproduced from Hayes et al. 2002). Brackets indicate bands, which are subdivided into at least three sub-bands as compared to the 1981 standardized international nomenclature system for the rabbit karyotype. The Y-chromosome is taken from a separate male metaphase spread

phylogeography (Biju-Duval et al. 1991; Branco et al. 2002), and to compare present-day animals and prehistoric animals by analyzing mtDNA extracted from bones found in Mesolithic and recent time archeological sites (Hardy et al. 1995). Two mtDNA sequences, referred to as B1 and B3–4 with 2% nucleotide divergence in their noncoding domains, can be simultaneously typed using two PCR-RFLP markers (Queney et al. 2002). One marker corresponds to a 565 bp fragment of the cytochrome b gene digested with *Alu*I, and the other marker corresponds to a 586 bp fragment of the noncoding region further digested by *Rsa*I and then sequenced, in order to evaluate the variability within each RFLP type (Bolet et al. 2000). Other strategies include direct sequencing of mtDNA fragments, such as a variable 233 bp fragment of the cytochrome b gene (Hardy et al. 1995), a 179–181 bp fragment of the control region (Branco et al. 2002), and a 700 bp fragment of the control region near the *tRNA-Pro* gene (Long et al. 2003).

RFLP and PCR-RFLP Markers

Major Histocompatibility Complex (MHC) Haplotyping

The rabbit is a valuable model for transplantation studies and, in that respect, a prospective diagnosis of allogenicity between donor and recipient is a prerequisite before a surgical operation. Due to the very small number of available allele-specific antirabbit MHC monoclonal antibodies (Boyer et al. 1995), no large-scale design of serological testing has been developed in rabbit, and thus the typing methods found in the literature are based on molecular marker polymorphism. RFLP markers have been described as relevant molecular tools to identify haplotypes in the class I and class II regions of the rabbit leukocyte antigen (RLA) complex in inbred rabbit lines (Marche et al. 1989) as well as outbred lines (Boyer et al. 1995). The probes have been designed from cloned RLA genes derived from class I and class II genes. Marche et al. (1989) set up a haplotyping method using two noncross-reacting class I probes referred to as 19-1-3' spanning exons 6–8 of a putative classical class I gene (*R9* gene) and a R27 probe corresponding to the 3' region of a single copy class I gene (*pR27* gene) together with three class II probes specific for DP-alpha, DQ-alpha, and DR-alpha genes. Prior to Southern blotting and hybridization, genomic DNA was digested with *Bam*HI for the 19-1-3' and DP-alpha probes, *Bgl*II for the R27 probe, *Pvu*II and *Eco*RI for the DQ-alpha probe, and *Eco*RI for DR-alpha probe. Haplotypes are numbered by class I and class II designations and the composites are given a numerical identification. A total of 13 distinct RLA haplotypes have been reported using this method (Marche et al. 1989). Boyer et al. (1995) refined this method for RFLP-based haplotyping by using similar class I probes and a DQ-alpha probe for the class II region. Genomic DNA was digested with *Bam*HI for the 19-1-3' probe, *Bgl*II for the pR27-3' probe, and *Pvu*II for the DQ-alpha probe. Using this method, 34 RFLP-based class I haplotypes and 16 RFLP-based class II haplotypes have been defined in two commercially available populations of New Zealand White rabbits. These methods have been extensively exploited for RLA haplotyping (F Drouet-Viard, personal communication) and in the European RESGEN research program on biodiversity (Bolet et al. 2000).

Immunoglobulin Genetic Diversity

Allelic diversity of immunoglobulins has been routinely inferred by serological tests (reviewed by Mage 1986), which very early on defined the two independently segregating *a* and *b* loci (Oudin 1956). These were later identified as the variable region of heavy chain (IGHV1) and the constant region of light kappa genes, respectively. The rabbit genome carries only one immunoglobulin gamma class. Serological tests have defined two other genetic markers for the heavy immunoglobulin chains, specifically *d* and *e* (Esteves et al. 2006), which map to the unique constant gene for the immunoglobulin gamma isotype (IGHG gene). The *d* locus corresponds to the hinge H and the *e* locus to the CH2 exon encoding a C-domain (Esteves et al. 2006). DNA typing methods based on sequencing and PCR-RFLP techniques have recently been set up to upgrade the serological methods for allele identification of IGH genes, thus providing additional typing tools in the absence of antisera (Esteves et al. 2002, 2004, 2006). At the *e* locus, the *e14* and *e15* allotypes have been defined, which correlate with an alanine-threonine exchange at amino acid position 309 (Appella et al. 1971; Lefranc et al. 2005). The PCR-RFLP typing designed for this locus consists in the amplification of a 553–557 bp fragment spanning the CH2 exon followed by a *Tha*I restriction enzyme digestion (Esteves et al. 2002). This typing distinguishes between the two allotypes and permits detection of heterozygotes. At the *d* locus, the *d11*, *d12*, and *d11a* allotypes have been identified (reviewed by Mage 1986; Lefranc et al. 2005). The *d11* and *d12* allotypes are associated with a methionine-threonine interchange at amino acid position 9 of the hinge exon. The DNA typing procedure for this locus consists in the PCR amplification of a 251 bp fragment spanning the entire hinge exon, flanking intron regions and part of the IGHG CH2 exon (Esteves et al. 2006), followed by a digestion with *Nla*III. Only *d11* and *d12* alleles have been found in tested animals, and this DNA-based method is suitable for the detection of homozygous and heterozygous loci (Esteves et al. 2006). Finally, DNA typing methods for the detection of allelic variation at the *a* locus corresponding to the variable region of the heavy chain (IgV$_H$*a*) have been based on RT-PCR and sequencing (Esteves et al. 2004). This typing method has allowed the identification of the *a4* lineage adding a new major allotypic IgVHa lineage to the already described *a1*, *a2* and *a3* lineages.

AFLP Markers

A first attempt to develop large-scale molecular genetic markers was based on amplified fragment-length polymorphism (AFLP) techniques (van Haeringen et al. 2001, 2002). The major advantage of this method is the possibility to develop hundreds of markers without any prior knowledge of the genome (Vos et al. 1995). In two reports, parents and progenies of inbred rabbit families including either dietary cholesterol susceptible (AX/JU) and dietary cholesterol-resistant (IIIVO/JU) rabbit strains were genotyped (van Haeringen et al. 2001, 2002). Initially, the authors used *Sse*I and *Mse*I with different primer combinations, resulting in high-density fingerprints containing more than 100 amplified fragments in the range of 50–700 bp. Two primer combinations referred to as S12-M61 (*Sse*I + AC/*Mse*I + CTG) and S13-M51 (*Sse*I + AG/*Mse*I + CCA) produced 21 and 23 clearly polymorphic AFLP markers based on the presence/absence of a specific band, respectively. In this report, the fingerprints were detected by radioactive labeling followed by electrophoretic separation, and heterozygosity of the markers was identified using the phosphor imaging software suitable for quantification of band intensities. Rabbits of several breeds were included in the study and the results confirmed the efficiency of the AFLP technique for genetic studies. Moreover, a Mendelian inheritance study of 12 AFLP markers and three markers mapping to the linkage group LG VI, including a coat color marker and biochemical markers Es-1 and Est-2, was conducted using rabbit families selected for dietary cholesterol susceptibility/resistance. In that study, AFLP markers D0Utr1 and D0Utr2 were assigned to LG VI (Fox 1994). In a second report, van Haeringen et al. (2002) refined the AFLP technique based on specificities of mammalian genomes not found in plant or bird genomes (Vos and Kuiper 1997) and used a combination of *Eco*RI and *Taq*I restriction enzymes, resulting in a significant increase in the level of polymorphism. Fluores-

cent methods for the detection of AFLP markers have improved the technology throughput, because fluorescence is more efficient than radioactivity to interpret data at heterozygous loci and second, because three primer combinations can be analyzed simultaneously using several fluorochromes. AFLP patterns ranging from 35 to 500 bp were scored; 226 of 1,039 total AFLP bands were polymorphic and 184 markers were tested for genetic linkage analyses. A male genetic map consisting of 103 AFLP markers distributed across 12 linkage groups was constructed (van Haeringen et al. 2002). The linkage group referred to as Chrom_1 was genetically linked to the *C* locus included in LG I (Fox 1994) that had been previously mapped to OCU1 (Korstanje et al. 2001b). The other 11 linkage groups were referred to as LG 1 to LG 11, with no chromosome anchoring (van Haeringen et al. 2002). This male map was estimated to cover 583 cM, which corresponds to an approximate 33% coverage of the rabbit genome (van Haeringen et al. 2002).

Microsatellites

Microsatellites are probably the most popular molecular genetic markers since their discovery in 1981 (Miesfeld et al. 1981; Spritz 1981). They have ubiquitous occurrences, are known to be quite polymorphic with a high number of alleles, and are interpreted as codominant markers. The PCR-based methods for microsatellite typing (Litt and Luty 1989; Tautz 1989; Weber and May 1989) have contributed to large-scale development of microsatellites. However, in contrast to AFLP markers, identification of microsatellites is a prerequisite, and the efficiency of a microsatellite-based genetic map for quantitative trait loci (QTL) mapping is based on the number and distribution of the microsatellites along the chromosomes. In that respect, the use of rabbit microsatellites for genetic mapping has been hampered by the lack of well-characterized markers. Although more than 200 microsatellites of various origins had been reported by 2003, most of them had not been anchored into existing linkage groups or cytogenetic maps. The majority of these microsatellites were recovered from the EMBL nucleotide database and included a first subset of 181 markers (van Lith and van Zutphen 1996), a second subset of three markers (Mougel et al. 1997), and a third subset of 23 markers (Korstanje et al. 2001a). An additional group of 15 markers was isolated from size-selected genomic libraries, including nine markers

(Rico et al. 1994; Surridge et al. 1997) and six markers (Mougel et al. 1997). Finally, 25 microsatellites were recovered from an OCU1-enriched library (Korstanje et al. 2001) and 50 microsatellites from OCU3- (2 microsatellites), OCU5- (9 microsatellites), OCU6- (12 microsatellites), OCU7- (11 microsatellites), OCU 12- (11 microsatellites), and OCU19- (5 microsatellites) enriched libraries (Korstanje et al. 2003). These microsatellites have been used for biodiversity studies (Mougel et al. 1997; Surridge et al. 1997; Bolet et al. 2000; Queney et al. 2001), parentage studies (Vicente et al. 2004), and genetic mapping by enrichment of linkage groups as well as anchoring and orientation onto rabbit chromosomes (Korstanje et al. 2001a, b, 2003). In spite of many efforts to produce microsatellites suitable for genetic mapping, a whole-genome microsatellite-based genetic map was still missing for the rabbit in the early 2000s. Thus, a program was launched to produce an integrated genetic and cytogenetic map for this species using microsatellites with a known cytogenetic position (Chantry-Darmon et al. 2005a, 2006). A total of 305 new microsatellites were reported, which included a subset of 164 microsatellites isolated from gene-containing bacterial artificial chromosome (BAC) clones previously mapped by FISH onto rabbit chromosomes (Chantry-Darmon et al. 2003, 2005b) and a subset of 141 microsatellites isolated from a size-selected rabbit genomic library (Chantry-Darmon et al. 2005a). Among the microsatellites isolated from the genomic library, 41 were further selected for BAC recovery and FISH mapping. The resulting cytogenetically anchored microsatellite map (Chantry-Darmon et al. 2005a) comprises 177 markers distributed on all the chromosomes except OCU21.

A higher proportion of $(TC)_n$ repeats compared to $(TG)_n$ repeats has been found for microsatellites recovered from gene-containing BAC clones (62%) (Chantry-Darmon et al. 2005a) or from the EMBL nucleotide database (50–72%) (van Lith and van Zutphen 1996; Mougel et al. 1997; Korstanje et al. 2001a). In contrast, a proportion of 78% of $(TG)_n$ repeat microsatellites were isolated from a randomly constructed genomic library (Chantry-Darmon et al. 2005a), in agreement with reports for most mammals (Lagercrantz et al. 1993), which have been further confirmed by the recent large-scale sequence analysis of the human (Lander et al. 2001), the mouse (Waterston et al. 2002), and the rat (Gibbs et al. 2004) genomes. Although these observations in rabbit need

to be confirmed, they do suggest a nonhomogeneous distribution of $(TC)_n$ and $(TG)_n$ repeats in the genome, as has been reported in *C. elegans* where $(TC)_n$ repeats are preferentially found in intergenic regions and $(AT)_n$ repeats within gene introns (review in Toth et al. 2000). In mammals, $(TG)_n$ repeats are often found associated with short interspersed elements (SINE) (review in Vaiman 2005), suggesting that a SINE-mediated microsatellite spreading within genomes may be a prevailing process. The most frequent known rabbit SINE elements are referred to as C repeats and comprise a tRNA-like sequence, a conserved central core, an A-rich tract, and a stretch of CT dinucleotides (Krane et al. 1991). These C repeats represent approximately 13% of the rabbit genome (Krane et al. 1991) and could play a major role in increasing the proportions of $(TC)_n$ repeats in this species.

The variability of rabbit microsatellites appears to be limited (Table 1), which could be related to extensive homoplasy (Bolet et al. 2000; Queney et al. 2001) because there is a contradiction between the high genetic diversity reported for domestic rabbit (Section Genetic Structure of Wild Rabbit Populations and 7.3.1) and the limited number of alleles at each locus. Several reports have confirmed the limited number of alleles for rabbit microsatellites, which is less than eight in most rabbits selected from Northern Europe, except microsatellites OCAS1CG (van Haeringen et al. 1996–1997), Sol03, and Sol33 (Rico et al. 1994; Surridge et al. 1997) that were reported to have more than ten alleles (Table 1). The average number of alleles for microsatellites is usually between three and four (Table 1).

To date, the number of rabbit microsatellites exceeds 400 (Table 1 and Fig. 4). Progress in sequencing the rabbit genome (http://www.ensembl.org/Oryctolagus_cuniculus/index.html) will help to increase this number. Moreover, many microsatellites that had already been isolated from EMBL nucleotide databases are embedded into genes (van Lith and van Zutphen 1996; Korstanje et al. 2001a,b). Enrichment of comparative mapping data between man and rabbit (Chantry-Darmon et al. 2003, 2005b) and the publication of reciprocal heterologous chromosome painting data (Korstanje et al. 1999) should help to anchor these gene-associated markers to rabbit chromosomes (c.f. human/rabbit comparative mapping data in Table 2). A few uncovered chromosomal segments remain, such as OCU5p, OCU15p, OCU20p, and OCU21. For OCU5p and OCU21q, the corresponding human chromosomal

Table 1 Summary of microsatellites characterized in rabbit

Chromosome	Arm	Microsatellite m	Accession no.	Associated gene	Cytogenetic position l	Allele no.	Linkage group
OCU1[v]	p	INRACCDDV0130[a]	AJ874471	–	1p33	3[j]	LG1a[j]
	p	INRACCDDV0278[a]	AJ874603	AQP7	1p31prox	3[j]	nl[n]
	p	INRACCDDV0236[a]	AJ874569	TJP2	1p21.2-p21.1[o]	1[j]	–
	p	INRACCDDV0259[a]	AJ874589	TJP2	1p21.2-p21.1[o]	7[j]	nl[n]
	p	INRACCDDV0269[a]	AJ874595	TJP2	1p21.2-p21.1[o]	2[j]	LG1a[j]
	p	INRACCDDV0328[a]	AJ874646	PSAT1	1p12	3[j]	nl[n]
	p	INRACCDDV0345[a]	AJ874661	PSAT1	1p12	2[j]	nl[n]
	p	D1Utr1[c]	AF389372	–	–	4[c]	LGI[c], lg1B[c], lg1C[c]
	p	D1Utr2[c]	AF389367	–	–	4[c]	LGI[c], lg1B[c], lg1C[c]
	p	D1Utr7[c]	AF389355	–	–	4[c]	LGI[c], lg1B[c]
	p	Sat13[b]	X99892	–	1p12[h]	5[b]	LGI[c], lg1C[c]
	p	D1Utr3[c]	AF389359	–	–	3[c]	LGI[c], lg1C[c]
	p	INRACCDDV0240[a]	AJ874573	DAPK1	1p11dist	–	–
	p	INRACCDDV0263[a]	AJ874593	DAPK1	1p11dist	–	–
	p	INRACCDDV0272[a]	AJ874598	PTCH	1p11dist[o]	2[j]	LG1a[j]
	p/q	INRACCDDV0204[a]	AJ874541	–	–	4[j]	LG1a[j]
	q	INRACCDDV0351[a]	AJ874666	MMP1	1q14	–	–
	q	INRACCDDV0320[a]	AJ874640	SLN	1q14	4[j]	LG1a[j]
	q	INRACCDDV0322[a]	AJ874641	SLN	1q14	2[j]	nl[n]
	q	INRACCDDV0252[a]	AJ874583	HTR3B	1q14	3[j]	nl[n]
	q	INRACCDDV0271[a]	AJ874597	HTR3B	1q14	2[j]	LG1a[j]
	q	Albino (C)	–	–	–		LG1a[j], LGI[c,k]
	q	D1Utr4[c]	AF389353	–	–	4[j], 4[c]	LG1a[j], LGI[c], lg1C[c]
	q	INRACCDDV0327[a]	AJ874645	RPS3	1q21.1	1[j]	–
	q	INRACCDDV0350[a]	AJ874665	RPS3	1q21.1	1[j]	–
	q	INRACCDDV0257[a]	AJ874587	PARVA	1q21.3dist[o]	1[j]	–
	q	INRACCDDV0169[a]	AJ874508	–	1q21.5	3[j]	LG1a[j]
	q	D1Utr5[c]	AF389357	–	–	2[c]	LGI[c], lg1C[c]
	q	D1Utr6[c]	AF389354	–	1q24-q26[h]	2[c]	LGI[c], lg1B[c], lg1C[c]

(Continued)

Table 1 (Continued)

Chromosome	Arm	Microsatellite m	Accession no.	Associated gene	Cytogenetic position l	Allele no.	Linkage group
OCU2	q	INRACCDDV0251[a]	AJ874582	SLC29A2	1q27prox[o]	–	–
	q	INRACCDDV0298[a]	AJ874620	–	1q27dist	2[j]	nl[n]
	p	INRACCDDV0192[a]	AJ874530	–	2p21.3dist	2[j]	nl[n]
	q	INRACCDDV0096[a]	AJ874439	–	2q14dist	3[j]	LG2a[i]
	q	INRACCDDV0173[a]	AJ874511	–	2q14-q15[o]	3[j]	nl[n]
	q	INRACCDDV0070[a]	AJ874414	TGFA	2q21.1[o]	4[j]	LG2a[i]
	q	INRACCDDV0077[a]	AJ874421	FSHR	2q22.2[o]	–	–
	q	INRACCDDV0358[a]	AJ874671	–	2q22.2-q22.3	–	–
OCU3	p	INRACCDDV0036[a]	AJ874398	CD14	3p21prox	5[j]	LG3a[i]
	p/q	D3Utr2[d]	AF421903	–	–	6[d], 5[j]	LGXI[d], LG3a[i]
	p/q	D3Utr1[d,g]	Z54345	FABP6	–	2[g]	LGXI[d,g]
	q	Sol33[e]	X94683	–	3q11[h]	15[e]	LGXI[d]
	q	INRACCDDV0225[a]	AJ874558	ASPH	3q14[o]	1[j]	–
	q	INRACCDDV0218[a]	AJ874552	ARFGEF1	3q14	1[j]	–
	q	Sat3[b]	J03744	PMP2	3q16[p,o]	7[b]	LGXI[d]
	q	D3Utr3[d]	AF421904	–	–	3[d]	LGXI[d]
	q	INRACCDDV0203[a]	AJ874540	–	3q22-q23	4[j]	nl[n]
OCU4	p	INRACCDDV0340[a]	AJ874657	NCOA6	4p13dist[o]	–	–
	p	INRACCDDV0068[a]	AJ874412	PRNP	4p13prox-p12[o]	1[j]	–
	p	INRACCDDV0314[a]	AJ874635	PRNP	4p13prox-p12[o]	–	–
	p	INRACCDDV0325[a]	AJ874643	PRNP	4p13prox-p12[o]	–	–
	p	INRACCDDV0333[a]	AJ874650	PRNP	4p13prox-p12[o]	3[j]	nl[n]
	q	INRACCDDV0022[a]	AJ874385	ERBB3	4q11dist[o]	2[j]	LG4b[i]
	q	INRACCDDV0040[a]	AJ874400	ERBB3	4q11dist[o]	5[j]	nl[n]
	q	INRACCDDV0090[a]	AJ874433	–	–	4[j]	LG4b[i]
	q	INRACCDDV0228[a]	AJ874561	–	–	2[j]	LG4a[i]
	q	INRACCDDV0182[a]	AJ874520	–	4q13	2[j]	LG4a[i]
	q	INRACCDDV0248[a]	AJ874579	PMCH	4q15.3prox[o]	4[j]	LG4b[i]

q	INRACCDDV0065[a]	AJ874410	IGF1	4q15.3prox[o]	–	–
q	INRACCDDV0100[a]	AJ874442	–	–	5[j]	nl[n]
OCU5[w]	–	–	–	–	–	–
p						
p/q	INRACCDDV0282[a]	AJ874606	–	–	3[j]	LG5a[j]
p/q	INRACCDDV0142[a]	AJ874482	–	–	3[j]	LG5a[j]
p/q	D5Utr4[d]	AF421907	–	–	2[d]	LGVI[d]
p/q	D5Utr3[d]	AF421905	–	–	4[d], 2[j]	LGVI[d], LG5a[j]
p/q	D5Utr2[d]	AF421913	–	–	3[d]	LGVI[d]
q	INRACCDDV0007	AJ874373	MT1A	5q12	–	–
q	INRACCDDV0012	AJ874376	MT1A	5q12	1[j]	–
q	D5Utr1[d-g]	X07790	MT1A	5q12[h]	4[g]	LGVI[d]
q	INRACCDDV0039[a]	AJ874399	LCAT	5q14	1[j]	–
q	INRACCDDV0211[a]	AJ874545	HAS3	5q14	4[j]	LG5a[j]
OCU6[w]	D6Utr1[g]	M74142	HBA1	6p14[q]	5[g]	lg6[d]
p	D6Utr2[d]	AF421922	–	–	4[d]	lg6[d]
p	D6Utr3[d]	AF421915	–	–	2[d]	lg6[d]
p	INRACCDDV0009[a]	AJ874374	DNASE1	6p14	4[j]	nl[n]
p	INRACCDDV0290[a]	AJ874613	–	6p14prox	2[j]	LG6a[j]
p	INRACCDDV0219[a]	AJ874553	MYH11	6p13-p12	–	–
p	INRACCDDV0187[a]	AJ874525	–	–	2[j]	LG6b[j]
p	INRACCDDV0213[a]	AJ874547	PRKCB1	6p13-p12	4[j]	LG6b[j]
p	INRACCDDV0224[a]	AJ874557	PRKCB1	6p13-p12	–	–
p	INRACCDDV0229[a]	AJ874562	PRKCB1	6p13-p12	–	–
p	INRACCDDV0238[a]	AJ874571	–	6p12	3[j]	nl[n]
p	D6Utr4[d]	AF421916	–	–	5[d], 3[j]	lg6[d], LG6a[j]
p	INRACCDDV0214[a]	AJ874548	EIF3S8	6p12prox	4[j]	LG6a[j]
p/q	INRACCDDV0292[a]	AJ874615	–	–	2[j]	LG6a[j]
p/q	INRACCDDV0127[a]	AJ874468	–	–	2[j]	LG6b[j]
p/q	D6Utr5[d]	AF421923	–	–	3[d]	lg6[d]

(Continued)

Table 1 (Continued)

Chromosome	Arm	Microsatellite m	Accession no.	Associated gene	Cytogenetic position l	Allele no.	Linkage group
	q	INRACCDDV0120[a]	AJ874462	–	6q12prox	–	–
	q	INRACCDDV0347[a]	AJ874663	EPO	6q12med	3[j]	nl[n]
OCU7[w]	p	D7Utr1[d,g]	M26312	TRB@	7p23[p]	2[g]	lg7[d]
	p	D7Utr6[d]	unavailable[y]	PODXL	–	3[d]	lg7[d]
	p	INRACCDDV0221[a]	AJ874555	GPR37	7p22-p21[o]	4[j]	nl[n]
	p	INRACCDDV0231[a]	AJ874564	GPR37	7p22-p21[o]	–	–
	p	INRACCDDV0311[a]	AJ874633	CALU	7p21prox	–	–
	p	INRACCDDV0042[a]	AJ874401	CFTR	7p11.3[o]	–	–
	p/q	D7Utr2[d]	AF421934	–		3[d]	lg7[d]
	p/q	D7Utr3[d]	AF421935	–		3[d]	lg7[d]
	p/q	D7Utr4[d]	AF421932	–		3[d]	lg7[d]
	p/q	D7Utr5[d]	AF421930	–		5[d]	lg7[d]
	q	INRACCDDV0323[a]	AJ874642	PROC	7q14	4[j]	nl[n]
	q	INRACCDDV0336[a]	AJ874653	PROC	7q14	–	–
	q	INRACCDDV0359[a]	AJ874672	CXCR4	7q15prox	–	–
	q	INRACCDDV0235[a]	AJ874568	ACTR3	7q15[o]	5[j]	nl[n]
	q	INRACCDDV0331[a]	AJ874649	C2ORF25	7q16prox	2[j]	nl[n]
	q	INRACCDDV0163[a]	AJ874502	–	7q21dist[o]	3[j]	nl[n]
	q	INRACCDDV0092[a]	AJ874435	–	7q25	4[j]	nl[n]
OCU8	p	INRACCDDV0317[a]	AJ874637	ORL1	8p12.3dist[o]	1[j]	–
	p	INRACCDDV0276[a]	AJ874602	–		2[j]	LG8a[j]
	p	INRACCDDV0074[a]	AJ874418	–		5[j]	LG8a[j]
	p	INRACCDDV0075[a]	AJ874419	SLC6A12	8p12.1[o]	–	–
	p	INRACCDDV0080[a]	AJ874423	SLC6A12	8p12.1[o]	3[j]	nl[n]
	p	INRACCDDV0087[a]	AJ874430	SLC6A12	8p12.1[o]	6[j]	LG8a[j]
	p	INRACCDDV0157[a]	AJ874497	–	8p11	5[j]	nl[n]
	q	INRACCDDV0341[a]	AJ874658	TPT1	8q13.3-q21	4[j]	LG8a[j]
	q	INRACCDDV0021[a]	AJ874384	SLC15A1	8q24	5[j]	LG8a[j]
OCU9	p	INRACCDDV0160[a]	AJ874500	–	9p13dist	–	–

	p	INRACCDDV0005[a]	AJ874371	*GPX1*	9p13	–	–
	p	INRACCDDV0184[a]	AJ874522	–	–	3[j]	LG9a[j]
	p	INRACCDDV0274[a]	AJ874600	*ITIH3*	9p13prox	4[j]	LG9a[j]
	p/q	INRACCDDV0146[a]	AJ874486	–	–	3[j]	LG9a[j]
	q	INRACCDDV0010[a]	AJ874375	*NPC1*	9q13	3[j]	LG9a[j]
	q	INRACCDDV0016[a]	AJ874380	*NPC1*	9q13	5[j]	nl[n]
	q	INRACCDDV0249[a]	AJ874580	*DSC3*	9q14.2	–	–
	q	INRACCDDV0344[a]	AJ874660	*MAPRE2*	9q14.2	4[j]	nl[n]
	q	INRACCDDV0239[a]	AJ874572	*GALNT1*	9q14.2	–	–
	q	INRACCDDV0273[a]	AJ874599	*GALNT1*	9q14.2	1[j]	–
	q	INRACCDDV0155[a]	AJ874495	–	9q15.1	3[j]	LG9a[j]
	q	INRACCDDV0200[a]	AJ874537	–	9q15.3	3[j]	LG9a[j]
		INRACCDDV0017[a]	AJ874381	*CYB5A*	9q17	2[j]	LG9a[j]
OCU10	p	INRACCDDV0006[a]	AJ874372	*ICAM5*	10p12	–	–
	p	INRACCDDV0001[a]	AJ874368	*CALR*	10p12	3[j]	nl[n]
	p	INRACCDDV0004[a]	AJ874370	*CALR*	10p12	–	–
	p	INRACCDDV0061[a]	AJ874407	*CALR*	10p12	2[j]	nl[n]
	q	INRACCDDV0138[a]	AJ874478	–	10q12prox	3[j]	nl[n]
	q	INRACCDDV0349[a]	AJ874664	*AOAH*	10q14dist	1[j]	–
	q	INRACCDDV0025[a]	AJ874388	*STK17A*	10q15prox[o]	–	–
	q	INRACCDDV0330[a]	AJ874648	*DLX5*	10q15dist[o]	–	–
	q	INRACCDDV0076[a]	AJ874420	*DLX5*	10q15dist[o]	1[j]	–
	q	D10Utr1[f,g]	unavailable[z]	*WAP*	10q16dist[r]	6[g]	nl[n]
	q	INRACCDDV0304[a]	AJ874626	*EGFR*	10q16ter	4[j]	nl[n]
	q	INRACCDDV0305[a]	AJ874627	*EGFR*	10q16ter	1[j]	–
OCU11	p	INRACCDDV0183[a]	AJ874521	–	11p11.1-p11.2	7[j]	LG11a[j]
	q	INRACCDDV0357[a]	AJ874670	*BXDC2*	11q13prox	2[j]	nl[n]
	q	INRACCDDV0108[a]	AJ874450	–	11q13prox	3[j]	nl[n]
	q	INRACCDDV0086[a]	AJ874429	*GHR*	11q13	3[j]	LG11a[j]
	q	INRACCDDV0237[a]	AJ874570	*NNT*	11q13-q14	4[j]	LG11a[j]

(Continued)

Table 1 (Continued)

Chromosome	Arm	Microsatellite m	Accession no.	Associated gene	Cytogenetic position l	Allele no.	Linkage group
OCU12[w]	p	INRACCDDV0072[a]	AJ874416	*DSP*	12p15prox	3[j]	nl[n]
	p	INRACCDDV0083[a]	AJ874426	*DSP*	12p15prox	–	–
	q	INRACCDDV0057[a]	AJ874404	*R19[u]*	12q11[t]	–	–
	q	INRACCDDV0055[a]	AJ874403	*NOTCH4*	12q11[t]	–	–
	q	INRACCDDV0088[a]	AJ874431	*DMA*	12q11[t]	–	–
	q	D12Utr1[g]	M22640	*RLADPA*	12q11[t]	4[g]	LGVII[k], lg12[d]
	q	INRACCDDV0318[a]	AJ874638	*CGA*	12q15	4[j]	nl[n]
	q	INRACCDDV0260[a]	AJ874590	*RNGTT*	12q15	3[j]	LG12a[j]
	q	INRACCDDV0201[a]	AJ874538	–	12q16	4[j]	LG12a[j]
	q	D12Utr2[d]	AY095444	–	–	3[d], 2[j]	LG12a[j], lg12[d]
	q	INRACCDDV0176[a]	AJ874514	–	12q23dist	2[j]	nl[n]
OCU13	p	INRACCDDV0215[a, s]	AJ874549	–	13p13	–	–
	p/q	INRACCDDV0106[a]	AJ874448	–	–	6[j]	LG13a[j]
	q	INRACCDDV0297[a]	AJ874619	–	13q21prox	3[j]	LG13a[j]
	q	INRACCDDV0346[a]	AJ874662	*CRP*	13q21prox[o]	1[j]	–
	q	INRACCDDV0027[a]	AJ874390	*CD1B*	13q21med[o]	3[j]	LG13a[j]
	q	INRACCDDV0139[a]	AJ874479	–	–	5[j]	LG13a[j]
	q	INRACCDDV0177[a]	AJ874515	–	–	3[j]	LG13a[j]
	q	INRACCDDV0137[a]	AJ874477	–	–	4[j]	LG13a[j]
	q	INRACCDDV0270[a]	AJ874596	*BCAS2*	13q22-q23	–	–
	q	INRACCDDV0342[a]	AJ874659	–	13q26-q27	4[j]	LG13a[j]
	q	INRACCDDV0202[a]	AJ874539	–	–	2[j]	nl[n]
	q	INRACCDDV0151[a]	AJ874491	–	–	2[j]	LG13a[j]
	q	INRACCDDV0014[a]	AJ874378	*SLC2A1*	13q31-q32	2[j]	LG13a[j]
	q	INRACCDDV0289[a]	AJ874612	–	–	3[j]	LG13a[j]
	q	INRACCDDV0291[a]	AJ874614	–	–	2[j]	LG13a[j]
	q	INRACCDDV0310[a]	AJ874632	*PTAFR*	13q33	3[j]	nl[n]
OCU14	p	INRACCDDV0230[a]	AJ874563	–	14p11dist	–	–
	q	INRACCDDV0082[a]	AJ874425	STAG1	14q11	2[j]	nl[n]

	q	INRACCDDV0082[a]	AJ874425	*STAG1*	14q11	2[j]	nl[n]
	q	INRACCDDV0131[a]	AJ874472	–	14q11	1[j]	–
	q	INRACCDDV0335[a]	AJ874652	*SIAH2*	14q13prox	–	–
	q	INRACCDDV0166[a]	AJ874505	–	14q13prox	3[j]	LG14a[j]
	q	INRACCDDV0081[a]	AJ874424	*FXR1*	14q17prox	3[j]	nl[n]
	q	INRACCDDV0079[a]	AJ874422	*AHSG*	14q17dist[o]	–	–
	q	INRACCDDV0313[a]	AJ874634	*HES1*	14q21prox	4[j]	nl[n]
	q	INRACCDDV0140[a]	AJ874480	–	14q23	3[j]	LG14a[j]
	q	INRACCDDV0241[a]	AJ874574	*TIAM1*	14q25	5[j]	LG14a[j]
	q	INRACCDDV0243[a]	AJ874576	*TIAM1*	14q25	–	–
	q	INRACCDDV0162[a]	AJ874501	–	–	2[j]	LG14a[j]
OCU15	p	–	–	–	–	–	–
	p/q	INRACCDDV0101[a]	AJ874443	–	–	3[j]	LG15a[j]
	p/q	INRACCDDV0125[a]	AJ874466	–	–	3[j]	LG15a[j]
	q	INRACCDDV0286[a]	AJ874609	–	15q11.3	3[j]	LG15a[j]
	q	INRACCDDV0288[a]	AJ874611	–	–	2[j]	LG15a[j]
	q	*Angora (L)*	–	–	–	–	LG15a[j], LGII[k]
	q	INRACCDDV0143[a]	AJ874483	–	15q21prox	4[j]	nl[n]
	q	INRACCDDV0294[a]	AJ874617	–	–	3[j]	LG15a[j]
	q	INRACCDDV0018[a]	AJ874382	*CSN1S1*	15q23	–	–
	q	OCAS1CG[f]	unavailable[z]	*CSN1S1*	15q23	11[f]	nl[n]
	q	Sat2[b]	M77195	*CSN1S1*	15q23	4[b]	nl[n]
	q	Sat4[b]	M33582	*CSN2*	15q23	5[b]	nl[n]
	q	INRACCDDV0032[a]	AJ874395	*CSN3*	15q23dist	4[j]	nl[n]
	q	INRACCDDV0035[a]	AJ874397	*CSN3*	15q23dist	5[j]	LG15a[j]
	q	INRACCDDV0303[a]	AJ874625	*SLC4A4*	15q23dist	2[j]	LG15a[j]
	q	INRACCDDV0307[a]	AJ874629	*SLC4A4*	15q23dist	–	–
	q	INRACCDDV0306[a]	AJ874628	*GC*	15q23dist	–	–
	q	INRACCDDV0044[a]	AJ874402	*ALB*	15q23	–	–
OCU16	p	INRACCDDV0148[a]	AJ874488	–	16p12.1	3[j]	LG16a[j]
	p/q	INRACCDDV0105[a]	AJ874447	–	–	4[j]	LG16a[j]

(Continued)

Table 1 (Continued)

Chromosome	Arm	Microsatellite m	Accession no.	Associated gene	Cytogenetic position l	Allele no.	Linkage group
	q	INRACCDDV0208[a]	AJ874544	HNRPU	16q13.1	–	–
	q	INRACCDDV0216[a]	AJ874550	HNRPU	16q13.1	–	–
	q	INRACCDDV0233[a]	AJ874566	HNRPU	16q13.1	–	–
	q	INRACCDDV0003[a]	AJ874369	PIGR	16q21med[o]	2[j]	nl[n]
	q	INRACCDDV0185[a]	AJ874523	–	16q23	3[j]	LG16a[i]
OCU17	p	INRACCDDV0152[a]	AJ874492	–	17p12	3[j]	nl[n]
	q	INRACCDDV0030[a]	AJ874393	CA12	17q11	–	–
	q	INRACCDDV0031[a]	AJ874394	CA12	17q11	4[j]	LG17a[i]
	q	INRACCDDV0172[a]	AJ874510	–	17q21dist	4[j]	LG17a[i]
	q	INRACCDDV0217[a]	AJ874551	GMFB	17q23prox	4[j]	–
OCU18	q	INRACCDDV0196[a]	AJ874534	–	18q12prox	2[j]	nl[n]
	q	INRACCDDV0280[a]	AJ874605	–	18q12prox	3[j]	nl[n]
	q	INRACCDDV0099[a]	AJ874441	–	18q12dist	4[j]	nl[n]
	q	INRACCDDV0119[a]	AJ874461	–	18q21.1	4[j]	LG18b[j]
	q	INRACCDDV0123[a]	AJ874464	–	–	2[j]	LG18b[j]
	q	INRACCDDV0188[a]	AJ874526	–	–	3[j]	LG18a[i]
	q	INRACCDDV0190[a]	AJ874528	–	18q22-q23	4[j]	LG18a[i]
	q	INRACCDDV0253[a]	AJ874584	MINPP1	18q23	–	–
	q	INRACCDDV0258[a]	AJ874588	MINPP1	18q23	–	–
	q	INRACCDDV0338[a]	AJ874655	FAS	18q23	–	–
	q	INRACCDDV0085[a]	AJ874428	CYP2C4	18q24-q31	1[j]	–
	q	D0Utr10[f,g]	M74203	CYP2C4	18q24[h]	3[f]	nl[n]
	q	INRACCDDV0168[a]	AJ874507	–	–	2[j]	LG18b[j]
	q	INRACCDDV0023[a]	AJ874386	CYP2C18	18q31prox[o]	3[j]	LG18b[j]
	q	INRACCDDV0029[a]	AJ874392	CYP2C18	18q31prox[o]	4[j]	nl[n]
	q	INRACCDDV0063[a]	AJ874409	CYP2C18	18q31prox[o]	–	–
	q	INRACCDDV0104[a]	AJ874446	–	–	3[j]	LG18b[j]
OCU19[w]	q	D19Utr2[d]	AF421952	–	–	4[d]	lg19[d]
	q	INRACCDDV0058[a]	AJ874405	ALOX15	19q12.3prox[o]	1[j]	–

	q	D19Utr1[f,g,d]	M33291	*ALOX15*	19q12[h]	4[f]	lg19[d]
	q	D19Utr3[d]	AF421951	–	–	3[d]	lg19[d]
	q	D19Utr4[d]	AF421949	–	–	3[d]	lg19[d]
	q	INRACCDDV0339[a]	AJ874656	*ENO3*	19q12.3prox[o]	1[j]	–
	q	INRACCDDV0212[a]	AJ874546	*NDEL1*	19q12.3prox[o]	–	–
	q	INRACCDDV0234[a]	AJ874567	*NDEL1*	19q12.3prox[o]	3[j]	LG19a[j]
	q	INRACCDDV0309[a]	AJ874631	*MYH3*	19q12.3prox	1[j]	–
	q	INRACCDDV0102[a]	AJ874444	–	–	3[j]	LG19a[j]
	q	INRACCDDV0094[a]	AJ874437	–	19q21prox	3[j]	LG19a[j]
	q	INRACCDDV0071[a]	AJ874415	*KRT12*	19q21med[o]	4[j]	LG19a[j]
	q	INRACCDDV0033[a]	AJ874396	*ITGB3*	19q21dist[o]	2[j]	LG19a[j]
	q	INRACCDDV0193[a]	AJ874531	–	–	3[j]	LG19a[j]
OCU20	p	–	–	–	–	–	–
	q	INRACCDDV0084[a]	AJ874427	*TGFB3*	20q12.3[o]	1[j]	nl[n]
OCU21	p/q	–	–	–	–	–	–
OCUX	p	INRACCDDV0232[a]	AJ874565	*SAT*	Xp15	–	–
	q	INRACCDDV0247[a]	AJ874578	*MSN*	Xq12prox	–	–
	q	INRACCDDV0275[a]	AJ874601	*MSN*	Xq12prox	–	–
	q	INRACCDDV0334[a]	AJ874651	*RPS4X*	Xq12	–	–
OCUY	p	INRACCDDV0326[a]	AJ874644	*SRY*	Yp12	2[j]	nl[n]
	q	–	–	–	–	–	–
Unknown[x]	–	Sat5[b]	X99887	–	–	6[b]	–
	–	Sat7[b]	X99888	–	–	5[b]	–
	–	Sat8[b]	X99889	–	–	4[b]	–
	–	Sat12[b]	X99891	–	–	4[b]	–
	–	Sat16[b]	X99890	–	–	3[b]	–
	–	Sol03[i]	X79189	–	–	5[i], 12[e]	–

(Continued)

Table 1 (Continued)

Chromosome	Arm	Microsatellite m	Accession no.	Associated gene	Cytogenetic position l	Allele no.	Linkage group
	–	Sol08[i]	X79217	–	–	5[i], 10[e]	–
	–	Sol28[i]	X79216	–	–	7[i], 7[e]	–
	–	Sol30[i]	X79215	–	–	4[i], 9[e]	–
	–	Sol44[e]	X94684	–	–	9[e]	–

[a] Chantry-Darmon et al. (2005a)

[b] Mougel et al. (1997)

[c] Korstanje et al. (2001b)

[d] Korstanje et al. (2003)

[e] Surridge et al. (1997)

[f] an Haeringen et al. (1996/97)

[g] Korstanje et al. (2001a)

[h] Zijlstra et al. (2002)

[i] Rico et al. (1994)

[j] Chantry-Darmon et al. (2006)

[k] Fox (1994)

[l] Cytogenetic position and description of the microsatellite have been published with the associated genes, except where indicated

[m] Most likely chromosomal assignment according to available cytogenetic, OCU-HSA comparative mapping and genetic data

[n] Not linked

[o] Refined cytogenetic localization (data not presented)

[p] Hayes et al. (2002)

[q] Original localization was 6q12 but OCU6 was initially inverted (Hayes et al. 2002). The most probable localization is 6p14, in agreement with HSA/OCU comparative mapping data

[r] Rogel-Gaillard et al. (2000)

[s] Initial association of this microsatellite with *FLJ11838* is no longer valid

[t] Rogel-Gaillard et al. (2001)

[u] Symbol R19 = RLA class I gene that was amplified using primers specific for the genomic clone 19-1 (GenBank accession number: K02819)

[v] Other unmapped microsatellites that were recovered using a chromosome 1-specific library (Korstanje et al. 2001b) exist

[w] Other unmapped microsatellites that were recovered using chromosome-specific libraries (Korstanje et al. 2003) exist

[x] Markers corresponding to the first characterized microsatellites

[y] *D7Utr6* is a polymorphic marker within the *PODXL* gene. No accession number exists for this marker, but primers were reported by Korstanje et al. (2003)

[z] No accession number exists for this microsatellite, but primers were reported by van Haeringen et al. (1996/97)

Table 2 Summary of genes mapped in rabbit (2008). Genes are listed according to their position on the human genome starting from HSA1pter to HSAY following the gene order given in human genome databases (Ensembl database: http://www.ensembl.org/index.html and UCSC Human Genome Browser: http://genome.cse.ucsc. edu/), with gene symbols and names (HUGO Gene Nomenclature Committee, website: http://www.gene.ucl.ac.uk/nomenclature/index.html), assignment or localization to rabbit chromosomes, and corresponding references

Gene	Gene name	Localization		Refer
		HSA	OCU	
ENO1	enolase 1, (alpha)	1p36.23	13q35	t
C1QB	complement component 1, q subcomponent, B chain	1p36.12	13q33	v
PTAFR	platelet-activating factor receptor	1p35.3	13q33	u
SLC2A1	solute carrier family 2 (facilitated glucose transporter), member 1	1p34.2	13q31-q32	s
PGM1	phosphoglucomutase 1	1p31.3	13q27	e and t
JAK1	Janus kinase 1 (a protein tyrosine kinase)	1p31.3	13q27	t
FUBP1	far upstream element (FUSE)-binding protein 1	1p31.1	13q26	t
EDG7	endothelial differentiation, lysophosphatidic acid G-protein-coupled receptor, 7	1p22.3	13q25d	t
AGL	amylo-1, 6-glucosidase, 4-alpha-glucanotransferase	1p21.2	13q24-q25	u
ADORA3	adenosine A3 receptor	1p13.2	13q23	t
BCAS2	breast carcinoma amplified sequence 2	1p13.2	13q22-q23	t
S100A4	S100 calcium-binding protein A4	1q21.3	13q21dist	t
SHC1	SHC (Src homology 2 domain containing) transforming protein 1	1q21.3	13q21med	v
CD1B	CD1b molecule	1q23.1	13q21med	t
CD1D	CD1e molecule	1q23.1	13q21med	u
CRP	C-reactive protein, pentraxin-related	1q23.2	13q21prox	t
MGST3	microsomal glutathione S-transferase 3	1q24.1	13q11-q21	t
CACYBP	Calcyclin-binding protein (alias SIP)	1q25.1	13p15prox-p14	t
PIGR	polymeric immunoglobulin receptor	1q32.1	16q21med	s
CENPF	centromere protein F, 350/400 ka (mitosin)	1q41	16q16-q21prox	u
GNPAT	glyceronephosphate O-acyltransferase	1q42.2	16q13.3	t
LYST	lysosomal trafficking regulator	1q42.3	16q13.3	u
HNRPU	heterogeneous nuclear ribonucleoprotein U (scaffold attachment factor A)	1q44	16q13.1	t
DTNB	dystrobrevin, beta	2p23.3	2q23	v
SLC5A6	solute carrier family 5 (sodium-dependent vitamin transporter), member 6	2p23.3	2q23	v
CDC42EP3	CDC42 effector protein (Rho GTPase binding) 3	2p22.2	2q22.3	u

(Continued)

Table 2 (Continued)

Gene	Gene name	Localization		Refer
		HSA	OCU	
FSHR	Follicle-stimulating hormone receptor	2p16.3	2q22.2	t
PELI1	pellino homolog 1 (Drosophila)	2p14	2q21.3prox	t
TGFA	transforming growth factor, alpha	2p13.3	2q21.1	s
GGCX	gamma-glutamyl carboxylase	2p11.2	2q14dist-q15	u
EIF5B	eukaryotic translation initiation factor 5B	2q11.2	2q13-q14prox	u
IL1B	interleukin 1, beta	2q13	2q14dist	t
ACTR3	ARP3 actin-related protein 3 homolog (yeast)	2q14.1	7q15	t
PROC	protein C (inactivator of coagulation factors Va and VIIIa)	2q14.3	7q14	u
LCT	Lactase	2q21.3	7q15prox	u
CXCR4	chemokine (C-X-C motif) receptor 4	2q21.3	7q15prox	u
ACVR2A	activin A receptor, type IIA (alias ACVR2)	2q22.3-q23.1	7q16prox	u
C2orf25	chromosome 2 open reading frame 25	2q23.2	7q16prox	u
PLA2R1	phospholipase A2 receptor 1, 180 kDa	2q24.2	7q16dist	u
TTN	Titin	2q31.2	7q21	t
TFPI	tissue factor pathway inhibitor (lipoprotein-associated coagulation inhibitor)	2q32.1	7q23.1	v
AOX1	aldehyde oxidase 1	2q33.1	7q23.3	v
MYL1	myosin, light chain 1, alkali; skeletal, fast	2q34	7q23.3-q24	u
ATIC	5-aminoimidazole-4-carboxamide ribonucleotide formyltransferase/IMP cyclohydrolase	2q35	7q25	t
SGPP2	sphingosine-1-phosphate phosphatase 2	2q36.1	7q25dist	v
UGT1A@	UDP glucuronosyltransferase 1 family, polypeptide A cluster	2q37.1	7q27	t
GHRL	ghrelin/obestatin preprohormone	3p25.3	9q17dist	v
RAF1	v-raf-1 murine leukemia viral oncogene homolog 1	3p25.1	9p13dist	u
SLC4A7	solute carrier family 4, sodium bicarbonate cotransporter, member 7	3p24.1	14p13dist	u
CRTAP	Cartilage-associated protein	3p22.3	14p13prox	v
PTHR1	parathyroid hormone receptor 1	3p21.31	9p13	p
GPX1	*glutathione peroxidase 1*	3p21.31	9p13	e and t
ACY1	aminoacylase 1	3p21.1	9	e

ITIH3	inter-alpha (globulin) inhibitor H3	3p21.1	9p13prox	t
ITIH4	inter-alpha (globulin) inhibitor H4 (plasma Kallikrein-sensitive glycoprotein)	3p21.1	9p13	t
MITF	microphthalmia-associated transcription factor	3p14.1	9p11	v
GBE1	glucan (1,4-alpha-), branching enzyme 1 (glycogen-branching enzyme, Andersen disease, glycogen storage disease type IV)	3p12.3-p12.2	14q23dist	t
CASR	**calcium-sensing receptor**	3q13.3-q21.1	14q11	m
RAB7	RAB7, member RAS oncogene family	3q21.3	9p13	t
TF	transferrin	3q22.1	14p11-q11	q
STAG1	stromal antigen 1	3q22.3	14q11	t
SIAH2	seven in absentia homolog 2 (Drosophila)	3q25.1	14q13prox	u
MYNN	myoneurin	3q26.2	14q15	t
TF-like*	(transferrin like in rabbit)	3?	14q15	v
FNDC3B	fibronectin type III domain containing 3B	3q26.31	14q15dist	u
FXR1	fragile X mental retardation, autosomal homolog 1	3q26.33	14q17prox	t
CLCN2	chloride channel 2	3q27.1	14q17dist	u
LIPH	lipase, member H	3q27.2	14q17dist	u
AHSG	alpha-2-HS-glycoprotein	3q27.3	14q17dist	s
HES1	hairy and enhancer of split 1, (Drosophila) (alias HRY)	3q29	14q21prox	t
APOD	apolipoprotein D	3q29	14q21	t
WDR1	WD repeat domain 1	4p16.1	2p25	u
UGDH	UDP-glucose dehydrogenase	4p14	2p21.3	u
CSN1S1	casein alpha s1	4q13.3	15q23	r and s
CSN2	casein beta	4q13.3	15q23	r
CSN3	casein kappa	4q13.3	15q23dist	t
SLC4A4	solute carrier family 4, sodium bicarbonate cotransporter, member 4	4q13.3	15q23dist	u
GC	group-specific component (vitamin D-binding protein)	4q13.3	15q23dist	u
ALB	Albumin	4q13.3	15q23	t
ART3	ADP-ribosyltransferase 3	4q21.1	15q23dist	t
PRKG2	protein kinase, Cgmp-dependent, type II	4q21.21	15q23prox	u
SPP1	secreted phosphoprotein 1 (osteopontin, bone sialoprotein I, early T-lymphocyte activation 1)	4q22.1	15q22-q23	u

Table 2 (Continued)

Gene	Gene name	Localization		Refer
		HSA	OCU	
PDLIM5	PDZ and LIM domain 5	4q22.3	15q22-q23	u
ADH1B	alcohol dehydrogenase IB (class I), beta polypeptide	4q23	15q21dist	u
TACR3	tachykinin receptor 3	4q24	15q21	t
IL2	interleukin 2	4q27	15q13prox	u
UCP1	uncoupling protein 1 (mitochondrial, proton carrier)	4q31.21	15q11.5	u
EDNRA	endothelin receptor type A	4q31.23	15q11.3-q11.4	t
TLR2	toll-like receptor 2	4q31.3	15q11.3	u
HAND2	heart and neural crest derivatives expressed 2	4q34.1	2p21.1prox	u
F11	coagulation factor XI (plasma thromboplastin antecedent)	4q35.2	2p13	t
CDH10	cadherin 10, type 2 (T2-cadherin)	5p14.2	11q12-q13	t
BXDC2	brix domain containing 2 (alias BRIX)	5p13.2	11q13prox	u
PRLR	prolactin receptor	5p13.2	11q13	q
GHR	growth hormone receptor	5p12	11q13	q and s
NNT	nicotinamide nucleotide transhydrogenase	5p12	11q13-q14	t
PPAP2A	phosphatidic acid phosphatase type 2A	5q11.2	11q15	t
CKMT2	creatine kinase, mitochondrial 2 (sarcomeric)	5q14.1	11p14dist	u
RASA1	RAS p21 protein activator (GTPase-activating protein) 1	5q14.3	11p14prox	t
CD14	CD14 molecule	5q31.3	3p21prox	t
HARS	histidyl-Trna synthetase	5q31.3	3p21prox	t
TCERG1	transcription elongation regulator 1	5q32	3p13dist	u
HAND1	heart and neural crest derivatives expressed 1	5q33.2	3p13prox	u
CYFIP2	cytoplasmic FMR1 interacting protein 2 (alias PRO1331)	5q33.3	3p13prox	t
ADRA1B	adrenergic, alpha-1B-, receptor	5q33.3	3p13prox	u
STK10	serine/threonine kinase 10	5q35.1	3p11	t
SLC34A1	**solute carrier family 34 (sodium phosphate), member 1 (alias NPT2)**	5q35.3	3p11	l and t
DSP	desmoplakin	6p24.3	12p15prox	t
RANBP9	RAN-binding protein 9	6p23	12p13	u
PRL	Prolactin	6p22.3	12p11	u

SLC17A1	**solute carrier family 17 (sodium phosphate), member 1 (alias NPT1)**	6p22.2	12p11	l
TNF	tumor necrosis factor (TNF superfamily, member 2)	6p21.33	12q11	o and s
HSPA1A	heat shock 70 kDa protein 1A	6p21.33	12q11	o
TNXB	tenascin XB	6p21.32	12q11	o
DRB1	major histocompatibility complex, class II, DR beta 1	6p21.32	12q11	o
DQA1	major histocompatibility complex, class II, DQ alpha 1	6p32.32	12q11	o
DQB1	major histocompatibility complex, class II, DQ beta 1	6p21.32	12q11	o
NOTCH4	Notch homolog 4 (Drosophila)	6p21.32	12q11	o
TAP1	transporter 1, ATP-binding cassette, subfamily B (MDR/TAP)	6p21.32	12q11	o
RP3-405J24.3*	cytochrome c oxidase subunit VIa polypeptide 1 pseudogene 2 (alias loc285849')	6p21.2	12q12-13prox	t
PGC	progastricsin (pepsinogen C)	6p21.1	12q11.3dist	u
BMP5	bone morphogenetic protein 5	6p12.1	12q12med	u
CGA	glycoprotein hormones, alpha polypeptide	6q15	12q15	u
RNGTT	RNA guanylyltransferase and 5'-phosphatase	6q15	12q15	t
REV3L	REV3-like, catalytic subunit of DNA polymerase zeta (yeast)	6q21	12q21	u
GJA1	gap junction protein, alpha 1, 43 kDa (connexin 43)	6q22.31	12q22	v
ARG1	arginase, liver	6q23.2	12q23	t
SOD2	**superoxide dismutase 2, mitochondrial**	6q25.3	12q14-q16	k
T	T, brachyury homolog (mouse)	6q27	12q25.3dist	u
FLJ20323*	Hypothetical protein FLJ20323	7p21.3	10q15dist	t
AQP1	aquaporin 1 (Colton blood group)	7p15.1	10q14prox	t
AOAH	acyloxyacyl hydrolase (neutrophil)	7p14.2	10q14dist	u
STK17A	serine/threonine kinase 17a (apoptosis-inducing)	7p13	10q15prox	t
PPIA	peptidylprolyl isomerase A (cyclophilin A)	7p13	10q16prox	t
WAP	wey acidic protein (unknown in man)	-	10q16dist	n
EGFR	epidermal growth factor receptor (erythroblastic eukemia viral (v-erb-b) oncogene homolog, avian)	7p11.2	10q16ter	u
POR	P450 (cytochrome) oxidoreductase	7q11.23	6q12med-dist	v
ZP3	zona pellucida glycoprotein 3 (sperm receptor)	7q11.23	6q12prox	v
PON1	paraoxonase 1	7q21.3	10q15dist	t

(Continued)

Table 2 (Continued)

Gene	Gene name	Localization		Refer
		HSA	OCU	
DLX5	distal-less homeobox 5	7q21.3	10q15dist	t
EPO	erythropoietin	7q22.1	6q12med	u
CFTR	cystic fibrosis transmembrane conductance regulator (ATP-binding cassette subfamily C, member 7)	7q31.2	7p11.3	t
GPR37	G protein-coupled receptor 37 (endothelin receptor type B-like)	7q31.33	7p22-p21	t
CALU	Calumenin	7q32.1	7p21prox	u
CALD1	caldesmon 1	7q33	7p21dist	u
TRB@	T cell receptor beta locus	7q34	7p23	q and s
PDIA4	protein disulfide isomerase family A, member 4 (alias ERP70)	7q36.1	7p21dist	u
PDGFRL	platelet-derived growth factor receptor-like	8p22	2p11-p12	t
SFTPC	surfactant, pulmonary-associated protein C	8p21.3	15q11	v
INTS10	integrator complex subunit 10	8p21.3	15q11.1	v
ADRA1A	adrenergic, alpha-1A-, receptor	8p21.2	2p21.1	v
TMEM66	transmembrane protein 66	8p12	2p11dist	v
POLB	polymerase (DNA directed), beta	8p11.21	2p11prox (1q12.3 chimeric)	v
SNAI2	snail homolog 2 (Drosophila)	8q11.21	3q12dist	t
ASPH	aspartate beta-hydroxylase	8q12.3	3q14	t
ARFGEF1	ADP-ribosylation factor guanine nucleotide-exchange factor 1(brefeldin A-inhibited) (alias BIG1)	8q13.2	3q14	t
JPH1	junctophilin 1	8q21.11	3q16	t
PMP2	peripheral myelin protein 2	8q21.13	3q16	q and s
EXT1	exostoses (multiple) 1	8q24.11	3q23prox	u
HAS2	hyaluronan synthase 2	8q24.13	3q23prox	t
MYC	v-myc myelocytomatosis viral oncogene homolog (avian)	8q24.21	3q23dist	t
ADFP	adipose differentiation-related protein	9p22.1	1p23	v
IFNA1	interferon, alpha 1	9p21.3	1p23	t
ACO1	aconitase 1, soluble	9p21.1	1p31prox	t
AQP7	aquaporin 7	9p13.3	1p31prox	t

GALT	*galactose-1-phosphate uridylyltransferase*	9p13.3	1		e
VCP	valosin-containing protein	9p13.3	1p31prox		u
TPM2	tropomyosin 2 (beta)	9p13.3	1p31		t
TJP2	tight junction protein 2 (zona occludens 2)	9q21.11	1p21.2-p21.1		t
PSAT1	phosphoserine aminotransferase 1 (alias PSA)	9q21.2	1p12		u
DAPK1	death-associated protein kinase 1	9q21.33	1p11dist		t
OGN	osteoglycin (osteoinductive factor, mimecan)	9q22.31	1p11dist		u
FBP1	fructose-1,6-bisphosphatase 1	9q22.32	1p11dist		t
PTCH1	patched homolog 1 (Drosophila)	9q22.32	1p11dist		t
CEL	carboxyl ester lipase (bile salt-stimulated lipase)	9q34.2	1p35dist		u
AKR1C3	aldo-keto reductase family 1, member C3 (3-alpha hydroxysteroid dehydrogenase, type II)	10p15.1	16p12.1prox		u
ITIH2	inter-alpha (globulin) inhibitor H2	10p14	16q11		t
ANXA8	annexin A8	10q11.22	18q12dist		u
DDX21	DEAD (Asp-Glu-Ala-Asp) box polypeptide 21	10q21.3	18q21.3dist		u
HK1	hexokinase 1	10q21.3	18q21		q
KCNMA1	potassium large conductance calcium-activated channel, subfamily M, alpha member 1	10q22.3	18q21		t
MINPP1	multiple inositol polyphosphate histidine phosphatase 1	10q23.31	18q23		t
FAS	Fas (TNF receptor superfamily, member 6) (alias TNFRSF6)	10q23.31	18q23		t
CYP2C18	cytochrome P450, family 2, subfamily C, polypeptide 18	10q23.33	18q31prox		t
CYP2C8	cytochrome P450, family 2, subfamily C, polypeptide 8 (alias CYP2C4)	10q23.33	18q24-q31		s
LOXL4	lysyl oxidase-like 4	10q24.2	18q31dist		u
GSTO1	glutathione S-transferase omega 1 ex GSTTLp28	10q25.1	18q31dist-q32		t
PNLIP	pancreatic lipase	10q25.3	18q33prox		u
FGFR2	fibroblast growth factor receptor 2	10q26.13	18q33prox		t
CYP2E1	cytochrome P450, family 2, subfamily E, polypeptide 1	10q26.3	18q33ter		u
HRAS	**v-Ha-ras Harvey rat sarcoma viral**	11p15.5	1q14-q21		g
IGF2	insulin-like growth factor 2 (somatomedin A)	11p15.5	1q27ter		s
ART1	ADP-ribosyltransferase 1	11p15.4	1q21.1-q21.2		t

(Continued)

Table 2 (Continued)

Gene	Gene name	Localization		Refer
		HSA	OCU	
HBB	**hemoglobin, beta**	11p15.4	1q14-q21	g
PARVA	parvin, alpha	11p15.3	1q21.3dist	t
PTH	**parathyroid hormone**	11p15.2	1q14-q21	g
LDHA	*lactate dehydrogenase A*	11p15.1	1	c
CAT	catalase	11p13	1q23	u
ACP2	*acid phosphatase 2, lysosomal*	11p11.2	1	c
LPXN	leupaxin	11q12.1	1q25dist	u
SCGB1A1	*secretoglobin, family 1A, member 1 (uteroglobin) (alias UGB)*	11q12.3	1	d
SLC29A2	solute carrier family 29 (nucleoside transporters), member 2	11q13.1	1q27prox	t
RBM4	RNA-binding motif protein 4	11q13.1	1q27med	t
RPS3	ribosomal protein S3	11q13.4	1q21.1	u
WNT11	wingless-type MMTV integration site family, member 11	11q13.5	1q21.1	u
TYR	tyrosinase (oculocutaneous albinism IA)	11q14.3	1q14-q15	u
MMP1	matrix metallopeptidase 1 (interstitial collagenase)	11q22.2	1q14	u
SLN	sarcolipin	11q22.3	1q14	u
HTR3B	5-hydroxytryptamine (serotonin) receptor 3B	11q23.2	1q14	t
HSPA8	heat shock 70 kDa protein 8	11q24.1	1q12.3	u
ETS1	v-ets erythroblastosis virus E26 oncogene homolog 1 (avian)	11q24.3	1q12.1dist	v
SLC15A4	solute carrier family 15, member 4 (alias PTR4)	11q24.32	21q14	v
SLC6A12	solute carrier family 6 (neurotransmitter transporter, betaine/GABA), member 12	12p13.33	8p12.1	t
OLR1	oxidized low-density lipoprotein (lectin-like) receptor 1	12p13.2	8p12.3dist	u
EMP1	epithelial membrane protein 1	12p13.1	8p12.3dist-p13	u
SSPN	sarcospan (Kras oncogene-associated gene)	12p12.1	8p14.1	t
KRT3	keratin 3	12q13.13	4q11prox	u
KRT18	keratin 18	12q13.13	4q11prox	u
TMEM4	transmembrane protein 4	12q13.3	4q11dist	t
LUM	lumican	12q21.33	4q15.1	u
PMCH	pro-melanin-concentrating hormone	12q23.2	4q15.3prox	t

IGF1	insulin-like growth factor 1 (somatomedin C)	12q23.2	4q15.3prox	s
HSP90B1	heat shock protein 90 kDa beta (Grp94), member 1 (alias TRA1)	12q23.3	4q15.2	v
DAO	D-amino-acid oxidase	12q24.11	21q12dist	t
NOS1	nitric oxide synthase 1 (neuronal)	12q24.22	21q14	s
GCN1L1	GCN1 general control of amino-acid synthesis 1-like 1 (yeast)	12q24.23-q24.31	21q12dist-q13	t
CRYL1	crystallin, lambda 1	13q12.11	8q11	u
ATP12A	ATPase, H+/K+ transporting, nongastric, alpha polypeptide	13q12.12	8q11	u
TPT1	tumor protein, translationally-controlled 1	13q14.12	8q13.3-q21	u
RB1	**retinoblastoma 1 (including osteosarcoma)**	13q14.2	8q21	k
DACH1	dachshund homolog 1 (Drosophila)	13q21.33	8q22prox	t
KLF5	Kruppel-like factor 5 (intestinal)	13q22.1	8q22prox	u
RBM26	RNA-binding motif protein 26 (alias C13orf10)	13q31.1	8q22dist	t
SLC15A1	solute carrier family 15 (oligopeptide transporter), member 1	13q32.3	8q24	t
F7	coagulation factor VII (serum prothrombin conversion accelerator)	13q34	8q26	s
NP	*nucleoside phosphorylase*	14q11.2	17	e
DHRS4	dehydrogenase/reductase (SDR family) member 4	14q11.2	17q21prox	t
COCH	coagulation factor C homolog, cochlin (Limulus polyphemus)	14q12	17q21dist	u
GMFB	glia maturation factor, beta	14q22.2	17q23prox	t
PPM1A	protein phosphatase 1A (formerly 2C), magnesium-dependent, alpha isoform	14q23.1	20q12.1	v
HIF1A	hypoxia-inducible factor 1, alpha subunit (basic helix-loop-helix transcription factor)	14q23.2	20q12.1	u
TGFB3	transforming growth factor, beta 3	14q24.3	20q12.3	s
CKB	**creatine kinase, brain**	14q32.33	20q13-q14	f
SNURF	SNRPN upstream reading frame	15q11.2	17q25prox	v
RYR3	ryanodine receptor 3	15q13.3-q14	17q15	v
RAD51	RAD51 homolog (RecA homolog, E. coli) (S. cerevisiae)	15q15.1	17q15	t
SLC28A2	solute carrier family 28 (sodium-coupled nucleoside transporter), member 2	15q21.1	17q15prox	t
CYP19A1	cytochrome P450, family 19, subfamily A, polypeptide 1	15q21.2	17q13prox	u
MAPK6	mitogen-activated protein kinase 6	15q21.2	17q13	v
MYO5A	myosin VA (heavy chain 12, myoxin)	15q21.2	17q13prox	u

(Continued)

Table 2 (Continued)

Gene	Gene name	Localization HSA	Localization OCU	Refer
LIPC	lipase, hepatic	15q22.2	17q11	s
CA12	carbonic anhydrase XII	15q22.2	17q11	t
PKM2	pyruvate kinase, muscle	15q23	17p14	v
RHCG	Rh family, C glycoprotein	15q26.1	17p11	t
SELS*	selenoprotein S (alias loc55829)	15q26.3	17q23dist	v
HBA1	**hemoglobin, alpha 1**	16p13.3	6p14 (6q12**)	i
DNASE1	deoxyribonuclease I	16p13.3	6p14	s
MYH11	myosin, heavy chain 11, smooth muscle	16p13.11	6p12-p13	t
PRKCB1	protein kinase C, beta 1	16p12.1	6p12-p13	t
EIF3S8	eukaryotic translation initiation factor 3, subunit 8, 110 kDa	16p11.2	6p12prox	t
SULT1A1	sulfotransferase family, cytosolic, 1A, phenol-preferring, member 1	16p11.2	6p12prox	t
PHKB	**phosphorylase kinase, beta**	16q12.1	5q12	j
FTS	fused toes homolog (mouse)	16q12.2	5q12	t
MT1A	metallothionein 1A (functional)	16q13	5q12	q and s
LCAT	lecithin-cholesterol acyltransferase	16q22.1	5q14	t
HAS3	hyaluronan synthase 3	16q22.1	5q14	t
DPEP1	dipeptidase 1 (renal)	16q24.3	5q16dist	u
ALOX15	arachidonate 15-lipoxygenase	17p13.2	19q12.3prox	q and s
ENO3	enolase 3 (beta, muscle)	17p13.2	19q12.3prox	t
TP53	tumor protein p53 (Li-Fraumeni syndrome)	17p13.1	19q12.3prox	t
NDEL1	nudE nuclear distribution gene E homolog like 1 (A. nidulans)	17p13.1	19q12.3prox	t
MYH3	myosin, heavy chain 3, skeletal muscle, embryonic	17p13.1	19q12.3prox	u
SREBF1	sterol regulatory element-binding transcription factor 1	17p11.2	19q12.1	t
NOS2A	nitric oxide synthase 2A (inducible, hepatocytes)	17q11.1	19q12.3dist	t
KRT12	keratin 12 (Meesmann corneal dystrophy)	17q21.2	19q21med	t
STAT5A	signal transducer and activator of transcription 5A	17q21.2	19q21med	t
ATP6V0A1	ATPase, H+ transporting, lysosomal V0 subunit a1	17q21.31	19q21med	t
ITGB3	integrin, beta 3 (platelet glycoprotein IIIa, antigen CD61)	17q21.32	19q21dist	t

TOB1	transducer of ERBB2, 1	17q21.33	19q21prox	u
ACE	angiotensin I-converting enzyme (peptidyl-dipeptidase A) 1	17q23.3	19q21-q22	q
TK1	**thymidine kinase 1, soluble**	17q25.3	19q22-q23	k
TYMS	**thymidylate synthetase**	18p11.32	9q12-q13	k
MRCL3*	myosin regulatory light chain MRCL3	18p11.31	9q13prox	u
NPC1	Niemann-Pick disease, type C1	18q11.2	9q13	t
DSC3	desmocollin 3	18q12.1	9q14.2	t
MAPRE2	microtubule-associated protein, RP/EB family, member 2	18q12.1	9q14.2	u
GALNT1	UDP-N-acetyl-alpha-D-galactosamine: polypeptide N-acetylgalactosaminyltransferase 1	18q12.2	9q14.2	t
ELAC1	elaC homolog 1 (E. coli)	18q21.2	9q15.1	u
CYB5A	cytochrome b5 type A (microsomal)	18q22.3	9q17	s
ICAM5	intercellular adhesion molecule 5, telencephalin	19p13.2	10p12	t
CALR	calreticulin	19p13.11	10p12	s
GPI	glucose phosphate isomerase	19q13.11	5p12prox	u
RYR1	ryanodine receptor 1 (skeletal)	19q13.2	5p12	t
FUT1(+2)	fucosyltransferase 1 (galactoside 2-alpha-L-fucosyltransferase, H blood group)	19q13.33	5p12	v
PRNP	prion protein (p27-30) (Creutzfeldt-Jakob disease, Gerstmann-Strausler-Scheinker syndrome, fatal familial insomnia)	20p13	4p13prox-p12	t
CST3	cystatin C (amyloid angiopathy and cerebral hemorrhage)	20p11.21	4p11	u
BCL2L1	BCL2-like 1	20q11.21	4p13med	u
NCOA6	nuclear receptor coactivator 6	20q11.22	4p13dist	u
GSS	glutathione synthetase	20q11.22	4p13dist	u
TIAM1	T-cell lymphoma invasion and metastasis 1	21q22.11	14q25	t
SERPIND1	serpin peptidase inhibitor, clade D (heparin cofactor), member 1	22q11.21	21q12prox	v
VPREB1	pre-B lymphocyte gene 1	22q11.22	21q12prox	v
IGLL1	immunoglobulin lambda-like polypeptide 1	22q11.23	21q12prox	u
FBXO7	F-box protein 7	22q12.3	4q15.3prox	u
DDX17	DEAD (Asp-Glu-Ala-Asp) box polypeptide 17	22q13.1	4q15.3prox	u
CPT1B	carnitine palmitoyltransferase 1B (muscle)	22q13.33	4q15.3dist	v

(Continued)

Table 2 (Continued)

Gene	Gene name	Localization		Refer
		HSA	OCU	
PHKA2	phosphorylase kinase, alpha 2 (liver)	Xp22.13	Xp15	t
ACOT9	acyl-CoA thioesterase 9 (alias ACATE2)	Xp22.11	Xp15	t
SAT1	spermidine/spermine N1-acetyltransferase 1	Xp22.11	Xp15	t
ZFX	zinc finger protein, X-linked	Xp22.11	Xp15	t
DDX3X	DEAD (Asp-Glu-Ala-Asp) box polypeptide 3, X-linked	Xp11.4	Xp13dist	u
RGN	regucalcin (senescence marker protein-30)	Xp11.3	Xp11	u
SMCX	Smcy homolog, X-linked (mouse)	Xp11.22	Xp11prox	t
MSN	moesin	Xq11.1	Xq12	t
RPS4X	ribosomal protein S4, X-linked	Xq13.1	Xq12	u
PHKA1	**phosphorylase kinase, alpha 1 (muscle)**	Xq13.1	Xq12	j
PGK1	*phosphoglycerate kinase 1*	Xq21.1	X	a and b
GLA	*galactosidase, alpha*	Xq22.1	X	a and b
SLC25A5	solute carrier family 25 (mitochondrial carrier; adenine nucleotide translocator), member 5	Xq24	Xq21.1prox	u
HPRT1	*hypoxanthine phosphoribosyltransferase 1*	Xq26.2	Xq23	b and s
G6PD	glucose-6-phosphate dehydrogenase	Xq28	X	a and b
SRY	sex-determining region Y	Yp11.31	Yp12	s
ZFY	zinc finger protein, Y-linked	Yp11.31	Y	h

[a] Cianfriglia et al. (1979)
[b] Echard et al. (1981)
[c] Echard et al. (1982)
[d] Gellin et al. (1983)
[e] Soulié and de Grouchy (1983)
[f] Mahoney et al. (1988)

[g] Xu and Hardison (1989)

[h] HGM10 (1989)

[i] Xu et al. (1991)

[j] Debecker et al. (1992)

[k] Lemieux and Dutrillaux (1992) [l] Kos (1996)

[m] Martin-DeLeon (1999)

[n] Rogel-Gaillard et al. (2000)

[o] Rogel-Gaillard et al. (2001)

[p] Martin-DeLeon et al. (2001)

[q] Zijlstra et al. (2002)

[r] Pauloin et al. (2002)

[s] Hayes et al. (2002)

[t] Chantry-Darmon et al. (2003) with some localizations refined after the publication

[u] Chantry-Darmon et al. (2005) with some localizations refined after the publication

[v] personal communication

* Nonapproved gene symbol

** Original localization was 6q12 but because OCU6 was initially inverted (Hayes et al. 2002). The most probable localization is 6p14, in agreement with HSA/OCU comparative mapping data

italic: Genes assigned by SCH analysis

bold italic: Genes assigned by SCH analysis and later confirmed by FISH

bold: Genes localized either by radioactive ISH or FISH with non-BAC probes

normal: Genes localized by FISH with BAC probes

segments are HSA19q and HSA12q24.1-24.3, respectively (Korstanje et al. 1999; Chantry-Darmon et al. 2003, 2005b), and will provide candidate genes for BAC recovery and microsatellite identification. In contrast, there is no obvious homology between OCU15p, OCU20p, or OCU21p and any human chromosomal segment (Korstanje et al. 1999), which suggests that these gaps will not be easily filled using a search based on available comparative mapping data.

The updated cytogenetically anchored genetic map is reported here and includes polymorphic microsatellites as well as those that are not polymorphic in the reference families (Fig. 4 and Table 1). Indeed, the set of available cytogenetically mapped microsatellites is a valuable reservoir of well-characterized potentially polymorphic markers. Moreover, rabbit microsatellites can be amplified using genomic DNA of other lagomorphs (Surridge et al. 1997), which means that a large set of microsatellites is probably available for genetic studies in lagomorphs other than the rabbit.

7.2.3
Physical Mapping Tools

To our knowledge, three BAC libraries are available for rabbit. The BAC library constructed by INRA (Rogel-Gaillard et al. 2001) contains 84,480 clones with an average insert size of 100–110 kb, for an approximate three-fold coverage of the rabbit haploid genome. Cloned DNA was extracted from lymphocytes of a male New Zealand rabbit homozygous for DRAdd-DQAbb in the RLA complex, according to RFLP marker typing (Han et al. 1992). The cloning vector is pBeloBAC11 and the restriction enzyme cloning site is *Hind*III. This library is referred to as LBAB and is maintained by the resource center for domestic animals managed by INRA (site at: http://www-crb.jouy.inra.fr/BRC/index.html). This BAC library can be screened upon request. A second BAC library has been published along with the physical mapping of the rabbit germline immunoglobulin heavy chain locus (Ros et al. 2004) and contains 100,000 clones with an average insert size of 124 kb, for a four-fold coverage of the haploid rabbit genome. Genomic DNA was extracted from an *a2b5* (heavy chain haplotype F-1) male rabbit and cloned into the pBeloBAC11 vector at the *Hind*III restriction enzyme

cloning site. This library, referred to as a2b5, has been spotted onto nylon filters for screening (Ros et al. 2004). The third library, referred to as LBNL-1 or LB1, was constructed at the Lawrence Berkeley National Laboratory using genomic DNA isolated from white blood cells of a New Zealand White rabbit. *Eco*RI-digested DNA was cloned into the pBACe3.6 vector. This library comprises a total of 123,648 clones with an average insert size of 175 kb, corresponding to an approximate seven-fold coverage of the rabbit haploid genome. This collection has been spotted onto nylon filters for screening (http://www-gsd.lbl.gov/cheng/LB1.html). Therefore, a 14× haploid genome is available for physical mapping thanks to the three rabbit BAC collections that are reported.

7.2.4
Sequencing Data

In October 2006, a total of 759,789 nucleotide ENTREZ records (http://www.ncbi.nlm.nih.gov) for rabbit were publically available. There is no focus specifically on the rabbit whole-genome sequence, but available sequencing data for this species are increasing. This is due, in part, to independent efforts from nonrelated laboratories and also because of at least two large-scale sequencing projects, namely, the ENCODE and the Mammalian Genome projects. These two projects aim at improving knowledge on the human genome by providing comparative sequencing data that will help identify coding and noncoding functional elements.

The ENCODE project (ENCyclopedia Of DNA Elements; http://www.genome.gov/10005107) is led by the NHGRI (National Human Genome Research Institute) and includes the rabbit. It aims at identifying all the functional elements present in the human genome using comparative sequence mapping data from 22 various vertebrate species (The ENCODE Project Consortium et al. 2004). A pilot study was started in 2003 and focused on the selection of 44 discrete regions of 0.5–2 megabases (Mb) encompassing about 1% of the human genome (30 Mb). These target regions of the ENCODE project have been selected according to gene density and level of nonexonic conservation with respect to the orthologous mouse genomic sequences. The UCSC (University of California at Santa Cruz) Genome Bioinformatics Group manages the official repository of sequence-related data for the ENCODE consortium and supports the coordination of data submission, storage, retrieval, and visualization (http://genome.ucsc.edu/ENCODE/). By October 2006, 225 of the 231 selected rabbit BAC clones covering all the ENCODE targets had been sequenced. A second phase of the project will consist in developing technology and a final phase should expand the pilot study to the whole human genome.

In 2004, the US National Institutes of Health (NIH) chose rabbit as one of the 16 mammalian species (rabbit, elephant, hyrax, tenrec, armadillo, sloth, hedgehog, shrew, microbat, megabat, cat, pangolin, guinea pig, squirrel, bushbaby, and treeshrew) included in the Mammalian Genome Project (http://www.broad.mit.edu/mammals/). The objective of the project was to produce low coverage (2×) of selected mammalian genomes to expand beyond the currently sequenced genomes, i.e., man, chimpanzee, mouse, rat, dog, cow, and pig. These 16 mammalian species have been selected in order to maximize the total branch length of the evolutionary tree over the last 100 million years of evolution and to provide sequencing data on biomedical models where possible. The rabbit genome project ID is number 12818 (http://www.intlgenome.org), and the two-fold coverage of the rabbit genome has been funded by the National Human Genome Research Institute (NHGRI) (http://www.ncbi.nlm.nih.gov). Shotgun sequencing was performed at the Broad Institute/Massachusetts Institute of Technology (BI/MIT) and led to the first publicly available assembly of the rabbit genome in May 2005. Owing to the fragmentary nature of the preliminary assembly of such low coverage genomes, a gene-building methodology relying on a whole-genome alignment (WGA) to the annotated human genome taken as a reference has been adopted. This WGA strategy has contributed to the arrangement of gene-scaffold superstructures in order to present complete genes; 7,268 such gene-scaffolds have been reported in August 2006 for rabbit (http://www.ensembl.org/Oryctolagus_cuniculus/index.html, Ensembl release 41, October 2006). A total of 495 known genes; 14,946 novel genes; 2,530 pseudogenes; 104,511 Genscan gene predictions; 210,256 gene exons; and 17,971 gene transcripts have been identified within 2,076,044,328 bp. The annotation of the rabbit genome sequence is still rudimentary and the rabbit scaffolds are not anchored onto

chromosomes. Identification of rabbit scaffolds containing gene-specific sequences may be achieved by searching similarities between rabbit anonymous genomic sequences and a gene sequence that is known in a related species such as man. Low coverage sequencing of the pika (*Ochotona princes*) genome will soon be completed, which should be very useful for annotating the rabbit sequence because the pika is closer to rabbit than man.

Although low coverage sequencing has proven efficient for recognizing features of the human genome shared across most mammals and for providing species-specific sequences, there is a growing interest in sequencing mammalian genomes at higher coverage than 2× (Kirkness et al. 2003). There are limitations to low coverage sequences, primarily because they provide only about 80% of the genome sequence, with several hundred thousand gaps remaining. In that respect, a request to expand the mammalian genome project toward a six to seven-fold genome coverage for ten species was presented in 2006 (http://www.genome.gov/Pages/Research/Sequencing/SeqProposals/2x-7x_promotion_seq.pdf). Three major objectives are mentioned, including (1) the identification of the "core" mammalian genome, (2) the increased use of medically relevant mammals, and (3) the reconstruction of the ancestral eutherian genome. The rabbit genome is considered as highly relevant for objectives (2) and (3). A new process to prioritize animal genomes for cloning and sequencing has been proposed by the NHGRI funded by the NIH (http://www.genome.gov/10002156), and the sequencing proposals and their processing can be found at http://www.genome.gov/10002154. In 2007, the NHGRI decided to fond the deeper coverage of the rabbit genome. A 7x coverage will be completed in 2008 by the Broad Institute.

7.2.5
Bioinformatics Tools

In contrast to other farm animals such as chicken, pig, ruminants, and horse, bioinformatics tools for genomics are still limited in the rabbit species, and the situation is referred to as "the forgotten rabbit"(Fadiel et al. 2005). A comprehensive list of informatics databases has been reported (Fadiel et al. 2005), with the addition of others. For genetic trait searches, the rabbit is referenced in the Online Mendelian Inheritance in Animals (OMIA) database (http://omia.angis.org.au/), which compiles information of inherited traits in animals and forms the domestic animal counterpart to the human OMIM database. For cytogenetic and genetic mapping, the INRA RabbitMap database provides data on chromosomal locations, genetic markers, and polymorphisms (http://dga.jouy.inra.fr/cgi-bin/lgbc/intro2.pl?BASE = rabbit). For comparative mapping data, the ICCARE (Interspecific Comparative Clustering and Annotation for EST) web server (Muller et al. 2004) is available (http://bioinfo.genopole-toulouse.prd.fr/Iccare/index.php). This tool compares expressed sequence tag (EST) and mRNA sequences from a species of interest to transcripts of a reference organism, which is man for mammalian genomes, and graphically presents the results, according to the chromosomal location of the genes on the reference organism. Thus, the ICCARE tool can summarize all rabbit ESTs and mRNAs that are in public databases with their corresponding positions on human chromosomes. In addition, this tool provides links to the rabbit sequences and BLASTN alignment against the human genome that indicates gene structure and highlights intron-exon boundaries. A direct link to software suitable for primer or overgo design has been created (Muller et al. 2004). For rabbit-sequencing data, the DDBJ/EMBL/GENBANK databases (http://www.ncbi.nlm.nih.gov/), the rabbit genome assembly (http://www.ensembl.org/Oryctolagus_cuniculus/index.html), and the UNIPROT database (http://www.expasy.uniprot.org/) devoted to proteins have been developed. Finally, for general knowledge on rabbit, the American House Rabbit Society (HRS) compiles articles and information on rabbit diseases and behavior written by veterinarians (http://www.hrschicago.org/articleslay.html).

7.2.6
Expected Tools and Development

Genomic resources for rabbits are still lacking, such as a radiation hybrid panel, large-scale sequencing of ESTs and full-length cDNAs, genetic markers such as single nucleotide polymorphisms (SNPs), and a coordinated project for polymorphism detection. However, research on the rabbit genome will benefit from strategies used for other genomes, and it may be possible to take advantage of high-throughput sequencing facilities that have been developed. The existing resources described in this chapter contribute significantly to the exploration of the rabbit genome and to biomedical research studies and QTL mapping that have been hampered by the lack of molecular tools for many years.

7.2.7
Genome Mapping

Cytogenetic Mapping in Rabbits

In contrast to other farm mammalian species (cattle, pigs, and to a lesser extent sheep, goats, and horses), very few mapping data were available for rabbit until the early 2000s. The first genes mapped in rabbits corresponded to groups of synteny, which are loci belonging to the same chromosome or chromosome fragment, using rodent × rabbit somatic cell hybrid (SCH) panels. These panels are composed of hybrid cell lines which retain various combinations of whole or fragmented rabbit chromosomes within the complete rodent chromosome complement and are tested for the presence or absence of a marker in each of the cell lines. If two markers cosegregate, they are considered as syntenic and constitute a synteny group, which can be assigned to a given chromosome if the panel has been cytogenetically characterized. For rabbits, two SCH panels were constructed in the late 1970s: a mouse × rabbit SCH panel in Italy (Cianfriglia and Pace 1979) and a Chinese hamster × rabbit SCH panel in France through an INSERM/INRA collaboration (Echard et al. 1982; Soulié and de Grouchy 1982). The first synteny group reported in rabbit is on the X-chromosome and contains *HPRT1, PGK1, G6PD,* and *GLA* (Cianfriglia et al. 1979; Echard et al. 1981). Since then, 18 other genes have been assigned using the SCH panels and reported in the literature (Echard et al. 1982; Gellin et al. 1983; Soulié and de Grouchy 1983; Medrano and Dutrillaux 1984). The chromosomal assignment of the first eight of these 18 genes was concordant with comparative human/rabbit mapping data (c.f. Table 2), i.e., *GALT, LDHA, ACP2,* and *UGB* (now *SCGB1A1*) to OCU1 (rabbit chromosome 1), *ACY1* and *GPX1* to *OCU9, PGM1* to OCU13, and *NP* to OCU17, while the chromosome assignment of the other ten genes has not been confirmed by comparative human/rabbit mapping data (Table 2), i.e., *IGKC* to OCU3, *TPI* (now *TPI1*), *GAPD* (now *GAPDH*) and *LDHB* to OCU4, *GUK* (now *GUK1*) and *MDH2* to OCU15, *IGH* (originally to OCU16), *TP* (now *ITPA*) and *PEPB* to OCU17, and *GSR* to OCU19.

After these initial gene assignments to rabbit chromosomes using SCH panels, a second phase in the construction of the rabbit gene map took place in 1985 to 1992 with the development of in situ hybridization on metaphase chromosome preparations using radioactive DNA probes. This technique permitted the first 11 regional localizations of genes: *CSN1S1* and *CSN2* on OCU12q24 (Gellin et al. 1985); *KRAS* on OCU16p11-q11 (Martin-DeLeon and Picciano 1988); *CKB* on OCU20q13-qter (Mahoney et al. 1988); *CKM* on OCU19q11-q12 (Martin-DeLeon et al. 1989); *HBB*, *PTH*, and *HRAS* on OCU1q14-q21, and *HBA1* on OCU6q12 (Xu and Hardison 1989, 1991); *PHKA1* on OCUXq12; and *PHKB* on OCU5q12 (Debecker and Martin-DeLeon 1992). Among these results, the localization of *CSNS1, CSN2, KRAS,* and *CKM* has been subsequently found to disagree with the human/rabbit comparative mapping data (Table 2), which may be due to difficulties in chromosome identification inherent to the isotopic in situ hybridization procedure. In fact, *CSN1S1* and *CSN2* have been relocalized more recently on OCU15q23 (Pauloin et al. 2002). Up until 1992, the rabbit gene map was still very poor with several chromosomes devoid of any assigned gene, i.e., OCU2, 8, 10, 11, 14, 18, 21, and Y.

A burst in cytogenetic gene mapping originated because of two developments:

- The use of nonradioactive DNA probes in the early 1990s for FISH, which permits more accurate localizations and easier identification of the hybridized chromosomes than in situ hybridization with radioactive DNA probes. In particular, the procedure developed by the French group in B Dutrillaux's laboratory (Viegas-Péquignot et al. 1989; Lemieux et al. 1992) has been applied extensively in rabbit. It combines hybridization and immunodetection of biotin-labeled probes with R-banding on BrdU incorporated chromosomes for their identification. This makes it possible to simultaneously visualize a green-yellow fluorescein hybridization signal and the red propidium iodide R-band pattern of chromosomes under a microscope equipped for fluorescence observations. This technique, initially developed for human chromosomes, has been successfully applied to other mammalian species, including sheep (Hayes et al. 1992) and rabbit (Rogel-Gaillard et al. 2000).

- The use of BAC clones in the late 1990s to construct genomic DNA libraries (Sect 7.2.3) resulted in large-sized DNA probes and thus intense hybridization signals, which can be easily detected with a fluorescence microscope.

The first FISH-mapped genes were localized either with short (1–4.5 kb) cDNA probes (Lemieux and Dutrillaux 1992; Kos et al. 1996) or with a 40 kb cosmid probe (Martin-DeLeon et al. 1999). Thereafter, rabbit BAC clones containing genes have been used in FISH-mapping experiments (Rogel-Gaillard et al. 2000, 2001; Martin-DeLeon et al. 2001; Zijlstra et al. 2002; Pauloin et al. 2002; Hayes et al. 2002; Chantry-Darmon et al. 2003, 2005b). FISH results on R-banded rabbit chromosomes using probes prepared from rabbit BAC clones containing *CALU* on OCU7p21prox and *NCOA6* on OCU4p13dist (Figs. 3a, b respectively).

With the advent of these technologies, the rabbit cytogenetic map progressed quickly with 301 additional gene assignments. In November 2006, there were 327 genes mapped on the rabbit genome (26 before 2000 + 301 after 2000–2006). These genes are listed in Table 2 with their official symbols and names (HUGO Gene Nomenclature Committee, http://www.gene.ucl.ac.uk/nomenclature/index.html), their localization on human chromosomes (Ensembl database: http://www.ensembl.org/index.html and UCSC Human Genome Browser: http://genome.cse.ucsc.edu/), their localization on rabbit chromosomes, and the corresponding references. These localizations are supported in most part by the comparative human/rabbit chromosome painting data (Korstanje et al. 1999). At least one gene has been localized to each rabbit chromosome arm, except the heterochromatic OCUYq, the small-sized p arm of OCU15, and the minute p arms of OCU20 and OCU21. It has been difficult to focus the search for genes for these autosomal regions due to the lack of any comparative human/rabbit mapping data (Table 2). The distribution of genes on the rabbit cytogenetic map is quite even along the chromosomes and the best-covered chromosomes are OCU1, 2, 6, 7, 8, 12, 13, 14, and 17. At present, about 45% of the bands at the 347-band

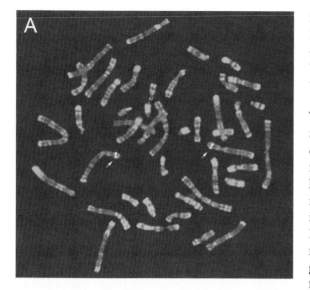

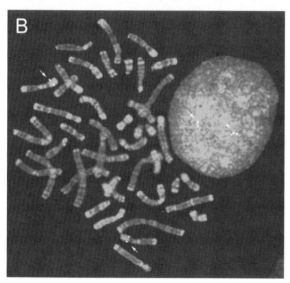

Fig. 3 Rabbit R-banded metaphase spreads hybridized with rabbit BAC clones. **a** *CALU* on OCU 7p21prox. **b** *NCOA6* on OCU4p13dist. Arrows indicate hybridization signals on chromosomes and nuclei

level (Hayes et al. 2002) carry at least one mapped gene, and most of these bands are R positive, in a large part because the R positive bands are gene-rich in mammals (Saccone et al. 1996). The resolution of the banded human karyotype is commonly about 850 bands (ISCN 1981), while that of the rabbit karyotype is now about 350 bands, which means that two genes located on separate but close bands in the human genome may be found on the same wide band in the rabbit genome. Therefore, in Table 2, many FISH localizations are specified "prox" (for proximal or nearer the centromere), "med" (in the middle), or "dist" (for distal or nearer the telomere) according to the position of the FISH signal within the band.

The Rabbit Linkage Map

Several studies containing rabbit linkage mapping data have been published. Castle (1924) reported the first coinherited pair of loci, including albinism (*a* locus) and brown pigmentation (*c* locus). The early rabbit maps included conventional biochemical, immunological, and morphological loci (reviewed by Fox 1994). The most recent review of rabbit linkage mapping (Fox 1994) included 10 autosomal linkage groups (LG) referred to as LG I to LG X, 11 identified chromosomes, and five X-linked loci, for a total of 69 assignments. From that time on, rabbit genome mapping developed slowly compared to that of other mammalian domestic species such as the cow and pig. Indeed, the first male genetic map based on AFLP markers was published in 2002 (van Haeringen et al. 2002) and the first microsatellite-based map was published in 2006 (Chantry-Darmon et al. 2006). While the rabbit linkage map is limited, the existing tools and maps are suitable for QTL mapping projects.

Different strategies have been reported to anchor linkage groups onto rabbit chromosomes. Of the first reported linkage groups (LG 1 to LG X), only LG I was cytogenetically anchored, with an assignment to chromosome 1 (Fox 1994). Later, linkage analysis of microsatellites led to an additional LG XI linkage group (Korstanje et al. 2001a). BAC clones containing markers or genes have been used for FISH mapping and have permitted orientation of LG I on rabbit chromosome (OCU)1, mapping and orientation of LG XI to OCU3, and anchoring of LG VI to OCU5 (Zijlstra et al. 2002). Microsatellites recovered from chromosome-specific libraries have expanded linkage groups anchored onto OCU1 (Korstanje et al. 2001b), OCU3, and OCU5 (Korstanje et al. 2003) and have been included into new linkage groups on OCU6, 7, 12, and 19 (Korstanje et al. 2003), leading to 15 distinct linkage groups of which seven are assigned to seven chromosomes. Chantry-Darmon et al. (2006) reported the most extensive work of cytogenetically anchored markers onto chromosomes, leading to 20 linkage groups that span 2,766.6 cM and cover 20 chromosomes excluding

OCU20, 21, and X. This map was built with 111 markers, i.e., 104 INRA microsatellites, five microsatellites previously identified by Korstanje et al. (2001b, 2003), and *angora* and *albino* morphological markers. In this map, linkage groups are referred to as LG and numbered according to the chromosomes (Chantry-Darmon et al. 2006). A summary of all linkage groups is presented in Fig. 4 and Table 1.

A male genetic map containing 103 AFLP markers organized into 12 linkage groups has also been reported (van Haeringen et al. 2002). This map covers only one-third of the rabbit genome, with only D0Utr1 and D0Utr2 markers assigned to LG VI (van Haeringen et al. 2001) on OCU5. Therefore, this map is not directly usable for comparative mapping.

Human–Rabbit Comparative Mapping

Comparative mapping is a strategy to transpose genomic information from one species to another species. Thus, information available for well-studied species, such as man and mouse, can be used for developing genome maps of less-studied species, such as rabbit. Comparative mapping is based on the conservation of genomes during evolution and contributes to the identification of conserved genomic segments among species and ancestral breakpoints. Comparative maps are essential for increasing marker density and identifying candidate genes for traits and diseases. Different approaches can be used to compare genomes, such as comparative cytogenetics, heterologous chromosome painting, comparative mapping based on data from somatic cell hybrid analyses, FISH on chromosomes, radiation hybrid mapping, linkage analyses, and whole-genome sequences. In rabbits, the approaches most often used are those with a low resolution at the level of the chromosome and chromosome bands, i.e., classical cytogenetics that compares R-banded human and rabbit chromosomes (in this Section), and molecular cytogenetics that uses FISH of DNA probes to chromosomes, which is more informative than classical cytogenetics. Three types of probes that can be hybridized to chromosomes include:

- Complex probes or chromosome paints, which are composed of a mixture of sequences corresponding to a unique chromosome or part of a chromosome sorted individually by cytometry or microdissection. Hybridization of such probes from species A

to chromosomes of species B detects conserved chromosome segments between the two species, hence the name heterologous chromosome painting. Korstanje et al. (1999) carried out human/rabbit bidirectional chromosome painting and their data are presented below.

- Probes representing a unique genomic sequence, which after hybridization determine the position of the target sequence on a given chromosome and at a precise chromosome band. These probes permit detailed comparisons of the localization of orthologous genes among species. In rabbit, gene-mapping data obtained by FISH (Sect. Human-Rabbit Comparative Mapping) have been exploited to compare the human and rabbit maps.

- Probes consisting of repeated DNA motifs, such as centromeric or telomeric DNA repeats, which serve as chromosome markers to study interphase nuclei. The only study in rabbits reporting the isolation, cloning, and characterization of centromeric satellite sequences described the existence of differences in the chromosomal distribution of satellite families across at least four distinct groups of chromosomes (Ekes et al. 2004).

In the mid-1990s, the number of orthologous genes mapped on human and rabbit chromosomes was too small to provide an overall correspondence between the two genomes. A major advance was the development of individual chromosome painting probes, which for the human genome have been commercialized since 1994 and for other animal species have been produced via collaborations with laboratories that have mastered chromosome-sorting technologies. Korstanje et al. (1999) reported global alignments of the human genome along the rabbit chromosomes and the rabbit genome along the human genome. They used human and rabbit chromosome-specific probes produced by bivariant fluorescence activated flow sorting of chromosomes that were then amplified by degenerate oligonucleotide primed PCR (DOP-PCR) and labeled by PCR in the presence of biotin-dUTP. Hybridization of each human chromosome paint to rabbit chromosome preparations and vice versa revealed the conserved chromosomal segments between these two species (Figs. 5a, b). This study showed that the conservation between these two genomes is extensive, with 42 conserved segments detected when the 23 rabbit chromosome paints were hybridized to human met-

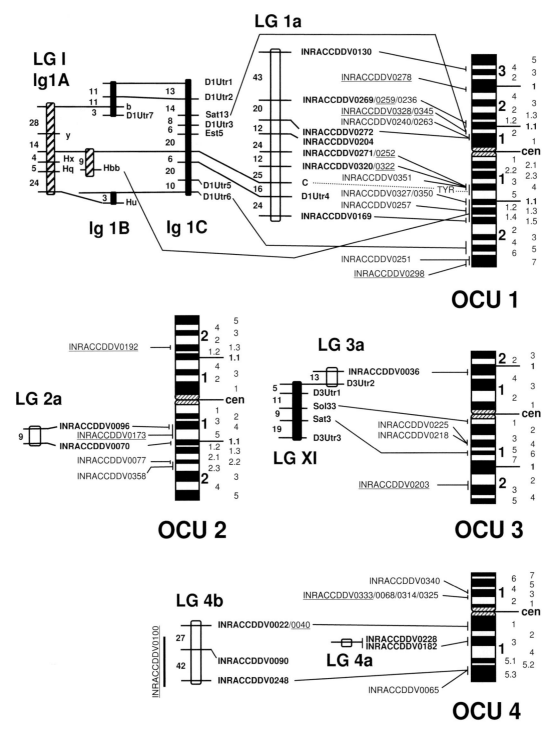

Fig. 4 Rabbit-integrated genetic and cytogenetic map. The linkage groups are represented on the *left side* of each R-banded idiogram. *Vertical bars* indicate the cytogenetic positions of microsatellites previously FISH-mapped (Chantry-Darmon et al. 2005b). Microsatellites included in linkage groups are in *bold letters*, polymorphic microsatellites not included in linkage groups are *underlined*, and microsatellites that were either not polymorphic in the tested families or for which the number of alleles were unknown are in *normal letters*. Genetic distances are estimated in Kosambi cM and shown to the *left* of the linkage groups. All microsatellites are summarized in Table 1. Linkage groups reported by Chantry-Darmon et al. (2006) are represented by *open boxes*, linkage groups reviewed in Fox (1994) are represented as *hatched boxes*, and linkage groups reported by Korstanje et al. (2001b, 2003) are represented as *large black bars* and are referred to as lg. The *INRACCDDV0100* and *INRACCDDV0202* microsatellites could not be precisely positioned in a linkage group; a bar is used to represent the most likely position for these microsatellites.

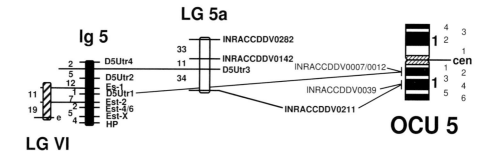

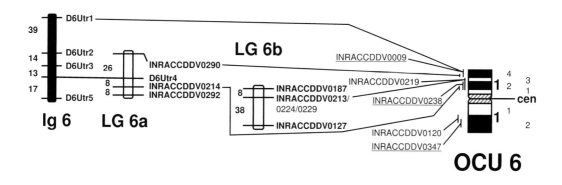

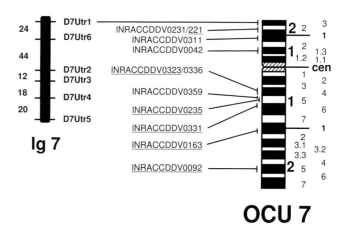

Fig. 4 (Continued)

aphase spreads and 38 conserved segments when the 24 human chromosome paints were hybridized to rabbit metaphase spreads. This is similar to reports for other animal species, including 44–58 segments for cattle (Hayes 1995; Solinas-Toldo et al. 1995; Chowdhary et al. 1996), 37 segments for pig (Goureau et al. 1996), 48 segments for sheep (Iannuzzi et al. 1999), and 43 segments for horse (Raudsepp et al. 1996). However, these studies do not provide an exhaustive view of the correspondences between genomes because of the low resolution of the technique that cannot detect conserved segments smaller than 10–20 Mb. However, chromosomal rearrangements, which have occurred since the divergence of human and rabbit lineages 65 million

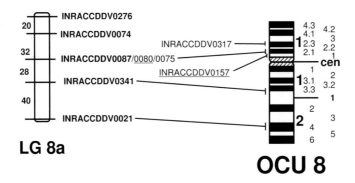

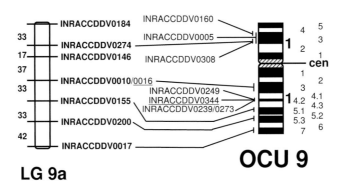

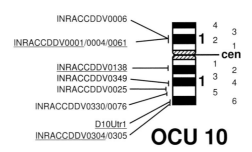

Fig. 4 (Continued)

of years ago, are mainly intrachromosomal, i.e., inversions. Indeed, inversions modify the banding pattern of chromosomes and the order of the genes without leading to complex reshuffling of the genome. In spite of the rearrangements that occurred in the genomes during evolution, the genomic context of the genes remains fairly conserved within mammalian species.

These human/rabbit heterologous chromosome painting data (Korstanje et al. 1999) were confirmed by 325 localizations available for rabbit genes and for which human orthologs have been identified (Table 2). Subsequent gene mapping data modified the position of some synteny junctions previously reported, as shown on Figs. 5a, b. Thus, the extent of some

Chapter 7 **Rabbit** 213

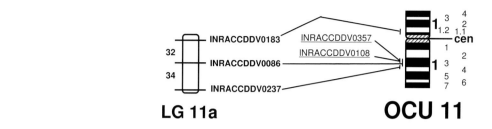

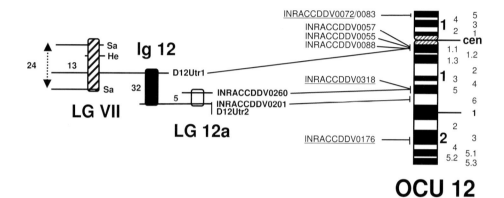

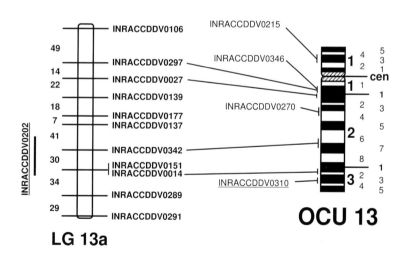

Fig. 4 (Continued)

conserved segments has been refined. In addition, two small conserved segments undetected by chromosome painting, i.e., one between human chromosome (HAS)22 and OCU4 and one between HSA8p and OCU15, have been identified. With the growing number of genes mapped to bands and sub-bands of rabbit chromosomes, it is possible to detect small inversions and rearrangements (Chantry-Darmon et al. 2005b). However, the overall picture of the human/rabbit comparative map is not yet complete, and many more genes need to be mapped in rabbit using higher-resolution techniques such as radiation hybrid mapping or even better high-coverage genome sequencing, to understand precisely the differences

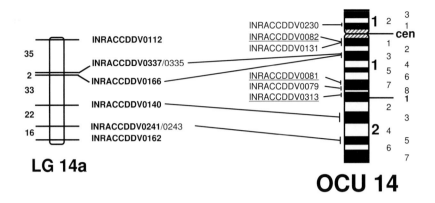

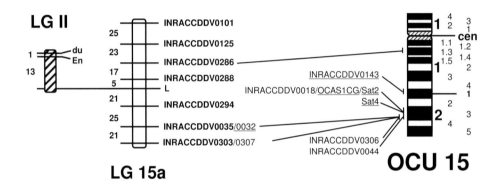

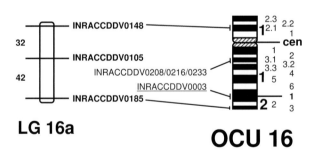

Fig. 4 (Continued)

in gene order and organization between the human and rabbit genomes.

Single-gene Traits and Quantitative Trait Loci (QTL)

Numerous heritable genetic mutations have been described (reviewed in Fox 1994). In addition, several QTL mapping projects and genetic association studies have been published. However, no positional cloning of a candidate gene has yet been reported in the rabbit.

QTLs and Candidate Genes for Atherosclerosis

Atherosclerosis is a complex histopathological process resulting from genetic and environmental influences.

Chapter 7 **Rabbit** 215

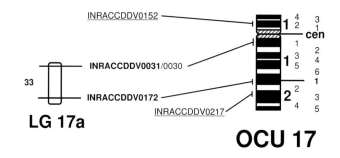

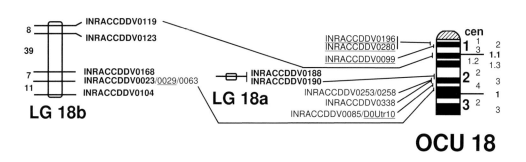

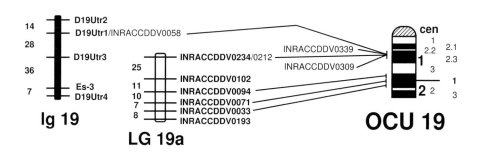

Fig. 4 (Continued)

Rabbit strains have been selected for genetic studies of atherosclerosis, such as the Watanabe rabbit (WHHL or Watanabe Heritable HyperLipidemic), which reproduces the pathology of human familial hypercholesterolemia type IIa, with the atherosclerotic process beginning in utero (Fan and Watanabe 2003). In order to study the genetic basis for susceptibility to atherosclerosis, rabbit strains have also been selected for high or low atherosclerotic response to diet-induced hypercholesterolemia (van Zutphen and Fox 1977; Thiery et al. 1995).

In the work by van Zutphen and Fox (1977), the AX/JU and IIIVO/JU inbred strains were selected as high and low responders, respectively, having high

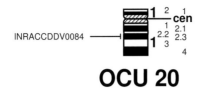

OCU 20

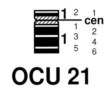

OCU 21

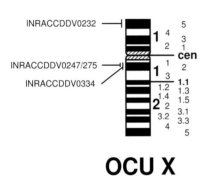

OCU X

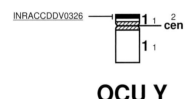

OCU Y

Fig. 4 (Continued)

and low basal serum HDL cholesterol levels corresponding to the basal serum HDL cholesterol level on a low-cholesterol control diet. Genetic mapping approaches have been carried out on backcross progeny IIIVO/JU × [IIIVO/JU × AX/JU] using AFLP markers (van Haeringen et al. 2001) and on F_2 intercrosses using microsatellite markers (Korstanje et al. 2001a). A significant genetic association was found between basal serum HDL cholesterol level and an AFLP marker mapping to LG VI (van Haeringen et al. 2001), and a significant genetic association was identified between the degree of atherosclerosis in aortas of male F_2 progeny and markers mapping to LG VI (Korstanje et al. 2001a). These results strongly suggest that a QTL for atherosclerosis susceptibility is found in LG VI that is cytogenetically anchored and orientated on OCU5 (Zijlstra et al. 2002; Chantry-Darmon et al. 2006). This QTL maps to a candidate region that is homologous to HSA16 and contains *lecithin-cholesterol acyltransferase (LCAT)* at HSA16q22.1 and *cholesteryl ester transfer protein (CETP)* at HSA16q21, both genes being involved in the metabolism of HDL cholesterol.

Thiery et al. (1995) created two rabbit strains with genetically determined high- (HAR) and low- (LAR) atherosclerosis responses to diet-induced hypercholesterolemia. Despite comparable cholesterol levels and similar lipoprotein profiles when fed a cholesterol-rich diet, the two strains differ in their susceptibility to develop aortic atherosclerosis. A suppression subtractive hybridization method was carried out on macrophages from the two strains, with focus on genes with increased expression in LAR macrophages. The *arginase 1 (ARG1)* gene was identified as a new candidate gene for the atherosclerosis resistance/susceptibility (Teupser et al. 2006) with a size polymorphism in *ARG1* mRNA, the shorter variant being associated with a higher enzymatic activity in LAR rabbits. The longer variant was detected in the macrophages of HAR rabbits and shown to be less stable than the shorter variant, which is probably due to the insertion of a 413 bp insertion of a C-repeat in the 3' untranslated region. The *ARG1* gene maps to HSA6q23. According to available comparative mapping data between rabbit and man, the *ARG1* gene cannot be considered as a candidate gene for the QTL that was mapped on OCU5.

Genetics of Susceptibility/Resistance to Papillomavirus-Induced Carcinogenesis

The European rabbit can be experimentally infected with cottontail rabbit papillomavirus (CRPV), and the resulting CRPV-induced warts may regress or evolve toward carcinoma. These lesions resemble those of genital intraepithelial neoplasia caused by human papillomaviruses (Breitburd et al. 1997; Campo 2002; Brandsma 2005). Because CRPV-induced warts either regress or

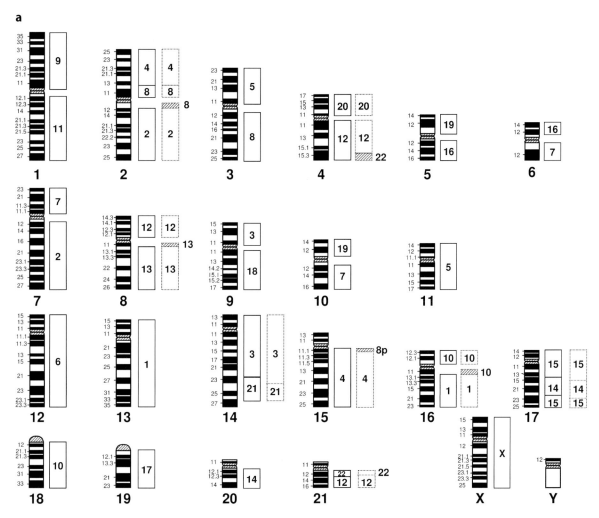

Fig. 5 Rabbit/human comparative genome maps. **a** Rabbit R-banded idiograms with blocks on the right (*full lines*) indicating correspondence with human chromosomes as determined by chromosome painting of rabbit chromosomes with human whole-chromosome paints (Korstanje et al. 1999) and revised blocks (*dotted lines*) accounting for FISH-mapping data from Table 2. **b** Human R-banded idiograms with blocks on the right (*full lines*) indicating correspondences with rabbit chromosomes as determined by chromosome painting of human chromosomes with rabbit whole-chromosome paints (Korstanje et al. 1999) and revised blocks (*dotted lines*) accounting for gene FISH-mapping data from Table 2

evolve into carcinomas in infected rabbits, genetic factors are likely to be involved in the persistence/regression of lesions. Polymorphism studies of the rabbit MHC revealed genetic associations between wart regression and a DR alpha *Eco*RI fragment, as well as an increased risk of malignant transformation for a DQ alpha *Pvu*II fragment (Han et al. 1992). These results suggest that susceptibility/resistance to the disease is genetically associated with the MHC class II region in rabbits. Many genetic studies have been carried out in human cohorts, and it has been confirmed that the MHC class II region contains a major genetic susceptibility locus for cervical cancer, and the class I region was excluded (Engelmark et al. 2004; Zoodsma et al. 2005).

Color

In mammals, pigmentation is mainly determined by the distribution of two distinct types of melanin: pheomelanin, which produces red/yellow phenotypes, and eumelanin, which produces dark phenotypes (Searle 1968). The final pigment color results from the relative amounts of eumelanin and pheomelanin that are primarily controlled by *extension* and

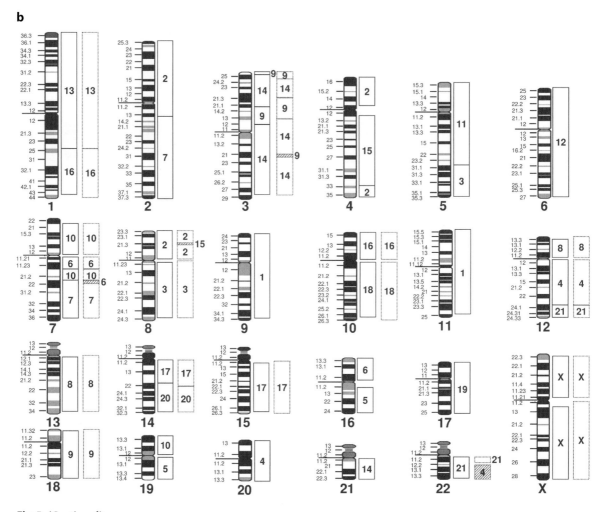

Fig. 5 (Continued)

agouti loci, which show epistatic interactions. The *extension* locus encodes the *melanocortin receptor 1* (*MC1R*) gene that maps to HSA16q24.3. In rabbit, *MC1R* has been assigned to linkage group LG VI (Fox 1994) which has been anchored onto OCU5q (Fig. 4). Five alleles have been reported in rabbit (Fox 1994) and are referred to as E^D (dominant black), E^S (steel, weaker version of E^D), E (normal gray, wild type), e^J (Japanese, mosaic distribution of black and yellow), and e (nonextension of black, coat yellow, white belly). The coding sequence of rabbit *MC1R* has been sequenced from 16 rabbits across 12 breeds with various coat colors (Fontanesi et al. 2006). Four alleles were detected, including two wild-type alleles differing by two synonymous SNPs, one allele with a 30 bp in-frame deletion (c.304_333del30) causing the loss of 10 amino acids that are part of the first extracellular loop, and one allele with a 6 bp in-frame deletion (c.280_285del6) resulting in the loss of two amino acids close to the extracellular end of the second transmembrane domain. The c.304_333del30 and c.280_285del6 act as recessive and dominant alleles, respectively (Fontanesi et al. 2006), and the c.280_285del6 allele corresponds either to E^S or E^D, which has thus been merged into allele E^{D-S} (Fontanesi et al. 2006). The *agouti* mouse gene maps to 89 cM on mouse chromosome (MMU)2, and its human ortholog, the *ASIP* gene, maps to HSA20q11.2-q12. The *agouti* locus has not yet been located in rabbit, but comparative mapping data between man and rabbit suggest that *agouti* maps to OCU4p (Table 2).

The *albino* locus has been genetically mapped to OCU1 (Fig. 4; Xu and Hardison 1989; Korstanje et al. 2001b; Chantry-Darmon et al. 2006), in agreement with the cytogenetic position of *tyrosinase* (*TYR*) on OCU1q14-q15 (Table 2). Mutations in the *TYR* gene are responsible for the *albino* phenotype in mouse (Beermann et al. 1990), and the *albino* phenotype can be rescued in rabbit by generating transgenic rabbits carrying a functional mouse *TYR* gene (Aigner et al. 1996; Brem et al. 1996). The rabbit *TYR* gene has been sequenced and *TYR* coding sequence mutations have been studied in eight rabbit strains varying in coat and eye pigmentation (Aigner et al. 2000). In that study, the German Giant and Belgian Hare stains that harbor the wild genotype with a complete pigmentation of coat and eye had nonstrain-specific coding sequence missense mutations. In contrast, strain-specific mutations were detected in *chinchilla*, *Californian* (adult animals showing coat pigmentation only at extremities), and *albino* animals (New Zealand White and ZIKA®-Hybrid). Colocalization of the *albino* and *TYR* loci is in agreement with these functional data.

Production and Fur Traits

Commercial rabbits are bred for meat and fur. A limited number of short reports suggest that performance in prolificacy, growth rate, meat, and carcass quality are at least partially controlled by inherited genetic factors, but no extensive QTL study of these traits has been published so far.

For meat quality, genetic factors are likely to be involved in intramuscular fat content and fatty acid percentages of meat and perirenal fat (Marin et al. 2006), as well as in sensory parameters that influence meat tenderness such as juiciness, hardness, or fibrousness (Hernandez et al. 2006).

For prolificacy, maternal lines have been selected (reviewed in Garreau et al. 2004) based on doe longevity (Sanchez et al. 2006a and b) and litter size (Piles et al. 2006). A divergent selection experiment for uterine capacity has been carried out, and results suggest the presence of a major gene affecting uterine capacity and a correlated trait, which is the number of implanted embryos (Argente et al. 2003). *Progesterone receptor* (*PRG*) (Peiro et al. 2006), *oviductin* (Merchan et al. 2006), and *tissue inhibitor of matrix metalloproteinase-1* (*TIMP1*) (Estellé et al. 2006) have been studied as candidate genes. SNPs within the *PRG*

promoter and exon 11 of *oviductin* identified previously (Merchan et al. 2005) were tested. Results suggest the exclusion of *oviductin* as a candidate gene (Merchan et al. 2006), but linkage disequilibrium was detected between polymorphisms in the *PRG* gene and litter size in the two divergent selected lines, as measured by the number of implanted embryos, total number of offspring born, and number of offspring born alive (Peiro et al. 2006). No polymorphism in *TIMP1* was found between the two selected rabbit lines, but real-time quantitative PCR indicated that the two lines expressed significantly different levels of *TIMP1* transcript. The three tested genes were chosen as functional candidate genes, not based on genetic mapping data. Other authors have reported a genetic association between litter size and the *casein* locus (Bösze et al. 2002), which maps to OCU15q23 (Table 2). A similar genetic association has recently been found between the *CSN3* and *CSN1S2* genes and litter size in Chinese Xinong Saanen dairy goats that are characterized by a high litter size (Chen et al. 2006). However, these results, which appear to be concordant in two distinct species, should be considered with caution due to the limited number of tested animals.

Genetic linkage has been detected between the *angora* locus and the *FGF5* gene (Mulsant et al. 2004), but no causative mutation within the rabbit *FGF5* gene was identified (Mulsant et al. 2004), in contrast to mouse (Hebert et al. 1994). However, genetic mapping of the *angora* locus to OCU15q (Chantry-Darmon et al. 2006) is in agreement with this linkage because the human *FGF5* gene maps to HSA4q21, which corresponds to OCU15q according to comparative mapping data (Table 2). Results strongly suggest that *FGF5* or a physically close gene or sequence is responsible for the *angora* phenotype.

Rabbits have been selected for the *rex* fur character, leading to the production of the Orylag® strain (Sect. Rabbit Fur and Wool Production). The mating of Orylag® rabbits with wild type rabbits has shown that the *rex* character segregates as an autosomal recessive trait, suggesting that it is a major gene mapping to a unique locus (JL Vrillon and H de Rochambeau, personal communication). Because successive inbreeding steps have been necessary to produce the Orylag® strain, it is likely that the Orylag® status is controlled by a more complex genetic determinism than *rex*. *Rex* and *rexoid* phenotypes have been reported in mouse, rat, cat, hamster, and guinea pig and are mostly characterized by curly whiskers as in rabbit and also a wavy coat found in other species is not reported in *rex* rabbits. The *rex* character was first described in mouse by Crew and Auerbach (1939), referred to as *Re,* with six phenotypic alleles that have been genetically mapped at position 58 cM on MMU11, a locus close to *Krt31c* (*keratin 31 complex*) (Mouse Genome Database, http://www.informatics.jax.org/). However, in the mouse, *Re* mutations behave as autosomal dominant mutations, and because the allelic phenotypes present a wavy coat in contrast to rabbit, it is anticipated that the locus on MMU11 is not a good candidate region for the rabbit *rex* mutation. In rat, hamster, cat, and guinea pig, *rex* mutations are dominant or recessive. Future studies should provide accurate descriptions of the species-specific *rex* mutations because they likely correspond to distinct phenotypes.

7.3
Future Scope of Work

The value of rabbit as a relevant animal model is presented in Section Production of Rabbits for Laboratory Stock and Pets, yet research projects in rabbits have been hampered because of a lack of genomic resources. With the recent development of molecular tools, sequencing data, gene mapping and cytogenetically anchored linkage data, it is anticipated that genomics-based studies using rabbits will increase over the next decade. In this context, some perspectives on rabbit research are given below.

7.3.1
Using Rabbits to Study the Domestication Process

Recent studies using molecular markers have greatly advanced our understanding of the genetic structure and evolution of rabbit populations. This, in turn, has contributed to a better knowledge of the process of rabbit domestication; we now know that domestic breeds are derived from a single subspecies (*O. c. cuniculus*) and have maintained most of the genetic diversity present in their wild ancestors. However, the essence of rabbit domestication is still elusive, and the most fundamental questions

related to the transformation of a wild rabbit into its domestic counterpart remain unanswered. The recent availability of a draft sequence of the rabbit genome together with novel studies focusing on the so-called domestication genes in a variety of species, from plants to animals, will be of great help to address these questions. In particular, the combination of (1) high levels of genetic diversity, (2) the availability of both wild and domestic populations, and (3) a strong phylogeographical structure of wild rabbit populations coupled with a well-known evolutionary history suggest that the European rabbit has enormous potential as a model species in domestication studies. Future research programs will focus on genes involved in the process of domestication and may pinpoint genomic regions that have been sculpted by artificial selection. These studies will ultimately advance our understanding of the molecular basis underlying morphological, physiological, and behavioral changes.

7.3.2
Using Rabbits to Study Color Patterns

Several research projects have been targeted toward the genetic mechanisms underlying color patterns in mammals. Rabbits represent a reservoir of color patterns because breeds have been highly selected and maintained according to size and color. In addition, the number of recorded breeds exceeds 60 (Section 7.1.4), each with characteristic color patterns.

7.3.3
Using Rabbits to Study Early Embryonic Development

Cloning and in vitro fertilization technologies have increased the need for research on embryo development and implantation, as well as on fetal development in mammals. The mouse has been considered a major mammalian model, but other mammals, including rabbits, are appropriate for studying early steps of embryo development and methylation dynamics (Shi et al. 2004; Haaf 2006). In mammals, the preimplantation embryo develops under a coordinate regulation of both maternal and zygotic information and undergoes a maternal to zygotic transition resulting in embryonic gene activation. This gene activation stage is due to a gradual decay of maternal information and an epigenetic methylation reprogramming of paternal and maternal genomes. Embryonic gene activation takes place after several cell cycles, i.e., at the 8–16 cell stage in rabbit (Manes 1973) and at the 4–8 cell stage in man (Braude et al. 1988), in contrast to mouse in which it occurs after a single cell cycle (Flach et al. 1982). These results suggest that the kinetics of transcriptional genome activation of the rabbit embryo is closer to man and most mammalian embryos than the mouse model (Christians et al.1994; Brunet-Simon et al.2001).A global genomic approach to analyze each developmental stage would be desirable, but microarray tools are still missing for the rabbit species. However, subtractive suppressive hybridization experiments have been carried out to study rabbit preimplantation embryos, and differentially expressed genes have been identified despite the scarcity of biological material (Bui et al. 2005). For embryo development studies, the use of rabbits should increase with the growing number of available genomic tools.

The rabbit is also suitable for studies of fetal growth using ultrasound techniques, paving the way for precise analysis of postimplantation development events mediated by microenvironment modifications.

7.3.4
Using Rabbits to Produce Embryonic Stem Cells and Validate Candidate Genes

Rabbits have been used routinely in transgenesis experiments (review in Fan and Watanabe 2003), but the identification of embryonic stem (ES) cells still represents a major challenge in this species. Putative rabbit ES cell lines were first isolated from the inner cell mass of blastocysts in high serum medium conditions (Graves and Moreadith 1993). Despite their capacity to form embryoid bodies in suspension cell cultures and to contribute to chimeras when injected into blastocysts (Schoonjans et al. 1996), these cell lines cannot be considered as true ES cells due to a high frequency of spontaneous mutations. Several attempts to isolate rabbit ES cells have been reported (Fang et al. 2006; Wang et al. 2007). Fang et al. (2006) isolated cells from the inner cell mass of blastocysts derived from embryos of various origins in serum-free medium conditions. The isolated cells proliferated in an undifferentiated state for a prolonged period, had a normal karyotype, expressed pluripotent markers, such as the *POU5F1/OCT-4* gene, produced embryoid bodies, and generated teratoma containing tissue types of all three germ layers. These cells are suitable for genetic engineering and have been successfully used to produce a live animal via nuclear transfer procedures (Fang et al. 2006).

However, because the capacity of these ES cell lines to produce germinal chimeras has not yet been demonstrated, they are considered as putative ES cells. It is anticipated that isolation of rabbit ES cells fulfilling all the appropriate criteria will be achieved very soon. Gene knock-out in rabbit ES cells will represent an alternative physiological model to the mouse model for functional tests and will help in some cases to validate candidate genes in a more relevant animal model.

7.3.5
Perspectives of the Rabbit as a Farm Animal

Resistance to Disease

Rabbits are bred in intensive conditions and are prone to infectious diseases that may be responsible for high mortality rates in breeding units. Antibiotherapy is a prophylactic strategy used to prevent bacterial-induced diseases. However, the international community, especially in Europe, is focusing on sustainable agriculture conditions, and consumers demand more and more that the use of antibiotics be reduced. Indeed, there is a need to better understand host-pathogen interactions, to address questions on the genetic susceptibility/resistance of the host to pathogens, and to introduce immune response parameters in future selection schemes, whenever possible. This problem is common to other species bred in intensive conditions, such as pigs and poultry. Diseases in rabbits depend on the country, but pasteurellosis is one of the most common and severe bacterial diseases, both in commercial breeding units and laboratory facilities (Coudert et al. 1999; reviewed by Kpodékon et al. 1999). Almost all rabbits are carriers of *Pasteurella multocida*, and the main source of infection is the normal carrier status of the animals themselves, with the major mode of transmission being from does to newborns. Most of the *Pasteurella multocida* strains that have developed do not cause disease in their host, but the balance between the rabbit's immune system and the bacterium may change and induce disease. Due to the chronic status of the disease, research on pasteurellosis and the search for rabbit lines genetically resistant to pathogenic strains is ongoing. Eady et al. (2003) estimated the genetic parameters of resistance to pasteurellosis in two French commercial rabbit populations. A statistical model was carried out to identify genetic and environmental factors of the disease trait based on routine observations. The heritability of pasteurellosis incidence was low but significantly different from zero. However, further studies are necessary to better estimate the genetic resistance to the disease. The same authors initiated a project in Australia, i.e., the CSIRO meat rabbit research project (Crusader), aimed at improving rabbit health and survival by selecting for pasteurellosis resistance. Based on weekly recordings of the disease incidence in grower rabbits, heritability of resistance to pasteurellosis was estimated, and this trait is included in the Crusader breeding program.

In late 1996 and early 1997, an emergent and severe gastrointestinal syndrome appeared in rabbit farms in the West of France and has become the main cause of mortality in rabbit farming, affecting broiler rabbits 6–14 weeks old (Duval 1998). The disease has spread very rapidly to other regions of France and to other European countries, including Spain, Portugal, Hungary, Belgium, the Netherlands, and Great Britain (Lebas and Coudert 1997; Jones and Duff 2001). The syndrome has been named Epizootic Rabbit Enteropathy (ERE) and is characterized by opportunistic pathogens (Coudert et al. 2003; Marlier et al. 2006) that hamper the understanding of the disease's origin. The etiologic agent has not been identified, but experimental transmission of ERE to specific-pathogen-free rabbits using infectious inoculums recovered from intestinal contents of diseased animals has been reported (Licois et al. 2005). The infectious inoculums were analyzed, and the results suggest that ERE is a bacterial disease and does not have a viral or parasitic etiology (Szalo et al. 2007). In contrast to other diseases, such as myxomatosis (Silvers et al. 2006), wild rabbits do not seem to be affected. The genetic variability of resistance to ERE has been studied and no clear conclusions

can be drawn despite encouraging results (Garreau et al. 2006).

In conclusion, disease resistance is a major investigation field in livestock production, and research programs are being set up worldwide in all species including rabbit.

Marker-assisted Selection

Classical genetic selection is based on a global genome value referred to as genetic value and does not take into account gene-specific effects. Despite high potential results for some characters, this approach is inefficient or too expensive for traits with low heritability or involving phenotypes that occur late in animal life or are expensive or difficult to measure. Therefore, it is of great interest to identify genes that have a major influence on the selected characters and to identify molecular markers that cosegregate with the selected traits, leading to marker-assisted selection (MAS). However, in contrast to other animals such as cattle, pig, or horse, the individual value of rabbits is low, which limits the economic interest of MAS in this species. The most interesting perspective would be marker-assisted introgression of genes for characters that are difficult to improve by classical genetic methods. As presented in Sect 7.3.5, genetic variability has been detected for production traits such as prolificacy, litter size, meat quality, and fat content. Genetic variability has also been reported in the occurrence of ERE (Garreau et al. 2006), and rabbits can be selected based on their susceptibility to a standard laboratory strain of myxoma virus (Best et al. 2000). QTLs have not been reported for rabbits, but these projects can be done by crossing rabbits harboring divergent phenotypes and analyzing segregation of the traits using the cytogenetically anchored genetic map and comparative mapping data with the well-referenced human genome. It is anticipated that most of the QTL studies will focus on rabbit prolificacy and pathologies. In the near future, a broader definition of breeding goals balancing productivity with improved functional traits such as health will create an integrated component of sustainable production systems.

Acknowledgments. Nuno Ferrand was partially supported by project POCTI/CVT/61590/2004, granted by FCT (Fundação para a Ciência e a Tecnologia), and CIBIO, University of Porto, Portugal.

References

Aigner B, Besenfelder U, Seregi J, Frenyo LV, Sahin-Toth T, Brem G (1996) Expression of the murine wild-type tyrosinase gene in transgenic rabbits. Transgenic Res 5:405–411

Aigner B, Besenfelder U, Müller M, Brem G (2000) Tyrosinase gene variants in different rabbit strains. Mamm Genom 11:700–702

Appella E, Chersi A, Mage RG, Dubiski S (1971) Structural basis of the A14 and A15 allotypic specificities in rabbit immunoglobulin G. Proc Natl Acad Sci USA 68:1341–1345

Arana A, Zaragoza P, Rodellar C, Amorena B (1989) Blood biochemical polymorphisms as markers for genetic characteristics of wild Spanish and domestic rabbits. Genetica 79:1–9

Argente MJ, Blasco A, Ortega JA, Haley CS, Visscher PM (2003) Analyses for the presence of a major gene affecting uterine capacity in unilaterally ovariectomized rabbits. Genetics 163:1061–1068

Ashworth LK, Carrano AV, Mayall BH (1979) DNA measurements of mitotic chromosomes of the rabbit, (*Oryctolagus cuniculus*). Cytogenet Cell Genet 24:68–71

Best SM, Collins SV, Kerr PJ (2000) Coevolution of host and virus: cellular localization of virus in myxoma virus infection of resistant and susceptible European rabbits. Virology 277:76–91

Beermann F, Ruppert S, Hummler E, Bosch FX, Müller G, Rüther U, Schütz G (1990) Rescue of the albino phenotype by introduction of a functional tyrosinase gene into mice. EMBO J 9:2819–2826

Biju-Duval C, Ennafaa H, Dennebouy N, Monnerot M, Mignotte F, Soriguer R, El Gaaïed A, El Hili A, Mounolou J-C (1991) Mitochondrial DNA evolution in lagomorphs: origin of systematic heteroplasmy and organization of diversity in European rabbits. J Mol Evol 33: 92–102

Bolet G, Monnerot M, Arnal C, Arnold J, Bell D, Bergoglio G, Besenfelder U, Bösze S, Boucher S, Brun JM, Chanteloup N, Ducourouble MC, Durand-Tardif M, Esteves PJ, Ferrand N, Hewitt G, Joly T, Koelh PF, Laube M, Lechevestrier S, Lopez M, Masoero G, Piccinin R, Queney G, Saleil G, Surridge A, van der Loo W, Vanhommerig J, Vicente JS, Virag G, Zimmermann JM (1999) A programme for the inventory, characterisation, evaluation, conservation and utilisation of European rabbit (*Oryctolagus cuniculus*) genetic resources. Animl Genet Resour Info 25:57–70

Bolet G, Brun JM, Monnerot M, Abeni F, Arnal C, Arnold J, Bell D, Bergoglio G, Besenfelder U, Bösze S, Boucher S, Chanteloup N, Ducourouble MC, Durand-Tardif M, Esteves PJ, Ferrand N, Gautier A, Haas C, Hewitt G, Jehl N, Joly T, Koelh PF, Laube T, Lechevestrier S, Lopez M, Masoero G, Menigoz JJ, Piccinin R, Queney G, Saleil G, Surridge A, van der Loo W, Vicente JS, Viudes de Castro MP, Virag GY, Zimmermann JM (2000) Evaluation and conservation of European rabbit (*Oryctolagus cuniculus*) genetic resources. First results and inferences. 7thWorld Rabbit Conf, Valencia, Spain, July 4–7, A:281–316

Bösze Z, Bolet G, Mészar Z, Virag G, Devinoy E (2002) Relation between litter size and kappa casein genotype in INRA rabbit lines. Proc 7th World Congr on Genet Appl to Livestock Prod, Belo Horizonte, Brazil, August 13–18, pp 8–10

Boyer MI, Bowen CV, Danska JS (1995) Restriction fragment-length polymorphism analysis of the major histocompatibility complex in New Zealand white rabbits. Transplantation 59(7):1043–1046

Braude P, Bolton V, Moore S (1988) Human gene expression first occurs between the four- and eight-cell stages of preimplantation development. Nature 332:459–461

Branco M, Ferrand N, Monnerot M (2000) Phylogeography of the European rabbit (*Oryctolagus cuniculus*) in the Iberian Peninsula inferred from RFLP analysis of the cytochrome b gene. Heredity 85:307–317

Branco M, Monnerot M, Ferrand N, Templeton A (2002) Postglacial dispersal of the European rabbit (*Oryctolagus cuniculus*) on the Iberian Peninsula reconstructed by nested clade and mismatch analyses of mitochondrial DNA genetic variation. Evolution 56:792–803

Brandsma JL (2005) The cottontail rabbit papillomavirus model of high-risk HPV-induced disease. Meth Mol Med 119:217–235

Breitburd F, Salmon J, Orth G (1997) The rabbit viral skin papillomas and carcinomas: a model for the immunogenetics of HPV-associated carcinogenesis. Clin Dermatol 15:237–247

Brem G, Besenfelder U, Aigner B, Müller M, Liebl I, Schütz G, Montoliu L (1996) YAC transgenesis in farm animals: rescue of albinism in rabbits. Mol Reprod Dev 44:56–62

Bruford MW, Bradley DG, Luikart G (2003) DNA markers reveal the complexity of livestock domestication. Nat Rev Genet 4:900–910

Brunet-Simon A, Henrion G, Renard JP, Duranthon V (2001) Onset of zygotic transcription and maternal transcript legacy in the rabbit embryo. Mol Reprod Dev 58:127–136

Bui LC, Léandri RD, Renard JP, Duranthon V (2005) SSH adequacy to preimplantation mammalian development: scarce specific transcripts cloning despite irregular normalisation. BMC Genomics 6:155

Callou C (2003) De la garenne au clapier. Étude archéozoologique du Lapin en Europe occidentale. Mémoires du Muséum National d'Histoire Naturelle 189 Paris, France

Campo MS (2002) Animal models of papillomavirus pathogenesis Virus Res 89:249–261

Castle WE (1924) On occurrence in rabbits of linkage in inheritance between albinism and brown pigmentation. Proc Natl Acad Sci USA 10:486–488

Cazenave PA, Benammar A, Sogn JA, Kindt TJ (1987) Immunoglobulin genes in the feral rabbit. In: Dubiski S (ed) The rabbit in Contemporary Immunological Research. Longman Sci & Tech, Wiley, New York, pp 148–163

Chan FPH, Sergovitch FR, Shaver EL (1977) Banding patterns in mitotic chromosomes of the rabbit (*Oryctolagus cuniculus*). Can J Genet Cytol 19:625–632

Chantry-Darmon C, Rogel-Gaillard C, Bertaud M, Urien C, Perrocheau M, Chardon P, Hayes H (2003) 133 new gene localizations on the rabbit cytogenetic map. Cytogenet Genom Res 103:192–201

Chantry-Darmon C, Urien C, Hayes H, Bertaud M, Chadi-Taourit S, Chardon P, Vaiman D, Rogel-Gaillard C (2005a) Construction of a cytogenetically anchored-microsatellite map in rabbit. Mamm Genom 16:442–459

Chantry-Darmon C, Rogel-Gaillard C, Bertaud M, Urien C, Perrocheau M, Hayes H, Chardon P, Hayes H (2005b) Expanded comparative mapping between man and rabbit and detection of a new conserved segment between HSA22 and OCU4. Cytogenet Genom Res 11:134–139

Chantry-Darmon C, Urien C, Rochambeau H, Allain D, Pena B, Hayes H, Grohs C, Cribiu EP, Deretz-Picoulet S, Larzul C, Save JC, Neau A, Chardon P, Rogel-Gaillard C (2006) A first generation microsatellite-based integrated genetic and cytogenetic map for the European rabbit (*Oryctolagus cuniculus*) and localization of angora and albino. Anim Genet 37:335–341

Chen H, Lan XY, Lei CX, Pan CY, Zhang RF, Zhang YD, Li RB (2006) Association of CSN3 and CSN1S2 genes with litter size in Chinese Xinong Saanen dairy goat. Proc 8th World Congr on Genet Appl Livestock Prod Belo Horizonte, Brazil, August 13–18, pp 2–15

Chesné P, Adenot PG, Viglietta C, Baratte M, Boulanger L, Renard J-P (2002) Cloned rabbits produced by nuclear transfer from adult somatic cells. Nat Biotechnol 20:366–369

Chowdhary BP, Fronicke L, Gustavsson I, Scherthan H (1996) Comparative analysis of the cattle and human genomes: detection of ZOO-FISH and gene mapping-based chromosomal homologies. Mamm Genom 7:297–302

Christians E, Rao VH, Renard J-P (1994) Sequential acquisition of transcriptional control during early embryonic development in the rabbit. Dev Biol 164:160–172

Cianfriglia M, Pace T (1979) Different chromosome banding procedures to distinguish between mouse and rabbit chromosomes in mouse-rabbit somatic cell hybrids. Bull Soc Italian Biol Sper 55:2208–2214

Cianfriglia M, Miggiano VC, Meo T, Muller HJ, Muller E, Battistuzzi G (1979) Evidence for synteny between the rabbit HPRT, PGK and G6PD in mouse x rabbit somatic cell hybrids. Human Gene Mapping 5. Cytogenet Cell Genet 25:142

Clutton-Brock J (1987) A Natural History of Domesticated Mammals. Cambridge University Press, Cambridge, UK

Clutton-Brock J (1992) Domestication of animals. In: Jones S, Martin R, Pilbeam D (eds) The Cambridge Encyclopedia of

Human Evolution. Cambridge University Press, Cambridge, UK, pp 380–385

Committee for Standardized Karyotype of *Oryctolagus cuniculus* (1981) Standard karyotype of the laboratory rabbit. Cytogenet Cell Genet 31:240–248

Coudert P, Rideaud P, Kpodekon M (1999) Pasteurellosis in the Rabbit: present situation. World Rabbit Sci 7:7

Coudert P, Jobert JL, Larour G, Guittet M (2003) Relation entre l'entéropathie épizootique du lapin (EEL) et l'infestation par les coccidies: enquête épidémiologique. Proc 10th Journées de la Recherche Cunicole, Paris, France, pp 239–241

Crew FAE, Auerbach C (1939) Rex: a dominant autosomal monogenic coat texture character in the mouse. J Genet 38:341–344

Debecker A, Martin-DeLeon PA (1992) Assignment of the rabbit genes for alpha (PHKA) and beta (PHKB) phosphorylase kinase subunits. Cytogenet Cell Genet 61:208–210

Doebley J (2004) The genetics of maize evolution. Annu Rev Genet 38:37–59

Douzery EJP, Huchon D (2004) Rabbits, if anything, are likely Glires. Mol Phylogenet Evol 33:922–935

Dutrillaux B, Viegas-Péquignot E, Couturier J (1980) [Great homology of chromosome banding of the rabbit (*Oryctolagus cuniculus*) and primates, including man (author's transl)]. Ann Genet 23:22–25 French

Duval ML (1998) Développement de l'entérocolite en France. ITAVI (ed). Séances d'actualité: l'Entérocolite Epizootique. Proc7th Journées de la Recherche Cunicole, Paris, France, pp 1–8

Eady SJ, Garreau H, Hurtaud J (2003) Heritability of resistance to bacterial infection in commercial meat rabbit populations. Proc 8th World Rabbit Congr, Puebla, Mexico, September 7–10, pp 51–56

Echard G (1973) Bandes chromosomiques de type G chez le lapin domestique (*Oryctolagus cuniculus*). Ann Genet 5:425–434

Echard G, Gellin J, Benne F, Gillois M (1981) The gene map of the rabbit (*Oryctolagus cuniculus* L.) I. Synteny between the rabbit gene loci coding for HPRT, PGK, G6PD and GLA: their localization on the X chromosome. Cytogenet Cell Genet 29:176–183

Echard G, Gellin J, Gillois M (1982) The gene map of the rabbit (*Oryctolagus cuniculus*L). II Analysis of the segregation of 11 enzymes in rabbit x hamster somatic cell hybrids: two syntenic groups, LDHB-TPI and LDHA-ACP2. Cytogenet Cell Genet 34:289–295

Ekes C, Csonka E, Hadlaczky G, Cserpan I (2004) Isolation, cloning and characterization of two major satellite DNA families of rabbit (*Oryctolagus cuniculus*). Gene 343:271–279

Engelmark M, Beskow A, Magnusson J, Erlich H, Gyllensten U (2004) Affected sib-pair analysis of the contribution of HLA class I and class II loci to development of cervical cancer. Hum Mol Genet 13:1951–1958

Estellé J, Sastre Y, Merchan M, Peiro R, Santacreu MA, Folch JM (2006) TIMP-1 as candidate gene for embryo survival in two divergent lines selected for uterine capacity in rabbits. Mol Reprod Dev 73:678–384

Esteves PJ, Alves PC, Ferrand N, van der Loo W (2002) Restriction fragment alleles of the rabbit IGHG genes with reference to the rabbit IGHGCH2 or e locus polymorphism. Anim Genet 33:309–311

Esteves PJ, Lanning D, Ferrand N, Knight KL, Zhai SK, van der Loo W (2004) Allelic Variations at the V_Ha Locus in Natural Populations of Rabbit (*Oryctolagus cuniculus*, L). J Immunol 172:1044–1053

Esteves PJ, Carmo C, Godinho R, van der Loo W (2006) Genetic diversity at the hinge region of the unique immunoglobulin heavy gamma (IGHG) gene in leporids (*Oryctolagus*, *Sylvilagus* and *Lepus*). Intl J Immunogenet 33:171–177

Fadiel A, Anidi I, Eichenbaum KD (2005) Farm animal genomics and informatics: an update. Nucl Acids Res 33:6309–6318

Fan J, Watanabe T (2003) Transgenic rabbits as therapeutic protein bioreactors and human disease models. Pharmacol Therap 99:261–283

Fang ZF, Gai H, Huang YZ, Li SG, Chen XJ, Shi JJ, Wu L, Liu A, Xu P, Sheng HZ (2006) Rabbit embryonic stem cell lines derived from fertilized, parthenogenetic or somatic cell nuclear transfer embryos. Exp Cell Res 312:3669–3682

Ferrand N, Branco M (2006) The evolutionary history of the European rabbit (*Oryctolagus cuniculus*): major patterns of population differentiation and geographic expansion inferred from protein polymorphism. In: Weiss S, Ferrand N (eds) Phylogeography of Southern European Refugia. Springer, Amsterdam, The Netherlands, pp 207–235

Flach G, Johnson MH, Braude PR, Taylor RAS, Bolton VN (1982) The transition from maternal to embryonic control in the 2-cell mouse embryo. EMBO J 1:681–686

Fontanesi L, Tazzoli M, Beretti F, Russo V (2006) Mutations in the melanocortin 1 receptor (MC1R) gene are associated with coat colours in the domestic rabbit (*Oryctolagus cuniculus*). Anim Genet 37:489–493

Ford CE, Pollock DL Gustavsson I (1980) Proceedings of the First International Conference for the Standardisation of Banded Karyotypes of Domestic Animals. University of Reading, England. 2nd-6th August 1976. Hereditas 92:145–162

Fox RR (1984) The rabbit as a research subject. Physiologist 27:393–402

Fox RR (1994) Taxonomy and genetics. In: Manning PJ, Ringler DH, Newcomer CE (eds) The Biology of the Laboratory Rabbits, 2nd edn. Academic Press, San Diego, pp 1–25

Garreau H, Piles M, Larzul C, Baselga M, de Rochambeau H (2004) Selection of maternal lines: last results and prospects. Proc 8th World Rabbit Congr, Pueblo, Mexico, September 7–10, pp 14–28

Garreau H, Licois D, Rupp R, de Rochambeau H (2006) Genetic variability of the resistance to Epizootic Rabbit Entheropathy (ERE). Proc 8th World Congr on Genet Appl Livestock Prod, Belo Horizonte, Brazil, August 13–18, pp15–28

Gellin J, Dalens M, Echard G, Hatey F (1983) Carte génique du lapin (*Oryctolagus cuniculus*): synténie entre les gènes utéroglobin, lactate deshydrogenase A et phosphatase acide 2. Genet Sel Evol 15:489–494

Gellin J, Echard G, Yerle M, Dalens M, Chevalet C, Gillois M (1985) Localization of the alpha and beta casein genes to the q24 region of chromosome 12 in the rabbit (*Oryctolagus cuniculus* L) by in situ hybridization. Cytogenet Cell Genet 39:220–223

Geraldes A, Rogel-Gaillard C, Ferrand N (2005) High levels of nucleotide diversity in the European rabbit (*Oryctolagus cuniculus*) SRY gene. Anim Genet 36:349–351

Geraldes A and Ferrand N (2006) A 7-bp insertion in the 3' untranslated region suggests the duplication and concerted evolution of the rabbit SRY gene. Genet Sel Evol 38:313–320

Geraldes A, Ferrand N, Nachman M (2006) Contrasting patterns of introgression at X-linked loci across the hybrid zone between subspecies of the European rabbit (*Oryctolagus cuniculus*). Genetics 173:919–933

Gibbs RA, Weinstock GM, Metzker ML, Muzny DM, Sodergren EJ, et al. (2004) Genome sequence of the Brown Norway rat yields insights into mammalian evolution. Nature 428:493–521

Gissi C, Gullberg A, Arnason U (1998) The complete mitochondrial DNA sequence of the rabbit, *Oryctolagus cuniculus*. Genomics 50:161–169

Goureau A, Yerle M, Schmitz A, Riquet J, Milan D, Pinton P, Frelat G, Gellin J (1996) Human and porcine correspondence of chromosome segments using bi-directional chromosome painting. Genomics 36:252–262

Graves KH, Moreadith RW (1993) Derivation and characterization of putative pluripotential embryonic stem cells from preimplantation rabbit embryos. Mol Reprod Dev 36:424–433

Gustavsson I (1972) Mitotic and meiotic chromosomes of the variable hare (*Lepus timidus* L), the common hare (*Lepus europaeus* Pall) and their hybrids. Hereditas 67:27–34

Haaf T (2006) Methylation dynamics in the early mammalian embryo: implications of genome reprogramming defects for development. Curr Top Microbiol Immunol 310:13–22

Hageltorn M, Gustavsson I (1979) Identification by banding techniques of the chromosomes of the domestic rabbit (*Oryctolagus cuniculus* L). Hereditas 90:269–279

Han R, Breitburd F, Marche PN, Orth G (1992) Linkage of regression and malignant conversion of rabbit viral papillomas to MHC class II genes. Nature 356:66–68

Hardy C, Casane D, Vigne JD, Callou C, Dennebouy N, Mounolou JC, Monnerot M (1994) Ancient DNA from

Bronze Age bones of European rabbit, (*Oryctolagus cuniculus*). Experientia 50:564–570

Hardy C, Callou C, Vigne JD, Casane D, Dennebouy N, Mounolou JC, Monnerot M (1995) Rabbit mitochondrial DNA diversity from prehistoric to modern times. J Mol Evol 40:227–237

Hayes H, Petit E, Lemieux N, Dutrillaux B (1992) Chromosomal localization of the ovine beta-casein gene by non-isotopic in situ hybridization and R-banding. Cytogenet Cell Genet 61:286–288

Hayes H (1995) Chromosome painting with human chromosome-specific DNA libraries reveals the extent and distribution of conserved segments in bovine chromosomes. Cytogenet Cell Genet 71:168–174

Hayes H, Rogel-Gaillard C, Zijlstra C, De Haan NA, Urien C, Bourgeaux N, Bertaud M, Bosma AA (2002) Establishment of an R-banded rabbit karyotype nomenclature by FISH localization of 23 chromosome-specific genes on both G- and R-banded chromosomes. Cytogenet Genome Res 98:199–205

Hebert JM, Rosenquist T, Gotz J, Martin GR (1994) FGF5 as a regulator of the hair growth cycle: evidence from targeted and spontaneous mutations. Cell 78:1017–1025

Hernandez P, Arino B, Pla M, Blasco A (2006) Comparison between rabbit lines for sensory meat quality. Proc 8th World Congr on Genet Appl Livestock Prod, Belo Horizonte, Brazil, August 13–18, pp 13–14

HGM10 (1989) Tenth International Workshop on Human Gene Mapping. Cytogenet Cell Genet 51:1–1148

Houdebine L-M (2002) Animal transgenesis: recent data and perspectives. Biochimie 84:1137–1141

Iannuzzi L, Di Meo GP, Perucatti A, Incarnato D (1999) Comparison of the human with the sheep genomes by use of human chromosome-specific painting probes. Mamm Genom 10:719–723

ISCN (1981) An International System for Human Cytogenetic Nomenclature-High Resolution Banding. Birth Defects. Original Article Series, 17, New York, March of Dimes Birth Defects Foundation, 1981; also in Cytogenet Cell Genet 31:1–23

Issa M, Atherton GW, Blank CE (1968) The chromosomes of the domestic rabbit, *Oryctolagus cuniculus*. Cytogenetics 7:361–375

Jones JR, Duff JP (2001) Rabbit epizootic enterocolitis. Vet Rec 149:532

Kirkness EF, Bafna V, Halpern AL, Levy S, Remington K, Rusch DB, Delcher AL, Pop M, Wang W, Fraser CM, Venter JC (2003) The dog genome: survey sequencing and comparative analysis. Science 301:1898–1903

Korstanje R, O'Brien PC, Yang F, Rens W, Bosma AA, van Lith HA, van Zutphen LF, Ferguson-Smith MA (1999) Complete homology maps of the rabbit (*Oryctolagus cuniculus*) and human by reciprocal chromosome painting. Cytogenet Cell Genet 86:317–322

Korstanje R, Gillissen GF, Kodde LP, Den Bieman M, Lankhorst A, Van Zutphen LF, van Lith HA (2001a) Mapping of microsatellite loci and association of aorta atherosclerosis with LGVI markers in the rabbit. Physiol Genom 6:11–18

Korstanje R, Gillissen GF, den Bieman MG, Versteeg SA, van Oost B, Fox RR, van Lith HA, van Zutphen LF (2001b) Mapping of rabbit chromosome 1 markers generated from a microsatellite-enriched chromosome-specific library. Anim Genet 32:308–312

Korstanje R, Gillissen GF, Versteeg SA, van Oost BA, Bosma AA, Rogel-Gaillard C, van Zutphen LF, van Lith HA (2003) Mapping of rabbit microsatellite markers using chromosome-specific libraries. J Hered 94:161–169

Kos CH, Tihy F, Murer H, Lemieux N, Tenenhouse HS (1996) Comparative mapping of Na+-phosphate cotransporter genes, NPT1 and NPT2, in human and rabbit. Cytogenet Cell Genet 75:22–24

Kpodékon M, Rideaud P, Coudert P (1999) Pasteurellose du lapin: revue. Revue de Médecine Vétérinaire 150:221–232

Krane DE, Clark AG, Cheng J-F, Hardison RC (1991) Subfamily relationships and clustering of rabbit C repeats. Mol Biol Evol 8:1–30

Kreider JW, Bartlett GL (1981) The Shope papilloma-carcinoma complex of rabbits: a model system of neoplastic progression and spontaneous regression. Adv Cancer Res 35:81–110

Lagercrantz U, Ellegren H, Andersson L (1993) The abundance of various polymorphic microsatellite motifs differs between plants and vertebrates. Nucl Acids Res 21:1111–1115

Lander ES, Linton LM, Birren B, Nusbaum C, Zody MC et al. (2001) Initial sequencing and analysis of the human genome. Nature 409:860–921

Lebas F, Coudert P, de Rochambeau H, Thébault RG (1997) The Rabbit – Husbandry, Health and Production. FAO Animal Production and Health Series No. 21, FAO Publ, Rome, Italy

Lebas F, Coudert P (1997) Entérocolite: les données récentes. Cuniculture 24:269–272

Lefranc MP, Pommié C, Kaas Q, Duprat E, Bosc N, Guiraudou D, Jean C, Ruiz M, Da Piédade I, Rouard M, Foulquier E, Thouvenin V, Lefranc G (2005) IMGT unique numbering for immunoglobulin and T cell receptor constant domains and Ig superfamily C-like domains. Dev Comp Immunol 29:185–203

Lemieux N, Dutrillaux B (1992) New gene assignments to rabbit chromosomes; implications for chromosome evolution. Cytogenet Cell Genet 61:132–134

Lemieux N, Dutrillaux B, Viegas-Péquignot E (1992) A simple method for simultaneous R- or G-banding and fluorescence in situ hybridization of small single-copy genes. Cytogenet Cell Genet 59:311–312

Li S, Chen X, Fang Z, Shi J, Sheng HZ (2006) Rabbits generated from fibroblasts through nuclear transfer. Reproduction 131:1085–1090

Licois D, Wyers M, Coudert P (2005) Epizootic Rabbit Enteropathy: experimental transmission and clinical characterization. Vet Res 36:601–613

Litt M, Luty JA (1989) A hypervariable microsatellite revealed by in vitro amplification of a dinucleotide repeat within the cardiac muscle actin gene. Am J Hum Genet 44:397–401

Liu GR, Miyamoto MM (1999) Phylogenetic assessment of molecular and morphological data for Eutherian mammals. Syst Biol 48:54–64

Long JR, Qiu XP, Zeng FT, Tang LM, Zhang YP (2003) Origin of rabbit (*Oryctolagus cuniculus*) in China: evidence from mitochondrial DNA control region sequence analysis. Anim Genet 34:82–87

Lopez-Martinez N (2006) The lagomorph fossil record and the origin of the wild rabbit. In: Hackländer K, Alves PC, Ferrand N, (eds) Lagomorph Biology: Ecology, Evolution and Conservation. Springer, Amsterdam, The Netherlands, pp 27–46

Mage R, Lieberman R, Potter M, Terry WT (1973) The immunoglobulin allotypes. In: Sela M (ed) The Antigens 1. Academic Press, New York, USA, pp 299–376

Mage RG (1986) Rabbit immunoglobulin allotypes. In: Weir DM, Herzenberg LA, Blackwell CC (eds) Handbook of Experimental Immunology, vol 3. Blackwell Scientific Publ, Oxford, pp 164–190

Mahoney CE, Picciano SR, Burton KM, Martin-DeLeon PA (1988) Regional mapping of the creatine kinase b (CKBB) gene in rabbit (*Oryctolagus cuniculus*) and man using a rat cDNA probe. Cytogenet Cell Genet 48:160–163

Manes C (1973) The participation of the embryonic genome during early cleavage in the rabbit. Dev Biol 32:753–759

Manning PJ, Ringler DH, Newcomer CE (eds) (1994) The Biology of the Laboratory Rabbits, 2nd edn. Academic Press, San Diego

Marche PN, Rebiere MC, Laverriere A, English DW, LeGuern C, Kindt TJ (1989) Definition of rabbit class I and class II MHC gene haplotypes using molecular typing procedures. Immunogenetics 29:273–276

Marin C, Arino B, Blasco A, Hernandez P (2006) Rabbit line comparison for lipid content, lipolytic activities and fatty acid composition of leg meat and perirenal fat. Proc 8th World Congr on Genet Appl Livestock Prod, Belo Horizonte, Brazil, August 13–18, pp 13–19

Marlier D, Dewree R, Lassence C, Licois D, Mainil J, Coudert P, Meulemans L, Ducatelle R, Vindevogel H (2006) Infectious agents associated with epizootic rabbit enteropathy: isolation and attempts to reproduce the syndrome. Vet J 172:493–500

Martin-DeLeon PA (1980) Location of the 18S and 28S rRNA cistrons in the genome of the domestic rabbit (*Oryctolagus cuniculus* L). Cytogenet Cell Genet 28:34–40

Martin-DeLeon PA, Picciano SR (1988) Localization of the *KRAS2* oncogene in the domestic rabbit (*Oryctolagus cuniculus*). Cytogenet Cell Genet 48:201–204

Martin-DeLeon PA, McLaughlin J, Mitzel E, Tailor S (1989) Mapping of the creatine kinase M gene to 19q11-q12 in the rabbit genome. Cytogenet Cell Genet 50:165–167

Martin-DeLeon PA, Canaff L, Korstanje R, Bhide V, Selkirk M, Hendy GN (1999) Rabbit calcium-sensing receptor (CASR) gene: chromosome location and evidence for related genes. Cytogenet Cell Genet 86:252–258 (Erratum in: Cytogenet Cell Genet 2000; 88: 168–169)

Martin-DeLeon PA, Piumi F, Canaff L, Rogel-Gaillard C, Hendy GN (2001) Assignment of the parathyroid hormone/parathyroid hormone-related peptide receptor (PTHR1) to rabbit chromosome band 9p14->p13 by fluorescence in situ hybridization. Cytogenet Cell Genet 94:90–91

Matthee CA, Van Vuuren BJ, Bell D, Robinson TJ (2004) A molecular supermatrix of the rabbits and hares (*Leporidae*) allows for the identification of five intercontinental exchanges during the Miocene. Syst Biol 53:433–447

Medrano L, Dutrillaux B (1984) Chromosomal location of immunoglobulin genes: partial mapping of these genes in the rabbit and comparison with Ig genes carrying chromosomes of man and mouse. Adv Cancer Res 41:323–367

Merchan M, Peiro R, Argente MJ, Agea I, Santacreu MA, Garcia ML, Blasco A, Folch JM (2006) Candidate genes for reproductive traits in rabbit: I. oviductin gene. Proc 8th World Congr on Genet Appl Livestock Prod, Belo Horizonte, Brazil, August 13–18, pp11–20

Merchan M, Peiro R, Estellé J, Sastre Y, Santacreu MA, Folch JM (2005) Candidate gene analysis in two lines of rabbits divergently selected for uterine capacity. Reprod Domest Anim 40:409

Miesfeld R, Krystal M, Arnheim N (1981) A member of a new repeated sequence family, which is conserved throughout eucaryotic evolution, is found between the human delta and beta globin genes. Nucl Acids Res 9:5931–5947

Misawa K, Janke A (2003) Revisiting the Glires concept-phylogenetic analysis of nuclear sequences. Mol Phylogenet Evol 28:320–327

Monnerot M, Vigne JD, Biju-Duval C, Casane D, Callou C, Hardy C, Mougel F, Soriguer R, Dennebouy N, Mounolou J-C (1994) Rabbit and man: genetic and historic approach. Genet Sel Evol 26(suppl 1):167s–182s

Mougel F, Mounolou JC, Monnerot M (1997) Nine polymorphic microsatellite loci in the rabbit, *Oryctolagus cuniculus*. Anim Genet 28:58–59

Muller C, Denis M, Gentzbittel L, Faraut T (2004) The Iccare web server: an attempt to merge sequence and mapping information for plant and animal species. Nucl Acids Res 32:W429–W434

Mulsant P, de Rochambeau H, Thébault RG (2004) A note on the linkage between the angora and Fgf5 genes in rabbits. World Rabbit Sci 12:1–6

Nei M (1987) Molecular Evolutionary Genetics. Columbia University Press, New York, USA

Nichols WW, Levan A, Hansen-Melander E, Melander Y (1965) The idiogram of the rabbit. Hereditas 53:63–76

Niederberger V (1989) Génétique et élevage du lapin rex: Historique, situation actuelle, perspectives d'avenir. Veterinary thesis. Ecole Nationale Vétérinaire d'Alfort, France

Oudin J (1956) L'"allotypie" de certains antigènes protéidiques du sérum. C R Hebd Seances Acad Sci. Comptes rendus de l'Académie des Sciences de Paris 242:2606–2608

Pauloin A, Rogel-Gaillard C, Piumi F, Hayes H, Fontaine ML, Chanat E, Chardon P, Devinoy E (2002) Structure of the rabbit alpha s1- and beta-casein gene cluster, assignment to chromosome 15 and expression of the alpha s1-casein gene in HC11 cells. Gene 283:155–162

Peiro R, Merchan M, Santacreu MA, Argente MJ, Garcia ML, Agea I, Folch JM, Blasco A (2006) Candidate genes for reproductive traits in rabbit: II. Progesterone receptor gene. Proc 8th World Congr on Genet Appl Livestock Prod, Belo Horizonte, Brazil, August 13–18, pp 11–13

Peterka M, Hartl GB (1992) Biochemical-genetic variation and differentiation in wild and domestic rabbits. On the significance of genetic distances, dendrograms and the estimation of divergence times in domestication studies. Zeitschrift für Zoologische Systematik und Evolutionforschung 30:129–141

Piles M, Garcia ML, Rafel O, Ramon J, Baselga M (2006) Genetics of litter size in three maternal lines of rabbits: repeatability versus multiple-trait models. J Anim Sci 84:2309–2315

Poulsen BS, Shibasaki Y, Ronne M (1988) The high-resolution R-banded karyotype of *Oryctolagus cuniculus* L. Hereditas 109:57–60

Queney G, Ferrand N, Marchandeau S, Azevedo M, Mougel F, Branco M, Monnerot M (2000) Absence of a genetic bottleneck in a wild rabbit (*Oryctolagus cuniculus*) population exposed to a severe viral epizootic. Mol Ecol 9:1253–1264

Queney G, Ferrand N, Weiss S, Mougel F, Monnerot M (2001) Stationary distributions of microsatellite loci between divergent population groups of the European rabbit (*Oryctolagus cuniculus*). Mol Biol Evol 18:2169–2178

Queney G, Vachot A-M, Brun J-M, Dennebouy N, Mulsant P, Monnerot M (2002) Different levels of human intervention in domestic rabbits: effects on genetic diversity. J Hered 93:205–209

Raudsepp T, Fronicke L, Scherthan H, Gustavsson I, Chowdhary BP (1996) Zoo-FISH delineates conserved chromosomal segments in horse and man. Chrom Res 4:218–225

Ray M, Williams TW (1966) Karyotype of rabbit chromosomes from leukocyte cultures. Can J Genet Cytol 8:393–397

Rico C, Rico I, Webb N, Smith S, Bell D, et al. (1994) Four polymorphic microsatellite loci for the European wild rabbit, *Oryctolagus cuniculus*. Anim Genet 25:367

Robinson TJ (1980) Comparative chromosome studies in the family Leporidae (*Lagomorpha*, Mammalia). Cytogenet Cell Genet 28:64–70

Robinson TJ, Elder FF, Chapman JA (1983) Karyotypic conservatism in the genus Lepus (order *Lagomorpha*). Can J Genet Cytol 25:540–544

Robinson TJ, Yang F, Harrison WR (2002) Chromosome painting refines the history of genome evolution in hares and rabbits (order *Lagomorpha*). Cytogenet Genom Res 96:223–227

Robinson TJ, Matthee CA (2005) Phylogeny and evolutionary origins of the *Leporidae*: a review of cytogenetics, molecular analyses and a supermatrix analysis. Mamm Rev 35:231–247

Rogel-Gaillard C, Zijlstra C, Bosma AA, Thepot D, Fontaine ML, Devinoy E, Chardon P (2000) Assignment of the rabbit whey acidic protein gene (WAP) to rabbit chromosome 10 by in situ hybridization and description of a large region surrounding this gene. Cytogenet Cell Genet 89:107–109

Rogel-Gaillard C, Piumi F, Billault A, Bourgeaux N, Save JC, Urien C, Salmon J, Chardon P (2001) Construction of a rabbit bacterial artificial chromosome (BAC) library: application to the mapping of the major histocompatibility complex to position 12q1.1. Mamm Genom 12:253–255

Rønne M, Gyldenholm AO, Storm CO (1993) The RBG-banded karyotype of *Oryctolagus cuniculus* at the 550-band stage. In Vivo 7:135–138

Rønne M (1995) The high-resolution RBG-banded karyotype of *Oryctolagus cuniculus*. Comparison between different classifications of landmarks and bands. In Vivo 9:239–245

Ros F, Puels J, Reichenberger N, van Schooten W, Buelow R, Platzer J (2004) Sequence analysis of 0.5 Mb of the rabbit germline immunoglobulin heavy chain locus. Gene 330:49–59

Rougeot J (1981) Origine et histoire du lapin. Le lapin: aspects historiques, culturels et sociaux. Colloque Société d'Ethnozootechnie, Paris, France. Ethnozootechnie 27

Saccone S, Caccio S, Kusuda J, Andreozzi L, Bernardi G (1996) Identification of the gene-richest bands in human chromosomes. Gene 174:85–94 (Erratum in: Gene 1999, 186:151)

Sanchez JP, Theilgaard P, Mingues C, Baselga M (2006a) Constitution and evaluation of a long-lived-productive maternal rabbit line. Proc 8th World Congr on Genet Appl Livestock Prod, Belo Horizonte, Brazil, August 13–18, pp 11–22

Sanchez JP, Korsgaard IR, Damgaard LH, Baselga M (2006b) Analysis of rabbit doe longevity using a semiparametric log-Normal animal frailty model with time-dependant covariates. Genet Sel Evol 38:281–295

Schroder J, van der Loo W (1979) Comparison of karyotypes in three species of rabbit: *Oryctolagus cuniculus, Sylvilagus nuttallii* and *S. idahoensis*. Hereditas 91:27–30

Searle AG (1968) Comparative Genetics of Coat Colour in Mammals. Logos Press, London, UK

Sharpe CR, Franco EL (1995) Etiology of Wilms' tumor. Epidemiol Rev 17:415–432

Shi W, Dirim F, Wolf E, Zakhartchenko V, Haaf T (2004) Methylation reprogramming and chromosomal aneuploidy in in vivo fertilized and cloned rabbit preimplantation embryos. Biol Reprod 71:340–347

Schoonjans L, Albright GM, Li JL, Collen D, Moreadith RW (1996) Pluripotential rabbit embryonic stem (ES) cells are capable of forming overt coat color chimeras following injection into blastocysts. Mol Reprod 45:439–443

Silvers L, Inglis B, Labudovic A, Janssens PA, van Leeuwen BH, Kerre PJ (2006) Virulence and pathogenesis of the MSW and MSD strains of Californian myxoma virus in European rabbits with genetic resistance to myxomatosis compared to rabbits with no genetic resistance. Virology 348:72–83

Simpson GG (1945) The principles of classification and a classification of mammals. Bull Am Mus Nat Hist 85:1–272

Solinas-Toldo S, Lengauer C, Fries R (1995) Comparative genome map of human and cattle. Genomics 27:489–496

Soulié J, de Grouchy J (1982) Of rabbit and man: comparative gene mapping. Hum Genet 60:172–175

Soulié J, de Grouchy J (1983) New gene assignments in the rabbit (*Oryctolagus cuniculus*). Comparison with other species. Hum Genet 63:48–52

Spritz RA (1981) Duplication/deletion polymorphism 5'- to the human beta globin gene. Nucl Acids Res 9:5037–5047

Stills HF Jr (1994) Polyclonal antibody production. In: Manning PJ, Ringler DH, Newcomer CE (eds) The Biology of the Laboratory Rabbits, 2nd edn. Academic Press, San Diego, USA, pp 435–448

Stock AD (1976) Chromosome banding pattern relationships of hares, rabbits, and pikas (order *Lagomorpha*). A phyletic interpretation. Cytogenet Cell Genet 17:78–88

Surridge AK, Bell DJ, Rico C, Hewitt GM (1997) Polymorphic microsatellite loci in the European rabbit (*Oryctolagus cuniculus*) are also amplified in other Lagomorph species. Anim Genet 28:302–305

Szalo IM, Lassence C, Licois D, Coudert P, Poulipoulis A, Vindevogel H, Marlier D (2007) Fractionation of the reference inoculum of epizootic rabbit enteropathy in discontinuous sucrose gradient identifies aetiological agents in high density fractions. Vet J 173:652–657

Tautz D (1989) Hypervariability of simple sequences as a general source for polymorphic DNA markers. Nucl Acids Res 17:6463–6471

Teupser D, Burkhardt R, Wilfert W, Haffner I, Nebendahl K, Thiery J (2006) Identification of macrophage arginase I as a new candidate gene of atherosclerosis resistance. Arterioscler Thromb Vasc Biol 26:365–371

The ENCODE Project Consortium et al. (2004) The ENCODE (ENCyclopedia Of DNA Elements). Science 306:636–639

Thebault RG, Allain D, de Rochambeau H, Vrillon JL (2000) Selection scheme and genetic improvement of Orylag for fur production. Proc 7th Intl Congr on Fur Anim Prod, Kastoria, Greece, August 13–15, 24:81–86

Thiery J, Nebendahl K, Rapp K, Kluge R, Teupser D, Seidel D (1995) Low atherosclerotic response of a strain of rabbits to diet-induced hypercholesterolemia. Arterioscler Thromb Vasc Biol 15:1181–1188

Toth G, Gaspari Z, Jurka J (2000) Microsatellites in different eukaryotic genomes: survey and analysis. Genom Res 10:967–981

Vaiman D (2005) DNA sequences of the human and other mammalian genomes. In: Rivinsky A, Graves M (eds) Mammalian Genomics. CABI, Wallingford, UK, pp 87–116

van den Hout H, Reuser AJ, Vulto AG, Loonen MC, Cromme-Dijkhuis A, van der Ploeg AT (2000) Recombinant human alpha-glucosidase from rabbit milk in Pompe patients. Lancet 356:397–398

van den Hout H, Reuser AJ, de Klerk JB, Arts WF, Smeitink JA, van der Ploeg AT (2001) Enzyme therapy for Pompe disease with recombinant human alpha-glucosidase from rabbit milk. J Inherit Metab Dis 24:266–274

van der Loo W (1987) Studies on the adaptive significance of the immunoglobulin alleles (Ig allotypes) in wild rabbit. In: Dubiski S (ed) The Rabbit in Contemporary Immunological Research. Longman Scientific & Technical, Wiley, New York, pp 164–190

van der Loo W, Arthur CP, Wallage-Drees M, Richardson B (1987) Non-random allele associations between unlinked protein loci: are the polymorphisms of the immunoglobulin constant regions adaptive? Proc Nat Acad Sci USA 84:3075–3079

van der Loo W, Ferrand N, Soriguer R (1991) Estimation of gene diversity at the b locus of the constant region of the immunoglobulin light chain in natural populations of European rabbit (*Oryctolagus cuniculus*) in Portugal, Andalusia and on the Azorean islands. Genetics 127:789–799

van Haeringen WA, den Bieman M, van Zutphen LF, van Lith HA (1996–1997) Polymorphic microsatellite DNA markers in the rabbit (*Oryctolagus cuniculus*). J Exp Anim Sci 38:49–57

van Haeringen WA, Den Bieman M, Gillissen GF, Lankhorst AE, Kuiper MT, Van Zutphen LF, van Lith HA (2001) Mapping of a QTL for serum HDL cholesterol in the rabbit using AFLP technology. J Hered 92:322–326

van Haeringen WA, Den Bieman MG, Lankhorst AE, van Lith HA, van Zutphen LF (2002) Application of AFLP markers for QTL mapping in the rabbit. Genome 45:914–921

van Lith HA, van Zutphen LF (1996) Characterization of rabbit DNA microsatellites extracted from the EMBL nucleotide sequence database. Anim Genet 27:387–395

van Zutphen LF, Fox RR (1977) Strain differences in response to dietary cholesterol by JAX rabbits: correlation with esterase patterns. Atherosclerosis 28:435–446

Vicente JS, Viudes de Castro MP, Lavara R, Moce E (2004) Study of fertilising capacity of spermatozoa after heterospermic insemination in rabbit using DNA markers. Theriogenology 61:1357–1365

Viebahn C, Stortz C, Mitchell SA, Blum M (2002) Low proliferative and high migratory activity in the area of Brachyury expressing mesoderm progenitor cells in the gastrulating rabbit embryo. Development 129:2355–2365

Viegas-Péquignot E, Dutrillaux B, Magdelenat H, Coppey-Moisan M (1989) Mapping of single-copy DNA sequences on human chromosomes by in situ hybridization with biotinylated probes: enhancement of detection sensitivity by intensified-fluorescence digital-imaging microscopy. Proc Natl Acad Sci USA 86:582–586

Vilà C, Seddon J, Ellegren H (2005) Genes of domestic mammals augmented by backcrossing with wild ancestors. Trends Genet 21:214–218

Vos P, Hogers R, Bleeker M, Reijans M, van de Lee T, Hornes M, Frijters A, Pot J, Peleman J, Kuiper M (1995) AFLP: a new technique for DNA fingerprinting. Nucl Acids Res 23:4407–4414

Vos P, Kuiper MTR (1997) AFLP analysis. In: Caetano-Anollés G, Gresshoff PM (eds) DNA Markers: Protocols, Applications and Overviews. Wiley, New York, pp 115–131

Wang S, Tang X, Niu Y, Chen H, Li B, Li T, Zhang X, Hu Z, Ji W (2007) Generation and characterization of rabbit embryonic stem cells. Stem Cells 25:481–489

Waterston RH, Lindblad-Toh K, Birney E, Rogers J, Abril JF, et al. (2002) Initial sequencing and comparative analysis of the mouse genome. Nature 420:520–562

Weber JL, May PE (1989) Abundant class of human DNA polymorphisms, which can be typed using the polymerase chain reaction. Am J Hum Genet 44:388–396

Weisbroth SH, Flatt RE, Kraus AL (1974) The Biology of the Laboratory Rabbit. Academic Press, Orlando

Xu J, Hardison RC (1989) Localization of the beta-like globin gene cluster and the genes for parathyroid hormone and c-Harvey-ras 1 to region q14-q21 of rabbit chromosome 1 by in situ hybridization. Cytogenet Cell Genet 52: 157–161

Xu J, Hardison RC (1991) Localization of the alpha-like globin gene cluster to region q12 of rabbit chromosome 6 by in situ hybridization. Genomics 9:362–365

Yerle M, Echard G, Gillois M (1987) The high-resolution GTG-banding pattern of rabbit chromosomes. Cytogenet Cell Genet 45:5–9

Zeuner FE (1963) A History of Domesticated Mammals. Hutchinson, London, UK

Zijlstra C, de Haan NA, Korstanje R, Rogel-Gaillard C, Piumi F, van Lith HA, van Zutphen LFM, Bosma AA (2002) Fourteen chromosomal localizations and an update of the cytogenetic map of the rabbit. Cytogenet Genom Res 97:191–199

Zoodsma M, Nolte IM, Schipper M, Oosterom E, van der Steege G, de Vries EGE, Te Meerman GJ, van der Zee AGJ (2005) Analysis of the entire HLA region in susceptibility for cervical cancer: a comprehensive study. J Med Genet 42:e49

CHAPTER 8

Dog

Dana S. Mosher, Tyrone C. Spady, and Elaine A. Ostrander (✉)

National Human Genome Research Institute, National Institutes of Health, 50 South Drive, MSC 8000, Building 50, Room 5334, Bethesda MD 20892-8000, USA, eostrand@mail.nih.gov

8.1
Introduction

For thousands of years man has relied on his closest companion, the domestic dog (*Canis familiaris*), for protection, work assistance, and companionship. The dog has been selectively bred to excel at certain tasks, including hunting, herding and drafting, and to possess phenotypic features typified by specific body size and shape, skull shape, leg length, and coat color. Today, there are more than 400 uniquely described breeds of domestic dog worldwide, exhibiting extremes in morphology, behavior, and disease susceptibility.

It has been said that there is a dog well suited for every person and for every way of life. Canines live comfortably in the city as companions and in the country as masters of a herd; each at ease in his surroundings. Indeed, dogs have touched every aspect of our lives as evidenced by their frequent appearance in art and literature (Fig. 1). The unique and long-lasting relationship between humans and dogs motivates studies aimed at deciphering the canine genome and to identifying variants responsible for traits of interest.

In the following, we highlight the many ways in which studies of canine population structure facilitate an understanding of both companion animal medicine and human health and biology. We describe the population structure of the modern domestic dog and its reported domestication from the gray wolf. We discuss the resources that have been developed for navigating the dog genome, including meiotic linkage, radiation hybrid (RH), comparative, and cytogenetic maps. A discussion of both the 1.5× survey sequence and a 7.5× whole-genome sequence is included, together with a primer on how that data have advanced our understanding of linkage disequilibrium (LD) and genome architecture in the dog.

All these resources have converged to facilitate mapping studies in the dog in the last three years. We describe recent progress in identifying single gene traits in the dog, as well as more recent studies aimed at mapping complex traits. Particular attention is paid to the identification of quantitative trait loci (QTL) controlling susceptibility to disease, as well as studies of morphology and behavior.

8.1.1
Dog Breeds

Dogs have been bred for a multitude of purposes. For instance, the Great Dane with its large frame and 30-in. height at the withers (shoulder) was originally bred in Germany to hunt wild boar (American Kennel Club 1998). By comparison, the Cavalier King Charles Spaniel was bred to be a lap dog and is only 12 in. tall and weighs about 15 pounds (lb; American Kennel Club 1998). Such selective breeding has resulted in a species more morphologically diverse than any other in the Canidae family (Wayne 1986a, b).

In general, gene flow between breeds is restricted by the "pedigree barrier." No dog can be registered as a particular breed unless both parents are registered members of the same breed (American Kennel Club 1998; Ostrander and Kruglyak 2000). Each dog breed, therefore, represents a closed breeding population. In the United States, breed registration is controlled by the American Kennel Club (AKC). Established in 1884, the AKC is composed of over 350 autonomous dog clubs that themselves oversee the activities of some 153 recognized breeds (American Kennel Club 1998). Parent clubs are responsible for setting the standard by which each breed is judged. The AKC breeds are organized into seven groups: sporting, hound, working, terrier, toy, herding, and

Genome Mapping and Genomics in Animals, Volume 3
Genome Mapping and Genomics in Domestic Animals
N.E. Cockett, C. Kole (Eds.)
© Springer-Verlag Berlin Heidelberg 2009

Fig. 1 This Egyptian limestone painting depicting a king accompanied by his dog as he hunts a lion is approximately 3,000 years old (1295–1069 BCE). It was excavated from western Thebes by Howard Carter for the Earl of Carnarvon and is now located at the Metropolitan Museum of Art in New York City, NY, USA

nonsporting (Fig. 2). The members of each group are often, although not always, similar in physical appearance, ancestry, and function (Fogel 1995; American Kennel Club 1998). Approximately 920,000 new puppies were registered in the United States in 2005 (http://www.akc.org/reg/dogreg_stats.cfm). The Labrador Retriever, Golden Retriever, and Yorkshire Terrier hold the top three slots in the numbers of dogs registered in 2005, with the German Shepherd Dog, Beagle, Dachshund, and Boxer also ranking high in popularity among registered dogs (http://www.akc.org/reg/dogreg_stats.cfm).

Dogs in the sporting group were originally developed to catch and retrieve birds (Yamazaki and Toyoharu 1995; American Kennel Club 1998). This group includes retrievers, spaniels, pointers, and setters, each of which is bred to assist in the hunt. By comparison, the hound group can be broadly divided into two groups, sight and scent hounds. Sight hounds, such as the greyhound, can spot prey from great distances and run the prey down using their incredible speed. Scent hounds, like the bloodhound, use their highly developed olfactory system to track down and

Fig. 2 Representatives of the seven breed groups defined by the American Kennel Club (AKC) (Sutter and Ostrander 2004). Breeds within each group share morphological characteristics, history, or function. Groups and the representative breed are shown: herding group (Cardigan Welsh Corgi), sporting group (German Shorthaired Pointer), hound group (Pharaoh Hound), working group (Mastiff), terrier group (Welsh Terrier), toy group (Maltese), and nonsporting group (Chow Chow)

eventually corner their prey (Fogel 1995; Yamazaki and Toyoharu 1995; American Kennel Club 1998).

The dogs of the working group are typified by guard dogs and described as powerful, brave, and tireless (Yamazaki and Toyoharu 1995). The Saint Bernard, German Shepherd Dog, and Akita all are classified as working dogs; interesting examples abound of these dogs' utility as a protector of both their owners and their property. For instance the Molossian dog, a predecessor of the present-day mastiff, fought alongside their human counterparts in ancient wars (Moody et al. 2006). Large dogs of great speed like the Irish Wolfhound were trained to "body slam" enemy messengers traveling by bike during World War II (Moody et al. 2006). The German Shepherd Dog was used in World War I, World War II, and the Vietnam War as lookouts and sentries, as well as to detect mines and tunnels (Moody et al. 2006). The number of dogs used by the US military is classified, but it is estimated that 1,000–2,000 dogs are currently deployed by the military in dozens of locations worldwide (Moody et al. 2006).

Dogs of the terrier group were bred for their ability to ferret out small prey, such as rodents. Many terriers, such as the Scottish Terrier and West Highland White Terrier, were developed in the United Kingdom, and nearly all possess strong limbs and a characteristically wedge-shaped head (Yamazaki and Toyoharu 1995; American Kennel Club 1998).

The dogs of the toy group were developed as indoor house pets or lap dogs. All the breeds in this group are small in stature, with some like the Chihuahua and Yorkshire Terrier being extremely tiny, each weighing no more than six and seven pounds, respectively (American Kennel Club 1998). Despite their small size these dogs show remarkable variation. For instance, there are at least six varieties of Chihuahua including the standard, miniature, curly coated, and long-haired varieties (Fogel 1995; Yamazaki and Toyoharu 1995; American Kennel Club 1998).

The herding group is composed of dogs bred for intelligence, rigorous obedience, and trainability. They make their livelihood by assisting herders in moving livestock (Yamazaki and Toyoharu 1995). The Australian Shepherd, Old English Sheepdog, and Border Collie are well-known breeds in this group. Although all are bona fide herders, different breeds use distinct behavior patterns to achieve their goals. The Border Collie has a hypnotic state that it uses to intimidate sheep into moving in the direction the herder requests. The Australian Shepherd, by comparison, will not stare intensely at one individual but will watch the entire herd (American Kennel Club 1998). They are in the category of herding dogs known as heelers who direct livestock by nipping at their heels.

The final group, termed nonsporting, includes a diverse set of breeds that due to their history, characteristics, or unique appearance are not included in the other six groups. The broad membership of this group includes, for instance, the Bulldog, Poodle, and Dalmatian (Yamazaki and Toyoharu 1995; American Kennel Club 1998).

8.1.2
Genetic Diversity and Dog Breeds

Although genetic diversity across dog breeds is high, diversity is relatively low within a breed (Parker et al. 2004). Many breeds were derived from a small number of founders, and once established a breed could not legitimately receive any new genetic information from dogs outside the breed. Genetic homogeneity within a breed is also increased by the "popular-sire effect" in which dogs most closely matching the breed standard in size, shape, and behavior are repeatedly bred. These restrictive breeding practices resulted in a loss of genetic diversity by effectively reducing the gene pool within that population (Ostrander and Kruglyak 2000).

In a genetic study of 85 breeds, Parker et al. (2004) and Parker and Ostrander (2005) observed that the overall nucleotide diversity in the dog is 8×10^{-4}, nearly equal to that of the human population. However, variation among breeds accounts for 27% of the total genetic diversity observed, in contrast to the 5–10% variation found between distinct human populations. In turn, the degree of homogeneity within a dog breed (94.6%) is much higher than that of distinct human populations (72.5%).

The genetic diversity within some breeds has been further reduced by fluctuations in their popularity as well as by catastrophic events such as the two world wars and the American depression. The world wars affected particularly breeders in Europe, who were unable to care for and maintain their kennels because of extensive food shortages. Larger dog breeds were particularly affected. In some cases, such as the Leonberger, the population was reduced to about six dogs from which the entire population that is alive today is thought to be derived (Ostrander and Kruglyak 2000). In the US, the Great Depression took a similar toll on the dog breeding community. With so many Americans out of work, maintaining kennels and developing breeds were simply not the priority. Dog breeds, therefore, offer unique opportunities for genetic studies because of their high degree of isolation, extreme population bottlenecks, and their well-documented genealogies (Ostrander and Kruglyak 2000; Parker and Ostrander 2005).

8.1.3
The Superfamily Canoidae

As reviewed elsewhere (Wayne and Ostrander 2004; Ostrander and Wayne 2005), dogs represent the earliest divergence in the superfamily Canoidae which includes familiar species such as bears, weasels, skunks, raccoons, as well as the pinnipeds (seals, sea lions, and walruses). The dog family (Canidae) has a diverse karyotype with chromosome numbers ranging from 36 to 78 (Ostrander and Wayne 2005). The domestic dog represents the extreme with 38 pairs of autosomes and the sex chromosomes.

Within the Canidae, there are three distinct phylogenetic groups (Wayne et al. 1987a, b, 1997). The first are the fox-like canids which includes the red fox, arctic, and fennec fox species (genus *Vulpes*, *Alopex*, and *Fennecus*, respectively). The second are the now familiar wolf-like canids which include the dog, wolf, coyote, and jackals, in the genus *Canis*. Also included are the African hunting dog and the dhole (genus *Cuon* and *Lycaon*, respectively). The last group features the South American canids including foxes (genus *Pseudolopex*, *Lycolopex*, *Atelocynus*), the maned wolf (genus *Chrysocyon*), and bushdog (genus *Speothos*). In addition, there are several canids that have no close living relatives and define distinct evolutionary lineages. Importantly, all members of the *Canis* genus can produce fertile hybrids, and several species may have genomes reflecting hybridization that has occurred in the wild (Wayne and Jenks 1991; Gottelli et al. 1994; Roy et al. 1994; Wilson et al. 2000; Adams et al. 2003).

Though modern populations of dogs can be found worldwide, there is strong evidence that they all share a common origin from Old World gray wolves (Leonard et al. 2002). Data supporting this hypothesis derive from several sources. In an analysis of 261 base pairs (bp) of mitochondrial sequence from dogs, wolves, and other wild canids, the dog and the wolf were found to have very similar mitochondrial profiles, differing by no more than 12 substitutions. The next closest wild canids, the coyote and the jackal, differed from the dog by 20 substitutions (Vila et al. 1997). In a separate study, analysis of exclusively New World breeds, such as the Xoloitzcuintli or Xolo breed (Mexican hairless dog), also confirmed a common Old World origin (Vila et al. 1999). In this latter study, Vila and collaborators compared the 261 bp mitochondrial sequence to a homologous 394 bp sequence in the Xolo. No unique haplotypes were found among the Xolo, and none was identical or similar to any of the sequences found in New World wolves (Vila et al. 1999).

Where did the original domestication event or events occur? This issue was addressed by Savolainen and colleagues (Savolainen et al. 2002) in a study of 654 dogs derived from worldwide populations. A similar phylogeny could be constructed from dogs of every geographic region, suggesting that a single domestication event from only one geographic region had predominated. The greatest diversity in haplotypes was found in East Asian dogs, further suggesting that the domestication event occurred in Asia.

8.1.4
Mitochondrial DNA (MtDNA) Analysis of Canids

A deeper discussion of the *Canis* genus is needed to fully appreciate the complex structure of modern dog breeds and to understand the timing of initial domestication events. Initial phylogenetic analysis of dog and wolf mitochondrial sequence has shown that dog haplotypes fall into four clades suggesting four founding lineages for the origin of the domestic dog (Vila et al. 1997). Clade I is exclusive to domestic dog haplotypes and contains 19 of the 27 dog haplotypes (70%). However, clades II through IV contain a mixture of dog and wolf haplotypes from a variety of locations worldwide. High nucleotide diversity in a dog-specific clade suggests a relatively ancient origin for dog domestication (Vila et al. 1997; Savolainen 2002). The level of nucleotide diversity present in clade I suggests that domestication occurred about ~135,000 years ago

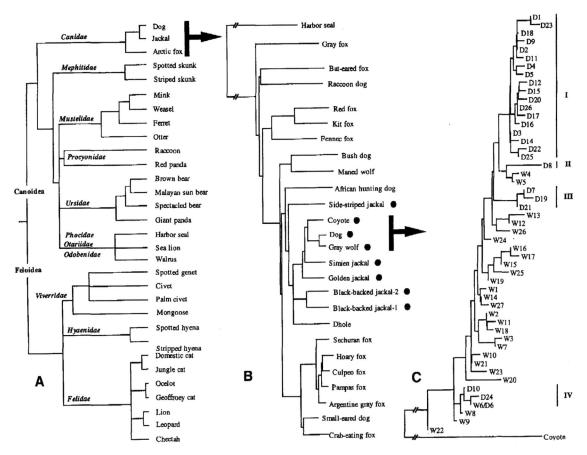

Fig. 3 Phylogenetic tree of the domestic dog (Vila et al. 1999). **a** The divergence of the superfamily Canoidae and the family Canidae is based on DNA hybridization data (Wayne et al. 1989). **b** Relationships between the members of the Canidae family were based on analysis of about 2,000 bp of mt DNA sequence (*mitochondrially encoded cytochrome b, mitochondrially encoded cytochrome c oxidase I and mitochondrially encoded cytochrome c oxidase II* genes) *Black circles* indicate species of the genus Canis (Wayne et al. 1997). **c** Neighbor-joining tree of wolf (*W*) and dog (*D*) haplotypes are based on DNA sequence of 261 bp of control region I mt DNA sequence (Vila et al. 1997). Dog haplotypes group into four clades marked I–IV. Position of the dog in the neighbor-joining tree was determined using 736 bp of *mitochondrially encoded cytochrome b* sequence (Wayne 1993)

(Vila et al. 1997). This time frame greatly exceeds that suggested by fossil evidence, which dates the origin of dogs to ~15,000 years before the present. It has been suggested that early dogs may not have been easily recognized as domesticated by archeological records because of their morphological similarity to gray wolves. The diagnostic change in phenotype from wolf to domestic dog began about 15,000 years ago and, according to Wayne and coworkers, may reflect the change from a hunter-gatherer to an agricultural society (Wayne et al., in press) (Fig. 3).

However, there are some recent phylogenetic analyses that show support for domestication dates closer to those predicted by archeological records (Savolainen 2002). In the study by Savolainen et al. (2002), evidence of six founding female lineages was discovered. Upon further analysis, individuals from East Asia exhibited the largest proportion of unique haplotypes supporting the East Asian origin of domestication. Careful examination of the three clades from East Asia displaying the largest number of haplotypes estimated the timing of domestication as ~15,000 years ago.

Finally, a study by Vila and coworkers used nuclear DNA to estimate the number of founding individuals needed to create the level of genetic diversity seen in present-day dog breeds (Vila et al. 2005). This was accomplished by examining the major histocompatibility (MHC) locus of the genome. The MHC region, also termed DLA for dog leukocyte antigen in canines, exhibits a high level of genetic diversity compared to neutral regions such as mtDNA. As a result, a large number of alleles are shared across closely related canine species, which can be useful in evolutionary studies to assess a comparatively longer time frame. Vila et al. (2005) found 42 alleles for the DRB MHC locus. By running simulations, they were able to determine that at least 21 founding individuals were needed to obtain the present-day level of variability. However, this is a minimum estimate. The statistic assumes that all individuals are heterozygous, of equal fertility, and that no allele loss occurs over time. In summary, while some controversy remains over the precise date of domestication, there is a clear consensus that the origin of the domestic dog was the result of multiple founding lineages in the region of East Asia.

8.1.5
The Domestic Dog Population

While mitochondrial DNA studies have been critical for elucidating the origins of the various Canidae, they have generally not been helpful in reconstructing breed origins or ascertaining relationships between breeds. To assess the recent evolution of domestic dogs, microsatellite loci have been used as they feature large number of alleles in the population. Initial efforts to undertake studies of canine population structure using microsatellites were of limited success because of the small number of dogs or breeds considered, or the use of neighbor joining trees as a primary analysis tool (Zajc et al. 1997; Koskinen and Bredbacka 2000; Irion et al. 2003; Koskinen 2003). Parker et al. (2004) and Parker and Ostrander (2005), however, have recently made significant strides in this area. Their approach involved the collection of DNA samples from at least five unrelated dogs from each of a large number of recognized breeds, the collection of genotyping data using a set of 96 CA-repeat microsatellite markers, and the analysis of the data using cluster-based algorithms.

In their analysis, Parker and colleagues (2004) found that individual breeds represented the smallest reproducibly definable cluster. Indeed, analysis of the same data set with the *Doh* assignment calculator suggested that dogs could be correctly assigned to their breed based on data from 96 markers, about 99% of the time.

Analysis of the same data using genetic distance trees revealed several distinct breed clusters. The most divergent group contained what is believed to be the most ancient dogs from the Arctic, Asia, Africa, and the Middle East. By comparison, the majority of modern breeds appeared to stem from a single node without significant phylogenetic structure. This group was termed the "European hedge" to indicate the recent origin and extensive hybridization that occurred between dog breeds during the nineteenth and twentieth centuries (Parker et al. 2004).

A genetic clustering algorithm from the computer program *Structure* (Pritchard et al. 2000) was used to explore the accuracy with which individual breeds were identified. *Structure* correctly assigned 335 dogs to 69 breed-specific clusters. Most clusters represented single breeds; however, the program could not easily distinguish approximately six obviously related pairs such as the Bernese Mountain Dog and Greater Swiss Mountain Dog, or Mastiff and Bullmastiff. The lack of resolution associated with these breeds was largely expected and indicative of their close history (Rogers and Brace 1995).

The data were also analyzed to determine if higher-order clusters, indicative of additional relationships between breeds, could be discerned. This was important because it defined relationships between breeds for the first time based on their molecular characterization, rather than historical information or lore. In addition, it now provides a guide for designing whole-genome scans aimed at mapping and defining variants for traits of interest.

The first distinct cluster to be defined included largely the breeds of Asian origin (Akita, Shar Pei, Lhasa Apso, Pekingese, Shiba Inu, Chow Chow, etc.), as well as some sled dogs and a few of the ancient hound breeds such as the Saluki. Not surprisingly, gray wolves grouped with this so-called "ancient cluster" as well (Fig. 4).

The next cluster comprised mastiff-type dogs including the Mastiff, Boxer, Bulldog, Miniature Bulldog, and Bullmastiff. Cluster 3 included working dogs such as the Collie and Shetland Sheepdog, together with a subset of the sight hounds, such as the Greyhound. The remaining group was reminiscent of dogs in the European hedge and comprised mostly spaniels, retrievers, and terriers. In order to define groups with finer resolution additional unpublished analyses by Parker and collaborators utilized more highly mutable tetra-nucleotide-based microsatellite markers (Francisco

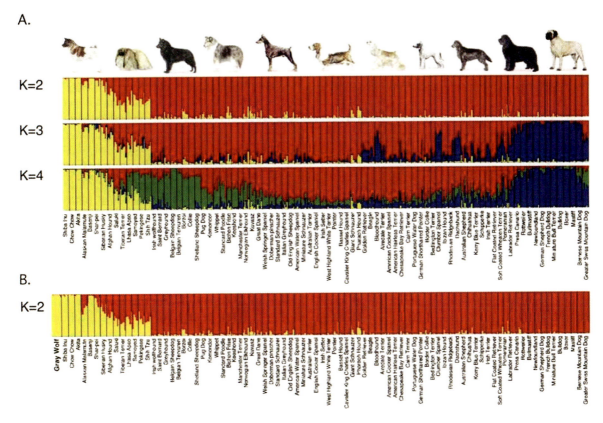

Fig. 4 Analysis of the population structure of 85 dog breeds (Parker et al. 2004). Data were generated using five unrelated individuals per breed and analyzed with the computer program, Structure. Each dog is represented by a *vertical line* and *thick black* lines separate breeds. K = number of clusters that are permitted in each analysis. Analysis reveals four distinct population clusters. Each cluster is represented by one color and the height of the color bar indicates a particular dog's relative membership into that cluster. Based on these data, domestic dog breeds cluster into four primary groups: Asian/Ancient group, Herding group, Hunting and Gun dog group, and Mastiff group. Representative breeds are indicated (Akita, Pekingese Belgian Sheepdog, Collie, Doberman Pinscher, Basset hound, American Cocker Spaniel, Bedlington Terrier, Flat-Coated Retriever, Newfoundland, and Mastiff). **a** Results shown represent averages over 15 runs of structure for each value of K. **b** Results as in **a**, but with inclusion of gray wolves which fall into Group 1, the Ancient/Asian Group. Data represents averages of five runs of structure at K = 2

et al. 1996) and less mutable markers based on single nucleotide polymorphisms (SNPs). In addition, many more breeds have been added to the analysis in an attempt to connect breeds that were impossible to relate previously (Parker et al. 2007).

Another promising approach toward reconstructing breed history has been described by Neff et al. (2004). Their analysis reconstructed the history of a set of related breeds using variation within the multidrug resistance gene (*ABCB1* formerly *MDR1*) and four closely linked microsatellite markers (Neff et al. 2004). A single *ABCB1* mutation was found to segregate in nine breeds that included seven herding breeds and two sighthound subgroups that were themselves related to at least one of the herding breeds. Indeed, the region right around *ABCB1* was identical by descent (IBD) in all nine breeds, suggesting the presence of a common ancestor for these breeds.

8.2
Molecular Genetics of Dogs

8.2.1
Canine Linkage and Radiation Hybrid Maps

Large numbers of microsatellite markers were isolated, characterized, and made available throughout 1990s to research groups working on the development of both meiotic linkage and RH maps (Ostrander et al. 1992, 1993, 1995; Francisco et al. 1996; Priat et al. 1998; Jouquand et al. 2000). A substantial number of gene sequences were similarly characterized (Jiang et al. 1998; Priat et al. 1998; Priat et al. 1999; Parker et al. 2001), facilitating the development of comparative maps.

The first meiotic linkage map of the dog, composed of 150 markers distributed on 30 linkage groups, was developed by typing markers on a set of 17 reference families and was published by the Ostrander group in 1997 (Mellersh et al. 1997). Shortly thereafter, second- and third-generation linkage maps were made public (Neff et al. 1999; Werner et al. 1999). In early 1999, a map of 341 markers distributed into 38 linkage groups with both a framework and a comprehensive map was described (Werner et al. 1999). The size of the canine genome was initially estimated from maximum-likelihood predictions based on the meiotic linkage map to be 2.7 Morgans in genetic distance (Neff et al. 1999). Estimates based on flow sorting of canine chromosomes agree well and suggest a physical size 2.8 megabases (Mb; Breen et al. 1999a). Predictions based on sequence analysis of euchromatic sequence derived from direct sequencing (discussed below) suggest a size of 2.3–2.4 Mb (Kirkness et al. 2003; Lindblad-Toh et al. 2005).

While the meiotic map proved useful for the mapping of some disease genes, it was clear that this was not an economical strategy for building a dense map and the community needed a much denser resource that integrated markers and genes. Indeed, meiotic linkage map development is a laborious and expensive process. It suffers from the additional limitation in that it does not allow for easy ordering of genes relative to anonymous microsatellite markers, unless informative markers are found within the gene. Therefore, investigators at the University of Rennes constructed a whole-genome canine radiation hybrid (WGRH) panel and in 1998 published a first-generation canine RH map (Priat et al. 1998; Vignaux et al. 1999). The RH panel, designated RHDF5000, consists of 126 hybrid clones and was the result of irradiation at 5,000 rads yielding a resolving power of 600 kb (Vignaux et al. 1999). Construction of the canine RH map was ongoing for the next five years and key features were summarized at periodic intervals (Breen et al. 2001, 2004; Guyon et al. 2003).

8.2.2
Comparative Maps

While the maps were useful independently, neither was maximally informative in the absence of a comparative map that linked the markers and genes localized using the RH panel or reference families to the human and mouse maps. To do this, a well-defined karyotype of the dog was needed. This was accomplished in 1996 (Switonski et al. 1996) and a now widely accepted DAPI banded ideogram of the dog was subsequently put forth (Breen et al. 1999b) (Fig. 5). In addition, a set of differentially labeled whole chromosome paint probes for reciprocal painting of dog and human chromosomes was developed (Langford et al. 1996), which allowed, for the first time, all canine chromosome arms to be assigned to human chromosomes and, conversely, all human chromosome arms to be assigned to dog chromosomes. Although the nomenclature was briefly in dispute, these results have subsequently been confirmed by independent studies (Yang et al. 1999; Sargan et al. 2000).

To facilitate the integration of all three maps with the human genome map, a set of 266 microsatellite-containing cosmid clones were developed (Breen et al. 2001). These were localized by both fluorescence in situ hybridization (FISH) and ordered on the meiotic linkage and RH maps. Because each cosmid was prescreened for the presence of a polymorphic microsatellite, these "anchor probes" (as the cosmid and its associated microsatellite are known) have been the cornerstones upon which the canine meiotic linkage map, RH, and cytogenetic map have been integrated. Selection of additional anchor clones near the centromere and telomere of each chromosome enabled researchers to orient the integrated groups on each chromosome and evaluate the extent of chromosomal coverage. The resulting map totaled 1,800 markers or about one marker every 1.2 Mb (Breen et al. 2001). It was followed quickly thereafter by a slightly denser map with a marker positioned every Mb (Guyon et al. 2003).

In 2004 efforts were focused on adding both genes and canine-specific bacterial artificial chromosomes (BACs) on the map (Breen et al. 2004). The latter were needed to facilitate positional cloning efforts recently initiated throughout the community. The resulting map featured a density of one marker every 900 kb and contained some 1,760 BACs localized

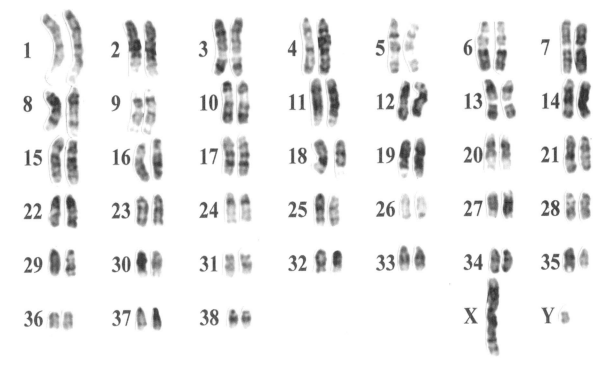

Fig. 5 The DAPI-banded karyotype of the domestic dog (Breen et al. 1999b). All 38 pairs of autosomes and the sex chromosomes are present. Chromosomes were placed in the karyotype following hybridization with chromosome-specific paint probes as described previously (Breen et al. 1999b)

to 1,423 unique positions, 804 of which were also localized by FISH. Excluding the Y-chromosome, the map featured an RH/FISH mapped BAC every 3.5 Mb and an RH mapped BAC-end, on average, every 2 Mb. For 2,233 markers, the orthologous human genes had been established, allowing the identification of the first 79 conserved segments (CS) between the dog and human genomes.

8.2.3
Survey Sequence and a Canine Gene Map

By 2003 it had become clear that while adding more markers to the map would be useful, the canine research community needed a dense gene map. To accomplish this we (Hitte et al. 2005) relied on a 1.5× survey sequence of the Standard Poodle generated by Celera and curated by Kirkness and colleagues (2003). The 1.5× sequence encompassed 6.22 million sequence reads. More than 650 million base pairs (>25%) of dog sequence aligned uniquely to the human genome, including fragments of putative orthologs for 18,473

of 24,567 annotated human genes. Assembly of the dog genome after 1.5× coverage yielded 1.9 million short contigs with an average length of 1.0 kilobase (kb) that could be ordered across only short regions of the genome with a mean length of 8.6 kb. At least 100 continuous bases from a minimum of one exon were present for nearly 85% of individual canine genes. Thus, while the 1.5x sequence was limited in terms of whole-genome coverage, it provided to be an excellent resource for selecting gene sequences to use in building a 9,000 rad high-resolution comparative map of the dog genome.

The gene-based markers selected for placement on the RH map were chosen exclusively from the aforementioned 1.5× poodle sequence. Specifically, we utilized a well-spaced set of 9,850 markers which corresponded to evenly spaced human genes, each of which was identified by their mutual-best Blast matches (Hitte et al. 2005). A newly constructed 9,000 rad panel was used for the mapping, which has a theoretical resolution of 200 kb (12,000 unique positions) (Hitte et al. 2005). In total, we mapped 10,348 gene-based markers which occupied 9,081 unique positions and were uniformly spaced across all canine chromosomes (CFAs) except the Y. In addition, we genotyped a set of 545 BAC-ends on both the newly created 9,000 rad panel and the originally described 5,000 rad panel (Vignaux et al. 1999) to connect the low (1,500)- and high (9,000)-resolution RH dog maps.

In this effort 264 conserved segments (CS), defined as those having two or more markers on the corresponding dog/human chromosomes, without interruption, comprising two to 332 markers were identified. These were defined by comparison of the canine map and the human genome sequence (NCBI Build 34) (Hitte et al. 2005). The number of conserved fragments identified by the RH map was higher than the 243 CS later identified by the CanFam 1.0 assembly (Lindblad-Toh et al. 2005). The larger number identified by the RH map could be explained by the possibly incorrect localization of about 20 singletons – that is genes that did not fall into any RH group.

Although the 7.5× high-quality draft sequence described later is now the primary reference map for the dog genome, we predict that the construction of dense RH maps may be critical for species where only a 2× sequence is currently planned (http://www.genome.gov/10002154) (Hitte et al. 2005). While 1–2× survey sequencing cannot provide a contiguous map of a genome, completion of the canine 10,000-gene RH map demonstrates how a collection of fragmentary survey-sequence data can be transformed into an accurate, high-resolution map of a mammalian genome. One additional utility of the 10,000 gene map has been its contribution to assembly of the 7.5x high-quality draft, as will be described now.

8.2.4
The 7.5x Canine Genome Sequence

In 2002, a white paper submitted by the canine genome community to request the 6× sequencing of the dog received approval from the National Human Genome Research Institute (NHGRI; www.genome.gov/Pages/Research/Sequencing/SeqProposals/CanineSEQedited.pdf). Sequencing of an anonymous female Boxer, chosen by Parker and Ostrander, began at the Broad Institute of MIT and Harvard in June 2003. The effort, which generated over 35 million sequence reads and a 7.5x sequence of the dog, was completed and the sequence released in July 2004 (Lindblad-Toh et al. 2005).

Key features of the sequence are well summarized by Lindblad-Toh et al. (2005). The assembled sequence spans most of the dog's 2.4 Gb and is derived from 31.5 million sequence reads (http://www.genome.ucsc.edu) (Lindblad-Toh et al. 2005). The sequence covers 99% of the eukaryotic genome. The quality of the assembly is high; 98% of the bases in the assembly have quality scores exceeding 40, corresponding to an error rate of less than 10^{-4} and comparable to the standard for the finished human sequence (Lander et al. 2001; Venter et al. 2001; Waterston et al. 2002; Gibbs et al. 2004). Half of the assembled bases (N50 contig size) are in contigs of 180 kb and the N50 supercontig size is 45.0 Mb, which is considerably longer than the mouse genome at a similar point in its assembly. Most canine autosomes comprise 1–3 supercontigs. The current gene count is listed as 19,120 with about 75% representing 1–1–1 orthologs among dog, human, and mouse and nearly all being unambiguous homologs of human genes. The assembly spans a total distance of 2.41 Gb.

The previously described maps proved very useful in assembling the sequence, as the assembly was anchored to chromosomes using both the RH and cytogenetic maps. About 97% of the assembled sequence was ordered and oriented on the chromosomes, showing an excellent agreement with the two maps. Indeed, there were only three discrepancies noted between the resources, and all three of these were resolved by obtaining additional FISH data from the sequenced boxer. The 3% of the assembly that could not be anchored consisted largely of highly repetitive sequence, including eight supercontigs of 0.5–1.0 Mb composed almost entirely of satellite sequence.

Several interesting features of genomic evolution were revealed through the analysis of the canine/human comparative sequence. First, the dog lineage has diverged more rapidly than the human lineage but at only half the rate of the mouse lineage. Interestingly, about 5.3% of the human genome contains functional elements that have been under purifying selection in both human and dog. Nearly all of these elements are confined to regions that have been retained in the mouse, indicating that they represent a common set of functional elements across mammals. In addition, 50% of the most highly conserved noncoding sequence in the genome shows striking clustering in ~200 gene-poor regions, most of which feature genes with key roles in establishing or maintaining cellular identity such as transcription factors or genes for neural development. As the assembly progresses and reaches higher levels of integrity, substantially more information about the evolution of the canine genome is expected.

8.2.5
Single Gene Traits

Dogs as Models for Human Disease

Modern dogs display a remarkably high frequency of genetic disease. It has been suggested that 46% of the genetic diseases reported in dogs occur predominantly or exclusively in one or a few breeds. To date, there are over 360 genetic disorders found in humans that have also been described in dogs (Patterson 1980, 1988). A detailed listing of over 1,000 canine diseases and their descriptions appears in the database of inherited diseases in dogs (IDID, http://www.vet.cam.ac.uk/idid) (Sargan 2004). This constitutes the largest set of naturally occurring genetic disorders in any nonhuman species (Patterson et al. 1982, 1988; Knapp and Waters 1997).

Recent interest in identifying genes for canine disorders has increased in both the veterinary and human health communities (Ostrander and Kruglyak 2000; Patterson 2000; Galibert and Andre 2002). The reasons are three fold. First, many diseases observed in humans are observed in excess in only a subset of dog breeds (Ostrander and Kruglyak 2000). This is not unexpected; as discussed, gene flow between dog breeds is restricted by the pedigree barrier. This limits the number of variant alleles that are likely to be responsible for even common diseases within a single breed, or group of breeds, as suggested by Parker et al. (2004).

Second, the high susceptibility to specific diseases in small numbers of breeds indicates enrichment for risk alleles in a subset of the population, which in turn facilitates mapping efforts. A small founding population, bottlenecks due to world wars or breed popularity shifts, and popular-sire effects all contribute to the utility of the dog system for mapping studies. It follows that within any given breed there is likely to be only a small number (or even one) disease allele of strong effect. Thus, dog breeds can be compared to geographically isolated human populations, such as the Bedouins or populations from Iceland or Finland. As a result, locus heterogeneity problems, which have in some cases halted the mapping of complex human

diseases such as heart disease, diabetes, deafness, blindness, autoimmune disorders, and cancer, may be solved through genetic mapping efforts undertaken in purebred dogs.

Finally, it is clear that many canine and human diseases are caused by mutations in the same genes. Indeed, of the more than 250 canine genetic diseases presently recognized, at least 215 have clinical and laboratory abnormalities that closely resemble a specific human genetic disease. For many of these diseases, it has already been shown that the same gene product is responsible for both the human and dog disease (Ostrander et al. 2000; Parker and Ostrander 2005). This suggests that the dog may be a useful model for the development of therapeutics. This has been demonstrated by the successful application of dogs for gene therapy studies (Howell et al. 1997; Acland et al. 2001; Mount et al. 2002; Ponder et al. 2002).

In the first case, gene therapy was used to treat the canine form of Leber congenital amaurosis (LCA), a disease that causes near total blindness in infancy and results from mutations in the gene *RPE65*. A naturally occurring animal model, the *RPE65-/-* dog, has been described (Aguirre et al. 1998). Using a recombinant adeno-associated virus carrying the wild-type gene, Acland and colleagues have been able to restore full sight to affected dogs (Acland et al. 2001). Similar success has been reported by Haskins and colleagues in the treatment of dogs with mucopolysaccharidosis VII using a retroviral vector expressing wild-type canine beta-glucuronidase (Ponder et al. 2002).

Linkage Disequilibrium and Identifying Variants for Single Gene Traits

The key to identifying variants for both simple and complex diseases lies with an understanding of the structure of the canine genome, particularly linkage disequilibrium (LD) (Hyun et al. 2003; Lou et al. 2003; Sutter et al. 2004; Lindblad-Toh et al. 2005). Linkage disequilibrium refers to the nonrandom association of two or more usually adjacent loci that segregate together through several generations. Specifically, LD is a measure of the distance that separates two loci before they can be considered independent. The most informative studies have been those of Sutter et al. (2004) and Lindblad-Toh et al. (2005), the latter performed as part of the dog genome-sequencing effort.

In an initial series of studies, Sutter and colleagues examined the extent of LD in five breeds with distinct breed histories; the Akita, Golden Retriever, Labrador

Retriever, Bernese Mountain Dog, and Pekingese. The average length of LD in these five breeds is approximately two Mb which is 40–100 times further than LD that typically extends in the human genome. The implications of this finding are important. While a typical whole-genome association study in humans requires about 500,000 SNPs (Kruglyak 1999), the same study in the dog requires only about 10–30,000 markers. For diseases found both in humans and in dogs such as heart disease, diabetes, cancer, and epilepsy, it may prove far easier to do the initial mapping study in dogs than in humans.

In addition, these investigators also found that the extent of LD varied over a near ten-fold range between breeds of dog (0.4–3.2 Mb) (Sutter et al. 2004). Thus, the selection of a particular breed for a mapping study over other available breeds could dramatically reduce the work load and expense associated with the study, and therefore, judicious selection of breeds for studies of common disorders is important. Finally, Sutter et al. (2004) demonstrated that haplotypes or blocks of alleles are frequently shared between dog breeds, a result that was not surprising given the recent evolution of the dog from the wolf (Vila et al. 1997; Savolainen et al. 2002). This means that a single map and a single nucleotide polymorphism (SNP) chip will be useful for mapping studies in any breed of dog.

These results were corroborated and greatly extended in a much larger study by Lindblad-Toh et al. (2005), using 10 breeds and nearly 1,300 SNPs. They showed that the within-breed polymorphism rate is about 1/1,500 bp, whereas it extends to about 1/900 bp when considering multiple breeds. Lindblad-Toh and colleagues suggest that whole-genome association studies in the dog should only require about 10,000 evenly spaced SNPs, which matches well with the conclusions of Sutter et al. (2004). These investigators further predict that as few as 100 cases and 100 unaffected controls will be needed to find a locus responsible for a simple Mendelian dominant trait with high penetrance and a low phenocopy rate when analyzing approximately 15,000 evenly spaced SNPs. Such criteria describe a theoretically perfect scenario, and the mapping of true complex traits is apt to present additional challenges. Their numbers also assumes a log-likelihood odds ratio (LOD score) score threshold of 5.0 (Table 1[b]).

As whole-genome studies are initiated to map loci of interest within single breeds, the question has been asked if there is sufficient polymorphism within a single breed to precisely localize disease variants. Such studies will likely require the comparison of data between closely related breeds that share an ancestral mutation. The large range of LD in dogs overall suggests that initial mapping studies will only localize a locus to a region of megabases (Mb). Comparison of the extent of LD between closely related breeds will be needed to reduce the region of interest from Mb to kb. Key to the success of such studies, however, is both a large number of informative SNPs for fine mapping and the identification of closely related breeds with the same phenotype. As described later, Lindblad-Toh and colleagues have undertaken the first of these and produced a SNP chip of 127,000 SNPs.

Identification of Genetic Loci for Single Gene Traits and PRCD LD

To date, many canine disease loci have been mapped (reviewed in: Ostrander and Kruglyak 2000; Sutter and Ostrander 2004; Switonski et al. 2004) including, but not limited to, disease loci controlling metabolic disorders (Yuzbasiyan-Gurkan et al. 1997; van De Sluis et al. 2002), blindness (Acland et al. 1978, 1998, 1999; Aguirre 1998; Aguirre et al. 1998), cancer (Jonasdottir et al. 2000; Lingaas et al. 2003), neurological disorders (Lingaas et al. 1998; Lin et al. 1999), hip dysplasia (Chase et al. 2004), osteoarthritis (Chase et al. 2005a, b), Addison's disease (Chase et al. 2006), and epilepsy (Lohi et al. 2005). For some of these, the disease-associated variant has been identified. In some cases, the availability of optimized pedigrees facilitated studies. In others, breeders and owners provided DNA samples for study. In aggregate, three important lessons can be derived from these studies.

First, several studies focused on important diseases in dogs and humans, and in doing so have illuminated our understanding of the genetics of human diseases. The most often-cited example is that of Lin et al. (1999) who showed that the sleep disorder narcolepsy is caused by mutations in the *hypocreten 2* receptor gene in the Doberman Pinscher. This work opened up the field of molecular biology of sleep and set in motion a series of experiments aimed at dissecting the pathways that are important in a variety of common sleep disturbances.

An additional example is provided by progressive rod cone degeneration (PRCD), a disease that is found in the poodle as well as several other breeds and is

Table 1[b] 100,000 sequence reads of each of nine dog breeds[a] and 20,000 sequence reads of the coyote and each of four types of wolves, compared to the 7.5× Boxer assembly and single nucleotide polymorphisms

Set number	Breed or species	Number of SNPs	SNP rate (one per × bases)
1	Boxer versus boxer	768,948	3,004 (observed)
			1,637 (corrected)
2	Boxer versus poodle	1,455,007	894
3a	Boxer versus breeds[c]		
	German shepherd	45,271	900
	Rottweiler	44,097	917
	Bedlington terrier	44,168	913
	Beagle	42,572	903
	Labrador retriever	40,730	926
	English shepherd	40,935	907
	Italian greyhound	39,390	954
	Alaskan malamute	45,103	787
	Portuguese water dog	45,457	896
	Total distinct SNPs	373,382	900
3b	Boxer versus Canids[d]		
	China gray wolf	12,182	580
	Alaska gray wolf	14,510	573
	India gray wolf	13,888	572
	Spanish gray wolf	10,349	587
	California coyote	20,270	417
	Total distinct SNPs	71,381	
3	Set of 3 total distinct SNPS	441,441	
Total	Total distinct SNPS	2,559,519	

[a] German shepherd, Rottweiler, Bedlington terrier, Beagle, Labrador retriever, English shepherd, Italian greyhound, Alaskan malamute, and the Portuguese water dog
[b] Taken from Lindblad-Toh et al. (2005)
[c] Based on ~100,000 sequence reads per breed
[d] Based on 1200,000 sequence reads per wolf

similar to human retinitis pigmentosa (RP) (Acland et al. 1998). The identification of a novel disease gene (*PRCD*) for PRCD sets the stage for both gene therapy and genetic testing studies in the dog (Zangler et al. 2006). Other studies of interest to human biologists included cancer, epilepsy, neurological disorders, Addison's disease, and osteoarthritis.

In addition to helping us understand the biology of human disease, these studies have suggested new mechanisms for genetic disease. For instance, Lohi et al. (2005) reported on the cause of canine epilepsy in the purebred miniature wirehaired dachshund. The disease manifests itself as autosomal recessive progressive myoclonic epilepsy (PME), which is similar to Lafora disease. The study presents strong evidence for the existence of repeat-expansion disease outside humans. Specifically, a canid-specific unstable dodecamer repeat in the *EPM2B* (*NHLRC1*) gene recurrently expands, causing a fatal epilepsy and

contributing to the high incidence of canine epilepsy (Lohi et al. 2005). The presence of the dodecamer repeat across canine breed barriers suggests that its origin predates dogs and it might, therefore, be present in related species. It is now of enormous interest to see if dodecamer repeats are responsible for any similar human disorders.

Finally, because of the sheer size of the canine families, some of the studies have allowed scientists to make progress in understanding the role of missense changes in disease susceptibility. For instance, canine hereditary multifocal renal cystadenocarcinoma and nodular dermatofibrosis (RCND) is an autosomal dominant form of renal cell cancer in the German Shepherd Dog that is characterized by uterine leiomyomas and skin lesions. A similar disease, called Birt–Hogg–Dube syndrome (BHD), was mapped in the late 1990s to chromosome 17p12-q11.2 in humans (Khoo et al. 2001; Schmidt et al. 2001). Because the dog disease had been previously localized to canine chromosome 5 in a region that corresponded to 17q22.1, it was likely that both diseases were due to a mutation in the same gene. This proved to be the case. The *BHD* gene encodes a protein called folliculin, and protein-truncating mutations in the gene account for a substantial number of all BHD cases. It is unclear whether the remainder of cases is caused by other genes or simply missense changes that cannot be ascribed with certainty to the disease. In the case of the dog, a single missense change in exon 7 of the canine BHD gene that altered a highly conserved amino acid sequence is clearly responsible for the disease (Lingaas et al. 2003). All unaffected German Shepherd Dogs tested lacked the change, whereas all affected dogs, regardless of their origin, carried the mutation. In addition, the mutation is homozygous lethal. In this case, dog studies were clearly able to identify a portion of the protein that is critical for function. Such studies would not have been possible in human genetics. We thus expect to turn increasingly to canine families to map genes of interest, as well as to understand the genetic mechanisms that are responsible for phenotypes of interest.

SINE Elements and Disease

Elements associated increasingly with canine disease and phenotype alterations are the short interspersed nuclear elements, or SINEs (Minnick et al. 1992; Bentolila et al. 1999; Vassetzky and Kramerov 2002). These retrotransposons are implicated in genome evolution and include several families of well-recognized repeats, such as Alu sequences in humans (Schmid 1996; Lander et al. 2001; Venter et al. 2001). In dogs, the major family of SINE elements is derived from a tRNA-Lys and is distributed throughout the genome at about 126 kb spacing (Coltman and Wright 1994; Bentolila et al. 1999; Kirkness 2006). The frequency of canine bimorphic SINE elements is 10- to 100-fold higher than what is observed in humans, due largely to expansion of a single subfamily termed SINEC_Cf in the canine lineage (Kirkness et al. 2003).

As with human Alu repeats, a surprising number of SINE elements seem to be located in positions that affect gene expression. A good example is the often-cited SINEC-Cf element inserted into intron 3 of the gene encoding the hypocretin receptor, resulting in the disease narcolepsy in the Doberman Pinscher (Lin et al. 1999). As mentioned earlier, these data were the first to link the hypocretin gene family to sleep disorders, and a large body of work on molecular biology of sleep has evolved from these initial studies. Likewise, the insertion of a SINE element into the canine *PTPLA* gene leads to multiple splicing defects causing an autosomal recessive centronuclear myopathy in the Labrador Retriever (Pele et al. 2005). Finally, a very recent study links merle coloring in several breeds to the insertion of a SINE element within the *SILV* gene, which has been shown to be important in mammalian pigmentation (Clark et al. 2006).

8.2.6
Complex Traits

Mapping QTLs for Morphology in the Portuguese Water Dog

The study of complex traits in the Portuguese Water Dog (PWD) represents one of the most progressive studies in canine genetics to date. There are about 10,000 living PWDs in the United States today, descended from a small number of dogs that derive primarily from two kennels (Braund and Miller 1986; Molinari 1993). Six dogs reportedly account for nearly 80% of the current gene pool. Not surprisingly, the breed is characterized by extensive consanguinity, ranging from 0 to 0.6 (Chase et al. 1999).

Several years ago Karen Miller, Gordon Lark, Kevin Chase, and collaborators at the University of Utah began a program termed the Georgie Project (www.georgieproject.com) in which they sought to collect extensive phenotypic data on as many living, registered PWDs as possible. In addition to DNA samples and detailed family and health history, they collected a set of 91 phenotypic measurements, derived principally from a set of five radiographs collected on each dog. To date over 1,100 dogs have been enrolled in the Georgie Project. Most dogs have been genotyped with a series of 500 microsatellite markers that span the genome at about 5 cM density (Francisco et al. 1996; Guyon et al. 2003; Clark et al. 2004).

The investigators used principal component (PC) analysis, which classifies variation of correlated traits into independent linear combinations to identify a large set of QTLs for body size and shape (Chase et al. 2002). Recall that PCs are phenotypes and as such are subject to genetic analysis, with the ultimate goal being to find the QTLs that control each PC. The analysis demonstrated that PC1 controls overall body size. PC1 is derived from 44 measurements that show strong correlation between the metrics of the skull, hip, and fore and hind limbs. PC2 demonstrated that the metrics of the pelvis were inversely correlated with those of the head and neck. Therefore, individuals that have a small pelvis and lumbar vertebrae will tend to have relatively large posterior faces, small anterior faces, and large attachment sites for jaw and neck muscles. In PC3, the metrics of the length of the skull and limbs were inversely correlated with the metrics of skull width and height, including those that define the volume of the cranium. This is illustrated by the tradeoffs between the Greyhound which has a long, narrow head, and long limbs compared to the Pit Bulls which is characterized by short limbs and a broad head and neck. In PC4, skull and limb lengths were inversely correlated with the metrics associated with the strength of the limb and axial skeletons. The thicker a bone is in cross-section, the greater its strength against failure, as in Pit Bulls, Rottweilers, and Bulldogs. However, longer, thinner bones are better adapted to hunting and racing dogs that require great speed, such as the Greyhound, Whippet, and Borzoi.

Initial analysis of these data highlighted an initial set of 44 QTLs on 22 chromosomes that were associated with these heritable skeletal phenotypes (Lark et al. 2006). Subsequent analysis has identified many more QTLs. Ongoing collaborations between the Ostrander lab and Gordon Lark, Kevin Chase, and collaborators are aimed at finding the specific causative variants for each principal component.

Sexual Dimorphism in the PWD

Among the most satisfying aspects of our work on the Portuguese Water Dog has been the ability to demonstrate how canine genetics has allowed scientists to unravel complex but nonadditive interactions between genetic loci, a problem which has proven difficult to approach using classical genetic methods. For example, 21% of the observed variation in skeletal size between PWDs results from differences between females and males. An interesting study published in late 2005 addressed the role of genetic loci controlling that variance in the PWD (Chase et al. 2005a, b). Using the same data set described earlier with the Georgie Project, Chase et al. (2005a, b) examined loci associated with sexual dimorphism. Male PWDs on average are 15% larger than females. There are, however, a small set of females, approximately 10%, that are as large as the largest males. Any genetic explanation of the overall size difference would have to account for the small cohort of very large females.

Analysis of 42 metrics of the pelvis, fore and hind limbs of 463 dogs demonstrated that PC1 (variation in overall body size) was significantly correlated with body mass ($r = 0.7$). A QTL-regulating size variation was found to be associated with a marker on canine chromosome 15 (CFA15) and associated with canine microsatellite FH2017 (Chase et al. 2005a, b). FH2017 alleles were associated with overall body size in the PWD and the effect was additive. In males, allele A (associated with larger dogs) was dominant (AA = AB > BB); whereas allele B (associated with smaller dogs) was dominant in females (BB = AB < AA) (Fig. 6). In addition, a marker in close proximity to the *CHM* locus on the X-chromosome has been shown to interact

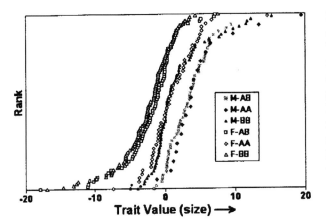

Fig. 6 Size variation (Principal Component 1 as originally described in Chase et al. (2005a, b)) in the Portuguese Water Dog (Chase et al 2005a, b). The *Y* axis indicates number of dogs and *X* indicates size, with smaller dogs on the *left* and larger on the *right*. Dogs are divided into groups based on sex (M or F) and on their genotypes at the FH2017 locus (AA, AB, or BB). Forty-two skeletal measurements from 463 dogs were used to calculate size. On average, females are smaller than males. For males, the larger haplotype was dominant (AA = AB > BB), while in females, the smaller haplotype was dominant (BB = AB > BB)

with the body size QTL associated with FH2017. The result is statistically significant ($p < 0.0001$).

Understanding the interactions of the various haplotypes to produce dogs of predicable size is not trivial. The analysis indicated that females that were homozygous for either allele at *CHM* on the X-chromosome, and possessed the FH2017 genotype for large size are, on average, as large as the largest males (Chase et al. 2005a, b). However, there was no effect in dogs that were heterozygous for the marker located on the X-chromosome. Those dogs were small and there was no segregation of FH2017 alleles for size in that background.

One of the most interesting aspects of this work will be determining if the results are applicable to other species. In nearly all species the male is larger than the female. Understanding the genetics that control such a fundamental observation across mammalian species would be very interesting and demonstrate the power of the canine system.

Mechanisms of Phenotypic Variation in the Dog

Very recently, efforts have been made to understand the apparent plasticity of the canine genome. Fondon and Garner (2004) propose that expansions and contractions of tandem repeats within coding sequences are a major source of phenotypic variation in dogs. As such, this mechanism serves to generate dogs with novel morphologies faster than would be otherwise predicted. To test their hypothesis, they sequenced 37 repeat-containing regions from 17 genes that were known or predicted to have a role in craniofacial development in dogs of 92 breeds. They found that the repeats in the dog were changing faster in terms of length than comparable repeats in humans. They also analyzed three-dimensional models of dog skulls from 20 breeds and some mixed breed dogs and found that variation in the number of repeats in the coding regions of the *Alx-4* (Aristaless-like 4) and *Runx-2* (Runt-related transcription factor 2) genes were associated with significant differences in limb and skull morphology.

The authors argue that the incremental effects of repeat length mutations would be an efficient way to generate the rapid yet morphologically conservative changes that distinguish various breeds of dog. If correct, this hypothesis would be in striking contrast to a commonly held view that variation arises largely from modification of gene regulatory sequences, such as transcriptional control elements.

Several avenues of experimentation are suggested by this work, including studies of additional genes and more phenotypic measures. However, the initial data provide a starting point for relating novel features of the canine genome to the repeated observation of the rapid creation of morphologically distinct breeds.

In a more recent study, Bjornerfeldt and collaborators (2006) propose that the domestication event itself had a profound effect on the dog's genome. They suggest that domestication resulted in an overall relaxation of selective constraints on the dog mitochondrial genome. They tested their hypothesis by sequencing the complete mtDNA from 14 dogs representing the four clades described previously by Vila et al. (1997) as well as from six wolves and three coyotes. They constructed a phylogenetic tree from the 23 complete mtDNA sequences and used maximum-likelihood approaches to estimate the rates of synonymous (dS) and nonsynonymous (dN) evolution in the mtDNA along each branch. Interestingly, they found no evidence that the overall mutation rate of mtDNA differs between dogs and wolves. However, they did find a significant difference between the average ratio of synonymous to nonsynonymous changes (dS/dN) in the mtDNA in branches of the tree associated with domestic dog versus the wolf, with the dog branches having a significantly higher ratio (Bjornerfeldt et al. 2006). This result was unexpected. Dogs represent an expanding population and one would therefore predict a greater probability of allele loss. The authors interpret their findings as suggestive of the relaxation of constraint on dog mtDNA. They further suggest that this relaxation contributes to the process of accumulating nonsynonymous mutations which in turn has led to an increase in nonlethal genetic changes throughout the dog genome. These genetic changes contribute to the phenotypic diversity observed in domestic dogs. Disease genes may be swept along as breeders select particular phenotypes for expansion. It will be interesting as more disease variants are identified to see if either of these hypotheses hold true.

Complex Disease in the Domestic Dog: The Example of Hip Dysplasia

Particularly challenging will be the identification of genes associated with complex diseases such as hip dysplasia, a common disease in dogs, affecting up to 50% of the large breeds. All dogs appear to have normal hips at birth, but within weeks to a year joint laxity becomes evident in affected individuals. The disease is recognized radiographically as subluxation of the femoral head from the acetabulum of the hip joint (Olsson et al. 1972; Riser 1973). The chronic uneven loading and remodeling eventually manifests as osteoarthritis, the hallmarks of which are observed histologically as synovitis, erosion of the articular cartilidge, edma, fibrosis of the joint capsule, and osteocyte formation. The disease is likely caused by both genetic (Smith et al. 2001; Bliss et al. 2002; LaFond et al. 2002; Todhunter et al. 2003a; Maki et al. 2004) and environmental factors (Kealy et al. 1997, 2000; Leighton et al. 1977; Smith et al. 2001; Todhunter et al. 2003a). The relative proportion of each is unknown and may, in fact, be different in different breeds. Two primary approaches have been taken in tackling this complex problem, as described later.

Investigators at the University of Utah have looked for a genetic association in a well-characterized group of 286 PWD, which have been genotyped using a set of 500 microsatellite markers. All dogs have been phenotyped for Norberg angle, a highly heritable and quantitative radiographic measure that is highly predictive of osteoarthritis. Analyzing the data separately for left and right hips, they report statistically significant associations for loci on the two ends of chromosome 1. These unlinked QTLs on CFA1 are located at 11.5 and 111.3 Mb, respectively (Chase et al. 2004). Of note, these two loci account for less than 25% of hip dysplasia. Also of interest, these investigators reproducibly found that laxity was greater in the left than the right hip, but the difference was not heritable. More loci clearly remain to be discovered, and both gene-gene and gene-environment interactions need to be considered. A third locus on a different chromosome (CFA3) was found to be associated with osteoarthritis (Chase et al. 2005a, b).

By comparison, Todhunter and colleagues have developed a large outcrossed pedigree of affected Labrador Retrievers crossed with unaffected Greyhounds (Todhunter et al. 1999, 2003a; Bliss et al. 2002). A variety of measures including age at detection of femoral capital epiphyseal ossification, distraction index, hip joint dorsolateral subluxation score, and hip joint osteoarthritis are being used in a genome-wide scan for classical linkage (Todhunter et al. 1999). While no gene has yet been found, pedigree analysis suggests that loci controlling these traits act additively and that the distraction index may be controlled by a single major locus (Todhunter et al. 2003a, b).

These studies illustrate two distinct methods for approaching a complex genetic problem in dogs using genetic analysis. The first makes use of the availability of large controlled populations with limited genetic diversity. The second demonstrates the ability to create crosses between populations showing extremes of phenotype. Comparison of the outcomes of the two approaches is likely to improve both the design of future studies and further our knowledge of this common and complex disease.

Poorly Understood Complex Traits: Aging and Behavior

The dog may be well suited for studies aimed at understanding the genetic contributions to aging and behavior. The dog is unique in that it possesses tremendous natural phenotypic variation in life expectancy and behavior in the context of highly structured breeds.

Studies of Aging in Dogs Longevity studies in humans have had limited success because of the high levels of genetic diversity and poor access to controls. Model organisms, however, have been useful in identifying several conserved biochemical pathways across a phylogenetically diverse sampling of species. Insulin signaling, stress resistance, cellular repair, chromosomal and nuclear structure, and caloric restrictions all influence lifespan (reviewed in: Warner 2003, 2005; Browner et al. 2004). The dog complements studies in other model organisms because of the dramatic variation in longevity observed across breeds, the extremes of which are demonstrated by the Irish Wolfhound and Papillion. The median lifespan of an Irish Wolfhound is only 7 years, whereas that of the Papillion is 16 years (Greer et al., 2007). Furthermore, several groups have documented an inverse relationship between body size and average lifespan in dogs (Li et al. 1996; Patronek et al. 1997; Galis et al. 2006; Greer et al., 2007). This is in contrast to the interspecific trend that is typically observed, where larger animals tend to be longer lived (Speakman et al. 2002). Because larger animals generally have slower metabolisms, differential metabolic rates have been cited as the causative agents for their longer life spans (Speakman et al. 2002, 2003). Thus, studies in the dog may expand our knowledge of canonical evolutionarily conserved biochemical pathways related to aging and reveal novel mechanisms that explain why the dog is unique with regard to the relationship between body size and longevity.

Recently, Canterberry et al. (2005) began studies of putative aging loci in the dog. Using a candidate gene approach they studied 54 genes, 26 of which were selected because of their roles in either increasing or decreasing longevity in other organisms. The remaining 28 genes were examined based upon their proximity to a marker, D4S1564, on human chromosome 4 (HSA4) (Puca et al. 2001). This region on HSA4 was chosen because previous work showed that D4S1564 correlated with an enhanced likelihood of siblings of centenarians achieving extreme old age (Puca et al. 2001). Close examination of the 54 genes by radiation hybrid mapping revealed that 45 genes mapped to regions in which there was conserved synteny between the canine and human genomes (Guyon et al. 2003). Of the 26 genes that are associated with aging in other organisms, two mapped to unexpected regions of the canine genome. Specifically, *FGF2*, which lies near the evolutionary breakpoint of HSA4, a region which maps to CFA32 and CFA19 (http://www-recomgen.univ-rennes1.fr/doggy.html) instead mapped to CFA8, which corresponds to HSA14 (Guyon et al. 2003). *SOD2*, predicted to be on CFA1, mapped to CFA19, in a region that has conservation of synteny with HSA2q21.1 (Guyon et al. 2003). These data result from gene paralogs of *FGF2* and *SOD2*. The gene paralogs result from gene duplication events that, in turn, are the foundation for gene families (Lynch and Conery 2000). Numerous gene paralogs that affect the aging process (Rikke et al. 2000) have been found in *C. elegans*. Therefore, although these gene paralogs were mapped inadvertently, the data may be pertinent to future studies of aging in the dog.

Behavior and the Dog Studies of behavior are certain to hold the most surprises. The three main promises of the canine system are to advance our understanding of mental illness, breed-specific differences in behavior, and personality and temperament traits. As with diseases like cancer, congenital heart defects, and cataracts, man and dog share numerous behavioral disorders. Obsessive compulsive disorders, anxiety disorders, narcolepsy, and epilepsy are all found in both humans and dogs. Much success has already been realized in studies of narcolepsy and epilepsy as described previously. Obsessive compulsive disorder (OCD) has been extensively described in the Bull Terrier where the disorder manifests as a fixation with a particular object or with an individual's own tail (tail chasing) (Moon-Fanelli and Dodman 1998). The Bull Terrier was selected by breeders for a desire to attack and hold on to prey and the obsessive-compulsive behavior could simply be their natural predatory behavior gone awry. The disorder is most likely heritable as it is seen in particular pedigrees and appears to be a spontaneous behavior rather than learned (Dodman et al. 1997). Of particular interest is the work by Moon-Fanelli and Dodman (1998) who report that Bull Terriers with OCD that were treated with clomipramide, a serotonin-reuptake inhibitor, had an overall decrease in compulsive behavior. This is the same as the response observed to similar drugs in humans with OCD (Moon-Fanelli and Dodman 1998). The breed specificity of canine OCD and its similarity to human OCD reinforces the value of the canine model for studying behavioral disorders.

The second area of research in dogs that is likely to revolutionize our understanding of behavioral genetics and mammalian evolution is uncovering the genetic basis of breed-specific behavioral variation. Individual breeds of domestic dog have been bred to perform a wide variety of highly specific jobs. As a consequence, domestic dogs exhibit enormous breed-specific variation in behavior. Not only is the dog's behavioral variation great, but the constellation of behaviors that are variable in the dog are also unique (tracking, herding, pointing, chasing, flushing, retrieving, etc.). Until recently, a lack of genetic resources and experimental approaches created a formidable barrier to extracting this goldmine of information. The genome sequence (Lindblad-Toh et al. 2005) and 64,000 SNP Affymetrix Genechip provide powerful experimental tools, while the in silico mapping, developed in the mouse (Pletcher et al. 2004), delineates a promising genetic approach to examining canine behavior.

One of the most exciting questions that the dog can address is how genes shape personality. Studies of aggressive behavior in Golden Retrievers are among the first to probe for the genes that govern personality and temperament traits. Variation in aggression among Golden Retriever lineages suggests that there is a significant genetic component to this behavior (Knol 1997). Van den Berg et al. (2006) demonstrate significant variation in aggressive behavior within a pedigree, a prerequisite for genetic interrogation by linkage analysis. Are we really poised to understand the genetics of loyalty, affection, and commitment? Time alone will tell. But one thing is sure - the dog will stand beside us as we ponder and debate these questions.

8.2.7
Resources

Genetic Disease Database

There are several resources available for canine geneticists that can facilitate their work. The first is a web-based compendium of information about genetic disease curated by Sargan (2004). The searchable web-based database (http://www.vet.cam.ac.uk/idid) catalogs diseases of dogs which are likely to be transmitted either wholly or partially through a genetic mechanism. It also lists the breeds in which they have been described and a short description of the disorder. When they are known, genetic and molecular genetic summaries are also included and the database connects to entry points in the literature for each disease.

Canine Map Resources

The canine RH and linkage maps continue to be important resources for the community. See http://genoweb.univ-rennes1.fr/Dogs/RH3270-page.html for a 1 Mb resolution RH map of the dog and http://genoweb.univ-rennes1.fr/Dogs/maquette2.html for links to other maps. The integrated map (Breen et al. 2004) provides valuable information about meiotic linkage distances, the RH map, and cytogenetic location. Reference families for contributing to the meiotic linkage map are freely available through Nestle Purina (Neff 1999; Mellersh 1997, 2000).

The RH map contains, as described earlier, about 10,000 gene sequences (Hitte et al. 2005), over a thousand BAC ends (Breen et al. 2004), and several hundred microsatellite markers. CRH_Server (http://idefix.univ-rennes1.fr:8080/Dogs/rh-server.html) is an online comparative and radiation hybrid mapping server for the canine genome (Hitte et al. 2004). RH mapping data can be submitted to investigators

at the University of Rennes through an easy-to-use web-based front end and locations of markers relative to others on the chromosome calculated. It also allows for dog/human genome mapping analysis. For determining the order of markers relative to genes in regions of the genome where the sequence is in question or where repeats have made sequence assembly difficult this resource (http://idefix.univ-rennes1.fr:8080/Dogs/rh-server.html) is important. In addition, considerably more information about several gene families, such as the olfactory gene set, is available through the RH maps than the assembled sequence (Quignon et al. 2003).

CanFam2.0

CanFam2.0 (June 2005), the current assembly of the canine genome, was generated by the Broad Institute of MIT/Harvard and Agencourt Bioscience. It contains approximately 2.5 billion bp. This sequence is based on 7.5× coverage of the dog genome, assuming a whole-genome sequence assembly size of 2.4 Gb. A variety of websites allows one to navigate the information effectively. The site, www.ensembl.org/Canis_familiaris/mapview?, gives scientists an excellent chromosome-by-chromosome view and provides information on length, known protein coding genes, novel protein coding genes, number of pseudogenes, mRNA genes, rRNA genes, shRNA genes, and more for each chromosome. The UCSC website, www.genome.ucsc.edu/cgi-bin/hg Gateway?clade=vertebrate&org=Dog&db=0&hgsid=4 2350241, provides a searchable genome browser gateway for obtaining detailed information about any position in the dog genome. Chromosomes can easily be searched one at a time, and detailed information about the size of each chromosome in base pairs including gaps are given on the website, www.genome.ucsc.edu/cgi-bin/hgTracks?hgsid=74674524&chromInfoPage= (Table 2).

Canine SNP Chips

Among the most promising resources recently made available is a SNP chip of some 127,000 SNPs on an Affymetrix platform (http://www.affymetrix.com/index.affx; http://www.broad.mit.edu/mammals/dog/snp/). Most of the SNPs were chosen by comparison of the two sequenced chromosomes of the 7.5× boxer sequence (Lindblad-Toh et al. 2005), the 1.5x

Table 2 Size of canine chromosome in base pairs based on the assembled sequence[a]

Chromosome	Length (bp) including gaps
Chr 1	125,616,256
Chr 2	88,410,189
Chr 3	94,715,083
Chr 4	91,483,860
Chr 5	91,976,430
Chr 6	80,642,250
Chr 7	83,999,179
Chr 8	77,315,194
Chr 9	64,418,924
Chr 10	72,488,556
Chr 11	77,416,458
Chr 12	75,515,492
Chr 13	66,182,471
Chr 14	63,938,239
Chr 15	67,211,953
Chr 16	62,570,175
Chr 17	67,347,617
Chr 18	58,872,314
Chr 19	56,771,304
Chr 20	61,280,721
Chr 21	54,024,781
Chr 22	64,401,119
Chr 23	55,389,570
Chr 24	50,763,139
Chr 25	54,563,659
Chr 26	42,029,645
Chr 27	48,908,698
Chr 28	44,191,819
Chr 29	44,831,629
Chr 30	43,206,070
Chr 31	42,263,495
Chr 32	41,731,424
Chr 33	34,424,479
Chr 34	45,128,234
Chr 35	29,542,582
Chr 36	33,840,356
Chr 37	33,915,115
Chr 38	26,897,727
Chr X	126,883,977
Chr Un	86,547,043
Chr M	16,727
Total	2,531,673,953

[a] www.genome.ucsc.edu/cgi-bin/hgTracks?hgsid=74674418&chromInfoPage=

poodle sequence (Kirkness et al. 2003), and 900,000 sequence reads generated (100,000 per dog) from each

of nine breeds representing the AKC groups (German Shepherd Dog, Rottweiler, Bedlington Terrier, Beagle, Labrador Retriever, English Shepherd, Italian Greyhound, Alaskan Malamute, and the Portuguese Water Dog). About 5% of SNPs were selected from sequence reads done on five wild canids including a coyote and four types of gray wolf.

Expression array chips are also available for the canine research community. One has been described by investigators at Nestle Purina and utilizes both RNA sequences contained in GenBank and a proprietary database of 350,000 ESTs (Holzwarth et al. 2005), using sequences isolated from more than 48 tissues. The final complement on the array includes sequences unique to GenBank (3,160), unique to the proprietary EST database (17,620), and present in both sources (1,996). Sequences with high or low abundance are represented. The majority of the sequences were found in multiple tissues (14,905), while 5,796 were found in only one tissue.

An additional expression array is also available from Affymetrix. The Affymetrix GeneChip® Canine Genome 2.0 Array (http://www.affymetrix.com/products/arrays/specific/canine_2.affx) enables researchers to simultaneously interrogate 18,000 *C. familiaris* mRNA/EST-based transcripts and over 20,000 nonredundant predicted genes. The sequence information for this array was selected from public data including the *C. familiaris* UniGene Build #11 (April, 2005), and mRNAs in GenBank up to April 15, 2005, as well as gene predictions derived from the 7.5× Boxer sequence (BROADD1 Prediction Set, Broad Institute), downloaded from EBI on May 9, 2005.

8.3
Future Scope of Work

Throughout our history, humans and dogs have enjoyed a unique and close relationship. Even Homer paid tribute to the dog's unwavering loyalty in *The Odyssey* (Homer 1996) when after traveling for 20 years, Odysseus returned to Ithaca disguised as a beggar. While his oldest friends failed to recognize him, his dog Argus, although crippled by age, recognized his master and wagged his tail in happiness at his master's return. Much of the drive in the field of canine genetics is because of this strong bond between man and dog.

Scientists in the companion animal community are facing enormous opportunities. Genomic resources are available to revolutionize the way companion animal health is carried out. Disease genes can be mapped, diagnostic tests developed, and breeding programs reprioritized to produce animals lacking the breed-specific diseases. Breed-specific diseases will truly fade from the genetic pool of individual breeds given enough generations and judicious mating schemes.

The most interesting findings in the next ten years will almost certainly be those associated with understanding the genetics of complex phenotypes. It will be fascinating to learn how legs become long or short, bodies squat or lean, and how head shapes become rounded or angular. These studies in turn will tell us even more about complex diseases such as cancer and heart disease, both of which are among the top killers of humans and dogs alike.

The last area to be explored in canine genetics is the study of behavioral traits. Research aimed at understanding the genetics of naturally occurring canine behaviors such as herding, pointing, and drafting is certain to create a new vocabulary for scientists studying behavioral genetics. Dogs also have much to offer scientists seeking to understand behavioral traits found in both humans and canines, and research into aberrant behaviors is certain to alter the way we think about human mental illnesses.

The dog has been a steadfast companion throughout our history, bred by humans to both assist us and to enrich our lives. It now seems the dog's next important contribution to our lives will be genetic one. It is the study of the dog genome that will allow us to understand complex traits exhibited not only in the dog, but also in his best friend.

Acknowledgments. We thank our colleagues Heidi Parker, Nathan Sutter, Pascale Quignon, Ed Giniger, Keith Murphy, and Gordon Lark for sharing ideas, data, and comments for this review. We thank the Intramural Program of the National Human Genome Research Institute and the American Kennel Club Canine Health Foundation who supported a portion of the work reviewed here. Finally we thank the many dog owners, breeders, and breed clubs for helping us to obtain DNA samples for these and other ongoing studies.

References

Acland, GM, et al. (1978) A novel retinal degeneration locus identified by linkage and comparative mapping of canine early retinal degeneration. Genomics, 1999. 59: 134–142

Acland GM, Ray K, Mellersh CS, Gu W, Langston AA, Rine J, Ostrander EA, Aguirre GD (1998) Linkage analysis and comparative mapping of canine progressive rod-cone degeneration (prcd) establishes potential locus homology with retinitis pigmentosa (RP17) in humans. Proc Natl Acad Sci USA 95:3048–3053

Acland GM, Ray K, Mellersh CS, Langston AA, Rine J, Ostrander EA, Aguirre GD (1999) A novel retinal degeneration locus identified by linkage and comparative mapping of canine early retinal degeneration. Genomics 59:134–142

Acland GM, Aguirre GD, Ray J, Zhang Q, Aleman TS, Cideciyan AV, Pearce-Kelling SE, Anand V, Zeng Y, Maguire AM, et al (2001) Gene therapy restores vision in a canine model of childhood blindness. Nat Genet 28:92–95

Adams JR, Leonard JA, Waits LP (2003) Widespread occurrence of a domestic dog mitochondrial DNA haplotype in southeastern US coyotes. Mol Ecol 12:541–546

Aguirre GD, Acland GM (1988) Variation in retinal degeneration phenotype inherited at the *prcd* locus. Exp Eye Res 46: 663–687

Aguirre G, Farber D, Lolley R, Fletcher RT, Chader GJ (1978) Rod-cone dysplasia in Irish Setter dogs: a defect in cyclic GMP metabolism in visual cells. Science 201:1133

Aguirre GD, Baldwin V, Pearce-Kelling S, Narfström K, Ray K, Acland GM (1998) Congenital stationary night blindness in the dog: common mutation in the RPE65 gene indicates founder effect. Mol Vis 4:23

American Kennel Club (1998) The Complete Dog Book. 19th Edn Revised edn. Official Publication of the American Kennel Club, Howell Book House, New York, USA

Bentolila S, Bach JM, Kessler JL, Bordelais I, Cruaud C, Weissenbach J, Panthier JJ (1999) Analysis of major repetitive DNA sequences in the dog (*Canis familiaris*) genome. Mamm Genom 10:699–705

Bjornerfeldt S, Webster MT, Vila C (2006) Relaxation of selective constraint on dog mitochondrial DNA following domestication. Genom Res 16:990–994

Bliss S, Todhunter RJ, Quaas R, Casella G, Wu R, Lust G, Williams AJ, Hamilton S, Dykes NL, Yeager A, et al (2002) Quantitative genetics of traits associated with hip dysplasia in a canine pedigree constructed by mating dysplastic Labrador Retrievers with unaffected Greyhounds. Am J Vet Res 63:1029–1035

Braund K, Miller DF (1986) The Complete Portuguese Water Dog. Howell Book House, New York, USA

Breen M, Langford CF, Carter NP, Holmes NG, Dickens HF, Thomas R, Suter N, Ryder EJ, Pope M, Binns MM (1999a)

FISH mapping and identification of canine chromosomes. J Hered 90:27–30

Breen M, Bullerdiek J, Langford CF (1999b) The DAPI banded karyotype of the domestic dog (*Canis familiaris*) generated using chromosome-specific paint probes. Chrom Res 7:401–406

Breen M, Jouquand S, Renier C, Mellersh CS, Hitte C, Holmes NG, Cheron A, Suter N, Vignaux F, Bristow AE, et al (2001) Chromosome-specific single-locus FISH probes allow anchorage of an 1800-marker integrated radiation-hybrid/linkage map of the domestic dog genome to all chromosomes. Genom Res 11:1784–1795

Breen M, Hitte C, Lorentzen TD, Thomas R, Cadieu E, Sabacan L, Scott A, Evanno G, Parker HG, Kirkness EF, et al (2004) An integrated 4249 marker FISH/RH map of the canine genome. BMC Genomics 5:1–11

Browner WS, Kahn A, Ziv E, Reiner A, Oshima J, Cawthon R, Hsueh W, Cummings S (2004) The genetics of human longevity. Am J Med 117:851–860

Canterberry SC, Greer KA, Hitte C, Andre C, Murphy KE (2005) Aging-associated loci in *Canis familiaris*. Growth Dev Aging 69:101–113

Chase K, Adler FR, Miller-Stebbings K, Lark KG (1999) Teaching a new dog old tricks: identifying quantitative trait loci using lessons from plants. J Hered 90:43–51

Chase K, Carrier DR, Adler FR, Jarvik T, Ostrander EA, Lorentzen TD, Lark KG (2002) Genetic basis for systems of skeletal quantitative traits: principal component analysis of the canid skeleton. Proc Natl Acad Sci USA 99:9930–9935

Chase K, Lawler DF, Adler FR, Ostrander EA, Lark KG (2004) Bilaterally asymmetric effects of quantitative trait loci (QTLs): QTLs that affect laxity in the right versus left coxofemoral (hip) joints of the dog (*Canis familiaris*). Am J Med Genet Part A 124:239–247

Chase K, Lawler DF, Carrier DR, Lark KG (2005a) Genetic regulation of osteoarthritis: A QTL regulating cranial and caudal acetabular osteophyte formation in the hip joint of the dog (*Canis familiaris*). Am J Hum Genet 135:334–335

Chase K, Carrier DR, Adler FR, Ostrander EA, Lark KG (2005b) Interaction between the X chromosome and an autosome regulates size sexual dimorphism in Portuguese Water Dogs. Genom Res 15:1820–1824

Chase K, Sargan D, Miller K, Ostrander EA, Lark K (2006) Understanding the genetics of autoimmune disease: two loci that regulate late onset Addison's disease in Portuguese Water Dogs. Intl J Immunogenet 33:179–184

Clark LA, Tsai KL, Steiner JM, Williams DA, Guerra T, Ostrander EA, Galibert F, Murphy KE (2004) Chromosome-specific microsatellite multiplex sets for linkage studies in the domestic dog. Genomics 84:550–554

Clark LA, Wahl JM, Rees CA, Murphy KE (2006) Retrotransposon insertion in SILV is responsible for merle patterning of the domestic dog. Proc Natl Acad Sci USA 103:1376–1381

Coltman DW, Wright JM (1994) Can SINEs: a family of tRNA-derived retroposons specific to the superfamily Canoidea. Nucl Acids Res 22:2726–2730

Dodman NH, Moon-Fanelli A, Mertens PA, Pfueger A, Stein DJ (1997) Veterinary Models of OCD. In: Hollander E, Stein DJ (eds) Obsessive-Compulsive Disorders: Diagnosis Etiology and Treatment. Marcel Decker, New York, USA, pp 99–143

Fogel B (1995) The Encyclopedia of the Dog. First Am edn. DK Publications, New York, USA

Fondon JW III, Garner HR (2004) Molecular origins of rapid and continuous morphological evolution. Proc Natl Acad Sci USA 101:18058–18063

Francisco LV, Langston AA, Mellersh CS, Neal CL, Ostrander EA (1996) A class of highly polymorphic tetranucleotide repeats for canine genetic mapping. Mamm Genom 7:359–362

Galibert F, Andre C (2002) The canine genome: alternative model for the functional analysis of mammalian genes. Bull Acad Natl Med 186:1489–1499; discussion 186:1499–1502

Galis F, Sluijs IVD, Dooren TJMV, Metz JAJ, Nussbaumer M (2006) Do large dogs die young? J Exp Zool B (Mol Dev Evol) pp 306B

Gibbs RA, Weinstock GM, Metzker ML, Muzny DM, Sodergren EJ, Scherer S, Scott G, Steffen D, Worley KC, Burch PE, et al (2004) Genome sequence of the Brown Norway rat yields insights into mammalian evolution. Nature 428:493–521

Gottelli D, Sillero-Zubiri C, Applebaum GD, Roy MS, Girman DJ, Garcia-Moreno J, Ostrander EA, Wayne RK (1994) Molecular genetics of the most endangered canid: the Ethiopian wolf *Canis simensis*. Mol Ecol 3:301–312

Greer KA, Canterberry SC, Murphy KE (2007) Statistical analysis regarding the effects of height and weight on life span of the domestic dog. Res Vet Sci 82:208–214

Guyon R, Lorentzen TD, Hitte C, Kim L, Cadieu E, Parker HG, Quignon P, Lowe JK, Renier C, Gelfenbeyn B, et al (2003) A 1-Mb resolution radiation hybrid map of the canine genome. Proc Natl Acad Sci USA 100:5296–5301

Hitte C, Derrien T, Andre C, Ostrander EA, Galibert F (2004) CRH_Server: an online comparative and radiation hybrid mapping server for the canine genome. Bioinformatics 20:3665–3667

Hitte C, Madeoy J, Kirkness EF, Priat C, Lorentzen TD, Senger F, Thomas D, Derrien T, Ramirez C, Scott C, et al (2005) Facilitating genome navigation: survey sequencing and dense radiation-hybrid gene mapping. Nat Rev Genet 6:643–648

Holzwarth JA, Middleton RP, Roberts M, Mansourian R, Raymond F, Hannah SS (2005) The development of a high-density canine microarray. J Hered 96:817–820

Homer (1996) The Odyssey. Penguin Group, New York, USA

Howell JM, Fletcher S, Kakulas BA, O'Hara M, Lochmuller H, Karpati G (1997) Use of the dog model for Duchenne muscular dystrophy in gene therapy trials. Neuromuscul Disord 7:325–328

Hyun C, Filippich LJ, Lea RA, Shepherd G, Hughes IP, Griffiths LR (2003) Prospects for whole genome linkage disequilibrium mapping in domestic dog breeds. Mamm Genom 14:640–649

Irion DN, Schaffer AL, Famula TR, Eggleston ML, Hughes SS, Pedersen NC (2003) Analysis of genetic variation in 28 dog breed populations with 100 microsatellite markers. J Hered 94:81–87

Jiang Z, Priat C, Galibert F (1998) Traced orthologous amplified sequence tags (TOASTs) and mammalian comparative maps. Mamm Genom 9:577–587

Jonasdottir TJ, Mellersh CS, Moe L, Heggebo R, Gamlem H, Ostrander EA, Lingaas F (2000) Genetic mapping of a naturally occurring hereditary renal cancer syndrome in dogs. Proc Natl Acad Sci USA 97:4132–4137

Jouquand S, Priat C, Hitte C, Lachaume P, Andre C, Galibert F (2000) Identification and characterization of a set of 100 tri- and dinucleotide microsatellites in the canine genome. Anim Genet 31:266–272

Kealy RD, Lawler DF, Ballam JM, Lust G, Smith GK, Biery DN, Olsson SE (1997) Five-year longitudinal study on limited food consumption and development of osteoarthritis in coxofemoral joints of dogs. J Am Vet Med Assoc 210:222–225

Kealy RD, Lawler DF, Ballam JM, Lust G, Biery DN, Smith GK, Mantz SL (2000) Evaluation of the effect of limited food consumption on radiographic evidence of osteoarthritis in dogs. J Am Vet Med Assoc 217:1678–1680

Khoo SK, Bradley M, Wong FK, Hedblad MA, Nordenskjold M, Teh BT (2001) Birt-Hogg-Dube syndrome: mapping of a novel hereditary neoplasia gene to chromosome 17p12–q11.2. Oncogene 20:5239–5242

Kirkness EF, Bafna V, Halpern AL, Levy S, Remington K, Rusch DB, Delcher AL, Pop M, Wang W, Fraser CM, et al (2003) The dog genome: survey sequencing and comparative analysis. Science 301:1898–1903

Kirkness EF (2006) SINEs of canine genomic diversity. In: Ostrander EA, Giger U, Lindblad-Toh K (eds) The Dog and Its Genome. Cold Spring Harbor Laboratory Press, Cold Spring Harbor, pp 209–219

Knapp DW, Waters DJ (1997) Naturally occurring cancer in pet dogs: important models for developing improved cancer therapy for humans. Mol Med Today 3:8–11

Knol BW (1997) Fear-motivated aggression in Golden Retrievers: no correlations with inbreeding. In: First Intl Conf on Vet Behav Med. Birmingham, UK

Koskinen MT, Bredbacka P (2000) Assessment of the population structure of five Finnish dog breeds with microsatellites. Anim Genet 31:310–317

Koskinen MT (2003) Individual assignment using microsatellite DNA reveals unambiguous breed identification in the domestic dog. Anim Genet 34:297–301

Kruglyak L (1999) Prospects for whole-genome linkage disequilibrium mapping of common disease genes. Nat Genet 22:139–144

LaFond E, Breur GJ, Austin CC (2002) Breed susceptibility for developmental orthopedic diseases in dogs. J Am Anim Hosp Assoc 38:467–477

Lander ES, Linton LM, Birren B, Nusbaum C, Zody MC, Baldwin J, Devon K, Dewar K, Doyle M, FitzHugh W, et al (2001) Initial sequencing and analysis of the human genome. Nature 409:860–921

Langford CF, Fischer PE, Binns MM, Holmes NG, Carter NP (1996) Chromosome-specific paints from a high-resolution flow karyotype of the dog. Chrom Res 4:115–123

Lark KG, Chase K, Carrier DR, Adler FR (2006) Genetic analysis of the canid skeleton: Morphological loci in the Portuguese Water Dog population. In: Ostrander EA, Giger U, Lindblad-Toh K (eds) The Dog and Its Genome. Cold Spring Harbor Laboratory Press, Cold Spring Harbor, pp 67–80

Leighton EA, Linn JM, Willham RL, Castleberry MW (1977) A genetic study of canine hip dysplasia. Am J Vet Res 38:241–244

Leonard JA, Wayne RK, Wheeler J, Valadez R, Guillen S, Vila C (2002) Ancient DNA evidence for Old World origin of New World dogs. Science 298:1613–1616

Li Y, Deeb B, Pendergrass W, Wolf N (1996) Cellular proliferative capacity and life span in small and large dogs. J Gerontol A Biol Sci Med Sci 51:B403–B408

Lin L, Faraco J, Li R, Kadotani H, Rogers W, Lin X, Qiu X, de Jong PJ, Nishino S, Mignot E (1999) The sleep disorder canine narcolepsy is caused by a mutation in the hypocretin (orexin) receptor 2 gene. Cell 98:365–376

Lindblad-Toh K, Wade CM, Mikkelsen TS, Karlsson EK, Jaffe DB, Kamal M, Clamp M, Chang JL, Kulbokas EJ III, Zody MC, et al (2005) Genome sequence, comparative analysis and haplotype structure of the domestic dog. Nature 438:803–819

Lingaas F, Aarskaug T, Sletten M, Bjerkas I, Grimholt U, Moe L, Juneja RK, Wilton AN, Galibert F, Holmes NG, et al (1998) Genetic markers linked to neuronal ceroid lipofuscinosis in English setter dogs. Anim Genet 29:371–376

Lingaas F, Comstock KE, Kirkness EF, Sorensen A, Aarskaug T, Hitte C, Nickerson ML, Moe L, Schmidt LS, Thomas R, et al (2003) A mutation in the canine BHD gene is associated with hereditary multifocal renal cystadenocarcinoma and nodular dermatofibrosis in the German Shepherd dog. Hum Mol Genet 12:3043–3053

Lohi H, Young EJ, Fitzmaurice SN, Rusbridge C, Chan EM, Vervoort M, Turnbull J, Zhao XC, Ianzano L, Paterson AD, et al (2005) Expanded repeat in canine epilepsy. Science 307:81

Lou XY, Todhunter RJ, Lin M, Lu Q, Liu T, Wang Z, Bliss SP, Casella G, Acland GM, Lust G, et al (2003) The extent and distribution of linkage disequilibrium in a multi-hierarchic outbred canine pedigree. Mamm Genom 14:555–564

Lynch M, Conery JS (2000) The evolutionary fate and consequences of duplicate genes. Science 290:1151–1155

Maki K, Janss LL, Groen AF, Liinamo AE, Ojala M (2004) An indication of major genes affecting hip and elbow dysplasia in four Finnish dog populations. Heredity 92:402–408

Mellersh CS, Langston AA, Acland GM, Fleming MA, Ray K, Wiegand NA, Francisco LV, Gibbs M, Aguirre GD, Ostrander EA (1997) A linkage map of the canine genome. Genomics 46:326–336

Mellersh CS, Hitte C, Richman M, Vignaux F, Priat C, Jouquand S, Werner P, Andre C, DeRose S, Patterson DF, et al (2000) An integrated linkage-radiation hybrid map of the canine genome. Mamm Genom 11:120–130

Minnick MF, Stillwell LC, Heineman JM, Stiegler GL (1992) A highly repetitive DNA sequence possibly unique to canids. Gene 110:235–238

Molinari C (1993) The Portuguese Water Dog. ELO-Publicidade, Portugal

Moody JA, Clark LA, Murphy KE (2006) Canine History and Breed Clubs. In: Ostrander EA, Giger U, Lindblad-Toh K (eds) The Dog and Its Genome. Cold Spring Harbor Laboratory Press, Cold Spring Harbor, pp 1–18

Moon-Fanelli AA, Dodman NH (1998) Description and development of compulsive tail chasing in terriers and response to clomipramine treatment. J Am Vet Med Assoc 212:1252–1257

Mount JD, Herzog RW, Tillson DM, Goodman SA, Robinson N, McCleland ML, Bellinger D, Nichols TC, Arruda VR, Lothrop CD Jr, et al (2002) Sustained phenotypic correction of hemophilia B dogs with a factor IX null mutation by liver-directed gene therapy. Blood 99:2670–2676

Neff MW, Broman KW, Mellersh CS, Ray K, Acland GM, Aguirre GD, Ziegle JS, Ostrander EA, Rine J (1999) A second-generation genetic linkage map of the domestic dog, *Canis familiaris*. Genetics 151:803–820

Neff MW, Robertson KR, Wong AK, Safra N, Broman KW, Slatkin M, Mealey KL, Pedersen NC (2004) Breed distribution and history of canine mdr1-1Delta, a pharmacogenetic mutation that marks the emergence of breeds from the collie lineage. Proc Natl Acad Sci USA 101:11725–11730

Olsson SE, Marshall JL, Story E (1972) Osteophytosis of the knee joint in the dog. A sign of instability. Acta Radiol Suppl 319:165–167

Ostrander EA, Kruglyak L (2000) Unleashing the canine genome. Genom Res 10:1271–1274

Ostrander EA, Wayne RK (2005) The canine genome. Genom Res 15:1706–1716

Ostrander EA, Jong PM, Rine J, Duyk G (1992) Construction of small-insert genomic DNA libraries highly enriched for microsatellite repeat sequences. Proc Natl Acad Sci USA 89:3419–3423

Ostrander EA, Sprague GF Jr, Rine J (1993) Identification and characterization of dinucleotide repeat (CA)n markers for genetic mapping in dog. Genomics 16:207–213

Ostrander EA, Mapa FA, Yee M, Rine J (1995) One hundred and one simple sequence repeat-based markers for the canine genome. Mamm Genom 6:192–195

Ostrander EA, Galibert F, Patterson DF (2000) Canine genetics comes of age. Trends Genet 16:117–124

Parker HG, Ostrander EA (2005) Canine genomics and genetics: running with the pack. PLoS Genet 1:e58

Parker HG, Yuhua X, Mellersh CS, Khan S, Shibuya H, Johnson GS, Ostrander EA (2001) Meiotic linkage mapping of 52 genes onto the canine map does not identify significant levels of microrearrangement. Mamm Genom 12:713–718

Parker HG, Kim LV, Sutter NB, Carlson S, Lorentzen TD, Malek TB, Johnson GS, DeFrance HB, Ostrander EA, Kruglyak L (2004) Genetic structure of the purebred domestic dog. Science 304:1160–1164

Parker HG, et al. (2007) Breed relationships facilitate fine-mapping studies: a 7.8-kb deletion cosegregates with Collie eye anomaly across multiple dog breeds. Genome Res. Nov; 17(11):1562–1571

Patronek GJ, Waters DJ, Glickman LT (1997) Comparative longevity of pet dogs and humans: implications for gerontology research. J Gerontol A Biol Sci Med Sci 52:B171–B178

Patterson DF (1980) A catalogue of genetic disorders of the dog. In: Current Veterinary Therapy VII. WB Saunders, Philadelphia, USA

Patterson D (2000) Companion animal medicine in the age of medical genetics. J Vet Intl Med 14:1–9

Patterson DF, Haskins ME, Jezyk PF (1982) Models of human genetic disease in domestic animals. Adv Hum Genet 12:263–339

Patterson DF, Haskins ME, Jezyk PF, Giger U, Meyers-Wallen VN, Aguirre G, Fyfe JC, Wolfe JH (1988) Research on genetic diseases: reciprocal benefits to animals and man. J Am Vet Med Assoc 193:1131–1144

Pele M, Tiret L, Kessler JL, Blot S, Panthier JJ (2005) SINE exonic insertion in the PTPLA gene leads to multiple splicing defects and segregates with the autosomal recessive centronuclear myopathy in dogs. Hum Mol Genet 14:1417–1427

Pletcher MT, McClurg P, Batalov S, Su AI, Barnes SW, Lagler E, Korstanje R, Wang X, Nusskern D, Bogue MA, et al (2004) Use of a dense single nucleotide polymorphism map for in silico mapping in the mouse. PLoS Biol 2:e393

Ponder KP, Melniczek JR, Xu L, Weil MA, O'Malley TM, O'Donnell PA, Knox VW, Aguirre GD, Mazrier H, Ellinwood NM, et al (2002) Therapeutic neonatal hepatic gene therapy in mucopolysaccharidosis VII dogs. Proc Natl Acad Sci USA 99:13102–13107

Priat C, Hitte C, Vignaux F, Renier C, Jiang Z, Jouquand S, Cheron A, Andre C, Galibert F (1998) A whole-genome radiation hybrid map of the dog genome. Genomics 54:361–378

Priat C, Jiang ZH, Renier C, Andr C, Galibert F (1999) Characterization of 463 type I markers suitable for dog genome mapping. Mamm Genom 10:803–813

Pritchard JK, Stephens M, Donnelly P (2000) Inference of population structure using multilocus genotype data. Genetics 155:945–959

Puca AA, Daly MJ, Brewster SJ, Matise TC, Barrett J, Shea-Drinkwater M, Kang S, Joyce E, Nicoli J, Benson E, et al (2001) A genome-wide scan for linkage to human exceptional longevity identifies a locus on chromosome 4. Proc Natl Acad Sci USA 98:10505–10508

Quignon P, Kirkness E, Cadieu E, Touleimat N, Guyon R, Renier C, Hitte C, Andre C, Fraser C, Galibert F (2003) Comparison of the canine and human olfactory receptor gene repertoires. Genom Biol 4:R80

Rikke BA, Murakami S, Johnson TE (2000) Paralogy and orthology of tyrosine kinases that can extend the life span of *Caenorhabditis elegans*. Mol Biol Evol 17:671–683

Riser W (1973) The dysplastic hip joint: It's radiographic and histologic development. J Am Vet Radiol Soc 14:35–50

Rogers CA, Brace AH (1995) The International Encyclopedia of Dogs. First Am Edn. Howell Book House, New York, p 496

Roy MS, Geffen E, Smith D, Ostrander EA, Wayne RK (1994) Patterns of differentiation and hybridization in North American wolflike canids, revealed by analysis of microsatellite loci. Mol Biol Evol 11:553–570

Sargan DR (2004) IDID: inherited diseases in dogs: web-based information for canine inherited disease genetics. Mamm Genom 15:503–506

Sargan DR, Yang F, Squire M, Milne BS, O'Brien PC, Ferguson-Smith MA (2000) Use of flow-sorted canine chromosomes in the assignment of canine linkage, radiation hybrid, and syntenic groups to chromosomes: refinement and verification of the comparative chromosome map for dog and human. Genomics 69:182–195

Savolainen P, Zhang YP, Luo J, Lundeberg J, Leitner T (2002) Genetic evidence for an East Asian origin of domestic dogs. Science 298:1610–1613

Schmid CW (1996) Alu: structure, origin, evolution, significance and function of one-tenth of human DNA. Prog Nucl Acid Res Mol Biol 53:283–319

Schmidt LS, Warren MB, Nickerson ML, Weirich G, Matrosova V, Toro JR, Turner ML, Duray P, Merino M, Hewitt S, et al (2001) Birt-Hogg-Dube syndrome, a genodermatosis associated with spontaneous pneumothorax and kidney neoplasia, maps to chromosome 17p11.2. Am J Hum Genet 69:876–882

Smith GK, Mayhew PD, Kapatkin AS, McKelvie PJ, Shofer FS, Gregor TP (2001) Evaluation of risk factors for degenerative joint disease associated with hip dysplasia in German Shepherd Dogs, Golden Retrievers, Labrador Retrievers, and Rottweilers. J Am Vet Med Assoc 219:1719–1724

Speakman JR, Selman C, McLaren JS, Harper EJ (2002) Living fast, dying when? The link between aging and energetics. J Nutr 132:1583S–1597S

Speakman JR, van Acker A, Harper EJ (2003) Age-related changes in the metabolism and body composition of three dog breeds and their relationship to life expectancy. Aging Cell 2:265–75

Sutter NB, Ostrander EA (2004) Dog star rising: the canine genetic system. Nat Rev Genet 5:900–10

Sutter NB, Eberle MA, Parker HG, Pullar BJ, Kirkness EF, Kruglyak L, Ostrander EA (2004) Extensive and breed-specific linkage disequilibrium in *Canis familiaris*. Genom Res 14:2388–2396

Switonski M, Reimann N, Bosma AA, Long S, Bartnitzke S, Pienkowska A, Moreno-Milan MM, Fischer P (1996) Report on the progress of standardization of the G-banded canine (*Canis familiaris*) karyotype. Committee for the Standardized Karyotype of the Dog (*Canis familiaris*). Chrom Res 4:306–309

Switonski M, Szczerbal I, Nowacka J (2004) The dog genome map and its use in mammalian comparative genomics. J Appl Genet 45:195–214

Todhunter RJ, Acland GM, Olivier M, Williams AJ, Vernier-Singer M, Burton-Wurster N, Farese JP, Grohn YT, Gilbert RO, Dykes NL, et al (1999) An outcrossed canine pedigree for linkage analysis of hip dysplasia. J Hered 90:83–92

Todhunter RJ, Bliss SP, Casella G, Wu R, Lust G, Burton-Wurster NI, Williams AJ, Gilbert RO, Acland GM (2003a) Genetic structure of susceptibility traits for hip dysplasia and microsatellite informativeness of an outcrossed canine pedigree. J Hered 94:39–48

Todhunter RJ, Casella G, Bliss SP, Lust G, Williams AJ, Hamilton S, Dykes NL, Yeager AE, Gilbert RO, Burton-Wurster NI, et al (2003b) Power of a Labrador Retriever-Greyhound pedigree for linkage analysis of hip dysplasia and osteoarthritis. Am J Vet Res 64:418–424

van De Sluis B, Rothuizen J, Pearson PL, van Oost BA, Wijmenga C (2002) Identification of a new copper metabolism gene by positional cloning in a purebred dog population. Hum Mol Genet 11:165–173

van den Berg L, Schilder MBH, de Vries H, Leegwater AJ, van Oost BA (2006) Phenotyping of aggressive behavior in Golden Retriever Dogs with a questionnaire. Behav Genet 36:882–902

Vassetzky NS, Kramerov DA (2002) CAN-a pan-carnivore SINE family. Mamm Genom 13:50–57

Venter JC, Adams MD, Myers EW, Li PW, Mural RJ, Sutton GG, Smith HO, Yandell M, Evans CA, Holt RA, et al (2001) The sequence of the human genome. Science 291:1304–1351

Vignaux F, Hitte C, Priat C, Chuat JC, Andre C, Galibert F (1999) Construction and optimization of a dog whole-genome radiation hybrid panel. Mamm Genom 10:888–894

Vila C, Savolainen P, Maldonado JE, Amorim IR, Rice JE, Honeycutt RL, Crandall KA, Lundeberg J, Wayne RK (1997) Multiple and ancient origins of the domestic dog. Science 276:1687–1689

Vila C, Maldonado JE, Wayne RK (1999) Phylogenetic relationships, evolution, and genetic diversity of the domestic dog. J Hered 90:71–77

Vila C, Seddon J, Ellegren H (2005) Genes of domestic mammals augmented by backcrossing with wild ancestors. Trends Genet 21:214–218

Warner HR (2003) Subfield history: use of model organisms in the search for human aging genes. Sci Aging Knowledge Environ 2003:RE1

Warner HR (2005) Longevity genes: from primitive organisms to humans. Mech Age Dev 126:235–242

Waterston RH, Lindblad-Toh K, Birney E, Rogers J, Abril JF, Agarwal P, Agarwala R, Ainscough R, Alexandersson M, An P, et al (2002) Initial sequencing and comparative analysis of the mouse genome. Nature 420:520–562

Wayne RK (1986a) Cranial morphology of domestic and wild canids the influence of development on morphological change. Evolution 40:243–261

Wayne RK (1986b) Limb morphology of domestic and wild canids: the influence of development on morphologic change. J Morphol 187:301–319

Wayne RK (1993) Molecular evolution of the dog family. Trends Genet 9:218–224

Wayne RK, Jenks SM (1991) Mitochondrial DNA analysis implying extensive hybridization of the endangered red wolf *Canis rufus*. Nature 351:565–568

Wayne RK, Ostrander EA (2004) Out of the dog house: the emergence of the canine genome. Heredity 92:273–274

Wayne RK, Nash WG, O'Brien SJ (1987a) Chromosomal evolution of the Canidae. I. Species with high diploid numbers. Cytogenet Cell Genet 44:123–133

Wayne RK, Nash WG, O'Brien SJ (1987b) Chromosomal evolution of the Canidae. II. Divergence from the primitive carnivore karyotype. Cytogenet Cell Genet 44:134–141

Wayne RK, Van Valkenburgh B, Kat PW, Fuller TK, Johnson WE, O'Brien SJ (1989) Genetic and morphological divergence among sympatric canids. J Hered 80:447–454

Wayne RK, Geffen E, Girman DJ, Koepfli KP, Lau LM, Marshall CR (1997) Molecular systematics of the Canidae. Syst Biol 46:622–653

Wayne RK, Leonard JA, Vila C (2006) Genetic analysis of dog domestication. In: Zeder MA, Bradley DG, Emshwiller E, and Smith BD (eds) Documenting Domestication: New Genetic and Archeological Paradigms. University of California Press, Berkeley, CA, USA (For the editors convenience, additional information on this book can be found at: http://www.ucpress.edu/books/pages/10279.php)

Wilson PJ, Grewal S, Lawford ID, Heal JNM, Granacki AG, Pennock D, Theberge JB, Theberge MT, Voigt DR, Waddell W, et al (2000) DNA profiles of the eastern Canadian wolf and the red wolf provide evidence for a common evolutionary history independent of the gray wolf. Can J Zool 78: 2156–2166

Yamazaki T, Toyoharu K (1995) Legacy of the Dog: The Ultimate Illustrated Guide to Over 200 Breeds. Chronicle Books, San Francisco

Yang F, O'Brien PC, Milne BS, Graphodatsky AS, Solanky N, Trifonov V, Rens W, Sargan D, Ferguson-Smith MA (1999) A complete comparative chromosome map for the dog, red fox, and human and its integration with canine genetic maps. Genomics 62:189–202

Yuzbasiyan-Gurkan V, Blanton SH, Cao Y, Ferguson P, Li J, Venta PJ, Brewer GJ (1997) Linkage of a microsatellite marker to the canine copper toxicosis locus in Bedlington terriers. Am J Vet Res 58:23–27

Zajc I, Mellersh CS, Sampson J (1997) Variability of canine microsatellites within and between different dog breeds. Mamm Genom 8:182–185

Zangler B, Goldstein O, Philp AR, Lindauer SJ, Pearce- Kelling SE, Mullins RF, Graphodatsky AS, Ripoll D, Felix JS, Stone EM, et al (2006) Identical mutation in a novel retinal gene causes progressive rod-cone degeneration in dogs and retinitis pigmentosa in humans. Genomics 88: 551–63

CHAPTER 9

Pig

Catherine W. Ernst(✉) and A. Marcos Ramos

Department of Animal Science, 3385 Anthony Hall, Michigan State University, East Lansing, MI 48824, USA, ernstc@msu.edu

9.1
Introduction

9.1.1
History and Description

The Eurasian wild boar, *Sus scrofa*, is the primary ancestor of domesticated pigs, and it is possible to identify at least 16 distinct *Sus scrofa* subspecies (Groves and Grubb 1993). These subspecies were originally scattered across Europe, Asia, and northern Africa, and they are differentiated by size, color, proportions, skull characters, and also chromosome number as a result of two Robertsonian translocations found in different geographic areas (Tikhonov and Troshina 1974). The various *Sus scrofa* subspecies exhibit extreme adaptability to food and climate conditions. They can tolerate temperature conditions from −50° to +50°C due to well-developed thermoregulation and nest building behavior, allowing them to thrive in harsh winters, tropical conditions, mountains, and semideserts. Wild boar population sizes remain relatively high in many locations despite pressures from humans and predators (Ruvinsky and Rothschild 1998).

Pigs were domesticated approximately 9,000 years ago (Epstein and Bichard 1984). Similar to other livestock species, the primary reason for the domestication of pigs was to provide human populations with a source of food protein. A 2005 study based on phylogenetic analyses of the sequence of a mitochondrial control region from 686 wild and domestic pig samples (Larson et al. 2005) provided evidence for multiple centers of domestication across Eurasia. This work showed that the basal lineages of *Sus scrofa* originated in Southeast Asia, and from this area dispersal of the species into India, East Asia, and

Western Europe occurred. Therefore, pigs appear to have been domesticated from local wild boar populations independently in several regions of Eurasia. In addition, mitochondrial diversity in European and Chinese pigs is consistent with population expansions that occurred 900,000 years ago for the Asian and European wild boar populations (Fang and Andersson 2006; Kijas and Andersson 2001).

After the initial events of domestication, dissemination of the species was relatively slow and generally limited to confined geographic regions. As a consequence, there was little exchange of genetic material between the different pig populations. In addition, because each region had specific practices regarding the way pigs were raised, as well as other unique socioeconomical factors, the phenotypic appearance of the pig populations also differed between regions. The most prominent illustration of these differences is the marked distinction between the European and Chinese breeds, in that the latter present wider and shorter heads, dished faces, short legs, and considerable fatness (Jones 1998). It was not until the 1700s that pig breeds from Asia were imported to Europe and used in the improvement of the European breeds by crossbreeding (Jones 1998).

These crossbreeding experiments were the basis for other exchanges of genetic material between different breeds. In addition, breeds also began to be selected for characteristics that were of interest for human populations. With the arrival and implementation of modern animal breeding techniques and eventually also artificial insemination in the late 1900s, the transfer of genetic material became commonplace. The impact that artificial selection had on the number and variety of pig breeds was considerable. A wide diversity of morphological and physiological characteristics exists today in various domestic breeds, resulting in extremes of lean, heavily muscled

Genome Mapping and Genomics in Animals, Volume 3
Genome Mapping and Genomics in Domestic Animals
N.E. Cockett, C. Kole (Eds.)
© Springer-Verlag Berlin Heidelberg 2009

European breeds and Chinese breeds with large litter sizes and high fat deposition. A large number of different pig breeds have been developed, and there are currently over 200 pig breeds distributed across all regions of the world (http://dad.fao.org/).

As a consequence of the domestication process and selections made by man, the morphology of the domestic pig differs from its wild boar ancestor. Typically, pigs have large heads with long snouts, small eyes, thick bodies, and short legs. Domestic pig breeds exhibit a range of coat colors from white to red to black, with various spotting patterns (Legault 1998). Hair coats also vary from coarse bristles to finer hairs. The numerous breeds of domestic pigs also display differences in mature size and ear shape. Several breeds have been selected for increased muscularity, which has altered the shape of the body.

The *Sus scrofa* genome is approximately 2.7 Gb, similar in size to that of the human genome. It consists of 18 pairs of autosomes and the sex (X and Y) chromosomes. Therefore, the swine karyotype includes 38 chromosomes ($2n = 38$), of which six pairs are acrocentric, while the remaining 12 pairs and the sex chromosomes are metacentric (Gustavsson 1988). There are also some wild boar populations located in Western Europe that display a $2n = 36$ karyotype (Fang et al. 2006). The sequencing of the pig genome is in progress (http://pre.ensembl.org/Sus_scrofa/index.html) and planned for completion in 2009.

9.1.2
Economic Importance

The pig is a species with significant economic importance to humans not only as a food source, but also for its value as a model organism for studying human health or as a potential source of organs for xenotransplantation, due to the extensive genome conservation and physiological similarities between humans and pigs. Pork is the major red meat consumed worldwide, and world pork consumption increased 27% between 1997 and 2005 to over 93 million metric tons (USDA 2006). In addition, other products can also be obtained from the pig, such as leather and components for human medical use such as heart valves. Modern pig production is primarily found in developed countries, but local pig breeds also play a very important role in the economies of developing

countries, in part because of their ability to feed on a variety of diets and their rapid reproductive rates. The top five pork-producing countries in 2005 were China, the European Union (25 countries of Europe), the USA, Brazil, and Canada (USDA 2006). World pork production increased 15.1% between 2000 and 2005 with the greatest percentage increases in Brazil, Vietnam, China, Russia, and Canada. Approximately 50% of the world pig population in 2005 was located in China, a country where per capita consumption is also among the highest in the world (33.8 kg in 2005). Despite its large population, until recently China has been almost self-sufficient in pork supplies. However, this situation is changing as a growing Chinese economy is leading to an increasing demand for pork.

Pork is an important part of the diets of people in many countries worldwide. It is a source of protein, iron, zinc, and B vitamins. Selection practices since the 1980s, especially in breeds descended from European populations, have emphasized increased lean and decreased fatness. In the USA, the United States Department of Agriculture (USDA) Agricultural Research Service (ARS) recently evaluated nutrient content for nine cuts of raw and cooked fresh pork. These data showed significant changes in nutrient content of some pork cuts over time with eight of the nine raw cuts found to be leaner than those previously analyzed, while protein content has remained unchanged (USDA 2007).

9.1.3
Breeding Objectives

Following domestication, different selection goals, as well as widely different production systems between Europe and Asia, led to pig breeds with very different characteristics. In more recent times, in order to meet worldwide pork demands, pig production requires fast growth, efficient feed utilization, high carcass merit and meat quality, as well as high levels of reproductive success among breeding animals, disease resistance, and increased survivability of young pigs (Rothschild 2004). Pork consumers, especially in Europe and the USA, are also making other demands including removal of antibiotic growth promoters from feed and altering facilities to promote better animal welfare. In addition, concerns regarding the environmental impact of swine waste have prompted producers to consider ways to reduce feed wastage and improve feed efficiency (Rothschild 2004).

The choice of the appropriate breeding goals is a key step in modern pig genetic improvement programs. It requires implementation of accurate phenotypic collection schemes, and it will impact the rate of genetic progress for the traits of interest. Breeding goals vary, but can generally be divided into two main categories, reproduction and production traits. Examples of reproductive traits include litter size, sow productivity, and longevity, while traits such as growth rate, food conversion, lean meat content, meat quality parameters, and fat quality are important production traits (Ollivier 1998). After a decision has been made regarding the choice of which breeding goals to prioritize, economic weights are assigned to each breeding goal. The genetic merit of each individual is then estimated using classical index selection and/or BLUP (Best Linear Unbiased Predictor) approaches.

The emphasis placed on various breeding objectives changes over time. For example, selection emphasizing growth rate and lean meat content that was practiced for several decades was very successful. However, a decline in pork quality was also observed. Consequently, pork quality traits (such as pH, marbling, and color) are receiving more attention in many breeding programs. Similarly for reproduction traits, litter size has long been an important breeding goal, and average litter size for several pig breeds has increased as a result of selection. However, a recent trend toward higher sow replacement rates has been observed, with reproductive failure being one of the factors responsible for this increase. In order to solve this problem, more attention is now being dedicated to traits such as sow longevity and sow replacement rates. In the future, additional traits may also be emphasized as new information becomes available from the swine genome and as some producers consider alternative production approaches such as traits related to disease resistance, animal welfare, or environmental sustainability.

9.2
Molecular Genetics of Pigs

9.2.1   Genetic Maps

The first genetic map of the pig, containing 36 genes assigned to 14 linkage groups, was published by Echard in 1984. More organized efforts aimed at

understanding the complexity, and organization of the entire pig genetic map did not begin until the 1990s. These efforts were centered in Europe where a multicountry initiative termed PiGMaP (Pig Gene Mapping Project) was established (Archibald et al. 1990), and in the USA where the USDA and several agricultural universities were also active. The initial pig linkage maps covered the pig genome only partially, since not all chromosomes were represented (Ellegren et al. 1994; Rohrer et al. 1994). However, linkage maps covering the entire genome were released a few years later (Archibald et al. 1995; Marklund et al. 1996; Rohrer et al. 1996). The markers included on the initial linkage maps were primarily microsatellite markers, although numerous type I (i.e., gene) markers were also assigned. These gene markers were typically single nucleotide polymorphisms (SNPs) that were genotyped by restriction fragment-length polymorphism (RFLP) analyses or single-strand conformation polymorphism (SSCP) analyses. Reference populations developed for constructing genetic maps made use of genetically divergent breeds, especially the European Wild Boar, Chinese Meishan and Large White, with F2 or backcross design strategies. Samples of DNA were collected in all generations and subsequently genotyped for a set of markers. Marker genotypes and pedigree information allowed map construction based on recombination frequencies. Linkage groups were anchored to chromosomes through physical mapping of some of the markers.

After the first linkage maps were published, several groups around the world began adding markers to these maps. Often, candidate genes were chosen based on their biological function and potential influence on a phenotypic trait. DNA panels of some pig population pedigrees, including the PiGMaP reference families, were made available to researchers and used to add a substantial number of genes to the linkage maps of all chromosomes. Updated maps and bioinformatics resources can be found at the US National Animal Genome Research Project web site (http://www.animalgenome.org/pigs/), the Roslin Institute, Scotland web site (http://www.thearkdb.org/), and at the National Center for Biotechnology Information (NCBI) web site (http://www.ncbi.nlm.nih.gov/projects/mapview/).

These initial maps laid the foundation for countless studies by providing relative locations for genetic markers in the pig genome and allowing for comparative mapping studies. The development of pig genetic maps was crucial for the large number of studies conducted to identify quantitative trait loci (QTL) for a variety of traits. Once identified, some of these QTL were then targeted for fine mapping, which resulted in even more markers being added to the pig genetic maps. As a consequence, the marker coverage and resolution of some regions of the pig genome dramatically increased. The determination of the complete pig genome sequence is expected to be finished in 2009. When complete, the pig genome sequence will represent the ultimate map and will be an extremely valuable resource for pig geneticists worldwide.

9.2.2
Physical and RH Maps

Concurrent with the release of the linkage maps, the first pig cytogenetic map was also published (Yerle et al. 1995). Physical maps of the pig genome developed in parallel to the linkage maps. The efforts to improve the porcine physical maps were greatly aided by the development of a pig × rodent somatic cell hybrid panel (SCHP; Yerle et al. 1996) and, at a later stage, by whole-genome radiation hybrid (RH) panels (Hawken et al. 1999; Yerle et al. 2002). These resources were very important for the development and refinement of the porcine physical maps because they were distributed to researchers worldwide and hundreds of markers were quickly added to the maps because mapping with these panels does not require the identification of polymorphic markers as is necessary for genetic mapping.

Initially, the SCHP panel was used as the main tool to improve the resolution of the pig physical maps. However, resolution obtained using the SCHP is limited, because markers could be assigned to a specific chromosomal region, but gene order and placement are not precisely determined with this resource. These problems were overcome with the release of a pig whole-genome radiation hybrid panel for high-resolution gene mapping in 1998 (Yerle et al. 1998), a resource that resulted from a collaboration established between INRA (France) and the University of Minnesota (USA). This panel, the INRA-University of Minnesota porcine radiation hybrid panel (IMpRH), includes 118 hybrids and has been made available to the community of researchers interested in

porcine gene mapping. The panel was characterized by Hawken and colleagues (1999), which added over 900 Type I and II markers to the panel. This effort was followed by the establishment of the IMpRH server and database (Milan et al. 2000b), a tool that allows researchers to map new markers relative to previously mapped markers. As a result, gene mapping in pigs significantly increased in the following years, with many researchers around the world using the IMpRH panel to map hundreds of genes and markers. Subsequently, a second porcine whole-genome radiation hybrid panel (IMNpRH2) was developed (Yerle et al. 2002). This panel was generated using a higher level of radiation (7,000-rad for IMpRH; 12,000-rad for IMNpRH2), which in turn allows an increased mapping resolution. These RH panels have been used in numerous comparative mapping studies (for example, Demeure et al. 2003; Martins-Wess et al. 2003; Liu et al. 2005).

The ultimate high-resolution map for any species is the complete sequence of its genome. The sequencing of the pig genome is currently underway and is expected to be completed in 2009 (http://pre.ensembl.org/Sus_scrofa/index.html). Some regions of the pig genome have already been sequenced, assembled, and annotated, including the MHC complex on SSC7 (Renard et al. 2006) and a portion of SSC17 (Hart et al. 2007).

9.2.3
Pig–Human Comparative Maps

In order for comparative maps to be created, similarities at the genome level between species need to be identified. In part because of the more advanced state of the human genome project, the initial comparative maps for the pig were human-porcine comparative maps. The first comprehensive comparative map between these two species was released in 1996 by Goureau and colleagues who were able to label 95% of the total length of the porcine chromosomes with human probes.

The initial sequence comparisons between species were made by analyzing the sequences available for each species genome, especially sequences derived from protein coding genes and type I loci. The comparisons made between human and pig allowed the identification of orthologous, paralogous, and syntenic genes between these species. Orthologous genes refer to genes that had a common origin, paralogous genes usually arise from duplication events and syntenic genes are located on the same chromosome. More recently, with the development of the radiation hybrid panels covering the pig genome and the increase in the number of pig specific sequences available, both from expressed sequence tag (EST) projects (Rink et al. 2006; Uenishi et al. 2007), bacterial artificial chromosome (BAC)-end sequences (Meyers et al. 2005), and shotgun sequences of the pig genome (Wernersson et al. 2005), the resolution of the human-pig comparative map has increased. Meyers et al. (2005) identified 51 conserved synteny segments between the human and pig genomes by RH mapping of BAC-end sequences, whereas Goureau et al. (1996) had previously revealed 38 conserved synteny segments using ZOO-FISH. Completion and annotation of the pig genome sequence will allow further refinement of the human-pig comparative map.

9.2.4
Quantitative Trait Loci

The objective of quantitative trait loci (QTL) mapping is to identify genomic regions associated with phenotypic variation in traits of economic importance. The first pig QTL, a major locus for fat deposition, was discovered in 1994 (Andersson et al. 1994). By 2007, over 110 QTL studies have been reported and QTL have been detected on all chromosomes, especially for traits related to growth, carcass composition, meat quality, and reproduction traits (Rothschild et al. 2007). In addition to these traits, QTL have also been identified for some stress response traits (Desautes et al. 2002), as well as a small number of QTL detected for traits related to disease resistance (Reiner et al. 2002; Wattrang et al. 2005) which are costly and difficult traits for which to obtain phenotypic measures.

Once pig genetic maps were constructed, researchers began using them to identify regions of the pig genome involved in regulation of the phenotypes of interest. The initial populations used for this purpose were usually experimental crosses (F2 or backcrosses) established by mating pigs of phenotypically divergent breeds. The typical strategy involved crossing commercial pig breeds with local or exotic breeds (e.g., Meishan, Iberian, or wild pig). Subsequent population development also included crosses of commercial breeds that exhibited phenotypic differences for traits of interest.

The size of these populations varied between studies, but many of the initial pig QTL scans used experimental crosses with sizes between 200 and 600 individuals. However, over time other studies were published that used larger numbers of individuals for QTL detection, facilitating discovery of QTL that control a smaller percentage of phenotypic variation. A common feature of all these studies was the use of microsatellite markers to initially generate chromosomal linkage maps. Marker coverage on these maps was not very dense, but nonetheless it was still sufficient to detect many major QTL present in the pig genome. The software that has commonly been used for QTL detection is QTL Express (Seaton et al. 2002), which uses a web interface where researchers load their data and perform the desired analyses. Another popular QTL mapping software is QXPAK (Perez-Enciso and Misztal, 2004), a versatile package that allows users to conduct QTL and other types of analyses. Results of these studies yield F-ratio curves such as those shown in Fig. 1.

The pig QTL mapping results obtained so far have detected a very large number of QTL for a variety of traits. To help researchers understand and integrate the results from various studies, a database comprising the majority of pig QTL was established (Hu et al. 2005; http://www.animalgenome.org/QTLdb/pig.html). By 2007 this database included over 1,675 QTL for over 280 traits. The number of QTL by trait categories, the top 20 traits, and the number of QTL per chromosome in 2007 are shown in Table 1. It can be seen from these summary statistics that the most common trait types for which QTL have been identified include anatomical measures, fatness, and growth. These categories tend to include traits of moderate to high heritability, and in addition these traits are relatively easy to measure. Therefore, most studies include measures for these traits even if they are not the primary focus of the study. It is also notable that the number of QTL reported across chromosomes is not equivalent even when accounting for chromosome size. While it is expected that QTL will be detected more frequently on some chromosomes than others due to differences in gene density, it should also be noted that many studies do not include a full genome scan, but rather target specific chromosomes on which QTL have already been reported.

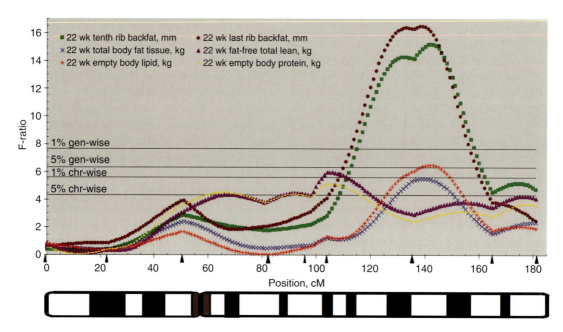

Fig. 1 F-ratio distributions for QTL for fat and protein traits on pig chromosome 6. The genetic map in centiMorgan units is shown on the X-axis with positions of genotyped markers denoted by *arrows*. An ideogram of chromosome 6 is shown for orientation

For example, numerous researchers have studied chromosome 4 because it was the first chromosome on which pig QTL were reported. Several QTL have been identified in multiple studies thus revealing similarity of QTL between populations, which may indicate genomic regions that have been under selection pressure for traits of interest to the pig industry (reviewed by Hu et al. 2005). As an example, several studies have detected QTL for backfat thickness on pig chromosomes 2, 4, 6, 7, and X.

The gene action of these QTL is mainly additive, but a number of dominant QTL have also been identified. In addition, imprinting also plays a role in the regulation of pig production traits, given the numerous imprinted pig QTL that have been identified (de Koning et al. 2000; Holl et al. 2004). More recently, the importance of epistasis has been emphasized because of results obtained in other species that demonstrated the importance of epistasis in QTL mapping, such as chickens (Carlborg et al. 2004), mice (Carlborg et al. 2005), and zebrafish (Wright et al. 2006). This research area is yet to be explored in the pig and can potentially explain some of the QTL detected so far. Moreover, there are also situations where the trait is regulated by more than one QTL on the same chromosome, such as the case for two pig carcass composition traits (carcass length and shoulder weight; Mercade et al. 2005). Overall, research in pig QTL mapping is evolving and moving toward more complex approaches and statistical models that may help in deciphering the complexity of the genetic regulation of economically important trait phenotypes.

Despite the large number of pig QTL identified so far, causative mutations or quantitative trait nucleotides (QTN) have been discovered for only a very limited number of these QTL. Examples of QTN identified in the pig include the *IGF2* (Van Laere et al. 2003) and *PRKAG3* (Milan et al. 2000a) genes. Presently, the genetic dissection of complex phenotypic traits, and the subsequent identification of pig QTN, still remains one of the biggest challenges in pig genomics. Future strategies to accelerate the discovery of these QTN will ideally include populations with larger sample sizes, more detailed phenotypes, and more sophisticated data analysis approaches in order to facilitate linking gene function to trait phenotypes.

9.2.5
Candidate Gene Discovery

The ultimate goal of pig genomics research is to identify the genes that control variation in economically important trait phenotypes. Numerous candidate genes have been studied either for their known

Table 1 Number of pig QTL (December 2006)[a]

Trait type	No. of QTL	Trait[b]	No. of QTL	Chromosome	No. of QTL
Anatomy	553	Backfat (average) thickness	89	1	203
Behavioral	22	Ham weight	48	2	157
Chemical	18	Average daily gain	46	3	66
Coat color	2	Loin eye area	45	4	226
Conductivity	25	Carcass length	40	5	56
Conformation	8	pH 24 H Post mortem (loin)	33	6	188
Defects	18	Lean percentage	31	7	194
Digestive organ	10	Dressing percentage	31	8	76
Disease resistance	7	Number of nipples	27	9	40
Endocrine	4	Backfat weight	27	10	37
Enzyme activity	1	Backfat at tenth rib	25	11	27
Fat composition	64	Backfat thickness, last rib	25	12	37
Fatness	403	Head weight	24	13	69
Feed conversion	8	Body weight, birth	23	14	64
Feed intake	16	Color L	23	15	61
Flavor	19	Loin and neck meat weight	23	16	11
Growth	224	Diameter of muscle fiber	22	17	32
Immune capacity	8	Carcass weight (cold)	21	18	22
Litter size	21	Backfat depth, last rib	19	X	106
Meat color	69	Body weight, slaughter	19	Y	1
Odor	5				
pH	58				
Reproductive organ	33				
Reproductive traits	9				
Stiffening	3				
Texture	65				

[a] Source of data: PigQTLdb (http://www.animalgenome.org/QTLdb/pig.html)

[b] Top 20 traits by number of QTL in database

functional significance for a trait or as positional candidates that correspond to chromosomal regions identified through QTL studies. Research involving candidate genes for growth, carcass composition, meat quality, reproduction, disease resistance, and coat color have been reported. In addition to determining the position of each gene in the pig genome, statistical analyses are also conducted in order to investigate the effect of each gene on the trait under study. However, even though the number of genes mapped to the pig genome is substantial, there is a relatively small number of genes for which the effect of the gene on the trait has been substantiated across populations, and an even smaller number of genes for which the QTN that causes the gene effect has been identified (Table 2).

The first pig gene for which the QTN was identified was the ryanodine receptor 1 (*RYR1*) gene, which is mapped to pig chromosome 6 and is a major gene affecting pork quality (Fujii et al. 1991). Coat color genes whose alleles have been determined include *KIT* (Johansson Moller et al. 1996; Marklund et al. 1998; Giuffra et al. 2002) and melanocortin 1 receptor or *MC1R* (Kijas et al. 1998, 2001). A polymorphism in the fucosyltransferase (*FUT1*) gene affects intestinal *E. coli* adherence (Meijerink et al. 2000) and selection for the favorable allele reduces postweaning diarrhea. Kim et al. (2000) reported associations of a missense variant of the melanocortin 4 receptor (*MC4R*) gene with fatness, growth, and feed intake traits. A G-to-A substitution in intron 3 of the insulin-like growth factor II (*IGF2*) gene was identified as the causative mutation for a muscle growth QTL on chromosome 2 (Van Laere et al. 2003). Milan et al. (2000a) reported a missense substitution in the protein kinase, AMP-activated, gamma 3 noncatalytic subunit (*PRKAG3*) gene in a QTL region on chromosome 15 which impacted skeletal muscle glycogen deposition and

Table 2 Pig genes for which quantitative trait nucleotide have been discovered

Gene	Trait	Reference(s)
FUT1	Intestinal *Escherichia coli* adherence	Meijerink et al. (2000)
IGF2	Muscle growth	Van Laere et al. (2003)
KIT	Dominant white color, hematopoiesis	Johansson Moller et al. (1996); Marklund et al. (1998); Giuffra et al. (2002)
LDLR	Hypercholesterolemia	Hasler-Rapacz et al. (1998)
MC1R	Coat color	Kijas et al. (1998); Kijas et al. (2001)
MC4R	Feed intake, growth, backfat	Kim et al. (2000)
PRKAG3	Glycogen content in skeletal muscle	Milan et al. (2000a, b); Ciobanu et al. (2001)
RYR1	Malignant hyperthermia	Fujii et al. (1991)

pork quality, and Ciobanu et al. (2001) subsequently identified additional *PRKAG3* polymorphisms affecting pork quality in other populations.

9.2.6
Marker-Assisted Breeding

Traditionally, genetic selection in the pig has been based on the calculation of estimated breeding values (EBVs) using phenotypic data. Although with this approach the genes involved in the regulation of the trait under selection are treated as an unknown "black box," the progress achieved with this method has been significant. This is due to the fact that genetic selection using pedigree and phenotypic information is able to accurately identify the animals that carry the superior genotypes for the trait. These animals are then selectively mated to generate progeny that are expected to maximize genetic gain. This approach has been used to advance genetic progress for a variety of pig production traits, and it also has some negative aspects associated with it. Perhaps the most important of these is the fact that selection for one trait may be associated with negative effects on other traits (Williams 2005). A classic example in pig production is the negative impact on pork quality caused by the intensive selection performed to increase leanness and decrease backfat thickness. The tremendous progress achieved in the field of molecular genetics has opened up new opportunities for using information from molecular markers in pig selection schemes. The use of genetic markers as part of the selection strategy in breeding programs is usually referred to as marker-assisted selection (MAS).

Different types of markers need to be considered when discussing MAS, including direct markers

(functional mutations for the trait), markers in population-wide linkage disequilibrium (LD) with the functional mutation, and markers in population-wide linkage equilibrium (LE) with the functional mutation (Dekkers 2004). Potentially, all these types of markers can be used in pig selection programs. However, there are differences between markers regarding the efficiency of the selection program and the genetic progress that can be obtained using each type of marker. The best-case scenario occurs when direct markers are used, because these polymorphisms represent the functional mutation for the trait (Dekkers 2004). Markers in close proximity (in LD) of the functional mutation can also be used, but as LD erodes with time the use of these types of markers in genetic selection can be problematic (Dekkers and Hospital 2002). The worst-case scenario for the use of markers in genetic selection regards the use of markers in LE with the causative mutation, because while in some families the marker allele will be associated with the favorable allele, this will not be the case for the entire population (Dekkers and Hospital 2002).

Ideally, a pig breeding program should incorporate both phenotypic and molecular information in order to maximize genetic gain. According to Dekkers (2004) and Lande and Thompson (1990) this can be achieved by tandem selection (an approach where individuals are first selected using molecular information followed by phenotype-based selection), index selection using a combination of molecular score and phenotype, and preselection on molecular score at a younger age followed by phenotypic selection at a later stage. All these strategies have advantages and disadvantages, and the application of each strategy will depend on the objectives of the breeding program. Several pig breeding companies have already implemented MAS in their breeding programs to variable extents, with PIC and Monsanto Choice Genetics (acquired by Newsham Genetics in 2007) being the two companies that have used molecular information most intensively. In contrast, there are fewer examples of marker-assisted introgression (MAI) in pig production. One successful example of MAI in pigs was the introgression of the favorable *RYR1* allele into a line of Pietrain pigs (Hanset et al. 1995), a breed known for increased muscularity, and also for elevated frequency of the unfavorable *RYR1* allele.

Despite the substantial number of genetic markers that could potentially be used in pig breeding schemes, the use of MAS has been somewhat limited. Nevertheless, MAS can be especially helpful for several economically important traits in pig production, such as disease resistance, pork quality, sow longevity, and sow fertility. Phenotypes for these traits are difficult and expensive to collect (disease resistance), are only collected after the animal reaches the end of its productive life (pork quality and sow longevity), or display low heritabilities (sow fertility). These traits are logical targets for MAS due to their clear economic values. However, Kanis et al. (2005) discuss societally important traits, such as animal welfare and environmental effects of pig production, that do not have a direct relationship on production costs or price of pork but to which citizens and consumers attach value. These researchers proposed a selection-index method to obtain proper weights for societally important traits in breeding plans. If markers favorably associated with traits of interest are identified, they can be used to accelerate genetic gain for these traits. However, the economic aspects and the relationship between costs and benefits of implementing a MAS program in pig selection also need to be considered because such programs must ultimately be economically feasible.

9.3
Functional Genomics

9.3.1
Expressed Sequence Tags

The number of pig ESTs significantly increased after 2000, and by 2007 there were over 623,000 pig ESTs deposited in the Genbank database (Tuggle et al. 2007). This marked development of the porcine transcriptome was a direct consequence of many studies

whose objective was the identification of porcine gene products (Fahrenkrug et al. 2002; Yao et al. 2002; Tuggle et al. 2003; Kim et al. 2006; Uenishi et al. 2007), including the analysis and assembly of over one million pig ESTs from 97 cDNA libraries (Gorodkin et al. 2007).

The discovery process for these ESTs included several pig tissues, derived from individuals displaying different physiological conditions and at different developmental stages. This will be important for later studies that will use the EST resources for annotation and assembly of the pig genome, for gene expression research, and also for positional cloning of QTL for economically important traits. Integration of such a large volume of data is also an essential detail to maximize its usefulness. For that purpose, several pig EST databases have been established, and many of these can be accessed through the US National Pig Genome Program web site (http://www.animalgenome.org/pigs/).

9.3.2
DNA Microarrays

The availability of a sufficient number of pig ESTs is an extremely important prerequisite for development of DNA microarray resources for the pig. DNA microarrays allow the simultaneous expression profiling of hundreds or thousands of genes. However, construction of such microarrays requires knowledge of a portion of the coding sequence for the genes to be spotted on the array. Some of the initial transcriptional profiling studies utilizing microarrays in the pig used custom-made cDNA arrays, usually following an EST discovery project, and the EST information was used for spotting clones from the cDNA libraries on the array (for example, Morimoto et al. 2003; Zhao et al. 2003). Early pig gene expression studies also included cross-species hybridization using human microarrays because of the more advanced resources available for the human genome (Moody et al. 2002; Gladney et al. 2004; Shah et al. 2004).

As the number of ESTs increased and microarray technologies improved, more advanced and comprehensive microarray resources were developed for the pig research community. A microarray containing over 23,000 transcripts was developed by Affymetrix (GeneChip® Porcine Genome Array) in 2004. Because this array was released at an early date, the annotation of the array was limited. A first-generation pig 70-mer oligonucleotide microarray containing over 13,000 oligonucleotides was developed in 2003 and has been used by many laboratories (Zhao et al. 2005). Like the Affymetrix array, this array also had limited annotation, because many features were not fully characterized. Efforts by several groups have improved the annotation and characterization of these arrays (Couture et al. 2006; Tsai et al. 2006). More recently, a porcine cDNA microarray containing nearly 27,000 elements was developed by a Danish group (Bendixen et al. 2006) and used to study the gene expression of porcine alveolar macrophages in response to lipopolysaccharide treatment (Jimenez-Marin et al. 2006).

Oligonucleotide microarrays are generally preferred over cDNA arrays due to their uniformity for hybridization efficiency and ease of probe synthesis. A second-generation pig 70-mer long oligonucleotide microarray, developed by several US institutions, overcomes the limitation of poor annotation of previous oligonucleotide microarrays because ESTs were compared to phylogenetically defined vertebrate proteins in order to develop the 20,400 probes on the array. This array, the Swine Protein Annotated Oligonucleotide Microarray (http://www.pigoligoarray.org/), is publicly available and proving to be a valuable resource for pig researchers (Fig. 2). The availability of advanced functional genomics resources allows pig researchers to explore a broad range of research questions related to pig gene expression.

9.4
Future Scope of Work

9.4.1
Swine Genome Sequencing

The ultimate map for any species is the complete and annotated genome sequence. Efforts toward sequencing the pig genome have used multiple approaches and have come from many research groups worldwide (Chen et al. 2007). Many groups have contributed EST sequences from various tissues as discussed previously, and, in addition, candidate genes have been sequenced. A Sino-Danish collaboration generated approximately 3.84 million shotgun sequences of the pig genome resulting in a 0.66X coverage of the pig genome (Wernersson et al. 2005) and sequence

of the swine major histocompatibility complex was reported in 2006 (Renard et al. 2006).

The Swine Genome Sequencing Consortium (SGSC) was formed in September 2003 to provide international coordination for sequencing the pig genome. The SGSC's mission is to advance biomedical research, and animal production and health through the development of DNA-based tools and products resulting from the sequencing of the swine genome (Schook et al. 2005; http://www.piggenome.org/). The pig whole-genome sequencing is being performed by The Wellcome Trust Sanger Institute using a clone-by-clone sequencing strategy based on the minimal tiling path of BAC clones (Chen et al. 2007). Additional BAC clones were selected to generate initial shotgun sequencing data, and BACs were selected from a different library for sequencing the Y chromosome. The swine genome sequence results are available at http://pre.ensembl.org/Sus_scrofa/index.html.

Several large SNP discovery projects are underway worldwide to complement the genome sequencing efforts. These include the Sino-Danish sequencing project which includes five different breeds (Duroc, Erhuanlian, Hampshire, Landrace, and Yorkshire) and efforts by the USDA in the USA. In addition, an SNP project was initiated by the INRA-Genescope in France in conjunction with the SGSC sequencing project. This effort includes plans to generate one million sequences from seven different breeds (Iberian, Landrace, Meishan, Minipig, Pietrain, Wild boar, and Yorkshire).

9.4.2
Expression QTL

Gene expression profiling using microarrays and genomic scans to identify chromosomal regions containing genes controlling important trait phenotypes have, until recently, been pursued as independent avenues of research. One direction has focused on variation in gene expression patterns, while the other has attempted to associate DNA sequence polymorphisms with phenotypic variation. However, within the past several years, significant interest has turned toward integration of these two research areas such that complex traits can now be studied in greater detail than is possible using either approach alone. Jansen and Nap (2001) proposed the term "genetical genomics" to describe the combination of expression

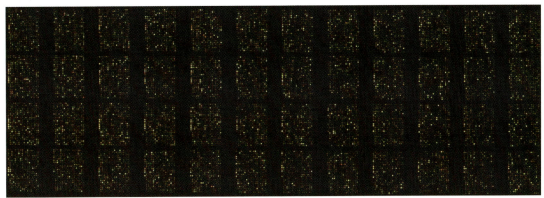
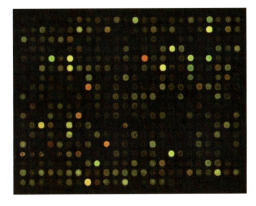

Fig. 2 Images of the Swine Protein Annotated Oligonucleotide Microarray (http://www.pigoligoarray.org/). **a** Full microarray depicting all 48 subarrays. **b** Expanded view of one of the subarrays. The array was hybridized with samples from skeletal muscle and uterus as part of a validation study

profiling and marker-based fingerprinting of individuals in a segregating population. Determination of mRNA transcript abundances for individuals in a segregating population makes it possible to treat the expression of individual genes as quantitative traits, designated by Schadt et al. (2003a, b) as expression QTL or eQTL. When the population has been screened for phenotypic traits of interest and trait QTL have been identified, the combined approach will help identify the gene(s) responsible for putative QTL. Thus, for each gene considered, QTL analysis will determine the regions of the genome influential for its expression and elucidate what portion of the variation in gene expression maps to the genes themselves (*cis*-acting factors), as opposed to other coregulated genomic locations (trans-acting factors). Unlike classic quantitative traits whose measures are far removed from the biological processes giving rise to them, the genetic linkages associated with transcript abundances afford a closer look at biochemical processes at the cellular level (Schadt et al. 2003b). This approach has tremendous potential for application in livestock populations including the pig (Haley and de Koning, 2006).

Application of the genetical genomics approach in livestock populations is illustrated in the diagram in Fig. 3. Phenotypic records, DNA, and RNA from target tissues are obtained from experimental resource populations. The DNA and phenotypic records are subjected to a classical QTL analysis, while the RNA is used for transcriptional profiling analyses. Individual mRNA abundances from the transcript profiles are then considered as quantitative phenotypes and they are incorporated into the genome scan. This integration will reveal cis- and trans-eQTL associated with the phenotypic traits of interest in the population,

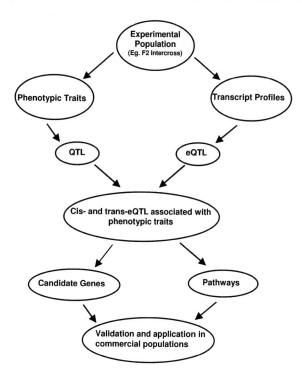

Fig. 3 Genetical genomics in livestock populations

However, completion of the pig genome sequence and efforts to develop a large SNP collection for pigs will aid in this effort. DNA microarray resources have also been developed for the pig, which are not only facilitating transcriptional profiling studies for many biological questions, but in addition the incorporation of gene expression data into QTL studies will further enhance efforts to identify the genes and QTN controlling traits of importance.

which will in turn facilitate not only identification of candidate genes at QTL, but also elucidation of biochemical pathways controlling phenotypic traits. Genes identified by this coordinated and synergistic approach could then be validated for incorporation into MAS strategies in commercial populations. To date, no results have been reported using this strategy in pigs, but some studies are underway and preliminary results from a project to identify eQTL for *longissimus dorsi* muscle genes were reported in the US at the National Swine Improvement Federation Conference in 2007 (Ernst et al. 2007).

9.4.3 Conclusion

The pig is an important species both as a food source and as a biomedical model for human biology. Significant progress has been made since the early 1990s to develop pig genetic, physical, and RH maps. In addition, numerous QTL have been discovered and candidate genes studied for their effects on traits of economic importance in the pig. Progress to identify the QTN underlying pig QTL has been more challenging.

References

Andersson L, Haley CS, Ellegren H, Knott SA, Johansson M, Andersson K, Andersson-Eklund L, Edfors-Lilja I, Fredholm M, Hansson I, et al (1994) Genetic mapping of quantitative trait loci for growth and fatness in pigs. Science 263:1771–1774

Archibald A, Haley CS, Andersson L, Bosma AA, Davies W, Fredholm M, Geldermann H, Gellin J, Groenen M, Gustavsson I, Ollivier L, Tucker EM, Van de Weghe A (1990) PiGMaP: an European initiative to map the porcine genome. Anim Genet 22 (Suppl. 1):82–83

Archibald AL, Haley CS, Brown JF, Couperwhite S, McQueen H, Nicholson D, Coppieters W, Van de Weghe A, Stratil A, Wintero AK, et al (1995) The PiGMaP consortium linkage map of the pig (*Sus scrofa*). Mamm Genome 6:157–175

Bendixen C, Gregersen V, Hedegaard J, Conley L, Hoj A, Panitz F (2006) Genetics of porcine gene expression. Proc 30th Intl Conf Anim Genet, Porto Seguro, Brazil, August 20–25, C444

Carlborg O, Hocking PM, Burt DW, Haley CS (2004) Simultaneous mapping of epistatic QTL in chickens reveals clusters of QTL pairs with similar genetic effects on growth. Genetical Res 83:197–209

Carlborg O, Brockmann GA, Haley CS (2005) Simultaneous mapping of epistatic QTL in DU6i × DBA/2 mice. Mamm Genome 16:481–494

Chen K, Baxter T, Muir WM, Groenen MA, Schook LB (2007) Genetic resources, genome mapping and evolutionary genomics of the pig (*Sus scrofa*). Int J Biol Sci 3:153–165

Ciobanu DC, Bastiaansen J, Malek M, Helm J, Woollard J, Plastow GS, Rothschild MF (2001) Evidence for new alleles in the protein kinase AMP-activated, subunit gene associated with low glycogen content in pig skeletal muscle and improved meat quality. Genetics 159:1151–1162

Couture O, Zhao X, Zhao SH, Recknor J, Lkhagvadorj S, Qu L, Nettleton D, Dekkers J, Tuggle C (2006) Improved annotation of the porcine Affymetrix GeneChip® and functional comparison to QIAGEN-NRSP8 oligonucleotide array data. Proc 30th Intl Conf Anim Genet, Porto Seguro, Brazil, August 20–25, E246

Dekkers JC (2004) Commercial application of marker- and gene-assisted selection in livestock: strategies and lessons. J Anim Sci 82 (E-Suppl):E313–328

Dekkers JC, Hospital F (2002) The use of molecular genetics in the improvement of agricultural populations. Nature Rev Genet 3:22–32

Demeure O, Renard C, Yerle M, Faraut T, Riquet J, Robic A, Schiex T, Rink A, Milan D (2003) Rearranged gene order between pig and human in a QTL region on SSC 7. Mamm Genome 14:71–80

Desautes C, Bidanel JP, Milan D, Iannuccelli N, Amigues Y, Bourgeois F, Caritez J, Renard C, Chevalet C, Mormede P (2002) Genetic linkage mapping of quantitative trait loci for behavioral and neuroendocrine stress response traits in pigs. J Anim Sci 80:2276–2285

Echard G (1984) The Gene Map of the Pig (*Sus scrofa domestica* L.). In: O'Brien SJO (ed) Genetic Maps, Book 3: A Compilation of Linkage and Restriction Maps. Cold Spring Harbor Laboratory Press, New York, USA pp 392–395

Ellegren H, Chowdhary BP, Johansson M, Marklund L, Fredholm M, Gustavsson I, Andersson L (1994) A primary linkage map of the porcine genome reveals a low rate of genetic recombination. Genetics 137:1089–1100

Epstein J, Bichard M (1984) Pig. In: Mason IL (ed) Evolution of Domesticated Animals. Longman, London, pp 145–162

Ernst CW, Steibel JP, Ramos AM, Tempelman RJ, Cardoso FF, Rosa GJM, Edwards DB, Bates RO (2007) Proc 32nd Ann Meet Genetics Sym, National Swine Improvement Federation. Kansas City, KS, USA, Dec 6–7 (in print).

Fahrenkrug SC, Smith TP, Freking BA, Cho J, White J, Vallet J, Wise T, Rohrer G, Pertea G, Sultana R, Quackenbush J, Keele J (2002) Porcine gene discovery by normalized cDNA-library sequencing and EST cluster assembly. Mamm Genome 13:475–478

Fang M, Andersson L (2006) Mitochondrial diversity in European and Chinese pigs is consistent with population expansions that occurred prior to domestication. Proc Royal Soc B 273:1803–1810

Fang M, Berg F, Ducos A, Andersson L (2006) Mitochondrial haplotypes of European wild boars with 2n = 36 are closely related to those of European domestic pigs with 2n = 38. Anim Genet 37:459–464

Fujii J, Otsu K, Zorzato F, de Leon S, Khanna VK, Weiler JE, O'Brien PJ, MacLennan DH (1991) Identification of a mutation in porcine ryanodine receptor associated with malignant hyperthermia. Science 253:448–451

Giuffra E, Törnsten A, Marklund S, Bongcam-Rudloff E, Chardon P, Kijas JM, Anderson SI, Archibald AL, Andersson L (2002) A large duplication associated with dominant white color in pigs originated by homologous recombination between LINE elements flanking KIT. Mamm Genome 13:569–577

Gladney CD, Bertani GR, Johnson RK, Pomp D (2004) Evaluation of gene expression in pigs selected for enhanced reproduction using differential display PCR and microarray. I. Ovarian follicles. J Anim Sci 82:17–31

Gorodkin J, Cirera S, Hedegaard J, Gilchrist MJ, Panitz F, Jorgensen C, Scheibye-Knudsen K, Arvin T, Lumholdt S, Sawera M, Green T, Nielsen BJ, Havgaard JH, Rosenkilde C, Wang J, Li H, Li R, Liu B, Hu S, Dong W, Li W, Yu J, Wang J, Staerfeldt HH, Wernersson R, Madsen LB, Thomsen B, Hornshoj H, Bujie Z, Wang X, Wang X, Bolund L, Brunak S, Yang H, Bendixen C, Fredholm M (2007) Porcine transcriptome analysis based on 97 non-normalized cDNA libraries and assembly of 1,021,891 expressed sequence tags. Genome Biol 8:R45

Goureau A, Yerle M, Schmitz A, Riquet J, Milan D, Pinton P, Frelat G, Gellin J (1996) Human and porcine correspondence of chromosome segments using bidirectional chromosome painting. Genomics 36:252–262

Groves CP, Grubb P (1993) The Eurasian Suids: *Sus* and *Babyrousa*. In: Oliver WLR (ed) Pigs, Peccaries and Hippos. IUCN, The World Conservation Union, Gland, Switzerland pp 107–111

Gustavsson I (1988) Standard karyotype of the domestic pig. Committee for the Standardized Karyotype of the Domestic Pig. Hereditas 109:151–157

Haley C, de Koning DJ (2006) Genetical genomics in livestock: potentials and pitfalls. Anim Genet 37(Suppl. 1):10–12

Hanset R, Dasnoi C, Scalais S, Michaux C, Grobet L (1995) Effets de l'introgression dons le genome Pietrain de l'allele normal aux locus de sensibilité a l'halothane. Genet Sel Evol 27:77–88

Hart EA, Caccamo M, Harrow JL, Humphray SJ, Gilbert JG, Trevanion S, Hubbard T, Rogers J, Rothschild MF (2007) Lessons learned from the initial sequencing of the pig genome: comparative analysis of an 8 Mb region of pig chromosome 17. Genome Biol 8:R168

Hasler-Rapacz J, Ellegren H, Fridolfsson AK, Kirkpatrick B, Kirk S, Andersson L, Rapacz J (1998) Identification of a mutation in the low density lipoprotein receptor gene associated with recessive familial hypercholesterolemia in swine. Amer J Med Genet 76:379–386

Hawken RJ, Murtaugh J, Flickinger GH, Yerle M, Robic A, Milan D, Gellin J, Beattie CW, Schook LB, Alexander LJ (1999) A first-generation porcine whole-genome radiation hybrid map. Mamm Genome 10:824–830

Holl JW, Cassady JP, Pomp D, Johnson RK (2004) A genome scan for quantitative trait loci and imprinted regions affecting reproduction in pigs. J Anim Sci 82:3421–3429

Hu ZL, Dracheva S, Jang W, Maglott D, Bastiaansen J, Rothschild MF, Reecy JM (2005) A QTL resource and comparison tool for pigs: PigQTLDB. Mamm Genome 16:792–800

Jansen RC, Nap JP (2001) Genetical genomics: the added value from segregation. Trends Genet 17:388–391

Jimenez-Marin A, Perez-Reinado E, Bendixen C, Conley L, Hedegard J, Martinez A, Garrido J (2006) Microarray analysis of lipopolysaccharide-treated porcine alveolar macrophages. Proc 30th Intl Conf Anim Genet, Porto Seguro, Brazil, August 20–25, C447

Johansson Moller M, Chaudhary R, Hellmén E, Höyheim B, Chowdhary B, Andersson L (1996) Pigs with the dominant white coat color phenotype carry a duplication of the KIT gene encoding the mast/stem cell growth factor receptor. Mamm Genome 7:822–830

Jones GF (1998) Genetic aspects of domestication, common breeds and their origin. In: Rothschild M, Ruvinsky A (eds) The Genetics of Pig. CABI, Oxon, UK pp 17–50

Kanis E, DeGreef KH, Hiemstra A, van Arendonk JAM (2005) Breeding for societally important traits in pigs. J Anim Sci 83:948–957

Kijas JMH, Andersson L (2001) A phylogenetic study of the origin of the domestic pig estimated from the near-complete mtDNA genome. J Mol Evol 52:302–308

Kijas JM, Wales R, Törnsten A, Chardon P, Moller M, Andersson L (1998) Melanocortin receptor 1 (MC1R) mutations and coat color in pigs. Genetics 150:1177–1185

Kijas JM, Moller M, Plastow G, Andersson L (2001) A frameshift mutation in MC1R and a high frequency of somatic reversions cause black spotting in pigs. Genetics 158:779–785

Kim KS, Larsen N, Short T, Plastow G, Rothschild MF (2000) A missense variant of the melanocortin 4 receptor (MC4R) gene is associated with fatness, growth and feed intake traits. Mamm Genome 11:131–135

Kim TH, Kim NS, Lim D, Lee KT, Oh JH, Park HS, Jang GW, Kim HY, Jeon M, Choi BH, Lee HY, Chung HY, Kim H (2006) Generation and analysis of large-scale expressed sequence tags (ESTs) from a full-length enriched cDNA library of porcine backfat tissue. BMC Genomics 7:36

de Koning DJ, Rattink AP, Harlizius B, van Arendonk JA, Brascamp EW, Groenen MA (2000) Genome-wide scan for body composition in pigs reveals important role of imprinting. Proc Nat Acad Sci USA 597:7947–7950

Lande R, Thompson R (1990) Efficiency of marker-assisted selection in the improvement of quantitative traits. Genetics 124:743–756

Larson G, Dobney K, Albarella U, Fang M, Matisoo-Smith E, Robins J, Lowden S, Finlayson H, Brand T, Willerslev E, Rowley-Conwy P, Andersson L, Cooper A (2005) Worldwide phylogeography of wild boar reveals multiple centers of pig domestication. Science 307:1618–1621

Legault C (1998) Genetics of colour variation. In: Rothschild M, Ruvinsky A (eds) The Genetics of Pig. CABI, Oxon, UK, pp 51–69

Liu WS, Eyer K, Yasue H, Roelofs B, Hiraiwa H, Shimogiri T, Landrito E, Ekstrand J, Treat M, Rink A, Yerle M, Milan D, Beattie CW (2005) A 12,000-rad porcine radiation hybrid (IMNpRH2) panel refines the conserved synteny between SSC12 and HSA17. Genomics 86:731–738

Marklund L, Johansson-Moller M, Hoyheim B, Davies W, Fredholm M, Juneja RK, Mariani P, Coppieters W, Ellegren H,

Andersson L (1996) A comprehensive linkage map of the pig based on a wild pig – Large White intercross. Anim Genet 27:255–269

Marklund S, Kijas J, Rodriguez-Martinez H, Rönnstrand L, Funa K, Moller M, Lange D, Edfors-Lilja I, Andersson L (1998) Molecular basis for the dominant white phenotype in the domestic pig. Genome Res 8:826–833

Martins-Wess F, Milan D, Drogemuller C, Vobeta-Nemitz R, Brenig B, Robic A, Yerle M, Leeb T (2003) A high resolution physical and RH map of pig chromosome 6q1.2 and comparative analysis with human chromosome 19q13.1. BMC Genomics 4:20

Meijerink E, Neuenschwander S, Fries R, Dinter A, Bertschinger HU, Stranzinger G, Vögeli P (2000) A DNA polymorphism influencing alpha(1,2)fucosyltransferase activity of the pig FUT1 enzyme determines susceptibility of small intestinal epithelium to *Escherichia coli* F18 adhesion. Immunogenetics 52:129–136

Mercade A, Estelle J, Noguera JL, Folch JM, Varona L, Silio L, Sanchez A, Perez-Enciso M (2005) On growth, fatness, and form: a further look at porcine chromosome 4 in an Iberian x Landrace cross. Mamm Genome 16:374–382

Meyers SN, Rogatcheva MB, Larkin DM, Yerle M, Milan D, Hawken RJ, Schook LB, Beever JE (2005) Piggy-BACing the human genome II. A high-resolution, physically anchored, comparative map of the porcine autosomes. Genomics 86:739–752

Milan D, Jeon JT, Looft C, Amarger V, Robic A, Thelander M, Rogel-Gaillard C, Paul S, Iannuccelli N, Rask L, Ronne H, Lundström K, Reinsch N, Gellin J, Kalm E, Le Roy P, Chardon P, Andersson L (2000a) A mutation in PRKAG3 associated with excess glycogen content in pig skeletal muscle. Science 288:1248–1251

Milan D, Hawken R, Cabau C, Leroux S, Genet C, Lahbib Y, Tosser G, Robic A, Hatey F, Alexander L, Beattie C, Schook L, Yerle M, Gellin J (2000b) IMpRH server: an RH mapping server available on the Web. Bioinformatics 6:558–559

Moody DE, Zou Z, McIntyre L (2002) Cross-species hybridisation of pig RNA to human nylon microarrays. BMC Genomics 3:27

Morimoto M, Zarlenga D, Beard H, Alkharouf N, Matthews BF, Urban JF Jr (2003) Ascaris suum: cDNA microarray analysis of 4th stage larvae (L4) during self-cure from the intestine. Exper Parasitol 104:113–121

Ollivier L (1998) Genetic improvement of the pig. In: Rothschild M, Ruvinsky A (eds) The Genetics of Pig. CABI, Oxon, UK, pp 511–540

Perez-Enciso M, Misztal I (2004) Qxpak: a versatile mixed model application for genetical genomics and QTL analyses. Bioinformatics 20:2792–2798

Reiner G, Melchinger E, Kramarova M, Pfaff E, Buttner M, Saalmuller A, Geldermann H (2002) Detection of quantitative trait loci for resistance/susceptibility to pseudorabies virus in swine. J General Virol 83:167–172

Renard C, Hart E, Sehra H, Beasley H, Coggill P, Howe K, Harrow J, Gilbert J, Sims S, Rogers J, Ando A, Shigenari A, Shiina T, Inoko H, Chardon P, Beck S (2006) The genomic sequence and analysis of the swine major histocompatibility complex. Genomics 88:96–110

Rink A, Eyer K, Roelofs B, Priest KJ, Sharkey-Brockmeier KJ, Lekhong S, Karajusuf EK, Bang J, Yerle M, Milan D, Liu WS, Beattie CW (2006) Radiation hybrid map of the porcine genome comprising 2035 EST loci. Mamm Genome 17:878–885

Rohrer GA, Alexander LJ, Keele JW, Smith TP, Beattie CW (1994) A microsatellite linkage map of the porcine genome. Genetics 136:231–245

Rohrer GA, Alexander LJ, Hu Z, Smith TP, Keele JW, Beattie CW (1996) A comprehensive map of the porcine genome. Genome Res 6:371–391

Rothschild MF (2004) Porcine genomics delivers new tools and results: this little piggy did more than just go to market. Genet Res Camb 83:1–6

Rothschild MF, Hu Z, Jiang Z (2007) Advances in QTL mapping in pigs. Int J Biol Sci 3:192–197

Ruvinsky A, Rothschild MF (1998) Systematics and evolution of the pig. In: Rothschild M, Ruvinsky A (eds) The Genetics of Pig. CABI, Oxon, UK pp 1–16

Schadt EE, Monks SA, Drake TA, Lusis AJ, Che N, Colinayo V, Ruff TG, Milligan SB, Lamb JR, Cavet G, Linsley PS, Mao M, Stoughton RB, Friend SH (2003a) Genetics of gene expression surveyed in maize, mouse and man. Nature 422:297–302

Schadt EE, Monks SA, Friend SH (2003b) A new paradigm for drug discovery: integrating clinical, genetic, genomic and molecular phenotype data to identify drug targets. Biochem Soc Trans 31:437–443

Schook LB, Beever JE, Rogers J, Humphrey S, Archibald A, Chardon P, Milan D, Rohrer G, Eversole K (2005) Swine Genome Sequencing Consortium (SGSC): a strategic roadmap for sequencing the pig genome. Comp Funct Genomics 6:251–255

Seaton G, Haley CS, Knott SA, Kearsey M, Visscher PM (2002) QTL Express: mapping quantitative trait loci in simple and complex pedigrees. Bioinformatics 18:339–340

Shah G, Azizian M, Bruch D, Mehta R, Kittur D (2004) Cross-species comparison of gene expression between human and porcine tissue, using single microarray platform–preliminary results. Clin Transpl 18 (Suppl 12):76–80

Tikhonov VN, Troshina AI (1974) Identification of chromosomes and their aberrations in karyotypes of subspecies of Sus scrofa L. by differential staining. Doklady Akademii Nauk SSSR 214:932–935

Tsai S, Cassady JP, Freking BA, Nonneman DJ, Rohrer GA, Piedrahita JA (2006) Annotation of the Affymetrix porcine genome microarray. Anim Genet 37:423–424

Tuggle CK, Green JA, Fitzsimmons C, Woods R, Prather RS, Malchenko S, Soares BM, Kucaba T, Crouch K, Smith C, Tack D, Robinson N, O'Leary B, Scheetz T, Casavant T, Pomp D, Edeal BJ, Zhang Y, Rothschild MF, Garwood K, Beavis W (2003) EST-based gene discovery in pig: virtual expression patterns and comparative mapping to human. Mamm Genome 14:565–579

Tuggle CK, Wang Y, Couture O (2007) Advances in swine transcriptomics. Int J Biol Sci 3:132–152

Uenishi H, Eguchi-Ogawa T, Shinkai H, Okumura N, Suzuki K, Toki D, Hamasima N, Awata T (2007) PEDE (Pig EST Data Explorer) has been expanded into Pig Expression Data Explorer, including 10 147 porcine full-length cDNA sequences. Nucl Acids Res 35(Database issue):D650–653

Van Laere AS, Nguyen M, Braunschweig M, Nezer C, Collette C, Moreau L, Archibald AL, Haley CS, Buys N, Tally M, Andersson G, Georges M, Andersson L (2003) A regulatory mutation in IGF2 causes a major QTL effect on muscle growth in the pig. Nature 425:832–836

United States Department of Agriculture, Foreign Agriculture Service (2006) Livestock and Poultry: World Markets and Trade. (http://www.fas.usda.gov/)

United States Department of Agriculture, Agriculture Research Service (2007) USDA Nutrient Data Set for Fresh Pork, Release 1.1 (http://www.ars.usda.gov/nutrientdata)

Wattrang E, Almqvist M, Johansson A, Fossum C, Wallgren P, Pielberg G, Andersson L, Edfors-Lilja I (2005) Confirmation of QTL on porcine chromosomes 1 and 8 influencing leukocyte numbers, haematological parameters and leukocyte function. Anim Genet 36:337–345

Wernersson R, Schierup MH, Jorgensen FG, Gorodkin J, Panitz F, Staerfeldt HH, Christensen OF, Mailund T, Hornshoj H, Klein A, Wang J, Liu B, Hu S, Dong W, Li W, Wong GK, Yu J, Wang J, Bendixen C, Fredholm M, Brunak S, Yang H, Bolund L (2005) Pigs in sequence space: a 0.66X coverage pig genome survey based on shotgun sequencing. BMC Genomics 6:70

Williams JL (2005) The use of marker-assisted selection in animal breeding and biotechnology. Revue Scientifique et Technique 24:379–391

Wright D, Butlin RK, Carlborg O (2006) Epistatic regulation of behavioural and morphological traits in the zebrafish (Danio rerio). Behav Genet 36:914–922

Yao J, Coussens PM, Saama P, Suchyta S, Ernst CW (2002) Generation of expressed sequence tags from a normalized porcine skeletal muscle cDNA library. Anim Biotech 13:211–222

Yerle M, Lahbib-Mansais Y, Mellink C, Goureau A, Pinton P, Echard G, Gellin J, Zijlstra C, De Haan N, Bosma AA, et al (1995) The PiGMaP consortium cytogenetic map of the domestic pig (Sus scrofa domestica). Mamm Genome 6:176–186

Yerle M, Echard G, Robic A, Mairal A, Dubut-Fontana C, Riquet J, Pinton P, Milan D, Lahbib-Mansais Y, Gellin J (1996) A somatic cell hybrid panel for pig regional gene mapping characterized by molecular cytogenetics. Cytogenet Cell Genet 73:194–202

Yerle M, Pinton P, Robic A, Alfonso A, Palvadeau Y, Delcros C, Hawken R, Alexander L, Beattie LB, Milan D, Gellin J (1998) Construction of a whole genome radiation hybrid panel for high-resolution gene mapping in pigs. Cytogenet Cell Genet 82:182–188

Yerle M, Pinton P, Delcros C, Arnal N, Milan D, Robic A (2002) Generation and characterization of a 12,000-rad radiation hybrid panel for fine mapping in pig. Cytogenet Genome Res 97:219–228

Zhao SH, Nettleton D, Liu W, Fitzsimmons C, Ernst CW, Raney NE, Tuggle CK (2003) Complementary DNA macroarray analyses of differential gene expression in porcine fetal and postnatal muscle. J Anim Sci 81:2179–2188

Zhao SH, Recknor J, Lunney JK, Nettleton D, Kuhar D, Orley S, Tuggle CK (2005) Validation of a first-generation long-oligonucleotide microarray for transcriptional profiling in the pig. Genomics 86:618–625

Index

Acrocentric chromosome. *See* Chromosome, acrocentric
Acrocephalus arundinaceus. See Great reed warbler
Amplified fragment-length polymorphism (AFLP), 23, 50, 118, 119, 125, 148, 160, 161, 176, 179, 208, 209, 216
African buffalo   19, 23
African collared dove   110
Agelaius phoeniceus. See Red-winged blackbird
Agouti   218
Albinism   87, 92, 99, 112, 114, 116, 149, 198, 208
Albino   113, 172, 173, 181, 209, 219
Albumen   84, 86
Alectoris. See Partridge, red-legged
Allelic variation   36, 150, 179
Allozyme   49, 52–55, 57, 61, 62, 64, 160
Alopochen aegyptiacus. See Goose, Egyptian
Anas   77, 78, 88, 111, 121
Anas platyrhynchos. See Duck, domestic; Duck, mallard
Angora   167, 172, 187, 209, 219
Animal model
– chicken   120
– common quail   120
– dog   241
– domestic duck   120
– Japanese quail   120
– pigeon   120
– rabbit   173, 175, 220, 221
– sheep   33, 34
– zebra finch   117–120
Anser albifrons. See Goose, white-fronted
Anser anser. See Goose, graylag
Anser cygnoides. See Goose, swan
Antibody(ies)   123, 174, 178
Atherosclerosis   174, 214–216
Autosomal gene   118, 119, 122, 144
Autosome. *See* Chromosome, autosome
Avian influenza   125, 147

BAC-end sequences   40, 158
BAC contig   6, 86, 105, 118, 119, 122, 158

BAC library. *See* Genomic libraries, BAC
BACPAC Resources Center   40, 157
Behavior
– cattle   17
– chicken   86
– dog   231, 233, 247, 251
– feather pecking   123
– foraging   123
– pig   263
– rabbit   206
– zebra finch   117–120
ß-lactoglobulin   8
Biochemical polymorphism(s)   57, 79, 101, 150, 179, 208
Biodiversity. *See* Genetic diversity
Bioinformatics   107, 204, 205, 206
Body size. *See* Growth traits
Body weight. *See* Growth traits
Booroola   36, 37
Branta canadensis. See Goose, Canada
Breeding industry   58, 85, 172
Breeding objectives   2, 22, 86, 144, 258, 259
Breeds
– cattle   1, 2, 4, 7–10
– dog   231–237, 241–244, 246–251
– chicken   75, 79–82, 85, 86, 110, 121, 122, 124, 125
– domestic duck   84, 110
– pig   257–259, 261, 266
– rabbit   165–173, 176, 178, 218, 220
– sheep   34, 40
– turkey   143, 160
– water buffalo   22, 23
Broilers   82, 83, 121, 122, 125
Brown kiwi   120
Brown-headed cowbird   118
Budgerigar. *See* Caged birds, budgerigar
Bulked segregant analysis   102

Caged birds
– budgerigar   110, 117, 118
– canary   110

– dove 110
– rose-ringed parakeet 78, 88, 115
– zebra finch 117–120
Cairina moschata. See Duck, Muscovy
California condor 118–120
Callipyge 37, 38
Candidate gene(s) 7, 8, 10, 36, 64, 102, 159, 160, 209, 214, 216, 219, 221, 248, 259, 262, 263, 268
Canidae 231, 234, 236
Canis 231, 234
Canis familiaris 231
Canoidae 234
Carcass 2, 5, 7, 33, 36, 37, 83–86, 120–122, 171, 219, 258, 261–263
Caribou 47, 56
Casein 8, 22, 50, 56, 176, 219
Cassowary, southern 78, 88, 117
Casuarius casuarius. See Cassowary, southern
cDNA 24, 39, 60, 107, 118, 151, 153, 154, 158–160, 207
cDNA library(ies) 124, 150, 265
Centromere 6, 19, 20, 25, 28, 175, 208, 238
Cervus 58
Chip. *See* Microarray
Cholesterol
– in eggs 84
– in meat 85
Chromosome(s)
- acrocentric 19-21, 34-36, 147
– autosome 19, 20, 24, 28, 34, 36, 39, 86, 107, 117, 175, 176, 234, 239, 240, 258
– banding 20, 21, 24–27, 34, 36, 117, 147, 175, 176, 207, 212
– breakpoint 101, 248
– fission 20, 21, 24–27, 34, 36, 61, 117, 147, 175, 176, 207, 212
– fusion 19–21, 34, 48, 61, 117, 157, 176
– inversion 24, 25, 97, 117, 119, 212, 213
– lampbrush 108
– macro- 86, 102, 107, 108, 110, 117, 119, 147, 157
– micro- 86, 99, 101, 105, 107, 108, 110, 117, 119, 147, 148, 157, 160
– map. *See* Linkage map
– number 61, 86, 119, 176, 234, 257
– painting 3, 28, 35, 108, 117, 157, 176, 180, 207, 209–213, 217, 238
– paints 108, 117–120, 238, 239
– R-banded 20, 21, 26, 35, 36, 175, 176, 207, 209
– rearrangement 24, 25, 28, 48, 61, 177, 119, 157, 176, 211–213, 238, 243
– sex 20, 23, 24, 34, 86, 88, 100, 107, 117, 118, 120, 147, 168, 175, 238, 239, 258
– translocation 19–21, 24, 34, 48, 61, 117, 257
– W 107, 108, 111, 113, 115, 116, 118, 119, 125, 147, 148
– Z 86, 88, 99, 101, 110, 118, 119, 122, 125, 126, 147, 151
Chronic wasting disease (CWD) 65–66

Chrysolophus pictus. See Pheasant, golden
Ciconiiformes 76, 78, 118
Cloning 174, 204, 205, 209, 214, 220
Coccidiosis 123, 166
Cockfighting 81
Colinus virginianus. See Quail, northern bobwhite
Color 37, 57, 79, 143, 144, 166, 173, 179, 217–220, 244, 258, 259, 263, 264
Columba livia. See Pigeon, rock
Comparative analysis 50, 67, 158
Comparative genomics 67, 106, 108
Comparative map 3, 4, 7, 24, 33, 117-118, 154–155, 213, 238, 239, 260, 261
Comprehensive map 238
Congenic lines 123
Conservation 23, 52, 53, 55, 57, 59, 60, 67 106, 108, 117–119, 125
Conserved synteny. *See* Synteny
Contig map. *See* BAC contig
Coturnix coturnix. See Quail, common
Coturnix japonica. See Quail, Japanese
Crane 88
Cross-species 58, 59, 117–119, 148, 154, 157, 158, 265
Cygnus olor. See Swan, mute
Cytogenetic map. *See* Physical map

Deer species (other) 47–49, 52, 54–56, 60, 66
Development biology 75, 86
Diapsids 75
Disease . *See also* Genetic disease(s); Infection 8, 33, 34, 37, 64, 66, 120, 160, 166, 173–176, 206, 209, 217, 221–222, 241–244, 246–249, 251
Disease resistance 33, 160, 256, 263, 265
– coccidiosis, *Eimeria maxima* 123
– *Escherichia coli* 123
– Marek's disease 120, 122
– *Mycobacterium butyricum* 123
– Newcastle disease 123
– QTL 7, 103, 120–124, 261
– *Salmonella enteritidis* 123
– selection 86, 122–123
Distribution of wild turkey 144, 153
Diversity 2, 47, 49, 52–54, 57, 58, 75, 80, 110, 124, 125, 160, 165, 167–169, 171, 176, 178, 180, 220, 235, 246, 247, 257
DNA content
– diploid 107
– haploid 86, 107, 204
DNA polymorphism 8, 101
DNA pooling 123
DNA probe(s) 60, 108, 151, 207, 209
Domestication 1, 34, 47, 48, 56, 75, 76, 79–81, 110, 143, 144, 166–171, 220, 231, 234–236, 246, 257, 258
Dominant trait 1, 61, 88, 102, 148, 160, 218, 220, 242

Dromaius novaehollandiae. See Emu
Duck
– domestic   84, 110, 118
– mallard   77, 78, 88, 121
– mulard   84
– Muscovy   75, 84, 110

Economic importance   1, 7, 21, 22, 33, 36, 48, 82, 83, 84, 100, 123, 125, 126, 145, 160, 166, 171, 173, 258, 261, 266, 268
Economic improvement   1, 3, 8
Economically important traits   3, 33, 85, 101, 120, 262, 265
Egg(s)   75, 81–86, 100, 120, 122, 124, 125, 145, 159
– age at first egg   122
– composition   86, 122
– persistency of lay   86
– production   83, 85, 86, 100, 120, 122
– quality   120, 122
– rate of lay   86
– shells   85, 86, 122, 124
– weight   86, 120–122
Egretta garzetta. See Little egret
Embryonic stem cells   9, 34, 174, 221
Emu   76, 85, 117, 118
ENCODE project   204, 205
Environmental factor(s)   75, 222, 247
Epistasis   10, 122, 262
Evolution   1, 25, 28, 49, 50, 61, 80–82, 85, 106–108, 110, 117–120, 157, 167, 168, 205, 209, 212, 220, 234, 236, 240–242, 244, 246, 248
Expressed sequence tags (ESTs)   3, 6, 39, 41, 61, 86, 103, 118, 119, 124, 148, 155, 158–159, 206, 250, 261, 265
Expression QTL   266–268

Fallow deer   48, 52, 57, 59
Fancy poultry   85
Fat
– abdominal   122
– backfat   262–264
– carcass   86, 122
– content   84, 171, 219, 222, 257–259, 261, 263
– distribution   79, 108
– domestic duck   84, 110, 118
– emu   85, 118
– genetic lines   85, 101, 120, 123
– in poultry meat   82, 83
– milk   5–8, 22
– perirenal   219
– skin   83–85, 122
Feather(s)
– emu   85, 123
– pinfeathers   144
– poultry   85, 143, 144, 148, 149
– ostrich   83–85, 107

Fecundity   36, 37, 81
Feed efficiency   120, 121, 160, 258
Feed intake   121, 122, 263-264
Fertility   2, 19, 33, 85, 86, 236, 265
Fibroblast(s)   102, 174
Ficedula albicollis. See Flycatcher, collared
Fluorescent in situ hybridization (FISH)   3, 20, 24, 25, 28, 61, 102, 105, 108, 117–119, 122, 157, 175, 207–209, 238–240
Flycatcher, collared   119
Food safety   84, 85, 123
Fossil   34, 52, 143, 165, 167, 235
Functional genomic resources   9
Functional genomics   9, 120, 123–124, 158, 265, 266
Fur   48, 166, 167, 171–173, 219

G-band karyotype   34, 117, 175, 176
Galliformes   75–77
Gallus   75–82
Gallus gallus. See Junglefowl, red
Gallus lafayettei. See Junglefowl, Ceylon
Gallus sonneratii. See Junglefowl, gray
Gallus varius. See Junglefowl, green
Gene expression   37, 38, 124, 158, 159, 244, 265–268
Gene flow   34, 56, 171, 231, 241
Gene order   3, 24, 25, 34, 35, 60, 106, 107, 119, 191, 214, 260
Gene pool   125
Genetic disease(s)   118, 241, 243, 249
Genetic distance   64, 87, 151, 153, 170, 210, 236, 238
Genetic diversity   2, 47–49, 52–54, 57, 58, 75, 80, 110, 124, 125, 160, 165, 167–169, 171, 176, 178, 180, 220, 233–236, 247, 257
Genetic map. *See* Linkage map
Genetic markers. *See* Markers
Genetic regulation   119, 122, 123, 262
Genetic resources   75, 85, 124, 125, 154, 157, 176, 248
Genetic selection. *See* Selection
Genetic susceptibility   65, 66, 123, 217, 221
Genetic variation   4, 6, 23, 48, 49, 53–57, 64–66, 80, 125, 150, 168
Genome   86, 100–103, 105–108, 110, 117–120, 122–126
Genome scan   7, 36, 112, 120, 121, 261, 267
Genome sequence   4, 7, 40, 50, 60, 67, 101–103, 105–110, 118, 119, 128, 154–158, 160, 204, 205, 209, 213, 231, 240, 241, 248, 249, 260, 261, 266, 268
Genome sequencing
– annotation   4, 33, 107, 206, 261, 265, 266
– shotgun   107, 158, 205, 261, 266
Genome size   88, 107
Genomic libraries
– BAC   4, 39, 40, 66, 86, 103, 105, 107, 117–120, 122 , 157, 158, 204, 207, 266
– chromosome-specific   208
– cosmid   118

- enriched 150, 151, 180
- fosmid 107, 118
- large-insert 33, 103, 108, 117, 118, 122
- YAC 4
Genotyping 4, 8, 10, 52, 54, 58, 59, 102, 123, 148, 153, 159, 236
Glires 166
Glycogen storage disease 37, 193
Golden eagle 88
Goose
- Canada 78
- Egyptian 78
- graylag 78, 88
- swan 78, 83, 88
- white-fronted 78
Great cormorant 78
Great reed warbler 119
Growth rate 2, 33, 64, 75, 83, 86, 123, 145, 160, 219, 259
Growth traits
- chicken 122, 124
- deer 63
- turkey 122
Guinea fowl, helmeted 78, 88
Gymnogyps californianus. See California condor

Haplotype(s) 41, 55, 168, 178, 204, 234, 235, 242, 246
Heritability 122, 222, 261
Heterogametic/homogametic sex 88, 107, 125
Heterogeneity 53–57, 107, 241
Heterosis 85
Horns 37
Hybridization
- Pere-David x red deer 61, 64
- red x wapiti 52, 54, 60
- reindeer 47–51, 56
- sika 47, 49, 52, 55–57, 59, 61
- white-tailed deer 47, 52–54, 57, 59, 61, 62, 65
Hybrids 22, 60, 62, 65, 80, 123, 234

Identical by descent (IBD) 237
Immune response 62, 122–124, 173, 221
Immunogenetics 86, 87
Immunoglobulin 167, 174, 176, 178–179, 204
Inbred lines/populations
- cattle 4, 7
- chicken 107, 122
- domestic duck 84, 110, 118
- rabbit 173, 178, 179, 215
- red junglefowl 79–82, 103, 110
- turkey 151, 157
Indigenous 1, 48, 49, 52, 56, 75, 82, 85, 143
Infection 85, 124, 174, 221
International Sheep Genome Consortium 39, 40

Interval mapping 5, 64, 65, 119, 123
Intervertebral disc degeneration 34
Isle of Rum 58, 61, 64
Junglefowl
- Ceylon 77, 78, 80
- gray 77, 78, 80
- green 78, 79
- red 77–82, 103, 107, 110, 123

Karyotype 19–21, 24–26, 34, 48, 61, 107, 108, 117, 125, 147, 175–176, 208, 221, 234, 238, 258

Laboratory lines/strains 9, 75, 171–173, 222
Lagomorpha 165, 166
Layers 110, 120–123
Leanness 264
Leporidae 165, 167, 176
Lesser rhea 78
Linkage analysis 3, 10, 102, 151, 155, 160, 208, 249
Linkage disequilibrium (LD) 123, 219, 231, 241–242, 264
Linkage map 3–5, 24, 28, 36, 37, 39, 40, 60, 64, 99, 102, 103, 105–107, 110, 118, 120–125, 148–155, 208, 238–240
Lipids 86
Little egret 78
Livability/viability 86, 159
Linkage group 3, 36, 60, 61, 64, 65, 89, 102, 105, 108, 118, 119, 122, 150, 151, 160, 179–190, 208, 210, 218, 238, 250
Longevity 7, 107, 210, 247, 248, 259, 265
Lophura nycthemera. See Pheasant, silver

Major histocompatibility complex (MHC) 23, 53, 56, 57, 62, 107, 122, 123, 157, 158, 171, 176, 178, 217, 235, 236, 260
- fitness 62
- parasite burden 57
- variation 49, 62
Mammalian Genome Project 204, 205
Map integration 102, 103, 118, 155
Mapping population 151, 153
Markers
- AFLP 23, 50, 61, 102, 118, 119, 125, 148, 160, 176, 179, 208, 209, 216
- DFP 120, 125
- microsatellites 3, 4, 7, 23, 24, 36, 39, 50, 52, 53, 57–61, 64, 67, 102, 108, 119–123, 125, 148, 150, 151, 153–155, 157, 168, 189, 176, 178, 204, 208, 209, 216, 236–238, 244, 245, 249, 250, 259, 261
- minisatellites 50, 120
- mtDNA 80, 82
- RAPD 57, 102, 125, 148
- RFLP 36, 50, 55, 59, 60, 102, 125, 151, 153, 176–178, 204, 259
- SNP. *See* Single nucleotide polymorphism (SNP)

– STS   102
– type I or coding sequences   24, 28, 49, 60, 61, 102, 108, 259, 260
– type II or anonymous   24, 28, 49, 60, 61, 102
Marker interval   6, 119, 153
Marker-assisted selection (MAS)   8, 100, 120, 222, 264, 265, 268
Meat   1, 2, 22, 33, 34, 36, 37, 48, 59, 144, 145, 147, 151, 159, 165–167, 171, 172, 219, 258, 259
– chicken   80, 85, 86, 120
– domestic duck   84, 118
– emu   85
– exports   145
– ostrich   84, 85
– quality   7, 120, 121, 144, 145, 159, 219, 222, 258, 259, 261, 265
– tenderness   8, 9
Megaloceros   52
Melopsittacus undulates. See Caged birds, budgerigar
Mendelian inheritance   61, 86, 167, 179, 242, 246, 247
Metabolism   8, 34, 107, 124, 158
MHC. *See* Major histocompatibility complex
Microarray(s)
– cDNA   9, 119, 242, 249, 250, 265, 266
– gene   124, 158, 159, 221, 248, 250, 265, 267
– SNP   4, 41, 242, 249, 268
Microchromosome(s). *See* Chromosome(s), micro
Microsatellite(s). *See* Markers, microsatellites
Migration   34, 75, 168
Milk   1–3, 6–8, 22, 23, 33, 34, 36, 48, 82, 174, 176
Milk production   2, 5, 6, 22, 23, 33, 36
Mitochondrial DNA (mtDNA)   1, 34, 49, 50, 52–55, 59, 80, 165, 167, 176–178, 234–236
Molothrus ater. See Brown-headed cowbird
Moose   47, 50, 52, 56, 57
Morphological traits/loci   49, 50, 62, 65, 79, 81, 101, 110, 208
mtDNA. *See* Mitochondrial DNA (mtDNA)
Muscle hypertrophy   7, 37–39, 150
Myopia   123
Myostatin   7, 37–39

National Center for Biotechnology Information (NCBI)   40, 158
New World vultures   120
National Human Genome Research Institute (NHGRI)   106, 107, 204, 205, 240
Nuclear transfer   34, 174, 221
Nucleolus organizer regions (NORs)   20, 21, 28, 119, 147, 157
Numida meleagris. See Guinea fowl
Nutritional value   82–85
Online Mendelian Inheritance (OMI)   33, 49, 206, 261

Origin/early history
– avian   118
– cattle   1
– chicken   76, 79–81
– deer   56, 57, 59, 60
– dog   231, 232, 234–236
– *Galliformes*   75–78
– *Gallus*   77, 79
– pig   257, 261
– rabbit   166, 171, 172
– sheep   34
– turkey   143, 144
– water buffalo   19, 20
Orthologs   119, 212, 239, 240
Orylag®   167, 172, 219
Osteoporosis   34, 85
Ostrich   83–85, 107
OVERGO (overgo)   103, 105, 118, 120, 158, 206

Papillomavirus   174, 216–217
Parasite resistance   36, 57, 62, 150, 180
Parentage   52, 55, 58–59
Partridge
– gray   78, 117
– red-legged   117
Passerines   118
Pavo cristatus. See Peafowl, Indian
Pavo muticus. See Peafowl, green
Peafowl
– green   117, 118
– Indian   78, 88
Pelican   83
Perdix perdix. See Partridge, gray
Pere-David's deer   60–62, 64, 65, 67
Peregrine falcon   88
Pet   166, 167, 169, 171–173, 220
Phalacrocorax carbo. See Great cormorant
Phasianus colchicus. See Pheasant, ring-necked
Phasianus versicolor. See Pheasant, green
Pheasant
– golden   78, 88, 112
– green   78, 112
– ring-necked   78, 88, 110, 112
– silver   78, 88, 117
Phenotypic trait   122, 148, 250, 262, 267, 268
Phenotypic variation   10, 39, 81, 246, 247, 261, 266
Phylogenetic tree /analysis   23, 49, 50, 52, 56, 80, 144, 167, 234, 236, 246, 257
Phylogeny   50, 54, 80
Physical map   33, 36, 101, 103, 105, 106, 119, 120, 155, 157, 158, 260
– chicken   86, 105, 118, 199
– deer   60, 61

– dog   231, 238–240
– domestic duck   118
– pig   260
– rabbit   176, 179, 180, 204, 206–210, 216, 219
– water buffalo   19, 23, 24, 26, 28
Pigeon, rock   78, 88
Polar overdominance   37, 38
Polyploidy   107
Positional cloning   41, 64, 122, 214, 238, 265
Poultry industry   75, 82, 83, 85, 86, 125
Prion   34, 50, 65
Productivity   75, 85, 86, 106, 120–122, 166, 222, 259
Pseudogenes   66, 108, 206, 249
Psittacula krameri. See Caged birds, rose-ringed parakeet
Pterocnemia pennata. See Lesser rhea

Quail
– common   78, 88, 110, 113, 121
– Japanese   78, 88, 110, 113, 117, 118, 120
– northern bobwhite   78
Quantitative trait loci (QTL)
– beef cattle   5, 7–8, 10
– chicken   103, 120–123
– dairy cattle   5–7
– deer   61–65
– dog   231, 244–247
– pig   260, 263, 266–268
– rabbit   179, 214–216, 291, 222
– sheep   33, 37
– turkey   153, 159
– water buffalo   28
Quantitative trait nucleotide (QTN)   262, 263, 268

R-banded chromosome. *See* Chromosome(s), R-banded
Radiation hybrid (RH) map   3, 4, 8, 24, 40, 209, 213, 231, 237–240, 248, 249
Radiation hybrid (RH) panel   3, 4, 7, 33, 39–40, 102, 155, 156, 206, 238, 260, 261
Recessive trait   219
Recombination   86, 102, 108
Red deer   47–56, 58–65
Red-winged blackbird   118
Reference/resource mapping population. *See* Mapping population
Regression analysis   121
Reindeer   47–52, 56, 59
Repetitive elements   80
Reproduction   33, 37, 61–66, 120, 173, 259, 261, 263
Reproductive traits   85, 106, 120
Rhea
– greater   78, 88
– lesser   78
Rhea americana. See Rhea, greater
Rib-eye muscling   37

River buffalo   19
– breeds   22, 23
– C-banding   20, 21
– chromosome abnormalities   20
– cytogenetic map   24, 26, 28
– milk composition   22
– R-banded karyotype   20, 21, 26
RNA interference (RNAi)   38, 124
Robertsonian translocations. *See* Translocation
Roe deer   47, 50–54, 59

Scrapie   37, 65
Seasonality   60–62, 64
Segregation   61, 110, 118, 151, 153, 222, 246
Selection   2–3, 5, 8, 10, 33, 48, 49, 57, 59, 62, 75, 85, 86, 100, 107, 110, 122, 144, 145, 159, 166, 167, 172, 191–222, 240, 242, 257–259, 262–265
Sequence annotation. *See* Genome sequencing, annotation
Serinus canaria. See Caged birds, canary
Sex-linked trait   86, 110, 135, 148
Sexual dimorphism   119, 125, 245
Sexual maturity   83, 86, 171
Short interspersed nucleotide element (SINE)   110, 180, 244
Sika   47, 49, 52, 55–57, 59, 61
Single nucleotide polymorphism (SNP)   4, 8, 9, 10, 40, 41, 50, 102, 107, 110, 119, 125, 148, 153, 154, 159, 160, 206, 216, 219, 237, 242, 248, 249, 250, 259, 266, 268
Single gene trait   37, 214
Skeletal trait   86
Skin   83–85, 122, 174, 243
Somatic cell hybrid (SCH)   3, 24, 36, 206, 209, 260
Spider lamb syndrome   37
Stem cell transplantation   34
Storks   88
Struthio camelus. See Ostrich
Survivability   258
Swamp buffalo   19, 20
– breeds   22, 23
– evolution   25
Swan, mute   110
Synteny   3, 24, 106, 109, 110, 119, 154, 206, 212, 248, 261

Taeniopygia guttata. See Caged birds, zebra finch
Taxonomy   47–48, 77, 78, 144
Telomere   19–21, 119, 147, 148, 208, 238
Toxicity   173, 174
Translocation   19–21, 24, 28, 34, 48, 61, 117, 257
Tumor   124, 174

Virtual sheep genome   33, 40, 41

Wapiti   47, 48, 51, 52, 54, 55, 58, 62, 65, 66
Water buffalo

– breeding objectives   22, 23
– economic importance   21, 22
– gene mapping   24–28
– molecular genetics   23, 24
Waterfowl   75, 84

White-tailed deer   47, 52–54, 57, 59, 61, 62, 65
Whole-genome scan. *See* Genome scan
Whole genome sequence. *See* Genome sequence
Wild turkey distribution   144, 153
Wool   33, 34, 36, 37, 40, 166, 171, 172